Biochemistry of Fruit Ripening

Edited by

G.B. Seymour
Horticulture Research International, West Sussex, UK

J.E. Taylor
Department of Physiology and Environmental Science,
University of Nottingham, UK

G.A. Tucker
Department of Applied Biochemistry and Food Science,
University of Nottingham, UK

CHAPMAN & HALL
London · Glasgow · New York · Tokyo · Melbourne · Madras

Published by
Chapman & Hall, 2–6 Boundary Row, London SE1 8HN

Chapman & Hall, 2–6 Boundary Row, London SE1 8HN, UK

Blackie Academic & Professional, Wester Cleddens Road, Bishopbriggs, Glasgow G64 2NZ, UK

Chapman & Hall Inc., 29 West 35th Street, New York NY10001, USA

Chapman & Hall Japan, Thomson Publishing Japan, Hirakawacho Nemoto Building, 6F, 1–7–11 Hirakawa-cho, Chiyoda-ku, Tokyo 102, Japan

Chapman & Hall Australia, Thomas Nelson Australia, 102 Dodds Street, South Melbourne, Victoria 3205, Australia

Chapman & Hall India, R. Seshadri, 32 Second Main Road, CIT East, Madras 600 035, India

First edition 1993

© 1993 Chapman & Hall

Typeset in 10/12 Palatino by EXPO Holdings, Malaysia

Printed in Great Britain at the University Press, Cambridge

ISBN 0 412 40830 9

A catalogue record for this book is available from the British Library

Library of Congress Cataloging-in-Publication data

Biochemistry of fruit ripening/edited by G. Seymour, J. Taylor, and
 G. Tucker. — 1st ed.
  p. cm.
  Includes index.
  ISBN 0-412-40830-9 (alk. paper)
  1. Fruit—Ripening. I. Seymour, G. (Graham) II. Taylor, J. (J.
 E.) III. Tucker, G. A. (Gregory A.)
 SB357.283.B56 1993
 664'.8—dc20   92-38655
                                                    CIP

Contents

Contributors

ELIZABETH A. BALDWIN
United States Department of
  Agriculture
Agricultural Research Service
South Atlantic Area Citrus
  and Sub-tropical Products
  Laboratory
600 Avenue S, NW
PO Box 1909
Winterhaven
Florida 33881
USA

COLIN J. BRADY
CSIRO
Division of Horticulture
North Ryde
New South Wales 2113
Australia

DONALD GRIERSON
Department of Physiology
  and Environmental
  Science
University of Nottingham
School of Agriculture
Sutton Bonington
Loughborough
LE12 5RD
UK

THE LATE NIGEL K. GIVEN

GRAEME HOBSON
Horticulture Research
  International
Worthing Road
LIttlehampton
West Sussex
BN17 6LP
UK

ANGELOS K. KANELLIS
Institute of Molecular Biology and
  Biotechnology
Foundation for Research and
  Technology
PO Box 1527
711 10 Heraklion
Crete

MICHAEL KNEE
Department of Horticulture
The Ohio State University
2001 Fyffe Court
Columbus
Ohio 43210
USA

CONCEPCION C. LIZADA
Post Harvest Training Research
  Centre
University of the Philippines at
  Los Banos College
Laguna
Philippines

KENNETH MANNING
Horticulture Research
  International
Worthing Road
Littlehampton
West Sussex
BN17 6LP
UK

BARRY McGLASSON
University of Western Sydney
Hawkesbury
Richmond
New South Wales
Australia

ROBERT E. PAULL
Department of Plant Molecular
  Physiology
College of Tropical Agriculture
  and Human Resources
University of Hawaii
3190 Maile Way
Honolulu
HI 96822
USA

KALLIOPI A. ROUBELAKIS-
  ANGELAKIS
Department of Biology
University of Crete
PO Box 1470
711 10 Heraklion
Crete

GRAHAM B. SEYMOUR
Horticulture Research
  International
Worthing Road
Littlehampton
West Sussex
BN17 6LP
UK

JANE E. TAYLOR
Department of Physiology and
  Environmental Science
University of Nottingham
School of Agriculture
Sutton Bonington
Loughborough
LE12 5RD
UK

GREGORY A. TUCKER
Department of Applied
  Biochemistry and Food Science
University of Nottingham
School of Agriculture
Sutton Bonington
Loughborough
LE12 5RD
UK

Preface

It is over 20 years since the publication of A.C. Hulme's two volume text on *The Biochemistry of Fruits and their Products*. Whilst the bulk of the information contained in that text is still relevant it is true to say that our understanding of the biochemical and genetic mechanisms underlying ripening have advanced tremendously in the intervening years. For instance three of the key advances have been the elucidaton of the pathway for ethylene synthesis, the application of molecular biology to begin to unravel the genetic basis of ripening and the manipulation of ripening by genetic engineering.

This book brings the study of the biochemistry of fruit ripening up to date. The introduction is designed primarily for those readers new to the field. It provides an overview of the biochemical pathways, whose actions combine to give the multitude of physiological changes that constitute ripening. It also provides an insight into the role of ethylene in initiating and co-ordinating these various diverse biochemical pathways. The subsequent chapters deal with individual, or groups of, fruit and cover all those of current commercial importance along with some of those with future potential as economically important crops. Emphasis is given to advances in knowledge made over the last two decades, and in particular to our current understanding of the molecular biology of ripening. The book contains contributions from several experts in the field and the editors express their sincere thanks to all the contributors for their efforts and perseverance.

A thorough understanding of the biochemical basis of ripening is essential if improved quality, storage and processing of fruit is to be achieved. Such improvements, by the application of genetic engineering, have already started and the scope for future advances is enormous. The prospects of producing fruit with enhanced shelf lives, better flavour and texture and higher nutritional value are good. However, our understanding of fruit biochemistry must be even further improved to assist in these aims. In particular aspects such as ethylene perception, the role of other plant growth regulators and the intricate details of the control of gene expression need to be investigated further.

In providing an overview of our understanding of fruit biochemistry this book should have much to interest the plant physiologist, biochemist and molecular biologist. We hope that it will stimulate further research especially in those key areas identified by the contributors.

Finally the editors would like to extend their thanks to those who helped with critical reading of parts of the final manuscript, particularly Graeme Hobson, Miguel Vendrell and Matthew Hills.

G.A.T
J.E.T
G.B.S

Chapter 1

Introduction

G. A. Tucker

Since fresh, or processed fruit, forms an important part of our diet, there is an ever-increasing demand, at least in western society, for both improved quality and extended variety of the fruit available. This demand ensures that fruits are commercially valuable crops. Commercially, trade is dominated by a relatively small number of fruit, those of greatest importance being grape, banana, citrus, pome (apples, pears), and tomato. However, western consumers are becoming more aware of exotic fruit, such as mango, and trade in this type of fruit is increasing rapidly.

Fruit are often attractive to the consumer because of their aesthetic qualities of flavour, colour and texture. However, they also provide essential nutrients and this quality attribute is assuming greater significance in the public eye. Fruit in general are poor sources of protein or fat, though avocado and olive represent obvious exceptions in the case of fat. In some diets fruit can represent major sources of energy. For example, plantains, with their high starch content, are a major staple food in many countries. In western societies, fruit are generally important in the diet for their vitamin and mineral content (Table 1.1). While these vitamin levels may not be high when compared to other sources, if the quantity of fruit consumed is also taken into account, then they assume, in several cases, prime importance. For instance tomato, orange and banana have been ranked 1st, 2nd and 6th, respectively, among all fruit and vegetables for their contribution of vitamins and minerals to the typical US diet (Wills *et al.*, 1989).

Fruit contain a very high percentage of their fresh weight as water. In tomato, for example, the water content is around 95% and may even be as high as 98%. Consequently, fruit exhibit relatively high metabolic activity when compared to other plant-derived foods such as seeds. This

Biochemistry of Fruit Ripening. Edited by G. Seymour, J. Taylor and G. Tucker. Published in 1993 by Chapman & Hall, London. ISBN 0 412 40830 9

Table 1.1   Vitamin content of fruit

Fruit	Vitamin A (mg/100g)	Vitamin C (mg/100g)	Folic Acid (mg/100g)
Apple	–	10	>5
Banana	0.1	20	10
Blackcurrant	–	200	>5
Mango	2.4	30	>5
Tomato	0.3	20	>5

metabolic activity continues postharvest and thus makes most fruit highly perishable commodities. It is this perishability, and inherent short shelf life, that presents the greatest problem to the successful trans-portation and marketing of fresh fruit. For the major commercial crops the postharvest problems have been largely solved by harvesting at the immature, or mature green stage, and/or by using refrigeration and controlled atmosphere storage. However, many other fruit cannot be handled successfully using these methods and this precludes the maximum commercial exploitation of these fruit. For instance, mangoes harvested at full maturity do not store well, but if harvested immature fail to ripen properly. Such exploitation would be of mutual benefit to both the producing and consuming countries. Solutions to the post-harvest problems of these fruit, as well as improvements in the handling procedures for the major established crops, may come from a better understanding of the biochemistry and molecular biology of fruit ripening.

The botanical definition of a fruit is 'a seed receptacle developed from an ovary'. This definition encompasses a very wide range of fruit types. The dupe fruit, as exemplified by apricot and peach, contain a single large seed surrounded by a fleshy mesocarp. Berry fruit types, tomato for instance, contain many small seeds within a gel contained in the locule of the fruit. Pome fruit and *Hesperidium* (citrus fruit) are again of completely different morphologies. These morphological variations can also be extended to include strawberry, pineapple and melon fruit type structures. Considering this wide range of fruit types, it is perhaps not surprising that they also differ, to a certain extent, in their respective metabolisms. However, the central biochemical pathways involved are common to all fruit and indeed very often to other plant tissues.

The aim of this introductory chapter is firstly, to provide the general reader with a broad outline of plant biochemistry and secondly, to apply this biochemistry more specifically to fruit tissues. More detailed infor-mation of the biochemistry of individual fruit is contained in the following specialized chapters. The biochemistry of fruit ripening has been reviewed most recently by Friend and Rhodes (1981), Brady (1987) and Tucker and Grierson (1987). For more general plant biochemistry the reader is referred to the recent texts by Dennis and Turpin (1990), and Anderson and Beardall (1991).

1.1 RESPIRATION AND ENERGY

Ripening requires the synthesis of novel proteins and mRNAs, as well as new pigments and flavour compounds. These anabolic processes require both energy and a supply of carbon skeleton building blocks. These are supplied in fruit, just as in other tissues by respiration.

While all fruit obviously carry out respiration there are marked differences in both the rates and patterns of change of this respiration between fruit. Fruit in general can be classified as either climacteric or non-climacteric on the basis of their respiration pattern during ripening (Fig. 1.1). Climacteric fruit display a characteristic peak of respiratory activity during ripening, termed the respiratory climacteric. This peak may correspond to optimum eating ripeness, or may precede or postdate this according to the fruit in question. Also the magnitude of the peak can vary enormously between fruit. It is interesting to note that in general, fruit with the highest respiratory rates, such as banana and avocado, also tend to ripen most rapidly and hence are most perishable. This has led to the regulation of respiration as a possible target for the biochemical manipulation of shelf life. In contrast, non-climacteric fruit simply exhibit a gradual decline in their respiration during ripening. There are however, still marked differences between fruit as regards the magnitude of their respiration rate. Again, for non-climacteric fruit the general correlation exists between high respiratory rate and short shelf life.

1.1.1 Respiratory substrates

The two major respiratory substrates found in fruit are sugars and organic acids. Some fruit, such as avocado, contain high levels of lipid, though this does not appear to be used as a respiratory substrate. The respiratory quotient in fruit varies but is normally in the region of 1, indicating that sugars are the predominant respiratory substrate in fruit. However, in many instances it is also apparent that metabolism of organic acids can account for a significant proportion of respiration (Tucker and Grierson, 1987). Both sugars and organic acids are found largely sequestered within the vacuole, and form a major contribution to the overall flavour of the fruit. However, they are presumably also released in a controlled manner from the vacuole, or alternatively a separate pool is maintained, and hence are available for respiration.

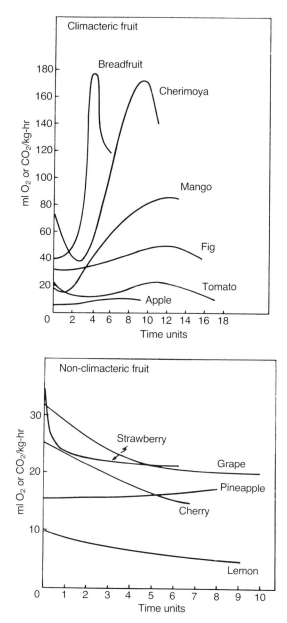

Fig. 1.1 Respiration patterns during the ripening of climacteric and non-climacteric fruit.

The respiratory pathways utilized by the fruit for the oxidation of sugars are those common to all plant tissues namely glycolysis, oxidative pentose phosphate pathway (OPP) and the tricarboxylic acid (TCA) pathway. These pathways are shown in Pathway Charts 1–3, located at the end of this chapter (pages 32–34). The increased respiration of sugars in climacteric fruit seems to be mediated largely by an increased flux through glycolysis. The contribution of the OPP is thought to be minimal but significant in preclimacteric fruit, and this contribution may decline during the climacteric (Tucker and Grierson, 1987). The respiratory capacity of preclimacteric fruit is thought to be sufficient to account for the maximum respiratory flux during the climacteric. For example, application of respiratory uncouplers to preclimacteric avocado fruit results in a rate of respiration equal to or greater than that at the climacteric peak (Millerd et al., 1953). Thus in the majority of climacteric fruit the increased respiration during the climacteric could arise from the de-inhibition of respiration.

Glycolytic control

The key regulatory steps in glycolysis are those mediated by phospho-fructokinase (EC 2.7.1.11; PFK) and pyruvate kinase (EC 2.7.1.40; PK), i.e. the conversion of fructose-6-P to fructose-1, 6-bisphosphate and phosphoenol pyruvate to pyruvate respectively. Both enzymes, when extracted from plant tissues, have been shown to be suceptible to a wide range of possible metabolic modulators (Turner and Turner, 1980). However, this does not imply that any or all of these are important regulators of activity *in vivo* (Copeland and Turner, 1987). The major effectors appear to be ATP and phosphoenol pyruvate, which act as inhibitors, and phosphate and potassium ions, which act as activators. How glycolytic flux is regulated during ripening is unknown. The level of ATP in fruit tends to increase during ripening (Solomos, 1983), as would be expected with such an increase in respiration; thus glycolysis is not regulated by a classical Pasteur effect. This is perhaps not surprising since plant PFK, unlike the mammalian counterpart, is relatively insensitive to inhibition by ATP (Turner and Turner, 1980). Levels of phosphoenol pyruvate tend to decline during ripening (Solomos, 1983) and both phosphate (Chalmers and Rowan, 1971) and potassium ions (Vickery and Bruinsma, 1973) are thought to increase. There is some evidence that PFK from banana fruit undergoes some form of structural alteration during ripening which alters its kinetics with respect to the substrate, fructose-6-phosphate, effectively rendering the enzyme more efficient (Salminen and Young, 1974). Also Isaacs and Rhodes (1987) reported the conversion of tomato PFK from an oligomeric to a monomeric form of the enzyme, which has been shown to be regulated by magnesium ions (Surendranathan *et al.*, 1990). However, since this change from oligomer to monomer does not correspond exactly with the kinetic change of the enzyme, its significance, if any, in terms of regulation is unclear.

The role of PFK activity in regulating glycolysis in plant tissue is further complicated by the existence of pyrophosphate-linked phosphofructokinase (EC 2.7.1.90; PPK) (Huber, 1986; Stitt, 1990) which catalyses a similar reaction to PFK. This enzyme also converts fructose 6-phosphate to fructose-1, 6-bisphosphate, but is reversible and uses pyrophosphate instead of ATP as a second substrate. The PPK activity, unlike that of PFK, can be regulated by levels of its key metabolic activator, fructose-2, 6-bisphosphate. Although PPK has been demonstrated to be important in controlling assimilate partitioning in leaves (Huber, 1986; Stitt 1990), its significance during fruit ripening is unclear. The enzyme has been found in fruit tissue and appears to undergo isoform changes during ripening of tomatoes (Wong *et al.*, 1990). Also, levels of fructose-2, 6-bisphosphate have been shown to increase in avocado fruit in response to ethylene treatment (Stitt *et al.*, 1986; Bennett *et al.*, 1987). The precise significance of these isoenzyme and activator changes is unknown. The PPK may play a role in regulating starch metabolism in fruit such as banana and this is dealt with in more detail in Chapter 3.

Malic acid

The major organic acid used for respiration appears to be malate. The sole utilization of malate as a respiratory substrate is mediated by malic enzyme, the activity of which is widespread in plant tissues. Activity in plants exists in several compartments, plastids (E1-Shora and ApRees, 1991) and in fruit a cytosolic NADPH-dependent malic enzyme (EC 1.1.1.40) and mitochondrial NAD-dependent malic enzyme (EC 1.1.1.39) have been demonstrated (Goodenough *et al.*, 1985). These enzymes catalyse the reductive decarboxylation of malate to pyruvate, allowing carbon from malate to be fed into the TCA cycle without any requirement for production of pyruvate by glycolysis (Fig. 1.2). Other organic acids such as citrate and succinate can feed directly into the TCA cycle. When malate is utilized as a respiratory substrate, the increased flux appears to be mediated by an increased activity of the malic enzyme (Moreau and Romani, 1982; Goodenough *et al.*, 1985). In the case of pome fruit this increased activity may occur as the result of enzyme synthesis (Frenkel *et al.*, 1968). More details of malate metabolism during ripening (in pome fruit) can be found in Chapter 11.

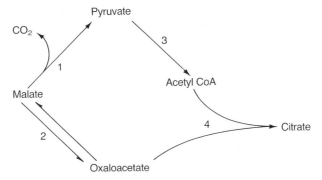

Fig. 1.2 Pathway for the oxidation of malate. The enzymes involved are (1) malic enzyme, (2) malate dehydrogenase, (3) pyruvate dehydrogenase, and (4) citrate synthase.

1.1.2 The fate of pyruvate

The pyruvate generated by glycolysis, OPP or malate metabolism is presumably fed into the TCA cycle. The control and utilization of the TCA cycle during ripening is poorly understood. The NAD(P)H produced by glycolysis, TCA or malic enzyme activity is oxidized by plant mitochondria and the energy used for ATP synthesis via oxidative phosphorylation. The oxidation of NAD(P)H by plant mitochondria is a complex process mediated by several membrane-bound dehydrogenases (Palmer and Moller, 1982). Synthesis of ATP by oxidative phosphorylation is linked to a complex system of electron carriers located on the inner membrane of the mitochondria (Douce *et al.*, 1987). This electron transport system and its links to NAD(P)H dehydrogenases is summarized in Pathway Chart 4 (page 35). Electron transport along these carriers is tightly coupled to the synthesis of ATP. Thus high ATP and low ADP levels, as occur in ripe fruit, would be expected to inhibit electron transport since no further ATP synthesis can occur. This in turn would inhibit the other respiratory pathways through a build up of NADH. However, respiration, in climacteric fruit, remains high despite this high level of ATP. It is possible that ATP is rapidly turned over during ripening. However, mechanisms for such turnover are unknown and the demand for ATP during ripening is insufficient to account for a large turn over (Solomos, 1983). Mitochondria can be uncoupled by a variety of compounds, including proton ionophores such as dinitrophenol. There is no evidence for any naturally occurring uncouplers in fruit, and mitochondria isolated from fruit during the climacteric have been shown to be tightly coupled with respect to ATP synthesis (Biale,

1974). One possible solution to this problem which has been studied is the use of alternative electron transport pathways in plant mitochondria.

Plant mitochondria, in general, differ from mammalian mitochondria by having the ability to bypass all or some of the so-called conservation sites in the conventional electron transport chain (see Pathway Chart 4). The Piericidin A-resistant pathway allows electrons to pass from the internal NADH dehydrogenase to ubiquinone, bypassing the first conservation site. A more common electron transport chain is the so-called cyanide-resistant pathway, which branches from the main cytochrome chain at ubiquinone thus allowing electrons to bypass the last two conservation sites. It is unclear whether or not fruit mitochondria possess Piericidin A resistance, though fruit in general do have the capacity for cyanide-resistant respiration. A correlation has been shown between ethylene-induced respiration and the presence of the cyanide-resistant alternative electron transport chain in a variety of tissues (Solomos and Laties, 1976). Since the climacteric is thought to occur as a response of the fruit to ethylene, several workers have attempted to determine the role, if any, of the alternate pathway during the respiratory climacteric. An increased capacity for cyanide-resistant respiration has been shown to occur in avocado mitochondria isolated from fruit at various stages during the climacteric (Moreau and Romani, 1982). The operation of the cyanide resistant-pathway in intact tissues is difficult to determine (Moller *et al.*, 1988). Some measurements using banana and avocado have suggested that while the capacity for such oxidation does exist in the mitochondria of these fruit, it is nonetheless not utilized during the climacteric (Theologis and Laties, 1978).

1.1.3 Role of the respiratory climacteric

The respiratory climacteric is common to a wide range of fruit yet its role, if any, in ripening is unclear. The increased levels of ATP may be required to drive other ripening events such as starch breakdown or nucleic acid, protein and pigment synthesis. However, the calculated energy demand in most fruit during ripening is much less than that produced during the climacteric (Solomos, 1983) and non-climacteric fruit ripen without any increase in respiration. Indeed, in some non-climacteric fruit respiration actually declines during ripening. The ripening of several fruit can be manipulated to separate the respiratory climacteric from other ripening events. For example, respiration and carotenoid biosynthesis can be separated in this way in cantaloupes (Pratt and Groeschl, 1968). Thus it is thought that respiration is neither dependent on, nor integrated with, the other ripening events. It is possible that respiratory increases result simply as a general response to the ethylene produced by climacteric fruit. This idea is supported by the fact that even non-climacteric fruit, which normally produce relatively low levels of ethylene, will respond to exogenous ethylene with increased respiration. Alternatively, Romani (1975) has suggested that mitochondria are autonomous and are simply responding to stress as the fruit tissue breaks down during ripening.

1.2 FLAVOUR CHANGES

Our perception of flavour relies on two senses, taste and smell. We can distinguish basically four different tastes these being sweet, sour, bitter and salt. In fruit it is primarily the sugars and organic acids which contribute to the fruit taste although the astringent nature of some fruit can be attributed to their phenolic and tannin content. However, the characteristic flavour of individual fruit is, also usually derived via our sense of smell and is due to the production of specific flavour volatiles. Thus the flavour of the fruit depends on the complex interaction of sugars, organic acids, phenolics and more specialized flavour compounds, including a wide range of volatiles.

1.2.1 Assimilation of sugars and organic acids

Sugars and organic acids are also used as respiratory substrates, as shown in the previous section, but are present in greater amounts than required simply for this involvement in energy generation within the fruit. Fruit differ in their relative contents of sugars and acids (Ulrich, 1970; Whiting, 1970). A representative list for some major fruit is given in Table 1.2. The most common sugars are fructose, glucose and sucrose and the most prevalent organic acids, malate and citrate.

Both sugars and organic acids originate from photosynthetic assimilates. Green fruit are capable of photosynthesis (Phan, 1970) but this is limited and the bulk of the assimilate is provided by the rest of the plant. However, fruits differ in how this assimilate accumulates during development and ripening. Some fruit have accumulated the bulk of their carbohydrate prior to the onset of ripening. This is either stored primarily as starch, as in banana, or as sugars, as in tomato. Fruit continue to accumulate sugar from the plant during ripening and in some cases (such as strawberry and grape) this accounts for a large part of their flavour. While this difference is of marginal importance for respiration (which uses only a relatively small fraction of the assimilate) it is of prime importance for flavour considerations. Fruit which accumulate their assimilate prior to ripening can be harvested at the mature green stage and still attain acceptable flavour on ripening. This is important when considering the need for early harvest to optimize shelf life.

Table 1.2 Sugars and acids in fruit

Fruit	Sugars			Acids	
	Glucose	Fructose	Sucrose	Major acid	Total acid
	(%fwt)	(%fwt)	(%fwt)		(μeq/100g)
Apple	2	6	4	Malate	3–19
Banana	6	4	7	Malate	4
Grape	8	7	6	Tartrate	2
Orange (juice)	2	2	5	Citrate	15
Peach	1	1	7	Malate\Citrate	4
Strawberry	3	2	1	–	10–18
Tomato	2	2	0	Citrate	–

Fruit which depend on the plant for assimilate during ripening fail to develop full flavour if harvested at the green stage; this is particularly evident in the case of strawberry and grape which are both commercially unacceptable if not 'vine ripened'. It can also be a consideration for fruit which can be commercially harvested at the green stage, such as tomato, since improved flavour can often be obtained by allowing fruit to initiate ripening prior to picking.

In general, levels of acids decline during ripening, presumably due to their utilization as respiratory substrates (Ulrich, 1970). During ripening sugar levels within fruit tend to increase (Whiting, 1970), due either to increased sugar importation from the plant or to the mobilization of starch reserves within the fruit, depending on the type of fruit and whether it is ripened on or off the plant. It is possible that some gluconeogenesis (from malate in particular) can occur during ripening. However, if this is occurring it would account for only a small percentage of the sugar accumulating in the fruit. The pathways for gluconeogenesis will not be dealt with here, but are covered in more detail in the section on organic acids in the grape (Chapter 6).

Starch mobilization

In many fruit the breakdown of starch to glucose, fructose or sucrose, is a characteristic ripening event. There are several enzymes in plant tissue capable of metabolizing starch (Presis and Levi, 1980; Steup, 1988); the pathways involved are shown in Pathway Chart 5 (page 36). α-Amylase (EC 3.2.1.1) hydrolyses the α (1–4) linkages of amylose at random to produce a mixture of glucose and maltose. In contrast, β-amylase (EC 3.2.1.2) attacks only the penultimate linkage and thus releases only maltose. The maltose produced in either case can be hydrolysed to glucose by the action of glucosidase (EC 3.2.1.20). Starch phosphorylase (EC 2.4.1.1) hydrolyses the terminal α(1–4) linkage to give glucose-1-phosphate, which can be converted to glucose-6-phosphate by the action of glucose phosphate mutase (EC 2.7.5.5).

All three starch degrading enzymes, α- and β-amylase and starch phosphorylase, have been identified in fruit. Not surprisingly the banana, with nearly 20% of pulp fresh weight as starch, is the most intensively studied fruit in this case; more detail is given in Chapter 3. In banana, the major activity is associated with α-amylase although the levels of the other two enzymes are not insignificant. All three enzymes occur in several isoenzyme forms and all three increase in activity during ripening to some extent (Tucker and Grierson, 1987). However, it has as yet been impossible to assign any definite mechanism for specific starch degradation during ripening.

In all cases the enzymes are active only against the linear glucose chains of amylose found in starch. They are unable to degrade the α(1–6) branch points also found in the amylopectin of starch. Starch granules in fruit are thought to resemble those in other tissues such as potato, cereal and mung beans (Kayisu and Hood, 1981). As such they are a mixture of amylose and amylopectin and must contain these branching linkages. Enzymes capable of attacking the branch points (debranching enzyme, EC 3.2.1.10) have been identified in several tissues including banana fruit (Garcia and Lajola, 1988).

The end products of starch degradation are either glucose or glucose-1-phosphate. Either can be converted to glucose-6-phosphate, by the action of hexokinase (EC 2.7.1.1) or glucose phosphate mutase respectively. Since starch is confined to the plastids of the fruit cells the degradation also occurs in this compartment. However, further utilization of the breakdown products of starch occur largely in the cytoplasm. In leaf tissue it is triose phosphates – primarily dihydroxyacetone, glyceraldehyde phosphate and glycerophosphates – which are transported across the chloroplast envelope following starch mobilization. How the products of starch degradation pass the envelope in fruit cells is not clear, but this could be either as six-carbon sugar phosphates (glucose-6-phosphate or fructose-6-phosphate) or more likely as triose phosphates, as in leaf tissue. Once in the cytoplasm, breakdown products could then either enter glycolysis to be respired, or converted back to glucose phosphate and fructose for use in the synthesis of sucrose (refer to Pathway Chart 5 at the end of this chapter).

Sucrose metabolism

Sucrose metabolism in plants has been reviewed by Akazawa and Okamoto (1980), though current data available on sucrose synthesis in fruit are minimal. In leaf tissue, the action of sucrose phosphate synthase (EC 2.4.1.14) is considered the major biosynthetic route for sucrose, the action of sucrose synthase (EC 2.4.1.13) being kinetically unfavourable for sucrose synthesis in leaf tissue (Avigad, 1982). However, levels of sucrose phosphate synthase in tissues other than leaves is very low, while that of sucrose synthase is more widely spread in plants (Avigad, 1982). Sucrose phosphate synthase has been reported in grape berries (Downton and Hawker, 1973) and in several other fruit (see later chapters), but, in general, control of sucrose synthesis in fruit is poorly understood. The breakdown of sucrose is probably mediated by the action of invertase (EC 3.2.1.26). This enzyme is widespread in fruit and often increases in activity during ripening. However, the activity is not always associated with increased metabolism of sucrose. For instance, in tomato this enzyme increases in activity during ripening, even though this fruit contains no sucrose store. Invertase activity may, however, be involved in the transport of sugar into the fruit from the parent plant. Invertase has been the subject of much investigation and a recent identification of a mRNA for acid invertase from tomato fruit (Endo *et al.*, 1990) may signal the beginning of a molecular biological approach to determining the exact function of this enzyme. Low levels of invertase activity have recently been linked to a sucrose accumulation trait in tomatoes (Yelle *et al.*, 1991).

1.2.2   Phenolics in fruit

Although sugars and acids constitute the major components in fruit flavour other, more specialized, compounds are also important (refer to the review on fruit flavour by Morton and McCleod, 1990).

To a certain extent, astringency is determined by phenolic compounds in the fruit. Although compared to many other plant tissues the levels of phenolics in fruit are relatively low, they can be quite significant, especially in determining the quality of fruit products. Several different phenolic compounds have been isolated from fruit, details of which may be found in the following chapters; general phenolic structures are shown in the Structure chart at the end of this chapter. Many phenolics are derived from phenylalanine via cinnamic and coumaric acids (Pathway Chart 6, p. 38). However, alternative biosynthetic routes exist, many of which are reviewed by Vickery and Vickery (1981).

One very important group of enzymes involved in the metabolism of phenolics is the polyphenoloxidases (PPO), of which there are basically two types: catechol oxidase (EC 1.10.3.1) and Laccase (EC 1.10.3.2) which catalyse the oxidoreduction of o-diphenols or p-diphenols, respectively, and have been classified together under the general name of monophenol monooxygenase (EC 1.14.18.1). The properties of these enzymes have been described by Mayer and Harel (1979) and their distribution in fruit reviewed by Mayer and Harel (1981). The enzymes are found in most, if not all fruit; however alterations during ripening are poorly documented and the role, if any, for this enzyme during ripening is not clear. The action of PPO is, however, an important factor in determing the quality of several processed fruit products. The mixing of PPO with its phenolic substrates when fruit tissue is disrupted during processing, accounts for some of the browning reactions occurring in both tissues and extracted juices.

1.2.3   Flavour 'volatiles'

Flavour in fruit is also dependent upon the synthesis of more specialized flavour compounds, which account for a very diverse range of fruit constituents and are described in more detail in the individual fruit chapters. Some examples of compounds important in determining fruit flavour are shown in the Structure Chart at the end of this chapter (page 42). Although several of these flavour compounds are complex, a large proportion are relatively simple molecules which being volatile account for fruit aroma. These 'flavour volatiles' are usually present at relatively low levels, often only as ppm, but are important since they are thought to provide the characteristic flavour and aroma of different fruit. The 'flavour volatile' profile of any fruit is usually very complex. For instance, analysis of the volatiles present in apple and orange fruit indicates at least 230 and 330 different compounds respectively (Van Straten, 1977). The nature of the volatiles involved is also very diverse and includes alcohols, aldehydes, esters and many other chemical groups (Nursten, 1970). While many of these volatile compounds are common to several fruit, the distribution of others may be more restricted. Nursten (1970) lists a wide range of fruit along with their respective major volatiles.

The biosynthetic pathways for such a wide range of volatiles is also obviously very diverse. The biosynthesis is further complicated by the fact that while some of these volatiles are synthesized in the intact fruit, others are produced only when the fruit tissue is macerated. The alcohols and aldehydes are presumably derived from the metabolism of their corresponding amino acids and oxo-sugars, although there is relatively little information available to confirm this. Another possible major biosynthetic pathway may be via the action of lipoxygenase (EC 1.13.11.12) on membrane lipids following maceration of the tissue. Such a system has been implicated in the formation of distinctive flavour notes in tomato tissue (Schrier and Gorens, 1981). Other flavour volatiles may be synthesized via the mevalonate/isoprene pathway as shown in Pathway Chart 6 (see page 38). The five-carbon hemiterpenoids as a group contain several volatile and aromatic compounds.

With such a wide range of volatiles it is difficult to assign precise roles for any particular compound in the development of fruit flavour and aroma. Much more work needs to be carried out in this area to identify those volatiles which have the major impact on fruit flavour. This is, however, a very difficult research area but progress is being made for apple and orange juice (Durr and Schobinger, 1981) and tomato fruit (Buttery et al., 1989). Further research is needed into the biosynthetic pathways of these key volatiles once identified, this being an obvious prerequisite to any manipulation of flavour by genetic engineering.

1.3   COLOUR CHANGES

Not all fruit change colour during ripening, for example many varieties of apple or pear remain green. However, in general colour change is associated with ripening and represents a key attribute, along with texture, for the determination of eating quality. Colour change can be brought about by the degradation of chlorophyll, which in turn unmasks previously present pigments, particularly β-carotene. Such a process occurs in the degreening of lemons. However, in most fruit this loss of chlorophyll is accompanied by the biosynthesis of one or more pigments, usually either anthocyanins or carotenoids. Although obviously coordinated during ripening, chlorophyll loss and pigment synthesis are not directly related or indeed inter-dependent. This is quite clearly demonstrated in the tomato ripening mutant greenflesh (gf), which has impaired chlorophyll degradation yet apparently normal carotenoid biosynthesis during ripening (Darby et al., 1977). The fruit end up as a dirty brown colour in the fully ripe state.

1.3.1   Chlorophyll degradation

The precise mechanism for chlorophyll degradation is unclear, but may involve both enzymic and chemical reactions. Chlorophyll is held tightly bound to the thylakoid membranes within the chloroplast. The first step in chlorophyll degradation would seem to be the 'solubilization' of this chlorophyll into the stroma, which may be brought about by enzymes capable of attacking the thylakoid membranes or the chlorophyll directly, but the mechanism is unknown. Once 'soluble', the chlorophyll can be oxidized chemically to the colourless purin and chlorin products.

1.3.2   Anthocyanins

Anthocyanins are a very diverse range of pigments localized within the vacuole of the plant cell (Timberlake, 1981). They can give rise to colours from red to blue and often occur in a wide range of types in individual fruit. For example, orange fruit are known to contain about 30 different anthocyanin-type pigments. The general structure of anthocyanidins and their glycoside derivatives, the anthocyanins, is given in Figure 1.3. A typical anthocyanin is cyanidin-3-galactoside, which is largely responsible for the colour of apple, blackberry, cherry and plum. Anthocyanins are derived from flavenoid compounds and as such are synthesized from the aromatic amino acid phenylalanine via the biochemical pathway shown in Pathway Chart 7 (see page 39). This pathway is discussed in detail by Timberlake (1981) and shown in more detail in Chapter 12. Two key enzymes in this pathway are phenylalanine ammonia lyase (EC 4.3.1.5; PAL), and flavanone synthase. These enzymes are known to be synthesized *de novo* in many plant tissues in response to UV light and mechanical damage. However, how they are regulated during fruit

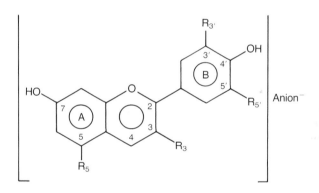

Anthocyanidins : R_3 = OH, R_5 = OH

		$R_{3'}$	$R_{5'}$
Pelargonidin	(Pg)	H	H
Cyanidin	(Cy)	OH	H
Peonidin	(Pn)	OCH$_3$	H
Delphinidin	(Dp)	OH	OH
Petunidin	(Pt)	OCH$_3$	OH
Malvidin	(Mv)	OCH$_3$	OCH$_3$

Anthocyanins: Pg, Cy, Pn, Dp, Pt, Mv with
R_3 = O–sugar or O–acylated sugar
R_5 = OH or O–glucose

Fig. 1.3   Structure of some fruit pigments.

ripening is unclear. Faragher and Chalmers (1977) found no direct correlation between pigment synthesis and PAL activity. Pigment synthesis in red cabbage seedlings is correlated more closely to flavanone synthase activity than to PAL (Hrazdina and Creasy, 1979). It is possible that synthesis or activation of these enzymes, or others within the pathway, respond to the plant growth regulator ethylene which is known to control other aspects of ripening.

1.3.3 Carotenoids

Carotenoid pigments are not as diverse as anthocyanins and are localized within the chloroplast which is then often termed a chromoplast. The structure of two typical carotenoids, β-carotene and lycopene, are shown in the Structure Chart (see page 42). The amount of carotenoids synthesized during ripening can be very large and this often leads to the formation of pigment crystals within the chromoplast; for a review of the conversion of chloroplasts to chromoplasts, refer to Tucker and Grierson (1987). This is especially true in tomato, where crystals of the predominant carotenoid pigment lycopene can be seen in the chromoplasts. Carotenoids are terpenoid compounds and as such derive from acetyl CoA via the mevalonic acid pathway (see Pathway Chart 6, page 38). The primary carotenoid produced from this pathway is phytoene and this is further metabolized to give the more familiar carotenoid pigments. Control of this pathway is again unclear. Many of the enzymes responsible for the biosynthesis of carotenoids have yet to be isolated, although there are several tomato-ripening mutants thought to be deficient in these enzymes (Darby *et al.*, 1977; Rick, 1980).

Carotenoids are normally synthesized in green tissue, a major product being β-carotene. However, in many fruit additional β-carotene and lycopene is synthesized during ripening. The pathway for carotenoid biosynthesis (Pathway Chart 8, p. 40) shows lycopene as the precursor of β-carotene. The accumulation of lycopene during ripening could therefore arise from either the inhibition of the final step in the production of β-carotene or to the initiation of a new biosynthetic route for lycopene. Several tomato-ripening mutants are known to be deficient, to varying degrees, in carotenoid biosynthesis during ripening (Rick, 1980; Frecknell and Pattenden, 1984) yet these mutations do not affect β-carotene synthesis in green tissue. Also high (30–35°C) temperatures are known to inhibit lycopene, but not β-carotene accumulation in tomato fruit (Goodwin, 1980). These observations would support the idea of a ripening-specific pathway for lycopene synthesis. The mevalonic acid pathway is central to the production of many other compounds – such as abscisic acid and gibberellin – which are likely to be essential throughout the development and ripening of the fruit. Any switch in carotenoid biosynthesis is therefore unlikely to occur at this level but probably occurs after the production of phytoene. However, exactly how, or if, fruit switch their carotenoid pigment production during ripening is unknown.

1.4   TEXTURE CHANGES

Most fruit soften during ripening and this is a major quality attribute that often dictates shelf life. Fruit softening could arise from one of three mechanisms: loss of turgor; degradation of starch; or breakdown of the fruit cell walls. Loss of turgor is largely a non-physiological process associated with the postharvest dehydration of the fruit, and as such can assume commercial importance during storage. Loss of water equivalent to about 5–10% of a fruit's fresh weight – although having little effect on the fruit's biochemistry – can render the fruit commercially unacceptable. Degradation of starch probably results in a pronounced textural change, especially in those fruit like banana, where starch accounts for a high percentage of the fresh weight. In general however, texture change during the ripening of most fruit is thought to be largely the result of cell wall degradation. Carbohydrate polymers make up 90–95% of the structural components of the wall, the remaining 5–10% being largely hydroxyproline-rich glycoprotein (HPRG). The carbohydrate polymers can be grouped together as cellulose, hemicelluloses or pectins.

1.4.1   Cell wall changes

Changes in cell wall structure during ripening have been observed under the electron microscope in many fruit, including avocado (Pesis *et al.*, 1978), pear (Ben-Arie *et al.*, 1979) and tomato (Crookes and Grierson, 1983). These changes usually consist of an apparent dissolution of the pectin-rich middle lamella region of the cell wall. At a biochemical level, major changes can be observed in the pectic polymers of the wall. During ripening there is a loss of neutral sugars, in most fruit this is predominantly galactose, but some loss of arabinose also occurs (Tucker and Grierson, 1987). These two sugars are the major components of the wall's neutral pectin. There are also major changes observed in the acidic pectin or rhamnogalacturonan fraction of the wall. During ripening there is an increase in the solubility of these polyuronides and in several cases these have been shown to become progressively depolymerized. (For a review of these wall changes, refer to Tucker and Grierson, 1987, and Fischer and Bennett, 1991.) The degree of esterification of the polyuronide fraction can also change during ripening; tomato polyuronide is about 75% esterified in green fruit and this declines to around 55% during ripening (Seymour and Tucker, unpublished). Conversely an increase in the degree of esterification of the soluble polyuronide from apple fruit has been reported (Knee, 1978).

Changes during ripening of cell wall components other than pectins have been poorly documented. It appears that levels of sugars commonly associated with either hemicellulose or cellulosic fractions remain constant throughout ripening, at least in tomato (Gross and Wallner, 1979). However, in tomato there is some evidence for a depolymerization of hemicelluloses during ripening (Huber, 1983a). Other poorly studied structural components are the cell wall proteins.

Much work has been carried out to identify enzymes in fruit responsible for these changes in the wall during ripening; this has been reviewed by Huber (1983b), Tucker and Grierson (1987), Brady (1987), and Fischer and Bennett (1991). However, such studies are hampered by an incomplete understanding of plant cell wall structure in general, and fruit wall structure in particular.

Cell wall structure

Keegstra *et al.* (1973) developed an early model of the plant primary cell wall in which cellulose fibrils, which were coated with hemicellulose, are found embedded in a matrix composed of pectin and protein (Fig. 1.4). This model has been developed further over the years and although the actual detailed structure of the various polymers and the nature of their interaction in the final three-dimensional wall is far from simple, this general model provides an adequate basis to investigate fruit softening. Several detailed reviews have been published on the fine structure of plant cell walls (Aspinall, 1980; Darvill *et al.*, 1980; McNeil *et al.*, 1984; Fry, 1986; Bacic *et al.*, 1988). Recent advances in microscopy have enabled the wall structure to be visualized to some extent (McCann *et al.*, 1990) and this confirms the lamellate structure of the primary wall.

Cellulose consists of linear chains of β(1–4)-linked glucose residues which aggregate together via hydrogen bonds to form fibrils. Hemi-celluloses are composed of a variety of polymers, the major ones in dicots being xyloglucans, glucomannans and galactoglucomannans (Aspinall, 1980). Pectins are also composed of a variety of polymers, the neutral pectins being arabinans, galactans or arabinogalactans, all of which can be structurally diverse (Aspinall, 1980). The acidic pectins are composed of rhamnogalacturonans and homogalacturonans. The former are complex molecules whose fine structure is far from understood (Darvill *et al.*, 1980). Basically rhamnogalacturonan consists of chains of α-(1–4)-linked galacturonic acid residues interspersed with rhamnose. The distribution of these rhamnose residues is unknown but may be either random or in organized clusters. Digestion of cell walls with enzymes releases discrete fragments of rhamnogalacturonans (Darvill *et*

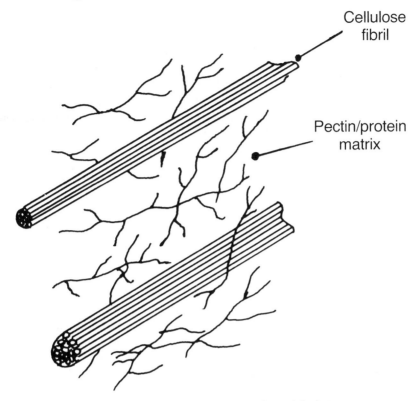

Cellulose
fibril

Pectin/protein
matrix

Fig. 1.4   Idealized cell wall model.

al., 1980; McNeil *et al.*, 1984) which indicates both some repeat structure and complexity in this molecule. The rhamnose is thought to provide attachment for arabinose or galactose side chains. The galacturonic acid residues can either be methylated or retain the free carboxyl group at the C-6 position, again distribution of these methylated residues within the polymers is unknown. A model structure for the pectin fraction of the cell wall is shown in Fig. 1.5.

It is thought that these polymers are held together in the three-dimensional cell wall by a variety of covalent and non-covalent bonds (Fry, 1986). Cellulose fibrils are held together by hydrogen bonds, and similar bonds account for the interaction of cellulose with hemi-celluloses. However, it has been postulated that in addition to simply coating the cellulose fibril, hemicellulose molecules may also form bridges between adjacent fibrils (Hayashi, 1989). The hemicellulose molecules are certainly of sufficient length for this, and if such linkages do occur then cleavage of these bridges could be a major cause of cell wall loosening and hence softening.

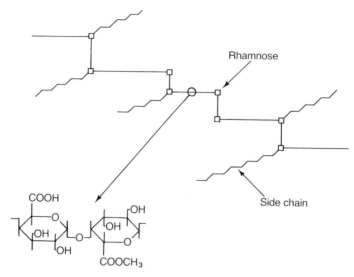

Fig. 1.5   Pectin structure. The side chains, branched polymers of galactose and arabinose, constitute the neutral pectin which is attached to the rhamnogalacturan acidic pectin polymers via the rhamnose residues.

Neutral side chains of galactose and/or arabinose are often linked to the acidic pectin backbone via covalent bonds between rhamnose and galactose residues, as shown in Figure 1.5. Not all of the available rhamnose residues are used, and it is likely that at least some neutral pectin remains unattached to any acidic pectin. Pectin polymers could also be covalently linked via diferulic acid bonds (Fry, 1986). These bonds are capable of linking together neutral pectins via their terminal galactose residues. These bonds have been identified in several plant tissues, but as yet not in fruit. If present in fruit cell walls they could account for covalent linkage between adjacent rhamnogalacturonan polymers by cross-linking of their neutral sugar side chains.

Pectin molecules can also be linked together non-covalently via a structure called the 'egg-box' (Grant *et al.*, 1973 Fig. 1.6). In this case calcium ions are chelated by regions of de-esterified galacturonic acid residues on adjacent polymers. Although this type of bond is relatively weak, the large number of possible bonds would be expected to hold adjacent polymers firmly together.

There are also possible links between the hemicellulose and pectin polymers, although these are much less defined. Neither is the role of the HPRG well understood; this protein may be covalently attached to the carbohydrate polymers or more likely it interacts in a non-covalent fashion. The HPRG is a strongly basic protein and thus its positively-

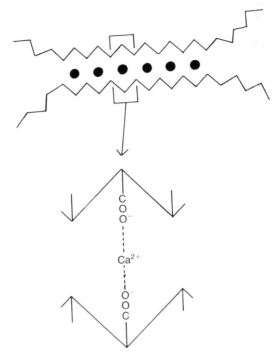

Fig. 1.6 The 'egg-box' model for the non-covalent linkage of adjacent acidic pectin polymers.

charged amino acids might be expected to interact with the negative charges found on de-esterified regions of the acidic pectins. The HPRG may interact with the other wall polymers in an even simpler manner. This protein is thought to form a mesh by the cross-linking of tyrosine residues on adjacent polypeptides by an isodityrosine bridge (Fry, 1986), the mesh physically entrapping the carbohydrate polymers and hence contributing to the integrity of the wall.

Most of the research on cell wall structure has been carried out on the primary walls from tissue other than fruit. However, it is generally thought that fruit cell walls have a three-dimensional structure similar to that of these other tissues. Fruit cell walls in general seem to have relatively high proportions of galacturonic acid, galactose and arabinose, suggesting that they are relatively rich in pectin. The structure of this pectin also seems to conform to the general rules as found for other tissues.

Considering the complex nature of the cell wall there are several possible mechanisms to account for the changes in wall polymers observed during ripening. Increased solubility of pectin could arise from disruption of the 'egg-box' structure or cleavage of either the rhamno-galacturonan backbone or the neutral sugar side chains. Similarly, breaking any bonds between pectin and hemicellulose could also give rise to increased solubility of polyuronide. Polyuronide depolymerization could also arise by cleaving either the rhamnogalacturonan backbone or the side chains. The loss of galactose and arabinose probably involves the action of exo-acting enzymes, in addition to the action of endo-acting enzymes, since these sugars, in most cases, are lost completely and are not usually recoverable as soluble polymers. One exception is the loss of arabinose in pear fruit (Ahmed and Labavitch, 1980).

Cell wall hydrolases

A wide range of cell wall hydrolases can be identified in fruit tissue (Huber, 1983b; Tucker and Grierson, 1987). Many of these are either constitutive throughout development and ripening, or are present in very low amounts. While neither of these observations rule out involvement of these enzymes in softening, it has meant that studies have tended to concentrate on the major wall hydrolases. The enzymes most studied are pectinesterase (EC 3.1.1.15), polygalacturonase (EC 3.2.1.15), β-(1–4) glucanase or cellulase (EC 3.2.1.4) and β-galactosidase (EC 3.2.1.23). Pectinesterase (PE) acts to remove the methyl group from the C-6 position of a galacturonic acid. Polygalacturonase (PG) hydrolyses the α(1–4) link between adjacent demethylated galacturonic acid residues. These two enzymes can act synergistically with PE, generating sites for PG action. However, the extent of this interaction *in situ* during ripening is unclear. Cellulase hydrolyses the β(1–4) link between adjacent glucose residues. This enzyme is assayed against the artificial substrate carboxymethylcellulose, and in most cases is totally inactive against native cellulose. Its trivial name of cellulase is, therefore, misleading. The natural substrate for this enzyme is unknown, but this is likely to be a hemicellulose polymer rather than cellulose itself. β-galactosidase activity is also often measured using artificial substrates. In many cases these enzymes are incapable of degrading native galactans and their natural substrate remains unknown. However, in several cases β-galactosidases from ripening fruit have been shown to attack native galactan polymers. These enzymes can be considered as true β-galactanases and could well be implicated in wall degradation.

These cell wall hydrolases are found in a wide variety of fruit (Tucker and Grierson, 1987). In fact PE and β-galactosidase seem to be ubiquitous, although the confirmation of the galactosidase activity as a true galactanase has been carried out only in a limited number of fruit. Both PG and cellulase are widespread, although in some fruit these enzymes are not detectable.

Many of the above enzymes tend to exist as several isoforms within the fruit tissue and this makes analysis of changes in activity during ripening difficult. Such changes have been reviewed (Huber, 1983b; Tucker and Grierson, 1987). In general, PE and β-galactosidase activities are present during the development of the fruit. During ripening, total PE activity can either decline, remain constant or increase depending on the fruit or extraction method used. Thus total PE activity in tomatoes has been reported to decline (Pressey and Avants, 1972), remain constant (Sawamura *et al.*, 1978) or increase (Tucker *et al.*, 1982) during ripening. These observations may be due to different cultivars being used in each case, since Smith *et al.* (1989a) have shown that PE changes in banana cultivars show similar marked variations during ripening. Changes in PE activity are also complicated by the presence of isoforms of this enzyme. Thus, again in tomatoes, a small increase in the total PE activity masks a much greater change in the pattern of isoenzyme activity during ripening (Tucker *et al.*, 1982). A further complication in PE activity may arise from the presence of endogenous enzyme inhibitors. A glycoprotein inhibitor of PE has been isolated from kiwifruit (Balestrieri *et al.*, 1990); however, the presence of such an inhibitor in fruit in general remains to be determined. The β-galactosidase activity in fruit tends to increase during ripening. Again, this usually occurs as several isoenzymes but, unlike the situation with PE, novel isoforms often appear during ripening. This has been demonstrated for tomatoes (Pressey, 1983) and for mango (Muda and Tucker, unpublished). Both cellulase and PG tend to be absent in green fruit, activity appearing only with the onset of ripening and tending to increase dramatically during ripening. Both of these enzymes tend to be predominantly endo-acting hydrolases; however, low levels of exo-PG are also found in several fruit and is the only form of this enzyme in apple (Bartley, 1978) and some peach cultivars (Pressey and Avants, 1978).

Considering the complexity of both wall structure and enzyme profiles it is perhaps not surprising that a mechanism for softening in fruit has not been fully elucidated. It is unlikely that a single enzyme is responsible for textural change and that this probably involves a complex interaction of enzyme activity with physicochemical changes in the wall. Also, it is becoming apparent that the softening of different fruit may proceed via different mechanisms, and this will be addressed for individual fruits in later chapters. One novel approach to elucidating the role of enzymes in wall degradation and softening is to employ antisense RNA technology. This approach is being used extensively in tomato fruit and is discussed in greater detail in Chapter 14. Using this technology, expression of PG in ripe fruit has been reduced to less than 1% of normal (Smith *et al.*, 1988). Analysis of the resultant fruit indicates that PG action is responsible for the polyuronide depolymerization observed during ripening, but that inhibition of this depolymerization is not sufficient in itself to prevent fruit softening (Smith *et al.*, 1990).

It remains to be determined exactly how the wall changes during ripening are brought about, but the application of molecular biology techniques should enable rapid advances in this area. Another aspect of softening that requires attention is to determine how exactly changes in wall structure result in actual softening. This again may be addressed using molecular biology techniques. A final problem is to understand how the softening process is regulated once initiated, since it is clear that the action of the wall hydrolases *in situ* must be severely restricted. It has been estimated, for instance, that there is sufficient PG and PE activity in tomato fruit to totally depolymerize the fruit pectin in a matter of hours. In fact, such degradation of the pectin can be seen *in vitro* following homogenization of the fruit, but this does not occur *in vivo* (Seymour *et al.*, 1987). There are several explanations for this, the localized conditions of pH, ionic strength or substrate accessibility in the wall may well be sufficient to inhibit enzyme activity. Alternatively, the enzymes may be sequestered within the wall and hence unable to carry out extensive degradation. Evidence for sequestration comes from investigations on the effect of silver ions on tomato wall enzymes and metabolism. Application of silver to ripening fruit prevents any further accumulation of PG within the wall; however, that PG which was present prior to silver treatment remains and can be extracted in an active form (Tucker and Brady, 1987). However, this same treatment has also been found to inhibit any further polyuronide solubilization or neutral sugar loss (Smith *et al.*, 1989b). Such sequestration may arise if enzymes are targetted to specific sites within the wall; such targetting could result in the spacial separation of enzymes and in limited polymer degradation. Recent studies have shown that pectin esterification, in developing root tissue, is spacially regulated both within the wall and between tissues (Knox *et al.*, 1990); whether such separation occurs in fruit is unknown.

1.5   CONTROL OF RIPENING

Ripening can be considered as a specialized stage of plant senescence. As such it was originally deemed that ripening was primarily a catabolic process in which cellular organization and control were breaking down (Blackman and Parija, 1928). However, over the last few decades, and in particular over the last few years, it has become apparent that ripening, like other plant senescent processes, is under strict genetic control. Early evidence for anabolic processes occurring during ripening came from *in vivo* radiolabelling studies to investigate turnover of proteins and nucleic acids during ripening. It was found that in many fruit, such as banana

(Brady and O'Connell, 1976) and tomato (DeSwardt *et al.*, 1973), protein synthesis continued during ripening. Similarly, nucleic acid synthesis continues (Richmond and Biale, 1967). More recently it has been shown that cell wall carbohydrate polymers also continue to be synthesized during ripening (Mitcham *et al.*, 1989). It is thus generally held that during ripening the processes of nucleic acid and protein synthesis continue, albeit at perhaps a reduced rate, of more significance was the realization that not only was protein synthesis continuing but that to a certain extent this was also being redirected. Thus analysis of mRNA and protein species during ripening of both avocado (Christoffersen *et al.*, 1982) and tomato (Rattanapanone *et al.*, 1978; Biggs *et al.*, 1986), showed the synthesis of distinct ripening-related proteins. This led to the concept of ripening as being controlled, at least partially, at the level of gene expression. The search is now underway to identify these 'ripening-specific proteins' and indeed some have already been identified, such as polygalacturonase (Grierson *et al.*, 1986), ACC synthase (Van der Straeten *et al.*, 1990) and ACC oxidase (Hamilton *et al.*, 1990). The rapidly expanding area of research on gene expression during ripening has been extensively reviewed (Brady, 1987; Grierson *et al.*, 1985, 1987, 1989, 1990; Spiers and Brady, 1991; Tucker and Grierson, 1987; Tucker, 1990). Also under investigation is the nature of the genetic controls which act to initiate and regulate ripening.

1.5.1 Genetic control of ripening

Considering the complex nature of fruit ripening, in which changes in flavour, texture and colour at least must be regulated and coordinated, it is perhaps not surprising to find the process is under genetic control. There are basically three major questions relating to our understanding of how the plant achieves this genetic control. Firstly, how is fruit ripening initiated? Secondly, how is the process once initiated, regulated? Thirdly, how are the various diverse biochemical changes coordinated throughout ripening? We cannot as yet provide precise answers to any of these questions, but they must rely on an interplay between the regulation of gene expression and enzyme activity.

Ripening, like other plant developmental processes, is probably under the control of plant growth regulators (McGlasson *et al.*, 1978; Bruinsma, 1983). All the five major growth regulators effect ripening in one way or another if applied exogenously. Thus auxin, gibberellin and cytokinin generally act to retard ripening, while ethylene and abscisic acid act to enhance the ripening process. Much work has been carried out in an attempt to monitor endogenous levels of these regulators during ripening. Although some correlations can be found, only ethylene is routinely found to be associated with ripening and hence has led to this plant growth regulator being considered the 'ripening hormone'. While it is undoubtedly true that ethylene plays a key role in ripening, the possible involvement of other hormones must not be ignored. Indeed, the most likely control mechanism for ripening probably resides in a complex interaction between ethylene and one or more other growth regulators. The possible involvement of auxin during ripening is covered in some detail in Chapter 12. Another group of regulatory compounds that may be implicated in the control of ripening are the polyamines (Saftner and Baldi, 1990).

Ethylene and ripening

Ethylene is produced by most plant tissues, albeit at a relatively low level of around $0.05\mu l. h^{-1}.gfwt^{-1}$. Climacteric fruit are characterized by a burst of ethylene production which occurs during ripening and most often precedes the respiratory climacteric. However, in some fruit it has been reported that this burst of ethylene either coincides with or, more rarely, follows, the respiratory climacteric. Like the respiratory climacteric, different fruit have different peak levels of ethylene production (Table 1.3). Non-climacteric fruit do not exhibit a burst of ethylene production, but simply a decline in production from the mature green to ripe stage of development, a pattern reminiscent of their respiration during ripening. However, different fruit do again have different levels of ethylene production (Table 1.3). Climacteric and non-climacteric fruit also appear to differ in the control of ethylene synthesis. The biosynthesis of ethylene in climacteric fruit is said to be autocatalytic. This is best demonstrated by the response of fruit to the ethylene analogue, propylene. Climacteric fruit exposed to propylene begin to synthesize ethylene in an autocatalytic manner; non-climacteric fruit, however, show no such response (McMurchie *et al.*, 1972). These

Table 1.3   Internal ethylene levels in ripening fruit

	Fruit	*Ethylene ($\mu l/l$)*
Climateric	Avocado	500
peak level	Banana	40
	Mango	3
	Pear	40
	Tomato	27
Non-climacteric	Lemon	0.1–0.2
steady state	Orange	0.1–0.3
	Pineapple	0.2–0.4

observations led McMurchie *et al.* (1972) to postulate two control systems for ethylene biosynthesis:

1. System 1, common to both climacteric and non-climacteric fruit, is responsible for both basal ethylene production, and as we shall describe later, the ethylene produced when the tissue is wounded.
2. System 2 is unique to climacteric fruit and is responsible for the autocatalytic production accompanying ripening in these fruit.

Increased biosynthesis of ethylene is not a phenomenon unique to the ripening of climacteric fruit. Most plant tissues will respond to wounding with an increase in ethylene production. In this respect both climacteric and non-climacteric fruit, at all stages of development, are no exception. Thus all fruit, if wounded, will respond with an increase in ethylene synthesis. However, even in unripe climacteric fruit this wound-induced ethylene synthesis is not considered to be autocatalytic in nature. The evidence for this comes from the fact that silver, which blocks ethylene perception, has no effect on the level of wound ethylene produced by green tomato fruit (Tucker and Grierson, 1987); thus it is considered that wounding results in the stimulation of System 1 of McMurchie *et al.* (1972).

Ethylene biosynthesis

It is generally thought that ethylene production in all plant tissues proceeds via a common biosynthetic pathway (Pathway Chart 9; see page 41) which was first established in apple fruit by Adams and Yang (1979) and has since been extensively reviewed (Yang, 1981; Yang *et al.*, 1985; Kende *et al.*, 1985). It has since been shown to exist in many other plant and fruit tissues. Many of the enzymic activities are thought to be constitutive. Thus the conversion of methionine to S-adenosyl methionine (SAM), which is used in other biochemical pathways such as methylation of galacturonosyl residues in pectin, is considered to be constant throughout the development and ripening of the fruit. Similarly, the enzymes involved in the sulphur reclamation half of the cycle are also considered as constitutive. The two key control enzymes for the bio-synthesis of ethylene are thus ACC synthase (EC 4.4.1.14) and the ethylene forming enzyme also referred to as ACC oxidase.

ACC synthase has been purified from a range of sources (Bleeker *et al.*, 1986) and shown to be a cytosolic enzyme with a molecular weight of around 50 kDa (Nakajimo *et al.*, 1988; White and Kende, 1990). The active site of this enzyme has been characterized (Yip *et al.*, 1990). ACC oxidase has only recently been solubilized in an active form, the activity being sensitive to oxygen (Ververdis and John, 1991). The cellular location of the ACC oxidase is still in question, but the enzyme can be isolated in association with vacuoles (Guy and Kende, 1984) or mitochondria (Djebar and Moreau, 1990). In all cases, ACC oxidase activity is severely inhibited by the application of uncouplers and ionophores. These results might imply that the enzyme is either membrane bound or is contained within a vesicle and is dependent on transmembrane potential for its activity *in situ*.

Since the enzyme activity can be solubilized, the requirement for transmembrane potentials *in vivo* may be linked to the transport of ACC across membranes. A cDNA isolated from a tomato ripening library has been shown to encode for ACC oxidase (Hamilton *et al.*, 1990). This cDNA codes for a protein with a molecular weight of around 35 000. When yeast were transformed with the cDNA they acquired the ability to convert ACC to ethylene (Hamilton *et al.*, 1991). It thus seems likely that this cDNA encodes for all of the catalytic components of the ACC oxidase system. The protein encoded by the ACC oxidase cDNA does not appear to resemble any known membrane-bound proteins. However, the presence of secondary regulatory subunits in the fruit, which could be membrane bound, cannot be ruled out.

During ripening, the levels of ACC are low in green fruit and accumulate rapidly and coincident with ethylene synthesis (Hoffman and Yang, 1980). This implies that ACC synthase may be a key enzyme in the control of ethylene synthesis. In post-climacteric fruit, levels of ACC remain high while ethylene production declines (Hoffman and Yang, 1980). This would indicate that ACC oxidase is becoming inactivated. The possibility of ACC oxidase as a controlling enzyme during the initial phase of ethylene synthesis must not be ignored, since the activity of this enzyme is known to increase during ripening and wounding. It is not considered as rate limiting, at least in the initial phase of ripening, but measurement of this enzyme's activity *in vivo* is difficult due to the need to wound the tissue. Since the levels of mRNA for ACC oxidase are very low in green fruit and rise dramatically at the onset of ripening (Smith *et al.*, 1986), this may indicate a more significant role for this enzyme than previously thought. The same biosynthetic pathway for ethylene synthesis operates in both climacteric and non-climacteric fruit. Presumably therefore, the differentiation into System 1 and 2 in these two types of fruit arises due to different control mechanisms acting on a common biosynthetic pathway. The precise nature of these control mechanisms is far from clear.

As well as responding to endogenous ethylene, fruit will also respond to exogenous ethylene. This is often an important consideration in fruit stores, since ethylene can accumulate from a wide range of sources. These include ripening or wounded tissue and also exhaust fumes from trucks or trollies. Both climacteric and non-climacteric fruit will respond to exogenous ethylene but this response is somewhat different in each case (Fig. 1.7). Non-climacteric fruit respond with an increased rate of respiration and a corresponding increase in their rate of ripening. This response is directly proportional to the level of exogenous ethylene, and dependent on its continuous presence. Green climacteric fruit also respond to ethylene with an increase in respiration and eventually by autocatalytic ethylene production. This results in a shortening of the shelf life of the fruit. A certain threshold of exposure is required to initiate this autocatalytic ethylene, but once started further response is not proportional to exogenous ethylene and is essentially irreversible.

Since both climacteric and non-climacteric fruit respond to exogenous ethylene, they must both possess a type of receptor mechanism. The nature of this receptor is poorly understood; it is known that ethylene

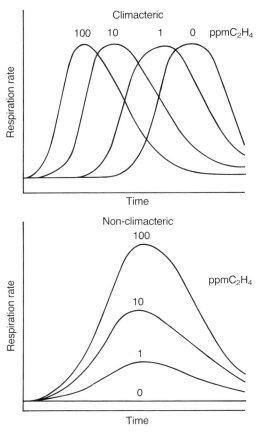

Fig. 1.7 Effect of exogenous ethylene on fruit respiration.

responses can be blocked by silver ions, and that this may be via an effect
of the silver on the receptor (Beyer, 1976; Hobson *et al.*, 1984). However,
the nature or location of this receptor remains unclear. It is also known
that a small proportion of the ethylene bound to plant tissues can be
metabolized to either ethylene oxide or ethylene glycol (Beyer, 1985).
However, no link between ethylene metabolism and the receptor or
ethylene action, has been established.

Another area in which information is sadly lacking is the nature of any
secondary hormone messenger systems for plants in general and fruit in
particular. Thus although calmodulin, cAMP and inositol triphosphates
– all proven secondary messengers in mammalian systems – have been
found in plants, there is no indication as yet of their possible role, if any,
in the signal transduction involved in the ethylene response.

It was originally thought that ethylene represented the trigger for
ripening. While this may be the case, it is not apparently a simple case
of increased ethylene causing ripening. In tomato fruit, changes in gene
expression are observable before any increase in ethylene levels
(Lincoln *et al.*, 1987). It is possible, therefore, that the initiation of
ripening is dependent on the interaction of ethylene with other growth
regulators. It was also originally thought that once initiated, ripening
would continue without any further role for ethylene. This is clearly
not the case, since it has been shown that the application of silver ions
to block ethylene perception will, as well as preventing the initiation of
ripening, also arrest subsequent ripening events, as judged by colour
and enzyme synthesis (Tucker and Brady, 1987). Thus ethylene
perception is required throughout ripening and this may indicate a role
for ethylene in the coordination of the process. However, ethylene is
not required for all ripening associated changes to occur. Fruit held
under controlled atmosphere storage conditions fail to produce normal
levels of ethylene. These fruit fail to change colour or soften, but do
exhibit the normal changes in sugars and acids seen during ripening
(Goodenough *et al.*, 1982). It is thus possible that only some ripening
events such as colour and texture changes are under the control of
ethylene, while others – including flavour development – may be
independent of ethylene. The separation of otherwise normally
coordinated ripening events can be demonstrated in several ways.
Thus, as we have already seen, respiration and colour changes in
cantaloupe can be temporarily separated (Pratt and Groeschl, 1968).
There is also a wide range of ripening mutants especially for tomato
(Rick, 1980, 1987) in which individual ripening events are inhibited,
without any apparent adverse effect on any other. These examples
illustrate the fact that ripening can be considered as a coordinated,
rather than as a linked, set of biochemical pathways.

1.6 CONCLUSIONS

Ripening, as we have seen, is best considered as a set of coordinated but otherwise loosely connected biochemical pathways. These pathways are most likely to be under hormonal control both for their initiation and coordination. However, we are far from understanding the complexity of this control system. It is also apparent that much of ripening is under strict genetic control. Ripening involves changes in gene expression and thus the synthesis of several novel enzymes, several of which have been identified, while many ripening-specific cDNAs for as yet unidentified enzymes have also been isolated. Most of the work on gene expression so far reported has been carried out in avocado and tomato fruit. More information on this rapidly advancing topic can therefore be found in Chapters 2 and 14, as well as in Chapters 12 and 13.

The advent of molecular biology techniques has provided a powerful tool to probe the biochemistry of fruit ripening. These techniques are being used to investigate the role of hydrolases in cell wall degradation and hence fruit softening. They can also be applied to investigations of other ripening events such as ethylene synthesis, pigment production and starch degradation. In particular, these techniques could be used to study the vexing question of how ethylene, or other hormone, perception results in altered gene expression during ripening. It is hoped that significant advances will therefore be made in the near future.

The application of molecular biology techniques also allows the **manipulation** of ripening, and this could have significant commercial advantages. Current storage of fruit usually employs refrigeration, either alone, or coupled with controlled atmospheres in which oxygen is maintained at a low level, and carbon dioxide content is increased. These techniques are expensive, often inefficient and can result in damage to the stored fruit. Manipulating ripening genetically to extend the shelf life would therefore have considerable commercial advantages. Such manipulation is currently restricted to tomato fruit (Kramer *et al.*, 1989; Schuch *et al.*, 1989; Smith *et al.*, 1990; Grierson *et al.*, 1990; Tucker, 1990). In this case, fruit softening has been influenced by the down regulation of the polygalacturonase gene. The resultant fruit have longer shelf lives and are more resistant to disease and cracking during transport. Their processing to pastes is also significantly improved. Ethylene synthesis can also be manipulated in tomato fruit (Hamilton *et al.*, 1990; Oeller *et al.*, 1991) and this again has a significant effect on the fruit's shelf life. In future, more aspects of ripening and other fruit may be manipulated to our advantage in this way; however, the major obstacle to achieving this aim remains a lack of fundamental knowledge of the biochemistry of the fruit ripening process.

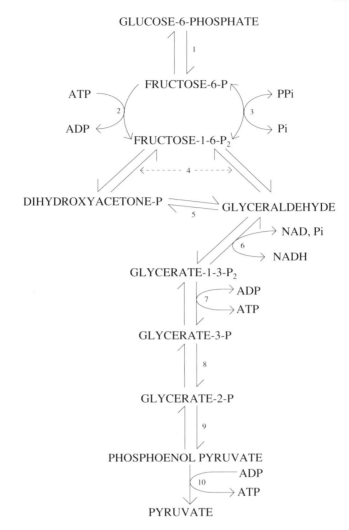

Pathway Chart 1   Glycolysis

Conversion of a six-carbon glucose to two molecules of pyruvate (3 carbons)

1   Hexose-phosphate isomerase (EC 5.3.1.9)
2   Phosphofructokinase (EC 2.7.1.11)
3   Phosphofructophosphotransferase (EC 2.7.1.90):
      A similar but irreversible conversion of fructose-1,6-bisphosphate to
      fructose-6-phosphate can be catalysed by fructose bisphosphatase (EC
      3.1.3.11) using ADP, instead of orthophosphate, as the phosphate
      receptor.
4   Aldolase (EC 4.1.2.13)
5   Triose-phosphate isomerase (EC 5.3.1.1)
6   Glyceraldehyde-3-phosphate dehydrogenase (EC 1.2.1.12)
7   Glycerate-3-phosphate kinase (EC 2.7.1.31)

8   Glycerate phosphate mutase (EC 2.7.5.4)
9   Enolase (EC 4.2.1.11)
10 Pyruvate kinase (EC 2.7.1.40)

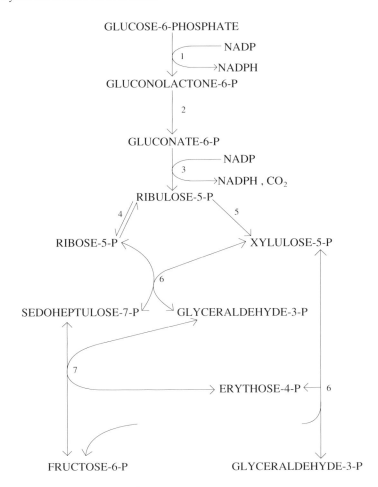

Pathway Chart 2   Oxidative pentose phosphate pathway

One complete turn of the pathway requires three glucose molecules to be converted to three five-carbon ribulose-5-Ps. One of these is then converted to ribose-5-P and the other two into xylulose-5-P. The completed pathway thus results in the conversion of three glucose into two fructose molecules.

1   Glucose-6-phosphate dehydrogenase (EC 1.1.1.49)
2   Phosphogluconolactonase (EC 3.1.1.31)
3   Gluconate-6-phosphate dehydrogenase (EC 1.1.1.44)
4   Ribose-5-phosphate isomerase (EC 5.3.1.6)
5   Ribulose-5-phosphate 3-epimerase (EC 5.1.3.1)
6   Transketolase (EC 2.2.1.1)
7   Transaldolase (EC 2.2.1.2)

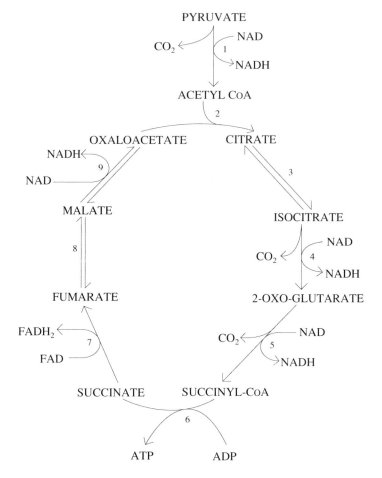

Pathway Chart 3   Tricarboxylic Acid (TCA) Cycle

1 Pyruvate dehydrogenase complex
      Pyruvate dehydrogenase (EC 1.2.4.1)
      Dihydrolipoyl transacetylase (EC 2.3.1.12)
      Dihydrolipoyl dehydrogenase (EC 1.6.4.3)
2 Citrate synthase (EC 4.1.3.7)
3 Aconitase (EC 4.2.1.3)
4 Isocitrate dehydrogenase (EC 1.1.1.42)
5 2-Oxoglutarate dehydrogenase (EC 1.2.4.2)
6 Succinate thiokinase (EC 6.2.1.4)
7 Succinate dehydrogenase (EC 1.3.99.1)
8 Fumarase (EC 4.2.1.3)
9 Malate dehydrogenase (EC 1.1.1.37)

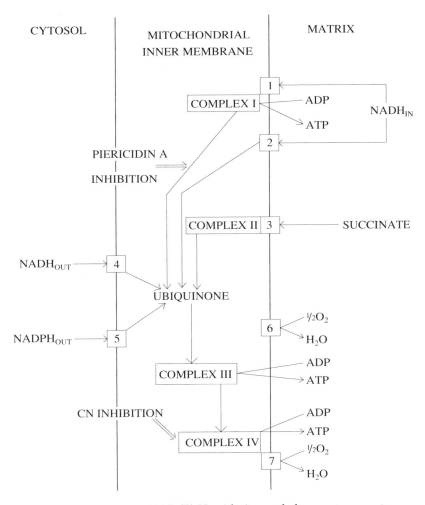

Pathway Chart 4   NAD (P) H oxidation and electron transport

Pathway indicates routes for electron transfer within the mitochondrial inner membrane. It should be noted that the mitochondrial outer membrane also contains NADH dehydrogenase activity which can feed electrons into the electron transport chain of the inner membrane, probably at the level of ubiquinone.

1   NADH dehydrogenase (EC 1.6.99.3)
2   NADH dehydrogenase (EC 1.6.99.3)
3   Succinate dehydrogenase (EC 1.3.99.1)
4   NADH dehydrogenase (EC 1.6.99.3)
5   NADPH dehydrogenase (EC 1.6.99.1)
6   Alternate oxidase
7   Cytochrome oxidase (EC 1.9.3.1)

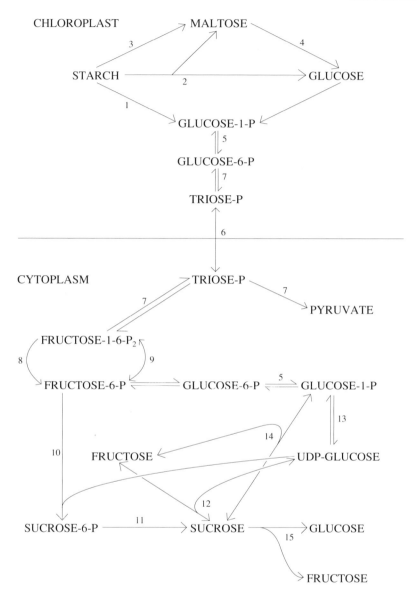

Pathway Chart 5   Starch degradation and sucrose synthesis

1   Starch phosphorylase (EC 2.4.1.1)
2   α-Amylase (EC 3.2.1.1)
3   β-Amylase (EC 3.2.1.2)
4   α-Glucosidase (EC 3.2.1.20)
5   Glucose phosphate mutase (EC 2.7.5.5)

6 Triose phosphate/phosphate transporter.
   It is possible that in some cases, export of starch
   degradation products from the chloroplast may occur via
   the transport of hexose phosphates.
7 Glycolysis/gluconeogenesis (see Pathway Chart 1)
8 Fructose bisphosphatase (EC 3.1.3.11)
9 Phosphofructophosphotransferase (EC 2.7.1.90)
10 Sucrose phosphate synthetase (EC 2.4.1.14)
11 Sucrose phosphate phosphatase (EC 3.1.3.24)
12 Sucrose synthase (EC 2.4.1.13)
13 UDP-glucose pyrophosphorylase (EC 2.7.7.9)
14 Sucrose phosphorylase (EC 2.4.1.7)
15 Invertase (EC 3.2.1.26)

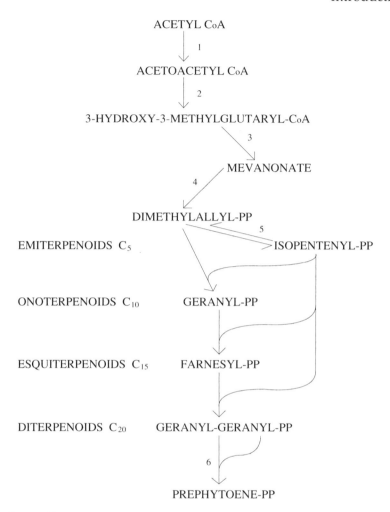

Pathway Chart 6   Carotenoid pigment biosynthesis: isoprenoid pathway

1   Acetyl CoA acetyl transferase (EC 2.3.1.9)
2   Hydroxymethylglutaryl– CoA synthase (EC 4.1.3.5)
3   Hydroxymethylglutaryl–CoA reductase (EC 1.1.1.88)
4   Mevalonate kinase (EC 2.7.1.36)
       Phospho mevalonate kinase (EC 2.7.4.2)
       Pyrophosphomevalonate decarboxylase (EC 4.1.1.33)
5   Isopentenyldiphosphate Δ -isomerase (EC 5.3.3.2)
6   Two molecules of geranyl–geranyl pyrophosphate undergo a tail-to-tail
       condensation.

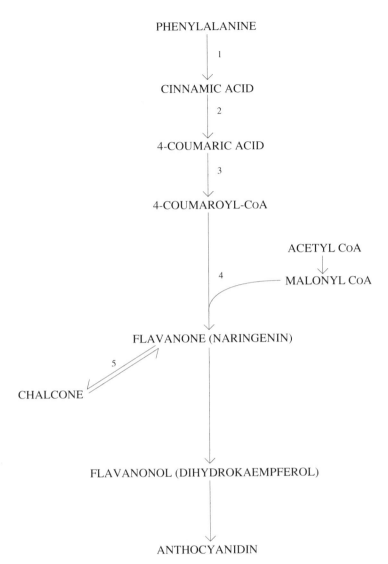

Pathway Chart 7   Anthocyanidin biosynthesis

Anthocyanins are synthesized by the addition of glycosyl residue(s) to the corresponding anthocyanidin. (See Fig. 1.3)

1   Phenylalanine ammonia lyase (EC 4.3.1.5)
2   Cinnamate 4 hydroxylase
3   4-coumarate:coenzyme A ligase (EC 6.2.1.12)
4   Flavanone synthase
5   Chalcone isomerase (EC 5.5.1.8)

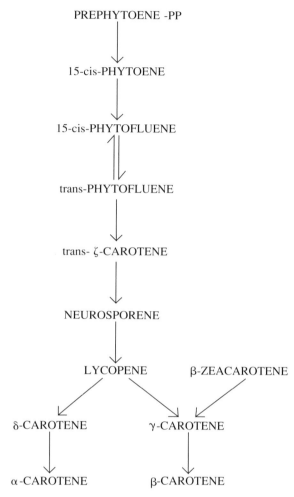

Pathway Chart 8   Carotenoid pigment biosynthesis: biosynthesis of lycopene and carotenes

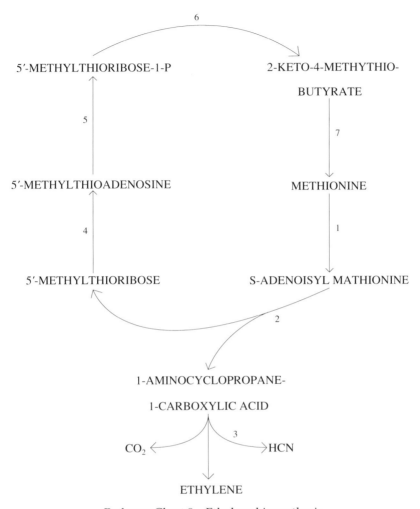

Pathway Chart 9   Ethylene biosynthesis

1   Methionine adenosyl transferase (EC 2.5.1.6)
2   ACC synthase (EC 4.4.1.14)
3   ACC oxidase
4   5′–methylthioadenosine nucleosidase
5   5′–methylthioribose kinase (EC 2.7.1.100)
6   This step is catalysed by at least three enzymes.
7   This step represents a transamination reaction with glutamine as the most
    efficient amino donor.

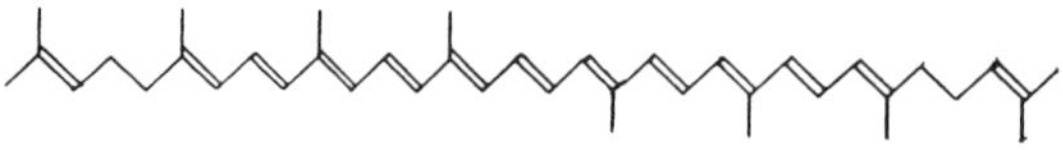

LYCOPENE

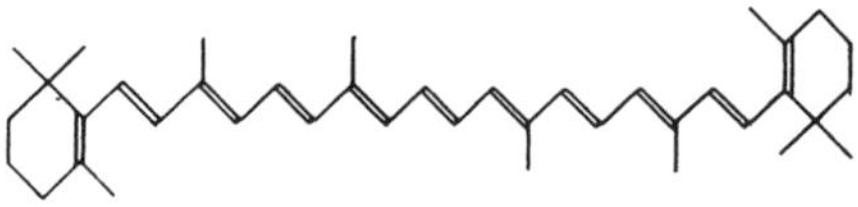

β–CAROTENE

CAROTENOIDS

GALLIC ACID 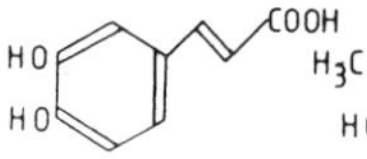   CAFFEIC ACID 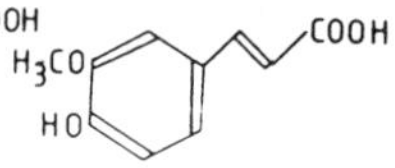   FERULIC ACID

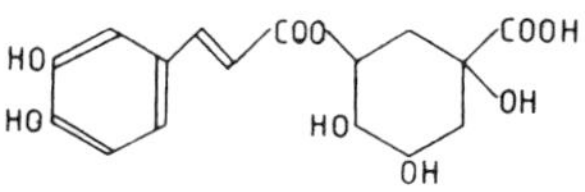

CHOROGENIC ACID 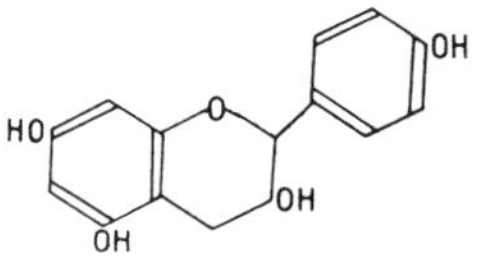   FLAVON–3–OL (CATECHINS)

PHENOLICS

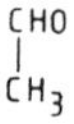

ETHANOL

ACETALDEHYDE   ETHYLACETATE

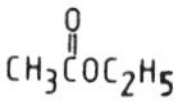

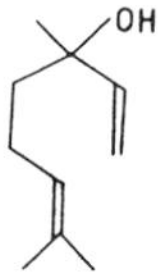

LINALOOL

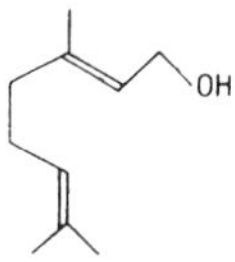

GERANIOL 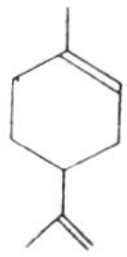   LIMONENE

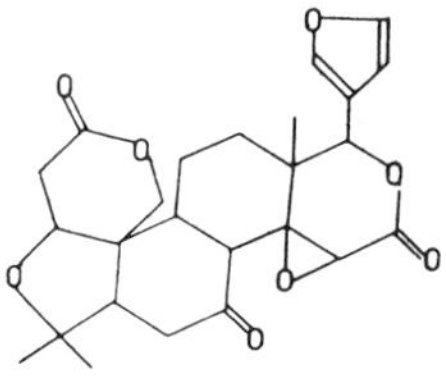

LIMONIN

FLAVOUR COMPOUNDS

Structure Chart

REFERENCES

Adams, D.O. and Yang, S.F. (1979) Ethylene biosynthesis: Identification of 1-aminocyclopropane-1-carboxylic acid as an intermediate in the conversion of methionine to ethylene. *Proceedings of the National Academy of Sciences, USA* **76**, 170–174

Ahmed, A.E. and Labavitch, J.M. (1980) Cell wall metabolism in ripening fruit. 1 Cell wall changes in ripening Bartlett pears. *Plant Physiology*, **65**, 1009–1013

Akazawa, T. and Okamoto, K. (1980) Biosynthesis and metabolism of sucrose. In *The biochemistry of plants – A comprehensive treatise*, (ed J. Preiss), Vol. 3, Academic Press, pp. 199–220

Anderson, J.W. and Beardall, J. (1991) *Molecular activities of plant cells. An introduction to plant biochemistry.* Blackwell.

Aspinall, G.O. (1980) Chemistry of cell wall polysaccharides. In *The biochemistry of plants – A comprehensive treatise*, (ed. J. Preiss), Vol. 3, Academic Press, pp. 473–500

Avigad, G. (1982) Sucrose and other disaccharides. In *Encyclopedia of plant physiology*, (eds F.A. Loewus, and W. Tanner), Vol. 13A, Springer-Verlag, pp. 217–347

Bacic, A., Harris, P.J. and Stone, B.A. (1988) Structure and function of plant cell walls. In *The biochemistry of plants – A comprehensive treatise*, (ed J. Preiss), Vol. 14, Academic Press, pp. 297–372

Balestrieri, C., Castaldo, D., Giovane, A., Quagliuolo, L. and Servillo, L. (1990) A glycoprotein inhibitor of pectinmethylesterase in kiwifruit. *European Journal of Biochemistry*, **193**, 183–187

Bartley, I.M. (1978) Exo-polygalacturonase of apple. *Phytochemistry*, **17**, 213–216

Ben-Arie, R., Kislev, N. and Frenkel, C. (1979) Ultrastructural changes in the cell walls of ripening apple and pear fruit. *Plant Physiology*, **64**, 197–202

Bennett, A.B., Smith, G.M. and Nichols, B.G. (1987) Regulation of climacteric respiration in ripening avocado fruit. *Plant Physiology*, **83**, 973–976

Beyer, E.M. Jr. (1976) A potent inhibitor of ethylene action in plants. *Plant Physiology*, **58**, 268–271

Beyer, E.M. Jr. (1985) Ethylene metabolism. In *Ethylene and plant development* (eds J.A. Roberts, and G.A. Tucker), Butterworths, pp. 125–138

Biale, J.B. (1974) Synthetic and degradative processes in fruit ripening. In *The postharvest biology and handling of fruit and vegetables*, (eds N.E. Haard and D.K. Salunkhe), AVI, pp. 5–18

Biale, J.B. and Young, R.E. (1971) The avocado pear. In *The biochemistry of fruit and their products*, (ed A.C. Hulme), Academic Press, Vol. 2, pp. 1–63

Biggs, M.S., Harriman, R.W. and Handa, A.K. (1986) Changes in gene expression during tomato fruit ripening. *Plant Physiology*, **81**, 395–403

Blackman, F.F. and Parija, P. (1928) Analytical studies in plant respiration. I: The respiration of a population of senescent ripening apples. *Proceedings of the Royal Society B. Biological Science*, **103**, 412–418

Bleeker, A.B., Kenyon, W.H., Somerville, S.C. and Kende, H. (1986) Use of monoclonal antibodies in the purification of 1-aminocyclopropane-1-

carboxylate synthase, an enzyme in ethylene biosynthesis. *Proceedings of the National Academy of Sciences, USA,* **83**, 7755–7759

Brady, C.J. (1987) Fruit ripening. *Annual Review of Plant Physiology,* **38**, 155–178

Brady, C.J. and O'Connell, P.B.H. (1976) On the significance of increased protein synthesis in ripening banana fruits. *Australian Journal of Plant Physiology,* **3**, 301–310

Bruinsma, J. (1983) Hormonal regulation of senesence, ageing, fading and ripening. In *Postharvest physiology and crop preservation,* ed. M. Lieberman, NATO ASI Series A, Plenum Press, 141–164

Buttery, R.G., Teranuhi, R., Flath, R.A. and Ling, C. (1989) Fresh tomato volatiles – composition and sensory studies. In *Flavour Chemistry Trends and Developments,* (eds R. Teranuhi, R.G. Buttery and R. Shahidi), ACS Symposium Series 388, pp. 213–222

Chalmers, D.J. and Rowan, K.S. (1971) The climacteric in ripening tomato fruit. *Plant Physiology,* **48**, 235–240

Christoffersen, R.E., Warm, E. and Laties, G.G. (1982). Gene expression during fruit ripening in avocado. *Planta,* **155**, 52–57

Copeland, L. and Turner, J.F. (1987). The regulation of glycolysis and the pentose phosphate pathways. In *The biochemistry of plants – A comprehensive treatise* (ed. D.D. Davies) Vol. 11, Academic Press, pp. 107–129

Crookes, P.R. and Grierson, D. (1983) Ultrastructure of tomato fruit ripening and the role of polygalacturonase isoenzymes in cell wall degradation. *Plant Physiology,* **72**, 1088–1093

Darby L.A., Ritchie, D.B. and Taylor, I.B. (1977) Isogenic lines of the tomato 'Ailsa craig'. Glasshouse Crops Research Institute, Annual Report, pp. 168–184

Darvill, A.G., McNeil, M., Albersheim, P. and Delmer, D.P. (1980) The primary cell walls of flowering plants. In *The Biochemistry of Plants,* (ed. N.E. Tolbert), Vol. 1, Academic Press, pp. 91–162

Dennis, D.T. and Turpin, D.H. (1990) *Plant Physiology, Biochemistry and Molecular Biology.* Longman Scientific and Technical.

DeSwardt, G.H., Swanepoel, J.H. and Dubenage, A.J. (1973) Relations between changes in ribosomal RNA and total protein synthesis and the respiratory climacteric in ripening pericarp tissue of tomato. *Zeitschrift fur Pflanzenphysiologie,* **70**, 358–363

Djebar, M.R. and Moreau, F. (1990) *In vitro* conversion of 1–amino cyclopropane–1–carboxylic acid to ethylene by apple mitochondria. *Plant Physiology and Biochemistry,* **28**, 523–530

Douce, R., Brouquiss, R. and Journet, E-P. (1987) Electron transfer and oxidative phosphorylation in plant mitochondria. In *The biochemistry of plants – A comprehensive treatise,* (ed. D.D. Davies), Vol. 11, Academic Press, pp. 177–213

Downton, W.J.S. and Hawker, J.S. (1973) Enzymes of starch metabolism in leaves and berries of *Vitis vinifera. Phytochemistry,* **12**, 1557–1563

Durr, P. and Schobinger, V. (1981) The contribution of some volatiles to the sensory quality of apple and orange juice odour. In *Flavour '81.* (ed. P. Schrier), Walter de Gruyter, pp. 179–193

El-Shora, H. and ApRees, T. (1991) Intracellular location of NADP-linked malic enzyme in C_3 plants. *Planta*, **185**, 362–367

Endo, M., Nakagawa, H., Ogura, N. and Sato, T. (1990) Size and levels of mRNA for acid invertase in ripe tomato fruit. *Plant Cell and Physiology*, **31**, 655–659

Faragher, J.D. and Chalmers, D.J. (1977) Regulation of anthocyanin synthesis in apple skin. III Involvement of phenylalanine ammonia lyase. *Australian Journal of Plant Physiology*, **4**, 133–141

Fischer, R.L. and Bennett, A.B. (1991) Role of cell wall hydrolases in fruit ripening. *Annual Review of Plant Physiology and Plant Molecular Biology*, **42**, 675–703

Frecknell, E.A. and Pattenden, G. (1984) Carotenoid differences in isogenic lines of tomato fruit colour mutants. *Phytochemistry*, **23**, 1707–1710

Frenkel, C., Klein, I. and Dilley, D.R. (1968) Protein synthesis in relation to ripening of pome fruit. *Plant Physiology*, **43**, 1146–1153

Friend, J. and Rhodes, M.J.C. (1981) *Recent advances in the biochemistry of fruit and vegetables*. Academic Press

Fry, S.C. (1986) Cross-linking of matrix polymers in the growing cell walls of angiosperms. *Annual Review of Plant Physiology*, **37**, 165–186

Garcia, E. and Lajola, A.M. (1988) Starch transformation during banana ripening. The amylase and glucosidase behaviour. *Journal of Food Science*, **53**, 1181–1186

Goodenough, P.W., Prossor, I.M. and Young, K. (1985) NADP–linked malic enzyme and malate metabolism in ageing tomato fruit. *Phytochemistry*, **24**, 1157–1162

Goodenough, P.W., Tucker, G.A., Grierson, D. and Thomas, T. (1982) Changes in colour, polygalacturonase, monosaccharides and organic acids during storage of tomatoes. *Phytochemistry*, **21**, 281–284

Goodwin, T.W. (1980) *The biochemistry of the carotenoids*. Vol. 1, Plants, Chapman and Hall, 2nd edition.

Grant, G.T., Morris, E.R., Rees, D.A., Smith, P.J.C. and Thom, D. (1973) Biological interactions between polysaccharides and divalent cations; The egg box model. *FEBS Letters*, **32**, 195–198

Grierson, D. (1985) Gene expression in ripening tomato fruit. *CRC Critical Reviews of Plant Science*, **3**, 113–132

Grierson, D., Maunders, M.J., Holdsworth, M.J. *et al.* (1987) Expression and function of ripening genes. In *Tomato Biotechnology*, (eds. D.J. Nevins and R.A. Jones) Alan Liss, New York, pp. 309–323

Grierson, D., Slater, A., Maunders, M. *et al.* (1985) Regulation of the expression of tomato ripening genes: the involvement of ethylene. In *Ethylene and Plant Development*, (eds. J.A. Roberts and G.A. Tucker), Butterworths, London, pp. 147–161

Grierson, D., Smith, C.J.S., Morris, P.C. *et al.* (1989) Manipulating fruit ripening physiology. In *Manipulation of Fruit*, (ed. C. Wright), Butterworths, London, pp. 387–398

Grierson, D., Smith, C.J.S., Watson, C.F. *et al.* (1990) Regulation of gene expression in transgenic tomato plants by antisense RNA and ripening-specific promoters. In *Genetic Engineering of Crop Plants*, (eds. G.W. Lycett and D. Grierson), Butterworths, London, pp. 115–125

Grierson, D., Tucker, G.A., Keen, J., Ray, J., Bird, C.R. and Schuch, W. (1986) Sequencing and identification of a cDNA clone for tomato poly-galacturonase. *Nucleic Acids Research*, **14**, 8595–8603

Gross, K.C. and Wallner, S.J. (1979) Degradation of cell wall polysaccharides during tomato fruit ripening. *Plant Physiology*, **63**, 117–120

Guy, M. and Kende, H. (1984) Conversion of ACC to ethylene by isolated vacuoles of *Pisum sativum* L. *Planta*, **160**, 281–285

Hamilton, A.J., Bouzayen, M. and Grierson, D. (1991) Identification of a tomato gene for the ethylene-forming enzyme by expression in yeast. *Proceedings of the National Academy of Sciences, USA*, **88**, 7434–7437

Hamilton, A.J., Lycett, G.W. and Grierson, D. (1990) Antisense gene that inhibits synthesis of the hormone ethylene in transgenic plants. *Nature*, **346**, 284–287

Hayashi, T. (1989) Xyloglucans in the primary cell wall. *Annual Review of Plant Physiology and Plant Molecular Biology*, **40**, 139–168

Hobson, G.E., Nichols, R., Davies, J.N. and Atkey, P. (1984) The inhibition of tomato fruit ripening by silver. *Journal of Plant Physiology*, **116**, 21–29

Hoffman, N.E. and Yang S.F. (1980) Changes of 1–aminocyclopropane–1–carboxylic acid content in ripening fruit in relation to their ethylene production rates. *Journal of the American Society for Horticultural Science*, **105**, 492–495

Hrazdina, G. and Creasy, L.L. (1979) Light-induced changes in anthocyanin concentration, activity of phenylalanine ammonia lyase and flavonone synthase and some of their properties in *Brassica oleracea*. *Phytochemistry*, **18**, 581–589

Huber, D.J. (1983a) Polyuronide degradation and hemicellulose modifications in ripening tomato fruit. *Journal of the American Society for Horticultural Science*, **108**, 405–409

Huber, D.J. (1983b) The role of cell wall hydrolases in fruit softening. *Horticultural Reviews*, **5**, 169–219

Huber, S.C. (1986) Fructose 2,6, bisphosphate as a regulatory metabolite in plants. *Annual Review of Plant Physiology*, **37**, 233–246

Isaacs, J.E. and Rhodes, M.J.C. (1987) Phosphofructokinase and ripening in tomato fruit. *Phytochemistry*, **26**, 649–653

Kayisu, K. and Hood, L.F. (1981) Molecular structure of banana starch. *Journal of Food Science*, **46**, 1894–1897

Keegstra, K., Talmadge, K.W., Bauer, W.D. and Albersheim, P. (1973) The structure of plant cell walls. III A model of the walls of suspension cultured sycamore cells based on the interactions of the macromolecular components. *Plant Physiology*, **51**, 188–196

Kende, H., Acaster, M.A. and Guy M. (1985) Studies on the enzymes of ethylene biosynthesis. In *Ethylene and plant development*, (eds. J.A. Roberts and G.A. Tucker), Butterworths, pp. 23–28

Knee, M. (1978) Metabolism of polymethylgalacturonate in apple fruit cortical tissue during ripening. *Phytochemistry*, **17**, 1257–1260

Knox, J.P., Linstead, P.J., King J., Cooper, C. and Roberts, K. (1990) Pectin esterification is spacially regulated both within cell walls and between developing tissues of root apices. *Planta*, **181**, 512–521

Kramer, M., Sheehy, R.E. and Hiatt, W.R. (1989) Progress towards the genetic engineering of tomato fruit softening. *Trends in Biotechnology*, **7**, 191–194

Lincoln, J.E., Cordes, S., Read, E. and Fischer, R.L. (1987) Regulation of gene expression by ethylene during *Lycopersicon esculentum* (tomato) fruit development. *Proceedings of the National Academy of Sciences, USA*, **84**, 2793–2797

Mayer, A.M. and Harel, E. (1979) Polyphenoloxidases in plants: A review. *Phytochemistry*, **18**, 193–215

Mayer, A.M. and Harel, E. (1981) Polyphenoloxidases in fruit – changes during ripening. In *Recent advances in the biochemistry of fruit and vegetables*, (ed. J. Friend, and M.J.C. Rhodes), Academic Press, pp. 159–178

McCann, M.C., Wells, B. and Roberts, K. (1990) Direct visualisation of cross-links in the primary cell wall. *Journal of Cell Science*, **96**, 323–324

McGlasson, W.B., Wade, M.L. and Adato, I. (1978) Phytohormones and fruit ripening. In *Phytohormones and related compounds – A comprehensive treatise*, Vol. 2, (eds. D. S. Letham, P.B. Goodwin and T.J.V. Higgins), Academic Press, pp. 447–493

McMurchie, E.J., McGlasson, W.B. and Eaks, I.L. (1972) Treatment of fruit with propylene gives information about the biogenesis of ethylene. *Nature*, **287**, 235–236

McNeil, M., Darvill, A.G., Fry, S. and Albersheim, P. (1984) Structure and function of the primary cell wall of plants. *Annual Review of Biochemistry*, **53**, 625–663

Millerd, A., Bonner, J. and Biale, J.B. (1953) The climacteric rise in fruit respiration as controlled by phosphorylative coupling. *Plant Physiology*, **28**, 521–531

Mitcham, E.J., Gross, K.C. and Ng, T.J. (1989) Tomato fruit cell wall synthesis during development and senescence. *Plant Physiology*, **29**, 477–481

Moller, I.M., Berezi, A., Van der Plas, L.H.W. and Lambers, H. (1988) Measurement of the activity and capacity of the alternate pathway in intact plant tissues. Identification of problems and possible solutions. *Physiologia Plantarum*, **72**, 642–649

Moreau, F. and Romani, R. (1982) Malate oxidation and cyanide insensitive respiration in avocado. *Plant Physiology*, **70**, 1385–1390

Morton, I.D. and McCleod A.J. (1990) *Food flavours*, part C. The flavour of fruits. Elsevier.

Nakajimo, N., Nakagawa, N. and Imaseki, H. (1988) Molecular size of wound induced ACC synthase from *Cucurbita maxima Duch* and change of translatable mRNA of the enzyme after wounding. *Plant Cell Physiology*, **29**, 898–998

Nursten, H.E. (1970) Organic acids. In *The Biochemistry of fruit and their products*, (ed. A. Hulme), Vol. 2, Academic Press, pp. 89–113

Oeller, P.W., Min-Wong, L., Taylor, L.P., Pike, D.A. and Theologis, A. (1991) Reversible inhibition of tomato fruit senescence by antisense RNA. *Science*, **254**, 437–439

Palmer, J.M. and Moller, I.M. (1982). Regulation of NAD(P)H dehydrogenases in plant mitochondria. *Trends in Biochemical Sciences*, **7**, 258–261

Pesis, E., Fuchs, Y. and Zaubermann, G. (1978) Cellulase activity and fruit softening in avocado. *Plant Physiology*, **61**, 416–417

Phan, C-T. (1970) Photosynthetic activity of fruit tissues. *Plant and Cell Physiology*, **11**, 823–825

Pratt, H.K. and Groeschl, J.A. (1968). In *Biochemistry and Physiology of plant growth substances*. (eds. F. Wightman and G. Setterfield) Runge Press, Ottawa, pp. 1293–1302

Presis, J. and Levi, C. (1980) Starch biosynthesis and degradation. in *The biochemistry of plants – A comprehensive treatise*, (ed. J. Preiss) Vol. 13, Academic Press, pp. 371–424

Pressey, R. (1983) β-galactosidases in ripening tomatoes. *Plant Physiology*, **71**, 132–135

Pressey, R. and Avants, J.K. (1972) Multiple forms of pectin methyl esterase in tomatoes. *Phytochemistry*, **11**, 3139–3142

Pressey, R. and Avants, J.K. (1978) Differences in polygalacturonase composition of clingstone and freestone peaches. *Journal of Food Science*, **43**, 1415–1417

Rattanapanone, N., Speirs, J. and Grierson, D. (1978) Evidence for changes in mRNA content related to tomato fruit ripening. *Phytochemistry*, **17**, 1485–1486

Richmond, A., and Biale, J.B. (1966) Protein and nucleic acid metabolism in fruit. I: Studies of amino acid incorporation during climacteric rise in respiration of the avocado. *Plant Physiology*, **41**, 1247–1253.

Richmond, A. and Biale, J.B. (1967) Protein and nucleic acid metabolism in fruit. II: RNA synthesis during respiratory rise of the avocado. *Biochimica et Biophysica Acta*, **138**, 625–627

Rick, C.M. (1980) Tomato linkage map. Tomato Genetics Cooperative Report, p. 30

Rick, C.M. (1987) Genetic resourse in *Lycopersicon*. In *Tomato Biotechnology* (eds. D.J. Nevins and R.A. Jones), Alan Liss, New York, pp. 17–26

Romani, R.J. (1975) Mitochondrial function and survival in relation to fruit ripening and the climacteric. *Center Nationale Research Symposium*, no. 238, pp. 229–233

Saftner, R.A. and Baldi, B.G. (1990) Polyamine levels and tomato fruit development: possible interactions with ethylene. *Plant Physiology*, **92**, 547–550

Salminen, S.O. and Young, R.E. (1974) Negative cooperativity of phospho-fructokinase as a possible regulator of ripening in banana fruit. *Nature*, **247**, 389–391

Sawamura, M., Knegt, E. and Bruinsma, J. (1978) Levels of endogenous ethylene, carbon dioxide and soluble pectin, and activities of pectinmethylesterase and polygalacturonase in ripening tomato fruit. *Plant Cell Physiology*, **19**, 1061–1069

Schrier, P. and Gorens, G. (1981) Formation of 'green-grassy' notes in disrupted plant tissues: characterisation of the tomato enzyme system. In *Flavour '81*, (ed. P. Schrier), Walter de Gruyter, pp. 495–507

Schuch, W., Bird, C.R., Ray, J. *et al.* (1989) Control and manipulation of gene expression during tomato fruit ripening. *Plant Molecular Biology*, **13**, 303–311

Seymour, G.B., Lasslett, Y. and Tucker, G.A. (1987) Differential effects of pectolytic enzymes on tomato polyuronide *in vivo* and *in vitro*. *Phytochemistry*, **26**, 3137–3139

Smith, C.J.S., Slater, A. and Grierson, D. (1986) Rapid appearance of an mRNA correlated with ethylene synthesis encoding protein of molecular weight 35,000. *Planta*, **168**, 94–100

Smith, C.J.S., Watson, C.F., Ray, J. *et al.* (1988) Antisense RNA inhibition of polygalacturonase gene expression in transgenic tomatoes. *Nature* **334**, 724–726

Smith, C.J.S., Watson, C.F., Morris, P.C., *et al.* (1990) Inheritance and effects on ripening of antisense polygalacturonase genes in transgenic tomatoes. *Plant Molecular Biology*, **14**, 369–379

Smith, N.J.S., Tucker, G.A. and Jeger, M.J. (1989a) Softening and cell wall changes in bananas and plantains. *Aspects of Applied Biology*, **20**, 57–66

Smith, R., Seymour, G.B. and Tucker, G.A. (1989b) Inhibition of cell wall degradation by silver(I) ions during ripening of tomato fruit. *Journal of Plant Physiology*, **134**, 514–516

Solomos, T. (1983) Respiration and energy metabolism in senescing plant tissues. In *Postharvest physiology and crop preservation*, (ed. M. Lieberman) NATO ASI Series A, Plenum Press, pp. 61–98

Solomos, T. and Laties, G.G. (1976) Effects of cyanide and ethylene on the respiration of cyanide-sensitive and cyanide-resistant plant tissues. *Plant Physiology*, **58**, 47–50

Speirs, J. and Brady, C.J. (1991) Modification of gene expression in ripening fruit. *Australian Journal of Plant Physiology*, **18**, 519–532

Steup, M. (1988) Starch degradation. In *The biochemistry of plants – A comprehensive treatise*, (ed. J. Preiss), Vol. 14, Academic Press, pp. 255–296

Stitt, M. (1990) Fructose 2,6 bisphosphate as a regulatory molecule in plants. *Annual Review of Plant Physiology and Plant Molecular Biology*, **41**, 153–185

Stitt, M., Cseke, C. and Buchanan, B. (1986) Ethylene-induced increase in fructose 2,6 bisphosphate in plant storage tissue. *Plant Physiology*, **80**, 246–248

Surendranathan, K., Iuer, M.G. and Nair, P.M. (1990) Characterisation of a monomeric phosphofructokinase from banana – role of magnesium in its regulation. *Plant Science*, **72**, 27–35

Theologis, A. and Laties, G.G. (1978) Respiratory contribution of the alternative path during various stages of ripening in avocado and banana fruit. *Plant Physiology*, **62**, 249–255

Timberlake C.F. (1981) Anthocyanins in fruit and vegetables. In *Recent advances in the biochemistry of fruit and vegetables*, (eds. J. Friend, and M.J.C. Rhodes), Academic Press, 221–247

Tucker, G.A. (1990) Genetic manipulation of fruit ripening. *Biotechnology and Genetic Engineering Reviews*, **8**, 133–159

Tucker, G.A. and Brady, C.J. (1987) Silver ions interrupt tomato fruit ripening. *Journal of Plant Physiology*, **127**, 165–169

Tucker, G.A. and Grierson, D. (1987) Fruit ripening. In *The Biochemistry of Plants – A comprehensive treatise* (ed. D.D. Davies), Vol. 12, Academic Press, pp. 265–318

Tucker, G.A., Robertson, N.G. and Grierson, D. (1982) Purification and changes in activity of tomato pectinesterase isoenzymes. *Journal of the Science of Food and Agriculture*, **33**, 396–400

Turner, J.F. and Turner, D.H. (1980) The regulation of glycolysis and the pentose phosphate pathway. In *The biochemistry of plants – A comprehensive treatise*, (ed. D.D. Davies), Vol. 2, Academic Press, pp. 279–316

Ulrich, R. (1970) Organic acids. In *The Biochemistry of fruits and their products*. (ed. A. Hulme), Vol. 1, Academic Press, pp. 89–118

Van der Straeten, D., Vanwiemeersch, L., Goodman, H.M. and Van Montagu, M. (1990) Cloning and sequencing of two different cDNAs encoding 1–amino cyclopropane–1-carboxylate synthase in tomato *Proceedings of the National Academy of Science*, **87**, 4859–4863

Van Straten, S. (1977) Volatile compounds in food. TNO. Zeist

Ververdis, P. and John, P. (1991) Complete recovery *in vitro* of ethylene forming enzyme activity. *Phytochemistry*, **30**, 725–727

Vickery, M.L. and Vickery, B. (1981) *Secondary plant metabolism*. Macmillan Press

Vickery, R.S. and Bruinsma, J. (1973) Compartmentation and permeability for potassium in developing fruits of tomato. *Journal of Experimental Botany*, **24**, 1261–1270

White, J.A. and Kende, H. (1990) ACC synthases. Are there charge and size variants in tomato? *Journal of Plant Physiology*, **136**, 646–652

Whiting, G.C. (1970) Sugars. In *The biochemistry of fruits and their products*, (ed. A.C. Hulme), Vol. 1, Academic Press, pp. 1–31

Wills, R.B.H., McGlasson, W.B., Graham, D., Lee T.H. and Hall, E.G. (1989) *Postharvest: an introduction to the physiology and handling of fruit and vegetables*. BSP Professional Books, Hong Kong

Wong, J.H., Kiss, F., Wu, M-X. and Buchanan, B.B. (1990) Pyrophosphate fructose–6 phosphate 1-phosphotransferase from tomato fruit. Evidence for change during ripening. *Plant Physiology*, **94**, 499–506

Yang, S.F. (1981) Ethylene the gaseous plant hormone and regulation of biosynthesis. *Trends in Biochemical Sciences*, **6**, 161–164

Yang, S.F., Liu, Y., Su, L., Peiser, G.D., Hoffman, N.E. and McKeon, T. (1985) Metabolism of 1-aminocyclopropane-1-carboxylic acid. In *Ethylene and plant development* (eds. J.A. Roberts, and G.A. Tucker), Butterworths, pp. 9–22

Yelle, S., Chetelat, R.T., Dorais, M., DeVerna, J.W. and Bennett, A.B. (1991) Sink metabolism in tomato fruit. IV Genetic and biochemical analysis of sucrose accumulation. *Plant Physiology*, **95**, 1026–1035

Yip, W-K., Dong, J-G., Kenny, J.W., Thompson, G.A. and Yang, S.F. (1990) Characterisation and sequencing of the active site of ACC synthase. *Proceedings of the National Academy of Science*, **87**, 7930–7934

Young, R.E. and Biale, J.B. (1967) Phosphorylation in avocado fruit slices in relation to the respiratory climacteric. *Plant Physiology*, **42**, 1357–1362

Chapter 2

Avocado

G.B. Seymour and G.A. Tucker

2.1 INTRODUCTION

The avocado fruit has been a major food for the people of Central America for, apparently, several thousand years (Bergh, 1976). The fruit is rich in unsaturated fats and vitamins and the flesh has more energy value than meat of equal weight. The avocado tree (*Persea americana Mill.*) belongs to the family Lauraceae. The avocado fruit is classified botanically as a berry comprising the seed and the pericarp, which is separated into (a) rind or exocarp, (b) flesh or mesocarp, and (c) the thin layer next to the seed coat, the endocarp. The exocarp consists of cuticle, epidermal, parenchyma and sclerenchyma or stone cells limiting the inner surface of the peel, while the mesocarp is composed of large iso-diametric lipid-containing parenchyma cells and is permeated by the vascular system (Valmayor, 1967; Biale and Young, 1971).

The avocado is now grown throughout most of the tropics or sub-tropics, but appears to have originated in Central America (Bergh, 1976). The commercial varieties are placed for horticultural purposes in one of three groups or races – West Indian, Guatemalan or Mexican. They are sometimes described as tropical, subtropical and semitropical on the basis of increasing cold hardiness and general climactic adaptation (Bergh, 1976; Knight, 1980). Fruit characteristics including size and skin texture vary considerably among the races (Knight, 1980). Prominent commercial varieties include Fuerte and Hass. Fuerte is a Guatemalan/Mexican hybrid and Hass originated from a Guatamalan seedling. Other commercially important varieties include Lula, Booth 8, Walden, Pollock (Bergh, 1976; Ahmed and Barmore, 1980). For further details on avocado breeding, refer to Bergh (1975; 1976).

Biochemistry of Fruit Ripening. Edited by G. Seymour, J. Taylor and G. Tucker. Published in 1993 by Chapman & Hall, London. ISBN 0 412 40830 9

Recently a considerable international trade in avocado fruits has developed. Total world production of avocados in 1990 was around 1.4 million metric tonnes, with major producing countries including Mexico, USA, Brazil, Dominican Republic, Indonesia, Zaire, South Africa and Israel.

Previous reviews on the physiology and biochemistry of the avocado fruit include Biale and Young (1971); Lewis (1978); Ahmed and Barmore (1988); Bower and Cutting (1988), the latter of which concentrates on the physiological aspects of the development and ripening of avocado. The present work is intended to cover the more biochemical aspects, and particularly recent developments on the role of gene expression in ripening avocados.

2.2   PHYSIOLOGY

Fruit growth in avocado follows a pattern similar to the development of that in other fruits, with rapid cell division at the early stages. However, in the avocado cell multiplication continues in the mature fruit and in general avocado fruit tend to continue growing while attached to the tree (Schroeder, 1953; Valmayor, 1967). Successful marketing of avocados requires the selection of mature fruit at harvest, since immature fruit may fail to develop the appropriate flavour or texture when ripe. However, unlike most other fruit avocados do not normally ripen until after harvest and can remain in a mature, but unripe condition until picked. In California, a minimum oil content of 8%, based on the fresh weight of the fruit, exclusive of the skin and seed, was used as the standard of maturity for all cultivars. Avocados having more than the minimum oil content may be lacking in organoleptic qualities, though raising the oil content standard might eliminate from the market avocado varieties or crops whose organoleptic qualities are adequate (Ahmed and Barmore, 1980). A strong relationship is apparent between avocado fruit growth and maturity and between oil content and dry weight. Consequently, the Californian minimum maturity index has been changed from oil content to percentage dry weight (Kader, 1992).

Ripening of avocado fruit may occur a few days after harvest. Ripening and softening can be delayed by precooling immediately after harvest. Some cultivars, e.g. Booth 1, Booth 8, Taylor, which are chilling tolerant can be stored for 4 to 8 weeks at 4.4°C. Other chilling-sensitive varieties store best at 13°C for a maximum period of 2 weeks. Ripening is also delayed by holding the fruit in low O_2 (2–5%) and high CO_2 (3–10%) conditions (Biale and Young, 1971; Kader, 1992). The biochemical effects of low oxygen atmospheres on avocado ripening are still poorly understood, but investigations to understand the mechanisms involved have been undertaken by Kanellis *et al.* (1989a, 1989b). The best ripening temperatures are between 15.5 and 24°C and ethylene may be used to stimulate ripening (Hardenburg *et al.*, 1986). Temperatures above about 30°C yield fruit with surface pitting and poor flavour (Lee and Young, 1984), while at 40°C normal ripening is inhibited (Eaks, 1978).

Factors which are responsible for delaying the onset of ripening in avocado fruits while they are attached to the tree are still unknown, but have received considerable attention. Early work (Biale and Young, 1971) indicated that there was a 'tree factor' translocated to the fruit and this was responsible for the inhibition of ripening. This hypothesis was based on girdling experiments which indicated that avocado fruit would not ripen if attached to a branch having functional leaves. To confirm these observations Tingwa and Young (1975a) undertook more comprehensive girdling and defoliation studies; results indicated that nearby leaves on a branch either had no effect, or accelerated the time of abscission and subsequent ripening. The conclusion was that the leaves did not appear to be a source of ripening inhibitor. However, it was observed that ripening occurred only after the fruit had abscissed from the peduncle, suggesting that the peduncle may be responsible for supplying such a factor. Inhibitory factors accumulated from the tree continue to exert their influence for a limited period after harvest, although this inhibition is inversely related to fruit maturity at harvest (Adato and Gazit, 1974). The possibility that these inhibitory factors are related to auxin or other plant hormones has been investigated by vacuum infiltration of avocado fruits with IAA, kinetin, abscisic acid and gibberellic acid. At low concentrations (1–10 μM), IAA delayed avocado ripening, but the other plant growth regulators had little effect (Tingwa and Young, 1975b; Adato and Gazit, 1976). The unidentified ripening inhibitor(s) may act partly by inhibiting ethylene biosynthesis at the onset of ripening (see page 59). In this respect it has been suggested that endogenous polyamines could also be involved in delaying avocado ripening. Polyamines have been shown to inhibit the production of ethylene in plant tissues and there is some evidence for an inverse relationship between polyamine levels and ethylene production. However, their role in avocado fruit ripening is still open to question (Winer and Apelbaum, 1986; Kushad *et al.*, 1988). Other factors that may regulate the onset of ripening include the internal calcium concentrations. Avocado fruit with low levels of calcium ripen more rapidly than those with higher levels of calcium (Eaks, 1985) and ripening in avocados can be delayed by vacuum infiltration of calcium (Wills and Tirmazi, 1982).

2.3 BIOCHEMISTRY

The onset of ripening in avocado is marked by a variety of biochemical changes including a large increase in ethylene production and respiration, texture changes and development of flavour components. The present review concentrates on the biochemical and molecular basis of these changes.

2.3.1 Respiration

Respiration rate of the fruit declines after harvest and low respiration rate defines a lag period between harvest and ripening. The avocado is a climacteric fruit and ripening is associated with a sharp increase in respiration and ethylene production (Fig. 2.1). The biochemical basis of the climacteric in avocado and other fruits is poorly understood. Avocado is one of the most metabolically active fruits yet studied (Biale and Young, 1971). The substrate for this metabolic activity is not clearly defined, but the respiratory quotient (RQ) – the ratio of CO_2 produced to O_2 consumed – remains at around 1 during the climacteric, indicating that the respiratory substrate during this period is carbohydrates rather than fat (Blanke, 1991). However, there are indications that some degradation of the fat reserve does occur during ripening (refer to page 61). This carbohydrate probably arises from the

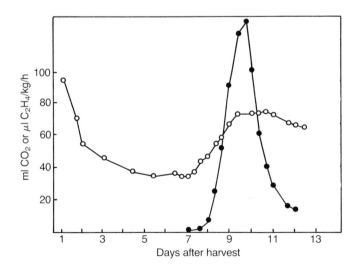

Fig. 2.1 Postharvest trends in carbon dioxide (o) and ethylene (•) production in an individual Fuerte avocado fruit. (Redrawn from Awad and Young, 1979.)

reserve of starch in the fruit, which falls from around 12 mg/g dry weight in unripe fruit to undetectable levels during ripening (Pesis *et al.*, 1978). Early studies on the biochemistry of respiration in ripening avocado were reviewed by Biale and Young (1971). Studies with inhibitors and isolated mitochondria suggested that the tricarboxylic acid cycle was operating during the climacteric to bring about oxidation of the respiratory substrate, and that this oxidation was coupled to the production of ATP, probably by electron transport through a cytochrome-mediated pathway (Biale and Young, 1971). Solomos and Laties (1974) reported that preclimacteric avocados under anaerobic conditions showed enhanced glycolysis and a similar pattern of enhanced utilization of glycolytic intermediates was apparent during the climacteric. This suggested that the carbon flux through glycolysis was enhanced during the respiratory increase in ripening avocados. Subsequent investigation of the regulation of enhanced glycolysis during ripening showed that conversion of fructose-6-phosphate to fructose 1, 6-bisphosphate is a rate limiting step in glycolysis. This reaction can be catalysed by ATP-dependent phospho-fructokinase (ATP–PFK) (EC 2.7.1.11) or pyrophosphate: fructose-6-phosphate phosphotransferase (PFP) (EC 2.7.1.90). In avocados, as in other fruit, e.g. bananas, levels of fructose-1,6-bisphosphate show a large (10-fold) increase during ripening (Solomos and Laties, 1974) and this may be brought about by the activation of ATP–PFK or PFP. Recently Bennett *et al.* (1987) measured the production of high energy phosphate (ATP) in climacteric avocados *in vivo* using ^{31}P nuclear magnetic resonance spectroscopy. They also examined the possible role of PFP and its activation by fructose-2,6-bisphosphate during the respiratory rise, finding large increases in ATP during ripening and confirming previous reports by Young and Biale (1967). An increase in fructose-2,6-bisphosphate which correlated temporally with the climacteric rise in respiration was also seen. Bennett *et al.* (1987) postulated that the climacteric rise might be mediated by enhanced PFP activity regulated by the presence of fructose-2,6-bisphosphate. However, recent studies in bananas cast some doubt on the accuracy of previously published measurements of fructose-2,6-bisphosphate and, in bananas at least, it appears that it is ATP – PFK and not PFP that is activated during the climacteric (Ball *et al.*, 1991).

The respiration and ripening of preclimacteric avocados can be stimulated by exposure to cyanide and this may indicate a role for the participation of the cyanide-resistant electron transport pathway in the climacteric in avocado and other fruits. Theologis and Laties (1978) investigated the contribution of this alternative pathway to respiration during ripening of intact avocado fruit and fruit slices using various inhibitors of the alternate pathway. These studies indicated that while present in avocado tissue, the alternate pathway does not appear to contribute to respiration in the absence of exogenous cyanide. These findings have been further supported by computer modelling, which on taking into account external oxygen tension, internal oxygen tension, intrinsic respiration rate, diffusion resistance and other factors, indicated that the cyanide-resistant oxidase does not appear to contribute appreciably to preclimacteric or climacteric respiration (Tucker and Laties, 1985). Moreau and Romani (1982) have presented evidence for a link between the action of NAD^+–malic enzyme (ME) (EC 1.1.1.39), cyanide-insensitive respiration and the climacteric in ripening avocado. ME converts malate to pyruvate in the mitochondria; the authors found that the rate of malate oxidation increases as ripening advances through the climacteric and the conversion of malate to pyruvate parallels the increase in mitochondrial O_2 uptake. ME in avocado appears to be linked to the cyanide-insensitive pathway and is therefore likely to function under conditions of low ATP demand. However, the role, if any, of the cyanide-resistant pathway in the climacteric is unresolved.

It has been proposed that the respiratory climacteric in avocado may represent maintenance metabolism of mitochondria in senescent fruit cells. In the ripening fruit cell increased permeability of intracellular membranes, e.g. the tonoplast, may expose mitochondria to stressful substances and there is evidence to suggest that to retain respiratory control, the mitochondria may respond with increased synthesis of ATP (Huang and Romani, 1991). As previously stated, there is a large increase in ATP synthesis during the climacteric (Bennett *et al.*, 1987) and the ADP/ATP ratio decreased with the transition from the preclimacteric to the climacteric, indicating a high energy charge, but low ATP demand. Thus the climacteric may not be initiated as the result of an increased energy requirement, but could indeed be a response to changes in the cytosol (availability of substrates, cofactors, activators and inhibitors) or the mitochondria (Blanke, 1991).

The avocado fruit may recycle some of the CO_2 produced during respiration using the enzyme phosphoenolpyruvate carboxylase (PEPC) (EC 4.1.1.31). This enzyme has been found in a number of fruit tissues and is concentrated in the seeds and perivascular tissue in avocado (Blanke, 1991).

2.3.2 Ethylene biosynthesis

In avocado, as in other climacteric fruit, the onset of ripening is marked by a large rise in ethylene production. Ethylene biosynthesis seems to be initiated in the distal end of the mature fruit, away from the pedicel and areas of greatest vascular concentration (Adato and Gazit, 1977). The precise role of ethylene is still uncertain, but it appears to be intimately involved in the initiation and coordination of ripening in fruits (Chapter 1). The rise in ethylene production accompanies the respiratory climacteric and ethylene biosynthesis in avocado follows the pathway from *S*-adenosyl-methionine (SAM) to 1-aminocyclopropane-1-carboxylic acid (ACC) to ethylene described by Adams and Yang (1979). Studies with Fuerte avocados (Hoffman and Yang, 1980) indicated that pre-climacteric fruit contained less than 0.05 $nmol.g^{-1}$ fresh weight of ACC, but these levels increased dramatically during the climacteric rise.

The time of onset of ethylene production in avocado appears to be carefully controlled. Fruit maturation apparently involves an increased ability to generate ethylene since mature, but not immature, fruit respond to exogenous ethylene or propylene by inducing endogenous ethylene production (Eaks, 1980). There is evidence that the rapid synthesis of ethylene by preclimacteric fruit is prevented by a lack of ACC. Sitrit *et al.* (1986) found that application of ACC to intact avocado fruit, attached or detached from the tree, resulted in increased ethylene production and ripening of these fruits. ACC synthase (EC 4.4.1.14) is the enzyme responsible for converting SAM to ACC. This enzyme shows only slight activity in preclimacteric avocado fruits, and inhibitor studies with avocado fruit discs indicate that the increase in the activity of this enzyme during ripening requires RNA and protein synthesis (Sitrit *et al.*, 1986). Thus regulation of the *de novo* synthesis of ACC synthase may play a major role in the initiation of ethylene biosynthesis and ripening, and repression of ACC synthase may be one of the factors preventing ripening on the tree. Other points in the biosynthetic pathway which may be important in this case include the conversion of ACC to ethylene by the ethylene forming enzyme (EFE) and malonylation of ACC to malonyl–ACC (MACC). Preclimacteric avocados contain significant levels of MACC, and this may play a role in regulating ethylene production in the preclimacteric stage (Sitrit *et al.*, 1986).

Changes in the fruit prior to the onset of ripening may reflect changes not only in the ability to synthesize ethylene, but also changes in ethylene sensitivity. Recently, Starrett and Laties (1991) showed that short pulses (24h) of exogenous ethylene or propylene on preclimacteric avocado fruit did not cause the immediate onset of ripening, but did stimulate some biochemical and molecular changes. EFE activity was sharply augmented by an ethylene pulse, although there was no change in ACC synthase activity and EFE activity declined after the pulse and did not rise again until ACC synthase was triggered at the onset of the climacteric. However, such ethylene pulses, when given 24 hours or more after picking, did reduce the length of the preclimacteric period. Hence, the authors suggest that the pulse must influence developmental phenomena other than ethylene synthesis *per se*, and in support of this proposition were able to find evidence for the up- and down-regulation of several genes in response to the ethylene pulse. These events may occur naturally in fruit during the preclimacteric period. However, while factors controlling the changes in ethylene sensitivity, and which are responsible for inhibiting ripening on the tree, are still unknown, the ability of molecular techniques to cast light on such difficult problems suggests that such events may not remain a mystery for much longer.

2.3.3 Lipids

The outstanding compositional feature of the avocado fruit is its high fat content, which can reach over 20% of the fresh weight in some cultivars (Mazliak, 1970; Biale and Young, 1971). The fat content increases during growth of the fruit. Changes during the ripening are much less marked than those during growth and development.

Avocado lipids can be divided into a number of fractions: (1) neutral lipids; (2) phospholipids; (3) glycolipids; (4) free fatty acids. The neutral lipid fraction constitutes about 96% of the total lipids at harvest time in mature Fuerte avocados and the majority of these neutral lipids are triglycerides (Table 2.1). The fatty acid composition of each lipid fraction is shown in Table 2.2. In each case, oleic, linoleic, palmitic and palmitoleic were the major acids present. There is apparently little difference in the fatty acid complement between mesocarp, endocarp and exocarp. The seed is low in fat, having about 1% on a fresh weight basis (Biale and Young, 1971).

Changes during development and ripening

The triglycerides represent the bulk of the storage lipids and these show the greatest change during development (Fig. 2.2), with much smaller

Table 2.1 Classes of lipids in the mesocarp of mature Fuerte avocado. (After Kikuta and Erickson, 1968.)

Class	% of fresh weight	% of total lipid
Neutral lipids		
Triglycerides	19.96	87.5
Diglycerides	1.29	5.7
Monoglycerides	0.78	3.4
Phospholipids	0.39	1.7
Free fatty acids	0.10	0.4
Others includes sterols, hydrocarbons	0.28	1.2

Table 2.2 Fatty acid composition of lipid fractions from the mesocarp of Fuerte avocado. (From Biale and Young, 1971).

Fraction	16:0	16:1	18:0	18:1	18:2	18:3	20:0	UN
Free fatty acid	20.3	9.7	0.4	43.7	22.5	3.0	–	0.4
Triglyceride	25.4	7.0	0.5	54.3	12.3	–	0.5	–
Diglyceride I	15.0	9.5	–	45.0	28.0	3.0	–	–
Diglyceride II	18.4	3.9	0.7	64.8	12.2	–	–	–
Glycolipid I	6.7	2.5	1.6	13.1	76.1	–	–	–
Monoglyceride	17.1	7.2	2.7	43.2	24.3	1.0	0.9	3.6
Glycolipid II	3.6	2.2	1.2	12.8	74.1	6.0	–	–
Phospholipid	16.9	4.4	3.3	20.5	36.1	9.8	–	9.0

16:0 (palmitic), 16:1 (palmitoleic), 18:0 (stearic), 18:1 (oleic), 18:2 (linoleic), 18:3 (linolenic), 20:0 (arachidic) and UN (unknown).

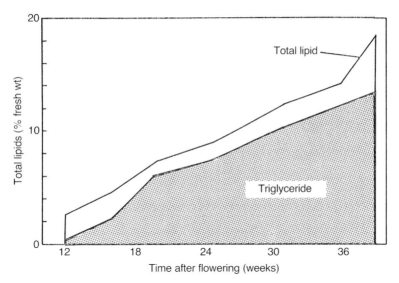

Fig. 2.2 Lipid content of the mesocarp of developing Lula avocado fruit. (After Gaydou *et al.*, 1987.)

changes being apparent in other fractions (Kikuta and Erickson, 1968). Overall changes in fatty acid components in the mesocarp of Fuerte and Lula avocados are shown in Figure 2.3. In Fuerte, one fatty acid in particular (oleic) increases dramatically during development. Kikuta and Erickson (1968) reported that in Fuerte avocados there were some changes in the lipids during ripening, including increases in the mono-glyceride and free fatty acid fractions and these may result from the degradation of the triglycerides. Thus the storage lipids may be involved in some way in the metabolic processes taking place during ripening.

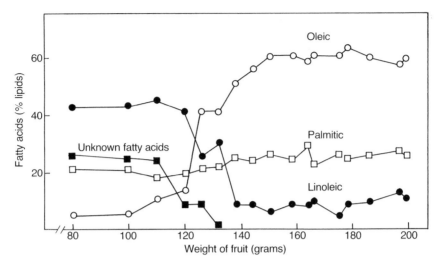

Fig. 2.3   Fatty acid composition of total lipids in Fuerte avocado fruit and changes during fruit development. (Redrawn from Kikuta and Erickson, 1968.)

Biosynthesis of triglycerides

Triglycerides are the main lipid fraction synthesized during avocado fruit development. They are fatty acid esters of glycerol where all three hydroxyl groups of glycerol are esterified by a fatty acid. The fatty acid composition of the major triglycerides of mature Fuerte avocados is shown in Table 2.3.

Early work on the biosynthesis of fat in avocado was reviewed by Mazliak (1970) and Biale and Young (1971). These and more recent studies on other plant species (Anderson and Beardall, 1991) indicate that the biosynthesis of fatty acids and subsequent storage of triglycerides in avocado probably proceeds by the pathway summarized in Figure 2.4. In avocado fruit, as in other tissues, the triglycerides are stored in oleosomes, which take up much of the space in the mature mesocarp cells (Fig. 2.5) (Platt-Aloia and Thomson, 1980; Anderson and Beardall, 1991). Processes by which the avocado triglycerides are broken down involve the action of lipases which liberate free fatty acids which are in

Table 2.3   Major triglycerides and their component fatty acids from Fuerte avocado. (From Gaydou *et al.*, 1987).

Triglyceride	Component fatty acids
Dioleyl palmitin	Oleic/Oleic/Palmitic
Triolein	Oleic/Oleic/Oleic
Dioleylpalmitolein	Oleic/Oleic/Palmitoleic
Linoleyl oleyl palmitin	Linoleic/Oleic/Palmitic
Linoleyl diolein	Linoleic/Oleic/Oleic

turn broken down to acetyl-CoA by β–oxidation. For a review of the composition of the cuticular wax and sterols of avocado fruit, refer to Ahmed and Barmore (1980).

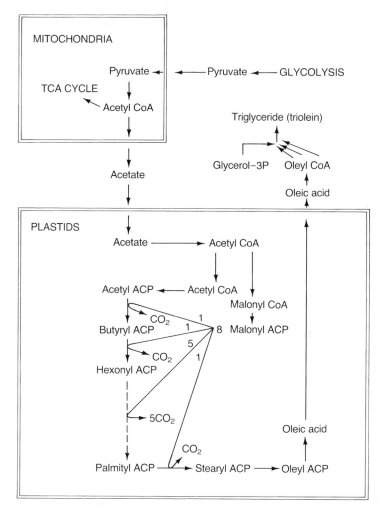

Fig. 2.4   Probable pathway for the biosynthesis of fatty acids and triglycerides in avocado fruit. It is now thought that the biosynthesis of fatty acids in plants occurs in the plastids. Acetyl CoA and malonyl CoA are attached to two separate molecules of acyl carrier protein (ACP) to give acetyl-ACP and malonyl-ACP. The acetyl-ACP is then involved in successive rounds of fatty acid synthase activity where additions of two-carbon C_2 units occur from malonyl-ACP. Eventually, after insertion of a double bond into stearyl-ACP by stearyl-ACP desaturase, oleate is formed from the action of oleyl-ACP thioesterase. Oleate is exported from the plastids and used as a base for the synthesis of other fatty acids and also in combination with glycerol-3-phosphate for the synthesis of triglycerides (Anderson and Beardall, 1991). Due to its relatively large size avocado has proved useful in the purification of enzymes of fatty acid synthesis (Shanklin and Sommerville, 1991).

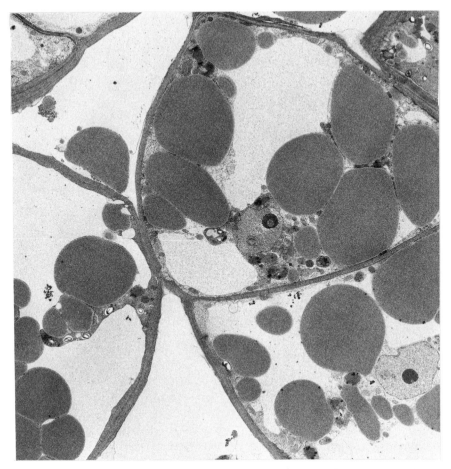

Fig. 2.5   Electron micrograph of a section of the mesocarp of a mature, unripe
avocado fruit (variety Hass). The large oil bodies (oleosomes) are present in the
cytoplasm. The cells contain the normal complement of cellular organelles –
nucleus, mitochondria, plastids and endoplasmic reticulum. These organelles all
remain intact throughout ripening. (Micrograph and accompanying information
kindly provided by Dr. K.A. Platt, University of California.)

Phospholipids

These are the predominant lipids in most cell membranes. The phospholipids are polar lipids consisting of fatty acid esters of an alcohol linked
via phosphate to a hydrophilic side chain. The phospholipids of Hass
avocado fruit have been separated into four fractions: (1) phosphatidic
acid; (2) phosphatidyl glycerol; (3) phosphatidyl ethanolamine; (4)
phosphatidyl choline. Analysis of the fatty acid composition of these
fractions has indicated that the main components are palmitic, oleic,

linoleic and linolenic acid (Ahmed and Barmore, 1980). Evidence for changes in membrane lipids during ripening in avocado fruit has been obtained by Meir *et al.* (1991), who observed increased formation of fluorescent lipid peroxidation products in the peel of avocado fruit prior to the onset of other changes associated with ripening. Lipid peroxidation involves free radical formation and yields lipofusein-like fluorescent compounds via chain reactions where polyunsaturated fatty acids act as the substrate, producing intermediates such as lipid hydroperoxides and eventually malondialdehyde. The pattern and time of onset of ripening was related to the appearance of these peroxidation products, and treatment with ethylene significantly increased their appearance.

2.3.4 Cell wall degradation

The biochemical basis of textural changes in avocado and other fruits is still incompletely understood, but probably involves changes in the structure of the fruits' cell walls. Light and electron microscopy studies show considerable degradation of avocado mesocarp cell walls during ripening (Platt-Aloia *et al.*, 1980; Dallman *et al.*, 1989). Enzymes which degrade cell wall polymers have been isolated from avocado fruit. Both cellulase (EC 3.2.1.4) and polygalacturonase (PG) (EC 3.2.1.15) activities have been reported to increase during ripening in avocado fruits (Fig. 2.6), while pectinesterase (EC 3.1.1.11) activity apparently declines during the same period. Also, there appears to be a close relationship between this increase in cell wall degrading activity and the rise in respiration and ethylene production (Awad and Young, 1979; Raymond and Phaff, 1965; Pesis *et al.*, 1978). Other putative wall degrading enzymes reported from avocado fruit include β-1, 4-D-endoxylanase and β-1, 4-D-exoxylanase. At present, the role of these enzymes in the cell wall degradation is not known and, in the case of cellulase, the *in vivo* substrate of this enzyme remains to be identified. However, the molecular biology of cellulase gene expression in avocado has received considerable attention (refer to section 2.4); current information on the biochemistry of these enzymes is described below.

Cellulase and polygalacturonase

Molecular studies have indicated that cellulase is synthesized at the onset of ripening in avocado fruit (Christoffersen *et al.*, 1984). Bennett and Christoffersen (1986) investigated the biosynthesis and processing of avocado cellulase. Using antibodies to the purified cellulase protein they found that immunoprecipitation of the cellulase translation product from avocado fruit RNA *in vitro* yielded a polypeptide of 54kDa. This represents the molecular weight of the unprocessed protein directly after

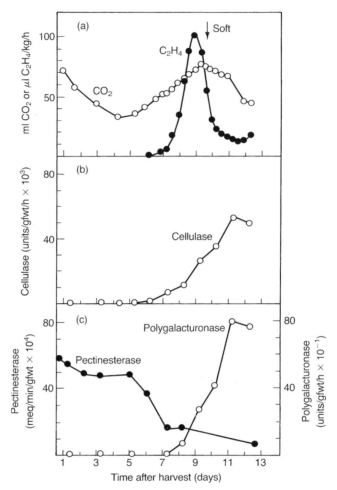

Fig. 2.6 Changes in cellulase, pectinesterase and polygalacturonase activity in relation to ethylene production and respiration in ripening Fuerte avocado fruit. (Redrawn from Awad and Young, 1979.)

translation. This polypeptide is, however, modified *in vivo*, because the mature cellulase protein is known to be glycosylated and has a molecular weight of 54.2kDa. Using deglycosylation agents and analysis of various molecular forms of cellulase purified from avocado fruit, the authors presented a scheme to describe the processing of this enzyme *in vivo* prior to secretion into the wall (Table 2.4). Further information of the processing of cellulase has come from other studies. Membrane vesicles were isolated from avocado mesocarp tissue by Dallman *et al.* (1989) and assayed for Golgi, endoplasmic reticulum and plasma membrane markers. A portion of each fraction was also analysed by electrophoresis and immunodetection for the presence of cellulase. Three molecular weight forms of cellulase were found between 50 and 55kDa, with a high and low molecular weight form associated with the endoplasmic reticulum and an intermediate weight form associated with the plasma membrane (see Table 2.4). Dallman *et al.* (1989) also localized cellulase in the fruit and within the mesocarp cells. Cellulase activity first appeared at the stylar end of the fruit and expanded outward and upward during the climacteric peak. Immunogold detection of cellulase by electron microscopy revealed the presence of the enzyme in the endoplasmic reticulum, plasmadesmata and cell wall. Cellulase also appeared to be present in the nucleus (Dallman *et al.*, 1989).

The exact substrate for cellulase *in vivo* remains to be demonstrated. However, attempts have been made to elucidate the substrate for this enzyme. The characteristics of purified avocado cellulase were studied by Hatfield and Nevins (1986). They found that only substrates containing (1-4)-β-glycosyl linkages were hydrolysed by the purified enzyme, although the enzyme will attack (1-3)(1-4)-β-D-glucans. Avocado cellulase did not appear to be effective at solubilizing cellulosic polymers contained within mature avocado cell walls. However, the

Table 2.4 Proposed relationship between various molecular forms of avocado fruit cellulase. (After Bennett and Christoffersen, 1986.)

Molecular size	Processing	Sub-cellular location
mRNA (~1800 bases) ↓ 54.0 kDa pre-protein	Translation	Cytoplasm/ endoplasmic reticulum
↓ 52.8 kDa mature polypeptide	Proteolytic processing and cleavage of signal sequence	
↓ 56.5 kDa secretory form	Glycosylation	Endoplasmic reticulum
↓ 54.2 kDa mature protein	Carbohydrate trimming	Golgi apparatus? Secretion could take place via the plasmodesmata*

*Refer to Dallman *et al.* (1989).

enzyme did release arabinose and galactose from the cell walls of unripe fruit. Thus the *in situ* substrate for cellulase remains unclear.

O'Donoghue and Huber (1991) have investigated the possibility that cellulase degrades the xyloglucan fraction in avocado cell walls. They found that this fraction shows, however, little change during ripening. PG activity has been reported to increase around three days after the first signs of cellulase activity in ripening avocados (Awad and Young, 1979), but PG levels are comparatively low in avocado as compared with for example tomato, where the levels are more than 10-fold greater (Hobson, 1962). Avocado fruit PG has been purified by Raymond and Phaff (1965) and shows the characteristics of an endoenzyme with a pH optimum of 5.5 in sodium acetate buffer. These authors also reported the presence of an endogenous inhibitor of PG, although this observation has not been confirmed by later studies (Awad and Young, 1979). Relatively little work appears to have been undertaken on changes in the pectin component during ripening, although Dolendo *et al.* (1966) noted that ripening was accompanied by an increase in water-soluble pectin and a decrease in the degree of pectin methylesterification from around 80 to about 50%.

2.3.5 Other changes

The skin of mature, 'harvest ripe' avocado fruit contains chloroplasts with prominent grana, while the yellow flesh contains only etioplasts with crystalline prolamella bodies. The pale green flesh has plastids of intermediate structure. Chlorophyll content ranges across the above tissues between 316 to 30 μg g^{-1} fresh weight (Cran and Possingham, 1973). Starch occurs in the plastids of mature unripe avocado fruit. A decrease in starch content from around 12 mg g^{-1} dry weight to undetectable levels occurs during ripening and is accompanied by a rise in amylase activity, but the presence of starch phosphorylase (EC 2.4.1.21) activity was not detected. The extractable amylase activity was found to have both α-and β-amylase (EC 3.2.1.1 and EC 3.2.1.2) reaction products (Pesis *et al.*, 1978). It is likely that this starch reserve is used as an energy substrate during ripening.

A slight increase in glucose and fructose content has been detected in ripening avocado fruit by Shaw *et al.* (1980), but these authors were unable to confirm earlier reports (Davenport and Ellis, 1959) of measurable sucrose levels. Avocado fruit also contain a number of unusual sugars (Biale and Young, 1971) including the seven-carbon sugar, D-manno-heptulose and the related seven-carbon alcohol, perseitol. Avocados are the richest natural source of D-manno-heptulose (Shaw *et al.*, 1980) at between 0.64 and 2.5% of the fresh pulp weight.

Postharvest browning of the mesocarp tissue of avocado is associated with the activity of the enzyme polyphenoloxidase (PPO) (EC 1.10.3.1).

PPO activity in avocados has been reported to consist of five fractions, of which a 28kDa fraction was the most active (Bower and Cutting, 1988). There is evidence that abscisic acid may stimulate PPO activity (Bower and Cutting, 1988; Cutting et al., 1990). Peroxidase (EC 1.11.1.7) activity has been reported in avocados and appears to decline during ripening (Zauberman et al., 1985). Other enzymes investigated during ripening include acid phosphatase (EC 3.1.3.2) and superoxide dismutase (EC 1.15.1.1). The activity of the former has been reported to increase in ripening avocados (Sacher, 1975), but there is little apparent change in superoxide dismutase activity during the climacteric (Baker, 1976).

2.4   GENE EXPRESSION DURING FRUIT RIPENING

Ripening of avocado, as with other fruit, is not simply a degradative process but involves the continued synthesis of both proteins and nucleic acids. Thus Richmond and Biale (1966) demonstrated that in avocado amino acid incorporation into proteins increased very early on in the respiratory climacteric, and coincident with this was an increased incorporation of ^{32}P into a presumed mRNA fraction (Richmond and Biale, 1967). This increased synthesis was found to be only transient, diminishing to preclimacteric levels by the time that respiration peaked. Similar increases in protein or nucleic acid synthesis early in ripening have been demonstrated in banana (Brady and O'Connell, 1976) and tomato (DeSwardt et al. 1973).

These observations suggest that ripening is under genetic control and as such may involve the expression of novel ripening-related genes. The advent of recombinant DNA techniques has allowed the nature of these specific genes, and their control, to be investigated. The most thoroughly investigated fruit in this area is the tomato (Chapter 14), and for a variety of reasons. Tomatoes are easy to grow and can provide several generations a year. There is a reasonably well-defined genetic map for the tomato (Stevens and Rick, 1986) and many ripening mutants. The tomato is a major commercial crop and can be readily transformed using derivatives of the T_i plasmid from Agrobacterium tumefaciens. However, it is evident that genetic studies must also be carried out on fruit other than tomato. While studies are underway on fruit such as apple, pear and strawberry, it is perhaps the avocado that has received the most attention after the tomato.

The easiest way to investigate possible novel gene expression is to isolate total protein from fruit at various stages of development and analyse this by either one- or two-dimensional gel electrophoresis. When proteins from unripe and ripe avocado fruit are compared in this way there is seen to be a considerable increase in polypeptides of molecular weights 39kDa and 32kDa (Christoffersen *et al.*, 1984) and 69 kDa, 55 kDa, 52 kDa, 41 kDa and 27 kDa (Percival *et al.*, 1991). The appearance of corresponding mRNAs during ripening can similarily be demonstrated by extracting total total mRNA from fruit at various stages of ripening and subjecting this to analysis by *in vitro* translation. Analysis of the resultant products has demonstrated the accumulation during ripening of avocado fruit of mRNAs encoding polypeptides of molecular weight 80 kDa, 36 kDa and 16.5 kDa (Christoffersen *et al.*, 1982) and 53 kDa, 40 kDa, 31 kDa and 20 kDa (Tucker and Laties, 1984). The latter study also found a series of mRNAs coding for polypeptides of molecular weight 25 kDa, 18 kDa and 17 kDa which all appear to decline during ripening. A summary of these changes is provided in Table 2.5, together with a similar summary of some results from tomato fruit. The dissimilarity of protein pattern changes during ripening of avocado and tomato is perhaps not surprising considering the widely different ripening patterns in these fruit. For instance, tomato undergoes a colour change during ripening. However, some similarities are to be expected and, as we shall see later, at least one of these novel proteins may encode the same enzyme in both fruit – namely the ethylene forming enzyme or ACC oxidase.

Table 2.5   Comparison of changes in gene expression during ripening of avocado and tomato fruit. All values are in kDa.

Proteins increasing during ripening		*mRNA increasing during ripening*[*]		
Avocado	Tomato[1]	Avocado	Tomato[1]	Tomato[2]
69	94	80	116	190
55	44	53	89	80
52	44	40	70	57
41	20	36	42	55
39	12	31	38	48
32		20	33	44
27		16.5	31	35
			29	20
			26	

[*]detected by *in vitro* translation
[1]Biggs *et al.* (1986)
[2]Grierson *et al.* (1985)

Christoffersen *et al.* (1982) observed that the major increase in new mRNA seemed to occur early in the climacteric rise and that total levels of mRNA appeared to decline after the climacteric peak of respiration. Tucker and Laties (1984) repeated and extended this observation by demonstrating a three-fold increase in both polysome prevalence and associated poly(A)$^+$ mRNA early in the respiratory climacteric, which then declined to a preclimacteric level at the respiratory peak. These authors also investigated in more detail the timing of appearance of novel mRNA species. Total mRNA was extracted from fruit at 10, 24 or 48 hours following the initiation of ripening with ethylene, translated *in vitro* and the products compared to those from unripe fruit. Although polysome prevalence and poly(A)$^+$ mRNA levels both increased dramatically during the first 10 hours following ethylene treatment, the *in vitro* translation pattern after 10 hours was very similar to that from unripe fruit, with very little evidence for the appearance of novel mRNA species. This suggests that the increase in poly(A)$^+$ over the first 10 hours following the initiation of ripening represents a generic increase in constitutive mRNA. The *in vitro* translation patterns from fruit at 24 and 48 hours after ethylene treatment showed a clear indication of novel gene expression when compared to unripe fruit. Tucker and Laties (1984) therefore postulated that the effect of ethylene on gene expression in avocado was biphasic; the first phase represents an initial increase in constitutive protein synthesis, and only later do the fruit enter phase two, in which novel ripening related proteins are synthesized.

2.4.1   Identification of ripening-related genes

One aim of the molecular biologist is to identify and isolate the key ripening-related genes and the enzymes for which they code. Again, most progress in this area has been in the tomato, where ripening-related specific cDNAs or genes have been identified for the cell wall hydrolases polygalacturonase and pectinesterase, and the ethylene biosynthetic enzymes ACC synthase and ACC oxidase, among others (Chapter 14). In avocado fruit, one target enzyme is cellulase, since this appears during ripening and is a predominant fruit enzyme which has been linked to softening of the fruit (refer to the section on cell wall degradation, page 65).

Avocado fruit cellulase has been purified and characterized (Awad and Lewis, 1980). The native enzyme has a molecular weight of 50–55kDa and the denatured protein a pI of 4.7. Polyclonal antibodies have been raised against the native enzyme and these have assisted greatly in the subsequent investigation of cellulase gene expression during avocado ripening. When total proteins from unripe or ripe fruit are analysed by one-dimensional denaturing gel electrophoresis there is no apparent increase in a protein band of a size corresponding to native cellulase enzyme (Tucker *et al.*, 1985)(Fig. 2.7). However, if the same gel is immunoblotted with cellulase antibody then there is clear evidence for the appearance of this protein during ripening (Fig. 2.7). The enzyme is obviously masked in the one-dimensional gel by constitutive proteins of similar molecular weight. Immunoblotting of a two-dimensional gel of total proteins from ripe avocado fruit with cellulase antiserum reveals at least three putative cellulase isoforms (Tucker *et al.*, 1985). These all have a molecular weight of around 53 kDa, but differ in their relative pIs which range from 5.6 to 6.2. Several reports have identified a 53 kDa *in vitro* translation product as that of cellulase by immunoblotting or immunoprecipitation (Christoffersen *et al.*, 1984; Tucker and Laties, 1984; Tucker *et al.*, 1985). When *in vitro* translation products from unripe and ripe fruit are compared, there is a clear increase in mRNA for a protein with a molecular weight of 53 kDa (Tucker *et al.*, 1985) (Fig. 2.7). This 53 kDa translation product is specifically precipitated with cellulase antibody (Tucker and Laties, 1984) (Fig. 2.8)

To study gene expression in more detail requires the isolation of ripening-specific cDNA clones and eventually genomic clones for ripening-specific genes. The first attempt to generate a cDNA library from avocado fruit was reported by Christoffersen *et al.* (1984). This was done by isolating poly(A)$^+$ mRNA from ripe Hass fruit and then cloning the corresponding cDNAs into the PstI site of pBR322. The resultant library of 330 clones was then differentially screened to identify

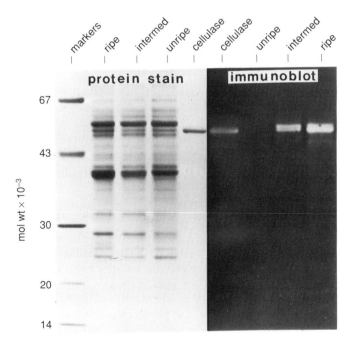

Fig. 2.7   Denatured SDS-gel electrophoresis of protein extracts from avocado fruit at various stages of ripening. Left: total proteins as stained with Coomassie blue. Right: the same gel after immunoblotting with cellulase-specific antibodies. (From Christoffersen *et al.*, 1984.)

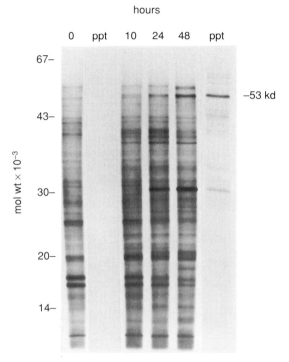

Fig. 2.8 *In vitro* translation products from avocado fruit polysomal poly-(A)+RNA before (o) and at specified times after treatment with ethylene. Cellulase antibody was used to precipitate antigen from time 0 and 48-hour samples. (From Tucker *et al.*, 1985.)

ripening-specific cDNAs. One such clone, pAV5 was subsequently shown by hybrid release translation to contain sequences complementary to the putative cellulase mRNA. Thus, hybrid selection using the pAV5 clone provided a mRNA from ripe fruit that on translation gave a polypeptide of 53 kDa which could be selectively precipitated with antibody to cellulase. The pAV5 clone was shown to have a cDNA insert of 640 base pairs and was found by Northern analysis to hybridize to a single mRNA of about 2000 bases. Thus, pAV5 probably represents about one third of the full mRNA sequence for cellulase.

Tucker *et al.* (1987) reported the construction of a much larger cDNA library from ripe avocado fruit. This was achieved by cloning into the PstI site of pUC18 and provided 1078 clones. This library was screened with pAV5 to try and isolate a full length cDNA for cellulase. Twelve clones were isolated with an average insert size of 1400 base pairs. Clone pAV363 contained the largest insert, this consisting of 2021 transcribed nucleotides and a poly-T tail of about 150 bases. Northern blots showed hybridization of pAV363 to a mRNA of about 2200 bases, suggesting that pAV363 may represent a full length cDNA clone for cellulase. The sequence of the cDNA insert in pAV363 contained an open reading frame of 1482 nucleotides which would encode the production of a peptide with a molecular weight of 54.1 kDa and with a predicted pI of 5.9. The deduced amino acid composition of this protein corresponds closely to that determined for native cellulase. The putative amino acid sequence from the pAV363 cDNA suggests the presence of a signal peptide consisting of the first 24 or 25 amino acids and having a molecular weight of about 2.7 kDa. This compares favourably with the 2.5 kDa signal sequence predicted by Bennett and Christoffersen (1986) from their studies of the processing of the native cellulase. Such a signal sequence is quite small when compared to other cell wall degrading enzymes investigated in tomato fruit. Thus polygalacturonase in tomato fruit has a 71 amino acid N-terminal extension (Chapter 14), and pectinesterase an even larger putative N-terminal signal sequence of 241 amino acids (Hall *et al.*, unpublished). The amino acid sequence predicted by pAV363 has two Asn-X-Ser regions. These sequences are potential glycosylation sites and native cellulase has indeed been shown to be a glycoprotein (Bennett and Christoffersen, 1986).

Analysis of genomic fragments by Southern blotting suggested a small gene family for cellulase may be present in avocado (Tucker *et al.*, 1987). This possibility was further investigated by generating and screening an avocado genomic library, which has been constructed using both λ-EMBL and λ-Charon 85 (Cass *et al.*, 1990). This library was screened with a cellulase cDNA probe pAVe6. The pAVe6 contains a 1800 base pair insert and has been described by Christoffersen (1987). After an initial screening using the entire pAVe6 insert, the isolated clones were rescreened with partial fragments of the insert corresponding to either the 3′ or 5′ specific non-translated regions of the mRNA. This allowed the genomic clones to be grouped into two classes, each corresponding to a separate gene which Cass *et al.* (1990) termed cel 1 and cel 2. The genomic sequence of one of these two genes, cel 1, has been determined and compared with that for the cellulase cDNA insert. Cel 1 is composed of eight exons, the sequence of which differs by only one base from the sequence of the pAVe6 insert. Since the sequence of the pAVe6 clone is identical to that found in all the other cellulase cDNAs – pAV5, pAV363 and pAVe8 – so far identified, it would appear that these are all derived from the cel1 gene and that this gene is the sole source of cellulase enzyme in the fruit. The partial sequence of the cel 2 gene shows about 81% homology with the cel1 gene. However, homology is greatest within the coding regions of the two genes (90%) and weaker in the non-translated regions (65%). There is no evidence for the expression of the cel2 gene in avocado fruit and its role, if any, in the plant is unclear. It may represent a pseudogene, but since the coding region is so highly conserved it could also represent a cellulase gene active in other tissues, such as the abscission zones of fruit or leaves. It is interesting to note that the avocado fruit cellulase has 64% and 50% homology at the nucleotide and amino acid level respectively with bean abscission zone cellulase (Tucker and Milligan, 1991). Also it has been demonstrated in tomato that fruit and leaf abscission zone polygalacturonases are probably the products of two completely separate genes (Taylor *et al.*, 1990).

The isolation of genomic fragments will allow the analysis of the 5′ and 3′ flanking regions of the cellulase gene which presumably are involved in the control of its ripening specific expression. The structure of a 1400 base pair 5′ flanking region of the cel1 gene has been determined (Cass *et al.*, 1990). This has been compared to sequences in the 5′ flanking regions of other ethylene-induced genes. Some similarity was observed with the ethylene-induced chitinase gene from bean (Broglie *et al.*, 1989) and to that of the E8 gene in tomato (Cordes *et al.*, 1989). It is interesting to note, however, that no similarity could be found with the 5′ flanking region of the tomato polygalacturonase gene. This polygalacturonase control region has been shown to direct the fruit specific expression of a reporter gene in transgenic tomato plants (Bird *et al.*, 1988; also Chapter 14).

It would thus seem that cellulase in avocado fruit is the product of a single gene switched on in response to ethylene. The mRNA corresponding to the cel1 gene increases 37-fold in response to propylene (Cass *et al.*, 1990) and mRNA corresponding to pAV5 increases at least 50-fold during normal ripening (Tucker *et al.*, 1987). The reason for several isoforms of native cellulase, as determined by two-dimensional gel electrophoresis, is unclear but these may arise from post-translational modification of the cellulase polypeptide.

2.4.2 Other ripening-specific cDNAs

Although most work has centred on the identification of the cellulase cDNAs and gene, progress towards the identification of other ripening specific cDNAs has been made. McGarvey *et al.* (1990) reported the nucleotide sequence of an avocado ripening-related cDNA pAVOe3. This 1151 base pair cDNA insert contains an open reading frame encoding a putative 320 amino acid protein of molecular weight 36.23 kDa and with a predicted pI of 4.49. This corresponds directly to one of the *in vitro* translation products identified earlier. The sequence of pAVOe3 showed a 72% identity with a tomato ripening-specific cDNA pTOM 13 (Holdsworth *et al.*, 1987). Subsequent experiments using pTOM 13 (reported in Chapter 14), have shown that this cDNA corresponds to the ethylene forming enzyme or ACC oxidase in tomato fruit. Presumably pAVOe3 corresponds to the same enzymic activity in avocado fruit, although this remains to be proven conclusively.

A further avocado ripening-specific cDNA clone, pAVOd8, has been putatively identified as coding for the fruit-related cytochrome P-450. Cytochrome P-450 is especially prevalent in the mesocarp of ripe avocado fruit (McPherson *et al.*, 1975) and has been purified and partially sequenced (O'Keefe and Leta, 1989). Using this amino acid sequence data, and that from several other cytochrome P-450 proteins that have been isolated, Bozak *et al.* (1990) have putatively identified pAVOd8 as coding for the fruit cytochrome P-450. Since this clone was itself isolated by differential screening of a cDNA library, cytochrome P-450 probably represents a ripening-induced protein. The specific role, if any, that cytochrome P-450 may play during ripening is unclear, since no endogenous substrates for enzymes that may use this cytochrome have been identified. However, considering the role of cytochrome P-450 in other tissues it may have a function in either flavour development or the metabolism of fruit phenolics.

The identification of further ripening-related clones will no doubt occur in the near future. In the meantime, the availability of cDNA and genomic clones specific for cellulase and other ripening-related proteins may eventually lead to the manipulation of avocado ripening, as described for tomato ripening in Chapter 14.

REFERENCES

Adams, D.O. and Yang, S.F. (1979) Ethylene biosynthesis: Identification of 1-aminocyclopropane-1-carboxylic acid as an intermediate in the conversion of methionine to ethylene. *Proceedings of the National Academy of Sciences, USA*, **76**, 170–174

Adato, I. and Gazit, S. (1974) Postharvest responses of avocado fruits of different maturity to delayed ethylene treatments. *Plant Physiology*, **53**, 899–902

Adato, I. and Gazit, S. (1976) Response of harvested avocado fruits to supply of indole-3-acetic acid, gibberellic acid and abscisic acid. *Journal of Agricultural and Food Chemistry*, **24**, 1165–1167

Adato, I. and Gazit, S. (1977) Changes in the initiation of climacteric ethylene in harvested avocado fruits during their development. *Journal of the Science of Food and Agriculture*, **28**, 240–242

Ahmed, E.M. and Barmore, C.R. (1980) Avocado. In *Tropical and subtropical fruits: composition, properties and uses* (eds S. Nagy and P.E. Shaw), AVI Publishing, Westport, CT, pp. 121–156

Anderson, J.W. and Beardall, J. (1991) *Molecular activities of plant cells. An introduction to plant biochemistry*. Blackwell Scientific Publications, Oxford

Awad, M. and Lewis, L.N. (1980) Avocado cellulase: extraction and purification. *Journal of Food Science*, **45**, 1625–1628

Awad, M. and Young, R.E. (1979) Post harvest variation in cellulase, polygalacturonase and pectinmethylesterase in avocado (*Persea americana* Mill cv Fuerte) fruits in relation to respiration and ethylene production. *Plant Physiology*, **64**, 306–308

Baker, J.E. (1976) Superoxide dismutase in ripening fruits. *Plant Physiology*, **58**, 644–647

Ball, K.L., Green, J.H. and apRees, T. (1991) Glycolysis at the climacteric of bananas. *European Journal of Biochemistry*, **197**, 265–269

Bennett, A.B. and Christoffersen, R.E. (1986) Synthesis and processing of cellulase from ripening avocado fruit. *Plant Physiology*, **81**, 830–835

Bennett, A.B., Smith, G.M. and Nichols, B.G. (1987) Regulation of climacteric respiration in ripening avocado fruit. *Plant Physiology*, **83**, 973–976

Bergh, B.O. (1975) Avocados. In *Advances in fruit breeding*, (eds. J. Janick and J.N. Moore), Purdue University Press, West Lafayette Indiana, USA.

Bergh, B.O. (1976) Avocado. In *Evolution of crop plants*, (ed. N.V. Simmonds), Longman, London

Biale, J.B. and Young, R.E. (1971) The avocado pear. In *The biochemistry of fruit and their products*, (ed. A.C. Hulme), Academic Press, Vol. 2, pp. 1–63

Biggs, M.S., Harriman, R.W. and Handa, A.K. (1986) Changes in gene expression during tomato fruit ripening. *Plant Physiology*, **81**, 395–403

Bird, C.R., Smith, C.J.S., Ray, J.A. *et al.* (1988) The tomato polygalacturonase gene and ripening-specific expression in transgenic plants. *Plant Molecular Biology*, **11**, 651–662

Blanke, M.M. (1991) Respiration of apple and avocado fruits. *Postharvest News and Information*, **2**, 429–436

Bower, J.P. and Cutting, J.G. (1988) Avocado fruit development and ripening physiology. *Horticultural Reviews*, **10**, 229–271

Bozak, K.R., Yu, H., Sirevag, R. and Christoffersen, R.E. (1990) Sequence analysis of ripening-related cytochrome P–450 complementary DNA from avocado fruit. *Proceedings of the National Academy of Sciences, USA*, **87**, 3904–3908

Brady, C.J. and O'Connell, P.B.H. (1976) On the significance of increased protein synthesis in ripening banana fruits. *Australian Journal of Plant Physiology*, **3**, 301–310

Broglie, K.E., Biddle, P., Cressman, R. and Broglie, R. (1989) Functional analysis of DNA sequences responsible for ethylene regulation of a bean chitinase gene in transgenic tobacco. *Plant Cell*, **1**, 599–607

Cass, L.G., Kirven, K.A. and Christoffersen, R.E. (1990) Isolation and characterisation of a cellulase gene family member expressed during avocado fruit ripening. *Molecular and General Genetics*, **223**, 76–86

Christoffersen, R.E. (1987) Cellulase gene expression during fruit ripening. In *Plant senescence: its biochemistry and physiology* (eds. W.W. Thompson, E.A. Nothnagel and R. Huffaker), American Society of Plant Physiology. Rockville, Maryland, USA, pp. 89–97

Christoffersen, R.E., Warm, E. and Laties, G.G. (1982). Gene expression during fruit ripening in avocado. *Planta*, **155**, 52–57

Christoffersen, R.E., Tucker, M.L. and Laties, G.G. (1984) Cellulase gene expression in ripening avocado (*Persea americana* cv Hass) fruit. The accumulation of cellulase mRNA and protein as demonstrated by cDNA hybridisation and immunodetection. *Plant Molecular Biology*, **3**, 385–392

Cordes, S., Deikman, J., Margorsian, C.L.J. and Fischer, R.L. (1989) Interaction of a developmentally regulated DNA binding factor with sites flanking two different fruit ripening genes from tomato. *Plant Cell*, **1**, 1025–1034

Cran, D.O. and Possingham, J.V. (1973) The fine structure of avocado plastids. *Annals of Botany*, **37**, 993–997

Cutting, J.G.M., Bower, J.P., Wolstenholme, B.N. and Hofman, P.J. (1990) Changes in ABA, polyphenoloxidase, phenolic compounds and polyamines and their relationship with mesocarp discolouration in ripening avocado (*Persea americana* Mill.) fruit. *Journal of Horticultural Science*, **65**, 465–471

Dallman, T.F., Thomson, W.W., Eaks, I.L. and Nothnagel, E.A. (1989) Expression and transport of cellulase in avocado mesocarp during ripening. *Protoplasma*, **151**, 33–46

Davenport, J.B. and Ellis, S.C. (1959) Chemical changes during growth and storage of the avocado fruit. *Australian Journal of Biological Sciences*, **12**, 445–454

DeSwardt, G.H., Swanepoel, J.H. and Dubenage, A.J. (1973) Relations between changes in ribosomal RNA and total protein synthesis and the respiratory climacteric in ripening pericarp tissue of tomato. *Zeitschrift für Pflanzenphysiologie*, **70**, 358–363

Dolendo, A.I., Luh, B.S. and Pratt, H.K. (1966) Relation of pectin and fatty acid changes to respiration rate during ripening of avocado fruits. *Journal of Food Science*, **31**, 332–336

Eaks, I.L. (1978) Ripening, respiration and ethylene production of "Hass" avocado fruits at 20° to 40° C. *Journal of the American Society for Horticultural Science*, **103**, 576–578

Eaks, I.L. (1980) Respiratory rate, ethylene production and ripening response of avocado fruit to ethylene or propylene following harvest at different maturities. *Journal of the American Society for Horticultural Science*, **105**, 744–747

Eaks, I.L. (1985) Effects of calcium on ripening, respiratory rate, ethylene production and quality of avocado fruit. *HortScience*, **110**, 145–148

Gaydou, E.M., Lozano, Y. and Ratovohery, J. (1987) Triglyceride and fatty acid composition of the mesocarp of *Persea americana* during fruit development. *Phytochemistry*, **26**, 1595–1597

Grierson, D., Slater, A., Maunders, M. *et al.* (1985) Regulation of the expression of tomato ripening genes: the involvement of ethylene. In *Ethylene and plant development*, (eds. J.A. Roberts and G.A. Tucker) Butterworth, pp. 147–161

Hardenburg, R.E., Watada, A.E. and Wang, C.Y. (1986) *The commercial storage of fruits, vegetables and nursery stocks*. USDA, ARS Handbook, 66

Hatfield, R. and Nevins, D.J. (1986) Characterisation of the hydrolytic activity of avocado cellulase. *Plant Cell Physiology*, **27**, 541–552

Hobson, G.E. (1962) Determination of polygalacturonase in fruits. *Nature*, **195**, 804–805

Hoffman, N.E. and Yang, S.F. (1980) Changes of 1-aminocyclopropane-1-carboxylic acid content in ripening fruit in relation to their ethylene production rates. *Journal of the American Society for Horticultural Sciences*, **105**, 492–495

Holdsworth, M.J., Bird, C.R., Ray, J., Schuch, W. and Grierson, D. (1987) Structure and expression of an ethylene related mRNA from tomato. *Nucleic Acids Research*, **15**, 731–739

Huang, L-S. and Romani, R.J. (1991) Metabolically driven self-restoration of energy-linked functions by avocado mitochondria. *Plant Physiology*, **95**, 1096–1105

Kader, A.A. (1992) *Postharvest technology of horticultural crops*. 2nd Edition. University of California, Division of Agriculture and Natural Resources.

Kanellis, A.K., Solomos, T., Mehta, A.M. and Mattoo, A.K. (1989a) Decreased cellulase activity in avocado fruit subjected to 2–5% O_2 correlates with lower cellulase protein and gene transcript levels. *Plant and Cell Physiology*, **30**, 817–823

Kanellis, A.K., Solomos, T. and Mattoo, A.K. (1989b) Hydrolytic enzyme activities and protein pattern of avocado fruit ripened in air and in low oxygen with and without ethylene. *Plant Physiology*, **90**, 257–266

Kikuta, Y. and Erickson, L.C. (1968) Seasonal changes of avocado lipids during fruit development and storage. *Californian Avocado Society Year Book*, **52**, 102–108

Knight, R. Jr (1980) Origin and world importance of tropical and subtropical fruit crops. In *Tropical and subtropical fruits*, (eds. S. Nagy and P.E. Shaw) AVI Publishing Inc., Westport, CT.

Kushad, M.M., Yelenosky, G. and Knight, R. (1988) Interrelationship of polyamine and ethylene biosynthesis during avocado fruit development and ripening. *Plant Physiology*, **87**, 463–467

Lee, S.K. and Young, R.E. (1984) Temperature sensitivity of avocado fruit in relation to C_2H_4 treatment. *Journal of the American Society for Horticultural Science*, **109**, 689–692

Lewis, C.E. (1978) The maturity of avocados – A general review. *Journal of the Science of Food and Agriculture*, 29, 857–866

Mazliak, P. (1970) Lipids. In *Biochemistry of fruits and their products*, (ed. A.C. Hulme), Vol. 1, Academic Press, London

McGarvey, D.J., Yu, H. and Christoffersen, R.E. (1990) Nucleotide sequence of a ripening-related complementary DNA from avocado fruit. *Plant Molecular Biology*, **15**, 165–168

Meir, S., Philosoph-Hadas, S., Zauberman, G., Fuchs, Y. and Akerman, M. (1991) Increased formation of fluorescent lipid-peroxidation products in avocado peel precedes other signs of ripening. *Journal of the American Society for Horticultural Science*, **116**, 823–826

McPherson, F.J., Markham, A., Bridges, J.W., Hartman, G.C. and Parke, D.V. (1975) Effects of preincubation *in vitro* with 3, 4-benzopyrene and phenobarbital on the drug metabolism system in the microsomal and soluble fractions of the avocado pear. *Biochemical Society Transactions*, **3**, 283–285

Moreau, F. and Romani, R. (1982) Malate oxidation and cyanide-insensitive respiration in avocado mitochondria during the climacteric cycle. *Plant Physiology*, **70**, 1385–1390

O'Donoghue, E.M. and Huber, D.J. (1991) Cell wall changes in ripening avocado fruit. *Hortscience*, **26**, 755 (abstract 524).

O'Keefe, D.P. and Leta, K.J. (1989) Cytochrome P-450 from the mesocarp of avocado (*Persea americana*). *Plant Physiology*, **89**, 1141–1149

Percival, F.W., Cass, L.G., Bozak, K.R. and Christoffersen, R.E. (1991) Avocado fruit protoplasts – a cellular model system for ripening studies. *Plant Cell Reports*, **10**, 512–516

Pesis, E., Fuchs, Y. and Zauberman, G. (1976) Cellulase activity and fruit softening in avocado. *Plant Physiology*, **61**, 416–419

Pesis, E., Fuchs, Y. and Zauberman, G. (1978) Starch content and amylase activity in avocado fruit pulp. *Journal of the American Society for Horticultural Science*, **103**, 673–676

Platt-Aloia, K.A. and Thomson, W.W. (1980) Aspects of the three-dimensional intracellular organization of mesocarp cells as revealed by scanning electron microscopy. *Protoplasma*, **104**, 157–165

Platt-Aloia, K.A., Thompson, W.W. and Young, R.E. (1980) Ultrastructural changes in the walls of ripening avocados: Transmission, scanning and freeze fracture microscopy. *Botanical Gazette*, **14**, 366–373

Raymond, D. and Phaff, H.J. (1965) Purification and certain properties of avocado polygalacturonase. *Journal of Food Science*, **30**, 266–273

Richmond, A. and Biale, J.B. (1966) Protein and nucleic acid metabolism in fruits. I. Studies of amino acid incorporation during climacteric rise in respiration of the avocado. *Plant Physiology*, **41**, 1247–1253

Richmond, A. and Biale, J.B. (1967) Protein and nucleic acid metabolism in fruits. II: RNA synthesis during respiratory rise of the avocado. *Biochimica et Biophysica Acta*, **138**, 625–627

Sacher, J.A. (1975) Acid phosphatase development during ripening of avocado. *Plant Physiology*, **55**, 382–385

Schroeder, C.A. (1953) Growth and development of the Fuerte avocado fruit. *Proceedings of the American Society for Horticultural Science*, **61**, 103–109

Shanklin, J. and Sommerville, C. (1991) Stearoyl-acyl-carrier protein desaturase from higher plants is structurally unrelated to the animal and fungal homologs. *Proceedings of the National Academy of Sciences of the USA*, **85**, 2510–2514

Shaw, P.E., Wilson, C.W. and Knight, R.J. Jr. (1980) High-performance chromatographic analysis of D-manno-heptulose, perseitol, glucose and fructose in avocado cultivars. *Journal of Agricultural and Food Chemistry*, **28**, 379–382

Sitrit, Y., Riov, J. and Blumenfield, A. (1986) Regulation of ethylene biosynthesis in avocado fruit during ripening. *Plant Physiology*, **81**, 130–135

Solomos, T. and Laties, G.G. (1974) Similarities between the actions of ethylene and cyanide in initiating the climacteric and ripening of avocados. *Plant Physiology*, **54**, 506–511

Starrett, D.A. and Laties, G.G. (1991) The effect of ethylene and propylene pulses on respiration, ripening advancement, ethylene-forming-enzyme and 1-aminocyclopropane-1-carboxylic acid synthase activity in avocado fruit. *Plant Physiology*, **95**, 921–927

Stevens, M.A. and Rick, C.M. (1986) Genetics and breeding. In *The tomato crop* (eds. J.G. Atherton and J. Rudich) Chapman & Hall, pp. 35–110

Taylor, J.E., Tucker, G.A., Lasslett, Y, *et al.* (1990) Polygalacturonase expression during leaf abscission of normal and transgenic tomato plants. *Planta*, **183**, 133–138

Theologis, A. and Laties, G.G. (1978) Respiratory contribution of the alternative path during various stages of ripening in avocado and banana fruits. *Plant Physiology*, **62**, 249–255

Tingwa, P.O. and Young, R.E. (1975a) Studies on the inhibition of ripening in attached avocado (*Persea americana* Mill.) fruits. *Journal of the American Society for Horticultural Science*, **100**, 447–449

Tingwa, P.O. and Young, R.E. (1975b) The effect of indole-3-acetic acid and other growth regulators on the ripening of avocado fruits. *Plant Physiology*, **55**, 937–940

Tucker, M.L. and Laties, G.G. (1984) Interrelationship of gene expression, polysome prevalence and respiration during ripening of ethylene and/or cyanide treated avocado fruit. *Plant Physiology*, **74**, 307–315

Tucker, M.L. and Laties, G.G. (1985) The dual role of oxygen in avocado fruit respiration: kinetic analysis and computer modelling of diffusion affected respiratory isotherms. *Plant Cell and Environment*, **8**, 117–127

Tucker, M.L. and Milligan, S.B. (1991) Sequence analysis and comparison of avocado fruit and bean abscission cellulases. *Plant Physiology*, **95**, 928–933

Tucker, M.L. and Christoffersen, R.E., Woll, L. and Laties, G.G. (1985) Induction of cellulase by ethylene in avocado fruit. In *Ethylene and plant development* (eds J.A. Roberts and G.A. Tucker), Butterworths, London, pp. 163–172

Tucker, M.L., Durbin, M.L., Clegg, M.T. and Lewis, L.N. (1987) Avocado cellulase nucleotide sequence of a putative full length complementary DNA clone and evidence for a small gene family. *Plant Molecular Biology*, **9**, 197–204

Valmayor, R.V. (1967) Cellular development of the avocado from blossom to maturity. *Philippine Agriculturist*, **50**, 907–976

Wills, R.B.H. and Tirmazi, S.I.H. (1982) Inhibition of ripening of avocado with calcium. *Scientia Horticulturae*, **16**, 323–330

Winer, L. and Apelbaum, A. (1986) Involvement of polyamines in the development and ripening of avocado fruits. *Journal of Plant Physiology*, **126**, 223–233

Zauberman, G., Fuchs, Y. and Akerman, M. (1985) Peroxidase activity in avocado fruit stored at chilling temperatures. *Scientia Horticulturae*, **26**, 261–265

Chapter 3

Banana

G. B. Seymour

3.1 INTRODUCTION

It is estimated that as many as 100 million people subsist on bananas and plantains as their main energy source (Rowe, 1981). Also the international trade in bananas is of considerable economic importance. The estimated world production of banana and plantain fruits in 1990 was about 71 million tonnes; bananas accounted for 46 million tonnes of which, in 1989, 8.1 million tonnes were exported (FAO, 1989; 1990).

Bananas are monocotyledons and belong to the family Musaceae. They are tree-like perennial herbs, two to nine metres tall, with an underground rhizome or corm, a pseudostem composed of leaf sheaths and a terminal crown of leaves through which an inflorescence emerges. Some seven to nine months after planting of a sucker (a shoot from the corm), an inflorescence is formed at the base of the pseudostem. About one month later, this inflorescence emerges through the centre of the leaf crown, and fruits may be suitable for harvest 90–150 days after inflorescence emergence. The banana fruit is classified as a berry, and in edible cultivars, vegetative parthenocarpy results in the formation of fruits with an edible seedless pulp in the absence of pollination. Almost all of the edible cultivated parthenocarpic bananas are derived from the wild diploid species *Musa acuminata* (A genome, 2n = 22) or by hybridization between this species and the wild diploid species *Musa balbisiana* (B genome, 2n = 22). The economically important cultivars are mainly triploids (2n = 33). Cultivars belonging to *Musa* AAA group include the Cavendish bananas, which form the basis of international trade. Other types of bananas which are of commercial importance, and are usually cooked prior to consumption, include plantains which are placed in *Musa* AAB group (Simmonds and Stover, 1987).

Biochemistry of Fruit Ripening. Edited by G. Seymour, J. Taylor and G. Tucker. Published in 1993 by Chapman & Hall, London. ISBN 0 412 40830 9

Bananas are plants of the tropical humid lowlands and are mostly grown between 40° north and south of the equator. Wild-seeded diploid forms of *Musa acuminata* have their centre of diversity in the Malaysian area and the evolution of the AAA triploids probably went on concurrently with the diploids, from which they originated, and in the same region. *Musa acuminata* cultivars (AA and AAA) were introduced to areas where the wild-seeded diploid *Musa balbisiana* is native such as India, the Philippines and New Guinea, and hybridization occurred to produce cultivars with various combinations of the A and B genomes (Simmonds, 1962; Simmonds, 1976).

To successfully market edible cultivated bananas, whether in the tropics or after export, requires control over the ripening process to ensure predictable ripening and good quality ripe fruit. Thus most bananas are harvested in the unripe state and may be induced to ripen by exposure to ethylene gas prior to sale. Ethylene is synthesized by ripening bananas, but unripe fruit can normally be initiated to ripen by exposure to an exogenous source of the gas. Bananas for export are shipped from the tropics under refrigeration at 12° to 14°C. On arrival, the fruit are commonly stored at 14° to 22°C under high humidity and exposed to ethylene gas at 100 to 1000 μl l^{-1} for 24 hours to initiate ripening (Hall, 1967; Watkins, 1974). In the tropics, refrigerated store rooms and ethylene gas may not be available and bananas are often stored at ambient temperature and initiated to ripen by other means, including exposure to fumes from charcoal fires and acetylene generated from calcium carbide (Seymour, 1985). (For a comprehensive treatment of the physiological aspects of banana ripening, refer to Marriott, 1980.)

Previous reviews on the biochemistry of banana ripening include those by Von Loesecke (1949), Palmer (1971), Marriott (1980) and Simmonds and Stover (1987). The majority of the studies on the biochemistry of banana ripening have been undertaken on cultivars of *Musa* AAA group and the information below describes work on cultivars in this genome group unless otherwise stated.

3.2   ETHYLENE PRODUCTION AND RESPIRATION

Unripe bananas show a constant, but low level of ethylene production (around 0.05 μl.kg^{-1}.h^{-1}) until the onset of ripening. Ethylene production then increases and this is followed by a rise in the rate of respiration (Burg and Burg, 1965a). During ripening peak ethylene production is normally reached while the rate of respiration is still increasing. Peak ethylene production by Cavendish bananas was reported to be around 3 μl.kg^{-1}.h^{-1} by McMurchie et al. (1972). As the rate of ethylene production declines, the rate of respiration reaches its maximum at around

125 mg CO_2. kg^{-1}. h^{-1} and then declines slightly, but remains at a high level (Palmer, 1971).

3.2.1 Ethylene and the initiation of ripening

Ethylene appears to be intimately involved in the initiation of ripening in bananas, as in other climacteric fruit, but its mode of action is unknown. Exposure of unripe (preclimacteric) bananas to low levels of ethylene (0.015 μl l^{-1}) will shorten the preclimacteric period, while ethylene at 1 μl l^{-1} for 24 hours will induce prompt initiation of ripening (Marriott, 1980). The time of onset of ripening in the absence of exogenous ethylene may be determined by a change in sensitivity to endogenous ethylene during maturation, and probably by a change in the capacity for ethylene production (Burg and Burg, 1965a; Peacock, 1972; McMurchie et al., 1972). These changes are still poorly understood.

Many fruits, including bananas, enter the climacteric phase soon after harvest, whereas if left on the plant they remain unripe for a longer period. These observations suggest that while attached to the plant an inhibitory factor may be supplied to the fruit which regulates the onset of ripening (Burg and Burg, 1965a,b). The regulation of ethylene sensitivity, ethylene synthesis and the onset of ripening may involve other phytohormones such as auxin and abscisic acid. Under certain circumstances auxins may act to retard the onset of banana ripening, while abscisic acid may induce its initiation (Vendrell, 1970, 1985; Vendrell and Dominguez, 1989). The role of these phytohormones in the control of ripening in intact banana fruit remains to be elucidated.

3.2.2 Ethylene biosynthesis

The pathway of ethylene biosynthesis in plant tissues is thought to follow the sequence of biochemical events detailed in Chapter 1 (page 27). The most thoroughly investigated portion of the pathway involves the steps linking S-adenosyl methionine (SAM), its conversion to 1-aminocyclopropane-1-carboxylic acid (ACC) and the subsequent production of ethylene from ACC. Work on the biosynthesis of ethylene in bananas indicates that the pathway involving SAM and ACC is operational in banana fruit. Preclimacteric bananas produce little ethylene and contain little ACC. At the onset of ripening ACC content, ethylene production and the capacity of produce ethylene are greatly enhanced, indicating an increase in the amount or activity of ACC synthase (EC 4.4.1.14) which converts SAM to ACC and the ethylene forming enzyme (EFE) which converts ACC to ethylene (Hoffman and Yang, 1980). Bananas show different patterns of ethylene production in the peel and pulp tissue during ripening. In preclimacteric fruit, peel and pulp ethylene production, ACC content and EFE activity are low and there was no difference between the tissues in the unripe state. However, at the onset of ripening pulp tissues showed much higher levels of ACC, but lower levels of EFE activity than the peel. Also, the pulp tissue was the principal source of ethylene production during ripening and peel tissue isolated from the pulp showed incomplete degreening (Vendrell and McGlasson, 1971; Ke and Tsai, 1988). Thus excess ethylene production by the pulp during ripening probably acts to bring ripening to a rapid and coordinated conclusion.

3.2.3   Respiration

Bananas and other climacteric fruit show a climacteric rise in respiration, the exact role of which is not understood. However, the metabolic events involved in the climacteric and factors regulating these processes have been intensively studied.

The climacteric rise in respiration is thought to reflect an increased flux of carbon through the glycolytic pathway (Young et al., 1974). The possibility that the enhanced rate of respiration in bananas resulted from utilization of the cyanide-resistant pathway was investigated by Theologis and Laties (1978), who concluded that the climacteric in bananas results from utilization of the cytochrome-mediated pathway.

An important control site in glycolysis is the enzyme phosphofructokinase (PFK) which catalyses the interconversion of fructose-6-phosphate to fructose 1,6-bisphosphate. Early studies on glycolytic intermediates in ripening bananas revealed a marked increase in the levels of fructose 1,6-bisphosphate during ripening, thereby indicating a role for PFK in the climacteric rise (Barker and Solomos, 1962). The enzyme was partially purified from bananas at various stages of ripeness by Salminen and Young (1975) and Nair and Darak (1981), who observed an increase in PFK activity during ripening, and a study of the kinetic properties of the enzyme suggested that this resulted from enzyme activation. The activation of PFK may result from the dissociation of the oligomeric form of the enzyme to monomers (Iyer et al., 1989a,b). The regulation of PFK activity in ripening bananas is still not fully understood, however investigations have concentrated on identifying regulatory factors observed in other tissues. Fructose 1,6-bisphosphate can be formed in plants by two types of PFK, the irreversible ATP-dependent PFK (EC 2.7.1.11) and the reversible pyrophosphate-dependent PFK (EC 2.7.1.90). The respective role of these enzymes in glycolysis is not yet well defined. In 1980, a new regulator of glycolysis, fructose 2-6-bisphosphate was discovered. In higher plants fructose 2-6-bisphosphate activates pyrophosphate-dependent PFK and inhibits cytoplasmic fructose 1,6-bisphosphatase (EC 3.1.3.11) (Van Schaftingen, 1987). Several groups have investigated the role of fructose 2-6-bisphosphate in regulating the climacteric rise in respiration in bananas (Mertens *et al.*, 1987; Beaudry *et al.*, 1987; Ball and Ap Rees, 1988; Beaudry *et al.*, 1989). Mertens *et al.* (1987) reported that the concentrations of fructose 1,6-bisphosphate and fructose 2,6-bisphosphate increased in parallel in ripening bananas, although the levels of fructose 2,6-bisphosphate were always much lower. Beaudry *et al.* (1987) observed a transient rise in fructose 2,6-bisphosphate concentration prior to the respiratory rise, but coincident with the increase in ethylene synthesis. The level of this metabolite then showed a more substantial rise one day after the initiation of the respiratory climacteric. However, Beaudry *et al.* (1987) stated that the timing of these changes did not causally implicate fructose 2,6-bisphosphate in the initiation of the respiratory climacteric through activation of pyrophosphate-dependent PFK. Also, data on the relative increase in the activity of pyrophosphate-dependent and ATP-dependent PFK indicated that during the climacteric, ATP-dependent PFK made a more substantial contribution. Beaudry *et al.* (1989) suggest that steady state levels of fructose 2,6-bisphosphate may exert their influence on the dynamic balance between sugar accumulation and glycolytic flux (Fig. 3.1) by exerting a relative preference for glycolysis under certain conditions during ripening, although they noted that fructose 2,6-bisphosphate concentrations did not appear to be tightly associated with absolute promotion of sugar accumulation and glycolysis. Also, extensive recovery experiments with fructose 2,6-bisphosphate failed to detect any marked change in fructose 2,6-bisphosphate during ripening in bananas (Ball and ApRees, 1988).

Very recently, Ball *et al.* (1991) reported a further study on glycolysis at the climacteric of bananas. Undertaking extensive precautions in the measurements of substrates and enzymes at the climacteric they concluded that the respiratory rise is accompanied by an increase in the maximum catalytic activity of ATP-dependent PFK, a fall in the content of hexose 6-phosphates and a rise in fructose 1,6-bisphosphate. There was no evidence that pyrophosphate-dependent PFK was primarily responsible for the increased flux from fructose 6-phosphate to fructose 1,6-bisphosphate at the climacteric. Also there was no marked change in fructose 2,6-bisphosphate at this time. Ball *et al.* (1991) observed a fall in the content of phosphoenolpyruvate as the rate of respiration rose. This they suggest, may indicate that activation of pyruvate kinase (EC 2.7.1.40) and/or phosphoenolpyruvate carboxylase (EC 4.1.1.31) is the initial response of glycolysis at the climacteric. The roles of pyrophosphate-dependent PFK and fructose 1,6-bisphosphatase in banana are unclear. They could be involved in the conversion of starch to sucrose (Beaudry *et al.*, 1987, 1989) (refer to Fig. 3.1). However, Ball *et al.* (1991) were unable to detect a cytosolic fructose 1,6-bisphosphatase in climacteric bananas

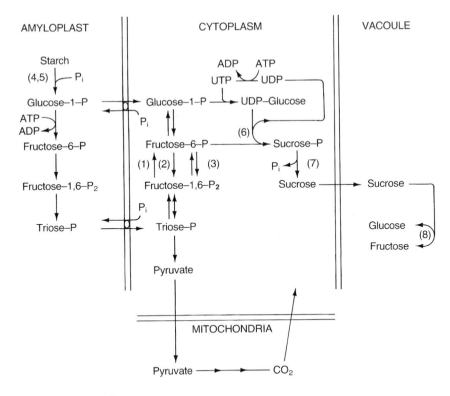

Enzymes

(1) Fructose 1,6-bisphosphatase
(2) ATP phosphofructokinase
(3) Pyrophosphate-dependent phosphofructokinase
(4) Phosphorylase
(5) α-amylase, β-amylase, α1,6-glucosidase
(6) Sucrose phosphate synthase
(7) Sucrose phosphate phosphatase
(8) Invertase

Fig. 3.1 Outline of biochemical pathways likely to be involved in respiration and carbohydrate metabolism in ripening bananas. (After Beaudry *et al.*, 1987, 1989; Hubbard *et al.*, 1990.)

and the observation that the climacteric was not accompanied by any marked change in pyrophosphate-dependent PFK, pyrophosphate or fructose 2,6-bisphosphate casts doubt on such a role for pyrophosphate-dependent PFK.

Both alcohol dehydrogenase (EC 1.1.1.1) and lactate dehydrogenase (EC 1.1.2.8) have been observed to increase during ripening in bananas (Hyodo *et al.*, 1983; Tan *et al.*, 1987) and ethanol accumulates at the overripe stage (Hyodo *et al.*, 1983).

3.3 CARBOHYDRATE METABOLISM

Starch forms about 20–25% of the fresh weight of the pulp of unripe bananas. During ripening this starch is degraded rapidly and the sugars sucrose, glucose and fructose accumulate; traces of maltose may also be present (Palmer, 1971). In the banana pulp sucrose is the predominant sugar, at least at the start of ripening, and its formation precedes the accumulation of glucose and fructose (Areas and Lajolo, 1981; Marriott *et al.*, 1981; Hubbard *et al.*, 1990). The peel tissue also contains starch, about 3% fresh weight, and appears to show similar changes in carbohydrate during ripening (Barnell, 1943). These characteristic patterns of carbohydrate metabolism can be altered under certain environmental conditions such as exposure to elevated temperatures during ripening, e.g. Lizana (1976) reported that at 40°C sucrose formation was suppressed, although the accumulation of glucose and fructose was still observed.

The biochemical basis of carbohydrate metabolism in ripening is now beginning to become apparent. The physical and chemical properties of banana starch have been investigated by Kayisu and Hood (1981), Lii *et al.* (1982) and Garcia and Lajolo (1988), who noted that banana starch has an amylose content of 16%, with granules generally of 20–60 μm in size. Enzymes for both hydrolytic and phosphorolytic breakdown of starch have been identified in banana. However, the exact contribution of each process to starch breakdown during ripening is still in question.

Starch degrading enzymes isolated from bananas are shown in Table 3.1, with α-amylase (EC 3.2.1.1), β-amylase (EC 3.2.1.2) and α1,6-glucosidase (EC 3.2.1.11) activity having all been reported to increase in

Table 3.1 Starch degrading enzymes isolated from bananas.

Enzyme	Mode of action	Reference
α-amylase (EC 3.2.1.1)	Endo, acting on α 1–4 glucose linkages	Young *et al.* (1974)
β-amylase (EC 3.2.1.2)	Exo, acting on α 1–4 glucose linkages at non-reducing end of substrate	Garcia and Lajolo (1988)
α 1,6-glucosidase (EC 3.2.1.11)	Attacks α 1–6 glucose Linkages of amylopectin	Garcia and Lajolo (1988)
Phosphorylase (EC 2.4.1.21)	Attacks the non-reducing end of polymer producing glucose-1-phosphate	Yang and Ho (1958), Areas and Lajolo (1981), Kumar and Sanwal (1982)

ripening bananas (Young *et al.*, 1974; Mao and Kinsella, 1981; Garcia and Lajolo, 1988). However, it has been noted that starch phosphorylase (EC 2.4.1.21) activity which is present in both unripe and ripe bananas and even in green bananas potentially has the activity required for starch breakdown during ripening (Areas and Lajolo, 1981). As previously mentioned, the primary product of starch breakdown in bananas appears to be sucrose, followed by hexose accumulation. Sucrose can be synthesized from glucose-1-phosphate by converting it to uridine diphosphate-D-glucose (UDP-glucose). This is then utilized in the reactions shown below:

$$\text{UDP-glucose + fructose 6-phosphate} \longrightarrow \text{Sucrose phosphate} \qquad 1$$
sucrose
phosphate synthase

$$\text{Sucrose phosphate + UDP} \longrightarrow \text{Sucrose} + P_i \qquad 2$$
Phosphatase

The biosynthesis of sucrose via sucrose phosphate synthase (EC 2.4.1.14) is the most likely pathway for sucrose synthesis in plants (Preiss, 1982) and recent evidence from Hubbard *et al.* (1990) indicates that sucrose synthesis in bananas takes this route, with a possible change in the kinetic properties of sucrose phosphate synthase during ripening. Their data also suggest that the hexose sugars arise from sucrose hydrolysis, perhaps by the action of acid invertase in the vacuole (Beaudry *et al.*, 1989). Acid invertase (EC 3.2.1.26) activity increases during ripening and Sum *et al.* (1980) have purified an invertase from ripe bananas. The conversion of starch to sucrose and sucrose turnover creates a very substantial demand for ATP, and sugar accumulation and respired carbon dioxide were highly correlated. Sucrose accumulation may therefore contribute causally to the respiratory climacteric in banana fruit by creating rapid adenylate turnover (Hubbard *et al.*, 1990) (refer to Fig. 3.1).

3.4   PIGMENT CHANGES

Over 60 years ago, Von Loesecke (1929) reported on pigment changes in the peel of ripening bananas. He observed a decrease in chlorophyll content from between 50–100 µg per gram fresh weight to almost zero in ripe fruit, while carotenoid levels (xanthophylls and carotenes) remained approximately constant at 8 µg per gram fresh weight. More recent studies have supported these early findings and added considerable detail, although the exact pathway of chlorophyll degradation in plants remains unknown.

Table 3.2 Quantitative changes in carotenoids in the peel of ripening bananas. (From Gross and Flügel, 1982.)

Carotenoid (μg per 10 g fresh weight)	Ripening stage			
	green	green–yellow	yellow–green	yellow
α-Carotene	27.4	12.2	16.4	20.0
β-Carotene	40.8	18.8	37.1	30.6
β-Carotene-5,6-epoxide	–	–	–	1.7
Unknown mixture	–	–	–	1.5
α-Cryptoxanthin	–	–	1.1	3.0
Cryptoxanthin	–	–	1.1	2.5
Cryptoxanthin-5,6-epoxide	–	–	–	1.5
Lutein	80.4	42.1	57.6	78.6
Isolutein a	5.4	2.7	5.3	1.9
Isolutein b	2.0	2.7	7.0	5.5
Antheraxanthin	2.8	1.4	2.6	2.9
Luteoxanthin	1.4	1.9	5.4	3.8
Violaxanthin a	19.0	3.4	8.8	4.0
Violaxanthin b	–	3.8	2.2	12.9
Neoxanthin	20.8	11.0	15.4	19.6
Total carotenoids	200.0	100.0	160.0	190.0

A number of studies (Gross *et al.*, 1976; Gross and Flügel, 1982) have examined the carotenoid composition of banana pulp and peel in some detail. Changes in the pattern of carotenoids in banana peel during ripening are shown in Table 3.2. Their data indicate that there is a reduction in total carotenoid content in the peel during the early stages of ripening followed by carotenoid biosynthesis at the yellow–green to yellow–ripe stage. Analysis of the carotenoid composition of the pulp of fully ripe bananas revealed the major types to be α-carotene (31%), β-carotene (28%) and lutein (56% diester : monoester : free 1 : 2 : 1) (Gross *et al.*, 1976).

3.4.1 Biochemical events

The biochemical events which regulate, and are responsible for, pigment changes in ripening bananas and other fruits are poorly understood. Some banana cultivars, e.g. Cavendish bananas, show a marked reduction in chlorophyll degradation during ripening at temperatures above about 24°C (Seymour *et al.*, 1987a) (Fig. 3.2). This is a comparatively low temperature for direct inhibition of 'chlorophyll degrading enzymes' and the effect does not appear to be related to capacity to synthesize ethylene or exchange of gases within the tissue (Seymour *et al.*, 1987b). Also, as

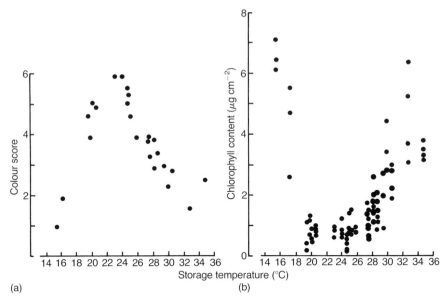

Fig. 3.2   Peel colour score (a) and chlorophyll content (b) of bananas ripened at temperatures of between 15°C and 35°C. (From Seymour *et al.*, 1987a.)

already stated, this failure to degreen occurs only in certain cultivars, with plantains degreening more rapidly at high temperature. Thus bananas may prove an interesting system by which to study the breakdown of chlorophyll in ripening fruits. Recently, Blackbourn *et al.* (1990a,b,c) have investigated the effect of high temperature on chlorophyll degradation in ripening bananas in more detail. They tested the hypothesis that elevated temperatures affect the activity of a chlorophyll bleaching system in banana thylakoids. The basis for this work lies in the discovery of an oxidative chlorophyll bleaching system requiring fatty acids identified in barley thylakoids (Martinoia *et al.*, 1982) and the postulate that thermal stability of membrane-associated chloroplast enzymes may be determined by changes in membrane lipid structure at high temperature (Raison, 1980). Their results show that banana thylakoids do possess a linolenate-dependent chlorophyll oxidase activity, though this activity is not directly inhibited by temperatures up to 40°C. Linolenic acid levels in the neutral lipid pool of banana peel were, however, much lower at elevated than at normal temperatures (35° rather than 20°C) and this could play a role in limiting chlorophyll oxidase activity. Also, studies on the physical distribution of chorophyll oxidase and its substrate in barley and related studies in banana indicate that in banana, chlorophyll retention at 35°C may result from the disruption of the normal spatial arrangement between the chlorophyll bleaching system and chlorophyll protein complexes. Differences between cultivars may reflect differences in membrane fluidity at elevated temperature. Thus the more 'fluid' thylakoid membranes of the relatively 'cold tolerant' Cavendish bananas may be unable to maintain appropriate organization of chlorophyll oxidases and chlorophyll protein complexes (Blackbourn *et al.*, 1990c).

3.5   CELL WALL CHANGES

Softening in fruits appears to be closely linked with changes in their cell wall structure. In bananas, the changes in texture of the fruit during ripening probably result from alterations in both cell wall structure and the degradation of starch. Preliminary visual examination of the cells from banana peel and pulp under the light microscope indicate wall modifications in the cells of the pulp, and to a lesser extent in the peel during ripening (Smith, 1989). In many fruits (Huber, 1983; Tucker and Grierson, 1987) the most apparent changes occur in the wall pectic polysaccharides, which may become more soluble and show a reduction in molecular weight. Comparatively little information is available for banana. Smith *et al.* (1989) found that the amount of pectin solubilized by acetate buffer from acetone-insoluble preparations of banana cell walls increased during ripening in both the pulp and the peel. Gel filtration and chromatography of pectic polyuronides solubilized from unripe and ripe fruit cell walls with acetate buffer containing the calcium chelator CDTA (*trans*-1,2-diaminocyclohexane-tetra acetic acid) indicated that these soluble pectins were depolymerized to some extent during ripening. Gross changes in banana cell wall composition were also examined by Smith (1989). The sugar composition of acetone-insoluble material and CDTA-soluble pectins from the pulp and peel of unripe and ripe bananas is shown in Tables 3.3 and 3.4, these data having been corrected for glucose derived from starch. However, the results are preliminary and require confirmation. Other data available on banana cell wall composition include that of Jarvis *et al.* (1988) who undertook a survey of the pectin content of nonlignified monocot cell walls and reported that bananas had a wall pectin content comparable to dicots.

3.5.1   Pectin degrading enzymes

Changes in pectin structure during banana ripening indicate that pectin degrading enzymes may be active in these cell walls. Pectin degrading enzymes have been identified from bananas. Markovic *et al.* (1975) reported the presence of endo- (EC 3.2.1.15) and exo- (EC 3.2.1.67) polygalacturonase in ripening bananas. Pectin methyl esterase (EC 3.1.1.11) activity has also been identified in banana tissue (Hultin and

Table 3.3 Neutral sugar and uronic acid composition (mol %) of acetone-insoluble residues from the pulp and peel of unripe and ripe bananas. (After Smith, 1989.)

Sugar	Pulp		Peel	
	unripe	ripe	unripe	ripe
Rhamnose	3.2	2.3	0.7	0.7
Arabinose	2.2	1.5	4.5	4.2
Xylose	3.9	2.2	5.8	9.6
Mannose	7.9	5.3	3.6	5.5
Galactose	4.2	2.2	3.6	5.0
Glucose	49.8	71.9	62.8	60.9
Uronic acid	28.8	14.7	19.1	13.1

Table 3.4 Neutral sugar and uronic acid composition (mol %) of CDTA-soluble pectins from the pulp and peel of unripe and ripe bananas. (After Smith, 1989.)

Sugar	Pulp		Peel	
	unripe	ripe	unripe	ripe
Rhamnose	5.7	2.4	8.1	3.0
Arabinose	2.4	2.0	4.7	7.9
Xylose	1.6	1.1	1.7	6.2
Mannose	0.7	3.2	0.9	0.9
Galactose	2.6	2.3	4.7	6.2
Glucose	7.9	5.49	3.0	1.2
Uronic acid	79.2	83.6	77.0	74.5

Levine, 1965; De Swardt and Maxie, 1967; Brady, 1976). Early reports (Hultin and Levine, 1965) indicated there were multiple forms of the enzyme in the pulp of banana fruit and that the activity of pectin methyl esterase increased during ripening. However, more recent work suggests that these changes in activity may have resulted from declining interference by endogenous phenolic compounds during ripening, with little actual change in enzyme activity being apparent (De Swardt and Maxie, 1967; Brady, 1976). Recently Smith (1989) using Cavendish fruit observed an increase in pectin methyl esterase activity in banana peel and a decrease in the activity of this enzyme in the pulp, with mixing experiments indicating that the activity recorded was not due to the presence of inhibitors or activators. Brady (1976) purified pectin methyl esterase about 200-fold from the pulp of climacteric bananas, finding it to be a small basic protein, with a molecular weight of about 30 kDa. Two iso-forms of pI 8.8–8.9 and pI 9.2–9.4 were resolved, although more detailed analyses are needed to confirm that these represent two distinct enzyme proteins. Both polygalacturonase and pectin methyl esterase may play a role in degrading pectic polysaccharides during ripening with pectin esterase removing methyl groups from galacturonic acid polymers prior to attack by endo-polygalacturonase. However, the extent of these events *in vivo* remains ill-defined at present. Enzymes which degrade wall components other than pectin have also been detected in ripening bananas. Smith (1989) reported the presence of cellulase activity in banana pulp.

Polypeptides which have the characteristics of glycoproteins, and some of which may be cell wall degrading enzymes, have been observed by two-dimensional electrophoresis to accumulate in ripening banana pulp (Fig. 3.3) (Dominguez-Puigjaner *et al.*, 1992). Two sets of poly-peptides were detected that increased substantially in ripe fruit. These polypeptides were characterized as glycoproteins by Western blotting and concanavalin A binding assays and antibodies to tomato polygalac-turonase cross-reacted with one of these sets of proteins (Fig. 3.4).

Cell wall degradation and loss of starch are likely to play a major role in textural changes in ripening bananas. Further studies are required to establish the relative importance of each of these factors, although pro-gress on the cell wall enzyme and compositional analysis is hampered by the high starch and phenolic content of bananas.

3.6 PHENOLIC COMPOUNDS

Banana fruit tissues, and particularly the peel, are rich in phenolics, such as 3,4-dihydroxyphenylethylamine and 3,4-dihydroxyphenylalanine. These compounds, when oxidized by the enzyme polyphenoloxidase (EC 1.10.3.1), are responsible for the rapid browning of banana tissues (Palmer, 1971). Studies on banana polyphenoloxidase indicate that the enzyme activity can be separated into a number of isoenzymes. Two distinct polyphenoloxidase fractions were isolated by Mowlah *et al.* (1982), where isoelectric focusing indicated major activities with pIs of 4.70 and 5.15 for the two fractions and molecular weights ranging from 38 kDa to 68 kDa and 40 kDa to 71 kDa respectively. Jayaraman *et al.* (1982, 1987) also identified two major fractions of polyphenoloxidase activity from bananas and examined the rates of enzymic browning and poly-phenoloxidase activity in a range of banana cultivars. Those cultivars which exhibited low browning rates had low polyphenoloxidase activity and a high ascorbic acid content. Studies on the substrate specificity of banana polyphenoloxidase have indicated that the system oxidizes specifically *o*-diphenols (Mayer and Harel, 1979). Thomas and Janave

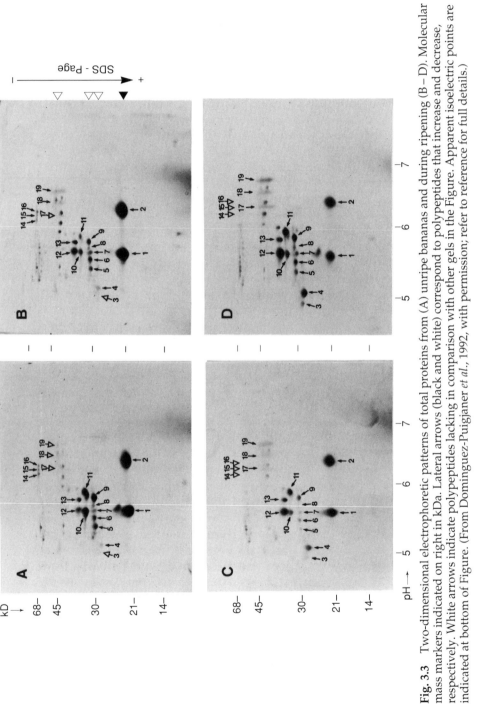

Fig. 3.3 Two-dimensional electrophoretic patterns of total proteins from (A) unripe bananas and during ripening (B – D). Molecular mass markers indicated on right in kDa. Lateral arrows (black and white) correspond to polypeptides that increase and decrease, respectively. White arrows indicate polypeptides lacking in comparison with other gels in the Figure. Apparent isoelectric points are indicated at bottom of Figure. (From Dominguez-Puigjaner *et al.*, 1992, with permission; refer to reference for full details.)

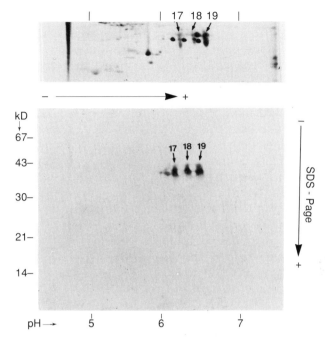

Fig. 3.4 Detection of a set of banana proteins cross-reacting with an antibody to tomato polygalacturonase. Total protein extracts from post-climacteric bananas were resolved by two-dimensional electrophoresis, blotted onto nitrocellulose and challenged with the antibody. The serum gave a specific immunoreaction with five polypeptides, some of which (17, 18, 19) corresponded with spots on Coomassie-stained gels (see Fig. 3.3). (From Dominguez-Puigjaner *et al.*, 1992, with permission; refer to reference for full details.)

(1986) reported banana isoenzymes with both mono-and diphenolase activities.

Peroxidases (EC 1.11.1.7) are also present in ripening bananas. These enzymes catalyse, in the presence of H_2O_2, the oxidation of substrates such as phenols, aromatic secondary and tertiary amines, leucodyes, ascorbic acid and certain heterocyclic compounds such as indoles (Haard and Tobin, 1971). Multiple forms of peroxidase have been reported in bananas and they appear to vary between clones (Jarret and Litz, 1986). The exact role of these enzymes in ripening fruit is unclear. Toraskar and Modi (1984) observed differing isoenzyme patterns between chill-injured and healthy banana fruit. An increase in cell wall-associated peroxidase activity in bananas has been reported to occur with the initiation of the climacteric rise in respiration (Haard, 1973).

Tannins are present in banana fruit tissue and particularly in the peel tissue. These compounds are generally defined as water-soluble phenolics having molecular weights between 500 and 3000 and having special properties, such as the ability to precipitate alkaloids and proteins (Haslam, 1981). In banana, the tannins appear to be confined mainly to the latex vessels of the pulp and peel and certain other small cells in the peel (Barnell and Barnell, 1945). Tannins are known to interact with salivary proteins and glycoproteins, rendering fruit astringent to taste, and the loss of astringency of banana during ripening may result from increased polymerization of tannins (Palmer, 1971; Haslam, 1981). The ability of tannins to bind proteins has long been recognized as a major problem to be overcome when extracting active enzymes from banana tissue (Young, 1965). Studies on the structure of banana tannins have been limited; however tannins of the proanthocyanidin (leucoan-thocyanadins) type have been reported. These phenolic compounds have a distinctive reaction with acids to give anthocyanidins (Haslam, 1981). Simmonds (1954) reported that the skin and pulp of edible bananas yielded the anthocyanidins delphinidin and cyanidin, with the former being predominant.

3.7   ORGANIC ACID AND AMINO ACID METABOLISM

The levels of organic acids in a fruit can markedly affect its taste. In banana the pulp shows an increase in acidity during ripening and the main organic acids present are malic, citric and oxalic (Palmer, 1971; Marriott, 1980). The astringent taste of unripe bananas is probably attributable at least partly to their oxalic acid content, which undergoes significant decarboxylation during ripening (Shimokawa et al., 1972), probably by the action of oxalate oxidase (EC 1.2.3.4) (Ueda and Ogata, 1976).

Enzymes involved in amino acid metabolism were studied in ripening banana fruit by Lal et al. (1974), who found that glutamate oxaloacetate transaminase and glutamate pyruvate decarboxylase (EC 2.6.1.2), enzymes involved in aspartate and alanine synthesis respectively, were at a maximum at a stage corresponding to the initiation of the climacteric. The protein content of bananas was found to remain approximately constant during ripening (Wade et al., 1972), and although the pattern of protein synthesis has been studied in ripening bananas, no conclusive evidence was obtained for any massive synthesis of new proteins to coincide with the onset of ripening (Brady and O'Connell, 1976). However, it may be possible to extend these investigations using molecular techniques.

3.8   PRODUCTION OF VOLATILE COMPOUNDS

A wide variety of volatile compounds emanate from ripe bananas; these include esters, alcohols, ketones, aldehydes and phenol esters. Esters account for about 70% of the volatile compounds and acetates and butyrates predominate within this fraction (Tressl and Drawert, 1973; Macku and Jennings, 1987). The exact contribution of each individual volatile to the characteristic flavour and aroma of bananas is not clear, but an early study (McCarthy *et al.*, 1963) indicated that the 'banana-like' flavour was due to amyl esters, while the distinctive 'fruity' note was attributable mainly to butyl esters. The production of most volatile components from bananas increases during ripening until the onset of peel browning, where production reaches a plateau or decreases (Tressl and Jennings, 1972; Macku and Jennings, 1987). Early reports suggested that acetate and butyrate esters were produced at rates that vary in a cyclic manner and that the two cycles were out of phase (Tressl and Jennings, 1972). However, these have not been confirmed by a more recent investigation (Macku and Jennings, 1987), which indicated that volatile production increases in a steady manner during ripening. The biosynthesis of the wide variety of banana volatiles appears to follow a few known metabolic pathways (Tressl and Drawert, 1973). Early work showed that possible precursors of the branched-chain esters and alcohols, which account for many of the major aroma compounds in bananas, were the amino acids leucine and valine. The quantity of these amino acids increased dramatically during ripening and along with the production of volatiles. Labelling of banana tissue slices with ^{14}C-labelled amino acids supported this hypothesis (Myers *et al.*, 1969; Tressl and Drawert, 1973). Another route for the production of volatiles is the metabolism of fatty acids, and labelled fatty acids are converted to esters by banana tissue slices (Tressl and Drawert, 1973). Other aroma compounds such as C_6 and C_9 aldehydes may be formed from linoleic and linolenic acid, again as shown from labelling studies (Tressl and Drawert, 1973).

3.9   LIPIDS

Goldstein and Wick (1969) observed that the total weight of lipid extracted from the banana peel and pulp did not change substantially during ripening, with lipids comprising around 1% dry weight of the pulp and 6.5% of the dry weight of the peel. They also reported that palmitic (16 : 0), oleic (18 : 1), linoleic (18 : 2) and linolenic acid (18 : 3) were the major fatty acids of the pulp and peel, with a tendency towards a loss of unsaturated fatty acids in both pulp and peel during ripening. More recent work has examined changes in the lipid fractions of the pulp and peel in greater detail. Wade and Bishop (1978) examined changes in the lipid composition of banana pulp during ripening. They noted that although in the pulp the relative proportions of neutral lipid, glycolipid and phospholipid did not change, the fatty acid content of these lipid fractions did alter. The changes were mainly confined to the phospho-lipid fraction, where they observed an increase in the proportion of linolenic acid and a decrease in the proportion of linoleic acid. They suggested that the increase in unsaturation of the phospholipid fraction may result in increased fluidity in cellular membranes, since increased fluidity is observed in liposomes prepared from banana phospholipids as ripening progresses. In the peel of ripening bananas Blackbourn *et al.* (1990b) reported a high level of galactolipid breakdown, probably due to the degradation of chloroplast thylakoid membranes during ripening. They observed a large loss of linoleic and linolenic acid from the galacto-lipid fraction; however, these losses were matched by increases in lino-lenic acid in the neutral fraction. These changes in fatty acids in the peel are probably related to the chloroplast–chromoplast transition during ripening and the degradation of chlorophyll.

3.10   OTHER CHANGES

Ribonuclease (EC 2.7.7.16) and acid phosphatase (EC 3.1.3.2) activities have been reported to increase markedly in relation to ripening in banana pulp (Hyodo *et al.*, 1981). In both unripe and ripe bananas acid phosphatase was present as nine isoforms. A similar number of isoforms were also observed by Kanellis *et al.* (1989). The role these changes in enzyme activity play in fruit ripening is not clear.

3.11   CONCLUDING REMARKS

The economic importance of bananas has stimulated intense research activity on the fruit's physiology and biochemistry. The work has allowed us to build up a picture of the exact sequence of metabolic events which occur in the ripening banana and sheds light on similar but less dramatic biochemical changes occurring in other fruits, such as the conversion of starch to sugar, or the respiratory climacteric. The high phenolic content of the fruit has hampered the quantitative isolation of unmodified enzymes from unripe pulp and peel tissue, and the high levels of starch and sugars in the fruit make cell wall analyses difficult and may reduce the efficiency of nucleic acid extraction. However, the potential for using molecular biology techniques to understand some of the biochemical events during banana ripening should be considerable once simple nucleic acid extraction techniques have been developed.

Commercially, the primary problems with the banana crop relate to difficulties in breeding disease-resistant clones with suitable yields. The future of such breeding programmes probably lies in genetic manipulation of tissue *in vivo* (Simmonds and Stover, 1987). Our understanding of the biochemical basis of ripening characteristics in banana fruits may allow us to incorporate especially desirable characteristics from the wide variety of banana fruit types available into disease-resistant and high yielding stock.

Acknowledgements

I would like to thank Dr Colin Brady for providing unpublished material and advice. Also, my thanks are due to Dr Miguel Vendrell for his comments on the completed manuscript.

REFERENCES

Areas. J.A. and Lajolo, F.M. (1981) Starch transformation during banana ripening. I. The phosphorylase and phosphatase behaviour in *Musa acuminata*. *Journal of Food Biochemistry*, **5**, 19–37

Ball, K. L. and Ap Rees, T. (1988) Fructose 2,6-bisphosphate and the climacteric in bananas. *European Journal of Biochemistry*, **177**, 637–641

Ball, K.L., Green, J.H. and Ap Rees, T. (1991) Glycolysis at the climacteric of bananas. *European Journal of Biochemistry*, **197**, 265–269

Barker, J. and Solomos, T. (1962) Mechanism of the 'climacteric' rise in respiration in banana fruits. *Nature*, **196**, 189

Barnell, H.R. (1943) Studies in tropical fruits. 14. Carbohydrate metabolism of the banana fruit during storage at 53°F. *Annals of Botany N. S.*, **9**, 1–22

Barnell, H.R. and Barnell, E. (1945) Studies in tropical fruits. XVI. The distribution of tannins within the banana and changes in their condition and amount during ripening. *Annals of Botany*, **9**, 77–99

Beaudry, R.M., Paz, N., Black, C.C. and Kays, S.J. (1987) Banana ripening: Implication of changes in internal ethylene and CO_2 concentrations, pulp fructose 2,6-bisphosphate concentration and activity of some glycolytic enzymes. *Plant Physiology*, **85**, 277–282

Beaudry, R.M., Severson, R.F., Black, C.C. and Kays, S.J. (1989) Banana ripening: Implication of changes in glycolytic intermediate concentrations, glycolytic and gluconeogenic carbon flux and fructose 2,6-bisphosphate concentration. *Plant Physiology*, **91**, 1436–1444

Blackbourn, H.D., Jeger, M.J., John, P. and Thompson, A.K. (1990a) Inhibition of degreening in the peel of bananas ripened at tropical temperatures. III. Changes in plastid ultrastructure and chlorophyll-protein complexes accompanying ripening in bananas and plantains. *Annals of Applied Biology*, **117**, 147–161

Blackbourn, H.D., Jeger, M.J., John, P., Teifer, A. and Barber, J. (1990b) Inhibition of degreening in the peel of bananas ripened at tropical temperatures. IV. Photosynthetic capacity of ripening bananas and plantains in relation to changes in the lipid composition of ripening banana peel. *Annals of Applied Biology*, **117**, 163–174

Blackbourn, H.D., Jeger, M.J. and John, P. (1990c) Inhibition of degreening in the peel of bananas ripened at tropical temperatures. V. Chlorophyll bleaching activity measured *in vivo*. *Annals of Applied Biology*, **117**, 175–186

Brady, C.J. (1976) The pectinesterase of the pulp of the banana fruit. *Australian Journal of Plant Physiology*, **3**, 163–172

Brady, C.J. and O'Connell, P.B.H. (1976) On the significance of increased protein synthesis in ripening banana fruits. *Australian Journal of Plant Physiology*, **3**, 301–310

Burg, S.P. and Burg, E.A. (1965a) Relationship between ethylene production and ripening in bananas. *Botanical Gazette*, **126** (3), 200–204

Burg, S.P. and Burg, E.A. (1965b) Ethylene action and the ripening of fruits. *Science*, **148**, 1190–1196

De Swardt, G.H. and Maxie, E.C. (1967) Pectin methylesterase in the ripening banana. *South African Journal of Agricultural Science*, **10**, 501–506

Dominiguez-Puigjaner, E., Vendrell, M. and Ludevid, M.D. (1992) Differential protein accumulation in banana fruit during ripening. *Plant Physiology*, **98**, 157–162

FAO (1989) Food and Agriculture Organisation of the United Nations. Trade Yearbook, Rome.

FAO (1990) Food and Agriculture Organisation of the United Nations. Production Yearbook, Rome.

Garcia, E. and Lajolo, A.M. (1988) Starch transformation during banana ripening. The amylase and glucosidase behaviour. *Journal of Food Science*, **53**, 1181–1186

Goldstein, J.L. and Wick, E.L. (1969) Lipid in ripening banana fruit. *Journal of Food Science*, **34**, 482–484

Gross, J., Carmon, M., Lifshitz, A. and Costes, C. (1976) Carotenoids of banana pulp, peel and leaves. *Food Science and Technology*, **9**, 211–214

Gross, J. and Flügel, M. (1982) Pigment changes in peel of the ripening banana (*Musa cavendish*). *Gartenbauwissenschaft*, **47**, 62–64

Haard, N.F. (1973) Upsurge of particulate peroxidase in ripening banana fruit. *Phytochemistry*, **12**, 555–560

Haard, N.F. and Tobin, L. (1971) Patterns of soluble peroxidase in ripening banana fruit. *Journal of Food Science*, **36**, 854–857

Hall, E.G. (1967) Technology of banana marketing. CSIRO *Food Preservation Quarterly*, **27** (2), 36–42

Haslam, E. (1981) Vegetable tannins. In *Biochemistry of Plants*, (ed. E.E. Conn) Vol. 7, Academic Press, London, New York

Hoffman, N.E. and Yang S.F. (1980) Changes of 1-aminocyclopropane-1-carboxylic acid content in ripening fruits in relation to their ethylene production rates. *Journal of the American Society for Horticultural Science*, **105**, 492–495

Hubbard, N.L., Pharr, D.M. and Huber, S.C. (1990) Role of sucrose phosphate synthase in sucrose biosynthesis in ripening bananas and its relationship to the respiratory climacteric. *Plant Physiology*, **94**, 201–208

Huber, D.J. (1983) The role of cell wall hydrolases in fruit softening. *Horticultural Reviews* **5**, 169–219

Hultin, H.O. and Levine, A.S. (1965) Pectin methylesterases of the banana. *Journal of Food Science*, **30**, 917–921

Hyodo, H., Ikeda, N., Nagatani, A. and Tanaka, K. (1983) The increase in alcohol dehydrogenase activity and ethanol content during ripening of banana fruit. *Journal of Japanese Society of Horticultural Science*, **52**, 196–199

Hyodo, H., Tanaka, K., Suzun, T., Mizukoshi, M. and Tasaka, Y. (1981) The increase in activities of acid phosphatase and ribonuclease during ripening of banana fruit. *Journal of Japanese Society of Horticultural Science*, **50**, 379–385

Iyer, M.G., Kaimal, K.S. and Nair, P.M. (1989a) Correlation between increase in 6-phosphofructokinase activity and appearance of three multiple forms in ripening banana. *Plant Physiology and Biochemistry*, **27**, 99–106

Iyer, M.G., Kaimal, K.S. and Nair, P.M. (1989b) Evidence for the formation of lower molecular forms of phosphofructokinase in ripening banana by dissociation of oligomeric form. *Plant Physiology and Biochemistry*, **27**, 483–488

Jarret, R.L. and Litz, R.E. (1986) Isoenzymes as genetic markers in bananas and plantains. *Euphytica*, **35**, 539–549

Jarvis, M.C., Forsyth, W. and Duncan, H.J. (1988) A survey of the pectic content of nonlignified monocot cell walls. *Plant Physiology*, **88**, 309–314

Jayaraman, K.S., Ramanuja, M.N., Dhakne, Y.S. and Vijayaraghavan, P.K. (1982) Enzymic browning in some banana varieties as related to polyphenol oxidase activity and other endogenous factors. *Journal of Food Science and Technology* (*India*), **19**, 181–186

Jayaraman, K.S., Ramanuja, M.N., Vijayaraghavan, P.K. and Vaidyanathan, C.S. (1987) Studies on the purification of banana polyphenoloxidase. *Food Chemistry*, **24**, 203–217

Kanellis, A.K., Solomos, T. and Mattoo, A.K. (1989) Visualisation of acid phosphatase activity on nitrocellulose filters following electroblotting of polyacrylamide gels. *Analytical of Biochemistry*, **179**, 194–197

Kayisu, K. and Hood, L.F., (1981) Molecular structure of banana starch. *Journal of Food Science*, **46**, 1894–1897

Ke, L.S. and Tsai, P.L. (1988) Changes of ACC content and EFE activity in peel and pulp of banana fruit during ripening in relation to ethylene production. *Journal of the Agricultural Association of China*, New Series **143**, 48–60

Kumar, A. and Sunwal, G.G. (1982) Purification and physicochemical properties of starch phosphorylase from young banana leaves. *Biochemistry*, **21**, 4152–4159

Lal, R.K., Garg, M. and Krishnan, P.S. (1974) Biochemical aspects of the developing and ripening banana. *Phytochemistry*, **13**, 2365–2370

Lii, C.Y., Chang, S.M. and Young, Y.L. (1982) Investigations of the physical and chemical properties of banana starches. *Journal of Food Science*, **47** 1493–1497

Lizana, L.A. (1976) Quantitative evolution of sugars in banana fruit ripening at normal to elevated temperatures. *Acta Horticulturae*, **57**, 163–173

Macku, C. and Jennings, W.G. (1987) Production of volatiles by ripening bananas. *Journal of Agricultural and Food Chemistry*, **35**, 845–848

Mao, W.W. and Kinsella, J.E. (1981) Amylase activity in banana fruit: properties and changes in activity during ripening. *Journal of Food Science*, **46**, 1400–1409

Markovic, O., Heinrichova, H. and Lenkey, B. (1975) Pectolytic enzymes from banana. *Collection Czechoslovakian Chemical Communications*, **40**, 769–774

Marriott, J. (1980) Bananas – physiology and biochemistry of storage and ripening for optimum quality. *CRC Critical Reviews of Food Science and Nutrition*, **13** (1), 41–88

Marriott, J., Robinson, M. and Karikari, S.K. (1981) Starch and sugar transformation during ripening of plantains and bananas. *Journal of Science, Food and Agriculture*, **32**, 1021–1026

Martinoia, E., Dalling, M.J. and Matile, P. (1982) Catabolism of chlorophyll: Demonstration of chloroplast localised peroxidative and oxidative activities. *Zeitschrift für Pflanzenphysiologie*, **107**, S269–S279

Mayer, A.M. and Harel, E. (1979) Polyphenol oxidases in plants. *Phytochemistry*, **18**, 193–215

Mertens, E., Marcellin, P., Van Schaftingen, E. and Hers, H.G. (1987) Effect of ethylene treatment on the concentration of fructose 2,6-bisphosphate and on the activity of phosphofructokinase. 2. Fructose-2,6-bisphosphatase in banana. *European Journal of Biochemistry*, **167**, 579–583

McCarthy, A.I., Palmer, J.K., Shaw, C.P. and Anderson, E.E. (1963) Correlation of gas chromatographic data with flavour profiles of fresh banana fruit. *Journal of Food Science*, **28**, 379–384

McMurchie, E.J., McGlasson, W.B. and Eaks, I.L. (1972) Treatment of fruit with propylene gives information about the biogenesis of ethylene. *Nature*, **237**, 235–237

Mowlah, G., Takano, K., Kamoi, I. and Obara, T. (1982) Characterisation of banana polyphenol oxidase (PPO) fractions with respect to electrophoretic gel filtration behaviour. *Lebensm.-Wiss U. Technol.*, **15**, 207–210

Myers, M.J., Issenburg, P. and Wick, E.L. (1969) Vapour analysis of the production by banana fruit of certain volatile constituents. *Journal of Food Science*, **34**, 504–509

Nair, P.M. and Darak, B.G. (1981) Identification of multiple forms of phosphofructokinase in ripening dwarf Cavendish banana. *Phytochemistry*, **20**, 605–609

Palmer, J.K. (1971) The Banana. In *The Biochemistry of Fruits and their Products*, (ed. A.C. Hulme), Vol. 2, Academic Press, London

Peacock, B.C. (1972) Role of ethylene in the initiation of fruit ripening. *Queensland Journal of Agricultural and Animal Sciences*, **29**, 137–145

Preiss, J. (1982) Regulation of the biosynthesis and degradation of starch. *Annual Reviews of Plant Physiology*, **33**, 431–454

Raison, J.K. (1980) In *The Biochemistry of Plants*, Vol. 4, Academic Press, London, New York

Rowe, P. (1981) Breeding an 'intractable' crop: bananas. In *Genetic Engineering for Crop Improvement*. Working Papers, The Rockefeller Foundation, New York

Salminen, S.O. and Young, R.E. (1975) The control properties of phosphofructokinase in relation to the respiratory climacteric in banana fruit. *Plant Physiology*, **55**, 45–50

Seymour, G.B. (1985) The effects of gases and temperature on banana ripening. Ph.D Thesis, University of Reading, UK

Seymour, G.B., Thompson, A.K. and John, P. (1987a) Inhibition of degreening in the peel of bananas ripened at tropical temperatures. I. Effect of high temperature on changes in the pulp and peel during ripening. *Annals of Applied Biology*, **110**, 145–151

Seymour, G.B., John, P. and Thompson, A.K. (1987b) Inhibition of degreening in the peel of bananas ripened at tropical temperatures. II. Role of ethylene, oxygen and carbon dioxide. *Annals of Applied Biology*, **110**, 153–161

Shimokawa, K., Veda, Y. and Kasai, Z. (1972) Decarboxylation of oxalic acid during ripening of banana fruit (*Musa sapientum* L.). *Agricultural and Biological Chemistry*, **36**, 2021–2024

Simmonds, N.W. (1954) Anthocyanins in bananas. *Annals of Botany, N.S.*, **18**, 471–482

Simmonds, N.W. (1962). *The Evolution of the Bananas*. Longman, London

Simmonds, N.W. (1976) *Evolution of Crop Plants*. Longman, London

Simmonds, N.W. and Stover, R.H. (1987) *Bananas*. Longman, London

Smith, N.J.S. (1989) Textural and biochemical changes during ripening of bananas. Ph.D Thesis, University of Nottingham, UK

Smith, N.J.S., Tucker, G.A. and Jeger. M.J. (1989) Softening and cell wall changes in bananas and plantains. In *Tropical Fruit – Technical Aspects of Marketing. Aspects of Applied Biology*, **20**, 57–65

Sum, W.F., Rogers, P.J., Jenkins, I.D. and Guthrie, R.D. (1980) Isolation of invertase from banana fruit (*Musa cavendishii*). *Phytochemistry*, **19**, 399–401

Tan, S.C., Ng, K.L., Ali, A.M., Othman, O. and Wade, N.L. (1987) Changes in the activities of alcohol and lactate dehydrogenases during modified atmosphere storage and ripening in air of banana fruit. *Asian Food Journal*, **3**, 138–143

Thomas, P. and Janave, M.T. (1986) Isoelectrofocusing evidence for banana isoenzymes with mono and diphenolase activity. *Journal of Food Science*, **51**, 384–387

Theologis, A. and Laties, G.G. (1978) Respiratory contribution of the alternate path during various stages of ripening in avocado and banana fruits. *Plant Physiology*, **62**, 249–255

Toraskar, M.V. and Modi, V.V. (1984) Peroxidase and chilling injury in banana fruit. *Journal of Agriculture and Food Chemistry*, **32**, 1352–1354

Tressl, R. and Jennings, W.G. (1972) Production of volatile compounds in the ripening banana. *Journal of Agricultural Food Chemistry*, **20**, 189–192

Tressl, R. and Drawert, F. (1973) Biogenesis of banana volatiles. *Journal of Agricultural Food Chemistry*, **21**, 560–565

Tucker, G.A. and Grierson, D. (1987) Fruit ripening. In *Biochemistry of Plants*. (eds P.K. Stumpf and E.E. Conn), Vol. 12, Academic Press, London, New York

Ueda, Y. and Ogata, K. (1976) Changes of oxalates in banana fruits during ripening. *Nippon Shokuhin Kogyo Gakkaishi*, **23**, 311–315

Van Schaftingen, E. (1987) Fructose-2, 6-bisphosphate. *Advances in Enzyme Related Areas of Molecular Biology*, **59**, 315–395

Vendrell, M. (1970) Relationship between internal distribution of exogenous auxins and accelerated ripening of banana fruit. *Australian Journal of Biological Sciences*, **23**, 1133–1142

Vendrell, M. (1985) Effect of abscisic acid and ethephon on several parameters of ripening in banana fruit tissue. *Plant Science*, **40**, 19–24

Vendrell, M. and Dominguez, M. (1989) Effect of auxins on ethylene biosynthesis in banana fruit. In *Biochemical and Physiological Aspects of Ethylene Production in Lower and Higher Plants*, (eds. Clijsters *et al.*), Kluwer Academic Publishers

Vendrell, M. and McGlasson, W.B. (1971) Inhibition of ethylene production in banana fruit tissue by ethylene treatment. *Australian Journal of Biological Sciences*, **24**, 885–895

Von Loesecke, H.W. (1929) Quantitative changes in the chloroplast pigments in the peel of bananas during ripening. *Journal of the American Chemical Society*, **51**, 2439–2443

Von Loesecke, H.W. (1949) *Bananas*. Interscience, London, New York

Watkins, J.B. (1974) Fruit ripening rooms. *Queensland Agricultural Journal*, **1**, 309–313

Wade, N.L. and Bishop, D.G. (1978) Changes in the lipid composition of ripening banana fruits and evidence for an associated increase in cell membrane permeability. *Biochimica et Biophysica Acta*, **529**, 454–464

Wade, N.L., O'Connell, P.B.H. and Brady, C.J. (1972) Content of RNA and protein of the ripening banana. *Phytochemistry*, **11**, 975–979

Yang, S.F. and Ho, H.K. (1958). Biochemical studies on post-ripening banana. *Journal of Chinese Chemical Society*, **5**, 71–85

Young, R.E. (1965) Extraction of enzymes from tannin bearing tissue. *Archives of Biochemistry and Biophysics*, **111**, 174–180

Young, R.E., Salminen, S. and Sornsrivichai, P. (1974) Enzyme regulation associated with ripening in banana fruit In *Facteurs et Regulation de la Maturation des Fruits. Colloq. Int. CNRS*, **238**, 271–279

Chapter 4

Citrus fruit

E. A. Baldwin

Citrus fruit are immensely popular worldwide for their flavour and nutrition. Citrus originated in south east Asia and is now grown in the tropical–subtropical belt from 40° latitude north to 40° latitude south, in both humid and arid regions. Citrus has an important place in world fruit production (Cooper and Chapot, 1977). Numerous reviews are available on various aspects of citrus fruit, including information on industry (Ward and Kilmer, 1989), production and harvesting (Samson, 1986), citrus products (Ting and Rouseff, 1986), postharvest (Salunkhe and Desai, 1986), nutrition (Ting and Attaway, 1971), chemistry and biochemistry (Kefford and Chandler, 1970), physiology (Erickson, 1968), and general (Nagy, *et al.*, 1977; Monselise, 1986).

4.1 COMMERCIAL IMPORTANCE OF CITRUS FRUIT

4.1.1 Production

World citrus fruit production was 47 million tonnes in 1989/90, of which oranges accounted for 80%. Total world orange production reached 39 million tonnes in 1989/90, 50% of which came from Brazil and the US (Fox, 1991). In addition to fresh fruit (sweet orange, mandarin, grapefruit, lemon, and lime), there are many processed citrus products. Frozen concentrated orange juice (FCOJ) is by far the most important processed product from citrus with most US (Florida) and Brazilian orange production being converted to this commodity.

Biochemistry of Fruit Ripening. Edited by G. Seymour, J. Taylor and G. Tucker. Published in 1993 by Chapman & Hall, London. ISBN 0 412 40830 9

4.1.2 Processed products

The popularity of FCOJ is waning, however, in favour of more convenient ready-to-serve juices 'made from concentrate' and 'not from concentrate'. The latter product has gained momentum due to the trend in
consumer preference for natural, i.e. minimally processed products (Fox,
1991). Other citrus products include fresh unpasteurized juice, processed
segments, pulp and molasses for cattle feed, pectin, essential oils and
flavonoids from the peel, citric acid from lemons and limes, and candied
peel and fruits from citron and kumquats, respectively (Samson, 1986).

4.2 TAXONOMY AND CULTIVARS

Citrus belongs to the family Rutaceae and subfamily Aurantioideae
(Samson, 1986) which contain an orange or lemon-like fruit classified as a
hesperidium or berry of special structure. These fruit are characterized
by a juicy pulp made of vesicles within segments. Only three genera in
this subfamily (*Citrus*, *Fortunella*, and *Poncirus*) produce edible juice
vesicles (Swingle and Reece, 1967). The scientific and common names of
commercially important citrus fruits and their complex array of hybrids
are given in Table 4.1.

4.2.1 Sweet orange

Sweet oranges are divided into three groups: blood, Navel, and common
oranges. Blood oranges have a pink to red colour in the flesh, juice and
rind. The Navel orange group has a small secondary fruit that is pushed
to the stylar end of the primary fruit giving a belly-button appearance.
All other cultivars belong to the common group of oranges. The most
important cultivar in the latter group is Valencia due to its adaptability,
abundant juice of excellent colour, good flavour, and paucity of seeds.
The maturation schedule of this cultivar results in two crops overlapping
on the tree. In general, oranges store well on the tree and therefore have
an extended harvest season. Other important cultivars include 'Shamouti'
(Israel), 'Hamlin' (Florida, US), 'Mosambi' (India), 'Pineapple' (Florida,
US), 'Pera' (Brazil), and 'Washington' Navel (California, US). 'Ruby' and
'Kuata' are the two main cultivars of the Blood orange group (Cooper and
Chapot, 1977; Samson, 1986).

4.2.2 Mandarin

Mandarin oranges (including tangerines) are divided into five groups:
Satsuma, King, Willowleaf (Mediterranean), common, and small fruited.
In general, these fruit have short on-tree life, resulting in a short season

Table 4.1 Botanical names of edible *Citrus*, *Citrus* relatives, and *Citrus* hybrids. (From Yokoyama and White, 1966; Samson, 1986; Mabesa, 1990).

Botanical name	Common name
Citrus sinensis	sweet orange
Citrus aurantium	sour orange
Citrus reticulata	mandarin (tangerine)
Citrus paradisi	grapefruit
Citrus grandis	pummelo (shaddock)
Citrus limon	lemon
Citrus medica	citron
Citrus aurantifolia	lime
Poncirus trifoliata	trifoliate orange
Fortunella margarita	kumquat

Common hybrids

tangor	=	mandarin	×	sweet orange
tangelo	=	mandarin	×	grapefruit
lemonime	=	lemon	×	lime
citrange	=	sweet orange	×	*Poncirus*
citrumelo	=	grapefruit	×	*Poncirus*
limequat	=	lime	×	kumquat
citrangequat	=	oval kumquat	×	'Rusk' citrange
calamansi	=	mandarin	×	kumquat

and the trees being subject to alternate bearing. Satsuma mandarins are the most cold tolerant, seedless, and have a navel. 'Ohari' and 'Uase' are important cultivars in Japan. Mandarin in the King group are suitable to the tropics and are grown in Indo-China. The common group of mandarins, known in the west as tangerines, includes 'Clementine', 'Dancy', and 'Ortanique' (Jamaica). 'Temple', a tangor, mandarin × orange hybrid, is also popular in the west (Cooper and Chapot, 1977; Samson, 1986). The ability to set parthenocarpic (seedless) fruit is greater in Satsuma mandarins than in the common group cultivars such as 'Clementine'. This ability was found to be associated with higher levels of active gibberellins (GA), a greater ability to catabolize abscisic acid (ABA) to conjugated ABA, and a lower ability to conjugate indoleacetic acid (IAA) during the early stages of fruit development (Talon *et al.*, 1990). Another link between hormone levels and seed set was observed when field treatments of 20 ppm GA or 150 ppm α-naphthaleneacetic acid (NAA) reduced the number of developing seeds in several mandarin-type cultivars (Feinstein *et al.*, 1975). Important tangelo cultivars (mandarin × grapefruit hybrids) include: 'Minneola', 'Orlando' (Florida, US), and

'Ugli' (Jamaica). Hybrids of tangerine ('Clementine')× tangelo produced 'Robinson', 'Page', 'Osceola', 'Nova', and 'Lee' (Florida, US) cultivars. The fifth group, which are the small fruited mandarins, include 'Cleopatra', which is important as a rootstock, and 'Calamondin', a hybrid of kumquat that is indigenous to the Philippines and used as a substitute for lemon (Cooper and Chapot, 1977; Samson 1986; Mabesa, 1990).

4.2.3 Grapefruit

Grapefruits are divided into two groups: white and pigmented cultivars. Both groups contain seedy and seedless types maturing in eight months to a year. 'Duncan' was a popular white-pigmented, seedy cultivar that has largely been replaced by 'Marsh', which is seedless. A pink flesh mutation of 'Marsh' named 'Thompson' and a sport of 'Thompson' with deeper colour called 'Redblush' (Samson, 1986) resulted in the current popular pigmented cultivars 'Star Ruby', 'Ruby Red', 'Rio', and 'Flame'. Some of the later mutations were induced by radiation.

4.2.4 Lemon and lime

Lemons are sensitive to both heat and cold. 'Eurika' and 'Lisbon' are important cultivars which are grown in Mediterranean-like climates for the fresh fruit market (Cooper and Chapot, 1977). Limes are more adapted to a tropical climate and are divided into two groups: acid (sour) and acidless (sweet) limes. The latter type is used primarily as a rootstock, although 'Palestine Sweet Lime' is popular for fresh consumption. Important acid cultivars include 'West Indian' (Mexican or Key) lime and the larger 'Tahiti' or 'Persian' lime (Samson, 1986).

4.2.5 Other citrus fruit

Sour orange is important as a rootstock and is processed for marmalade, drinks and essential oils. Pummelo (also called shaddock) fruits come in seedy, seedless, common, and pigmented cultivars and are popular in Asia (Samson, 1986).

4.3 HARVEST

Harvested citrus fruit are transported to packing houses for fresh market or are diverted to processing plants. The major forms of processed citrus products are ready-to-use packaged juice or concentrated juice that is stored and ultimately packaged in concentrate form (Ward and Kilmer, 1989).

4.3.1   Determination of maturity

Citrus fruits cannot be picked immature for after-ripening since the fruit contain little starch and are non-climacteric. Unfortunately, colour is not a good indicator of maturity, so percentage soluble solids (°Brix, of which 70–85% is composed of sugars) and titratable acid of the juice are used as a maturity index (Salunkhe and Desai, 1986). A Brix/acid ratio ranging from 8 to 10 is generally accepted as a measure of minimum maturity, while a ratio of 10 to 16 is considered to be of acceptable quality. If the fruit remain on the tree, the Brix continues to increase and the acid diminishes, until eventually overripe and unpleasantly sweet fruit reach a Brix/acid ratio of 20 or more (Samson, 1986).

4.3.2   Abscission

Hand-picking is still a large expense in the cost of producing citrus fruit (Cooper and Chapot, 1977). The tenacity with which mature citrus fruit hang on the tree (a fruit removal force of 9 kg is common) precludes the use of mechanical harvesters. Although much of the fruit can be removed in this manner, too much remains on the tree to make this practice economically feasible. Conversely, flowers and young fruitlets abscise easily, reducing potential yields (Erickson and Brannaman, 1950, 1960). For this reason there has been much interest in enhancing or retarding citrus fruit abscission at different points during fruit development and maturation.

Auxins and gibberellins

Abscission is thought to be controlled by plant growth regulators; either promoted by ethylene or retarded by auxin and GA. High endogenous levels of GA were associated with retardation of fruit separation and early fruit drop was reduced by the application of auxin in the form of 2,4,-dichlorophenoxyacetic acid (2,4-D), and GA to trees (Kefford and Chandler, 1970).

Ethylene

Ethylene promotes abscission in citrus in association with cellulase (Basiouny and Biggs, 1976) and polygalacturonase (Riov, 1974; Basiouny and Biggs, 1976) activation. Citrus is a nonclimacteric fruit in that only low basal levels of ethylene are produced throughout development (Rasmussen, 1975). There are no rapid chemical or physical changes that occur once the fruit is detached from the tree (Salunkhe and Desai, 1986). Citrus peel is capable, however, of producing substantial amounts of ethylene, as has been shown with peel disc explants (Riov *et al.*, 1969;

Evensen *et al.*, 1981; Baldwin and Biggs, 1983), whole fruit on trees
sprayed with abscission chemicals (Biggs, 1971; Kossuth *et al.*, 1979) or
harvested whole fruit injected with fungal cell wall-degrading enzymes
or cell wall fragments (Baldwin and Biggs, 1988). Both albedo and
flavedo portions of the peel can produce ethylene (Hyodo, 1977; Baldwin
and Biggs, 1983) via the methionine pathway (Biggs and Baldwin, 1981;
Riov and Yang, 1982b). Ethylene production by citrus flower petals,
flavedo, and leaves was reported to be autocatalytic and/or autoinhi-
bitory. Autoinhibition was due to suppression of ACC synthase activity
(Riov and Yang, 1982b; Zacarias *et al.*, 1990) and to increased ACC-
malonyltransferase activity (reduces availability of ACC for ethylene
production) (Liu *et al.*, 1985). In aged orange peel discs ethylene production was inhibited by aminoethoxyvinylglycine (AVG), putrescine and
spermine which stimulated incorporation of methionine into spermine
(Even-Chen *et al.*, 1982). This indicated a control point in the ethylene
(senescence-promoting) – polyamine (antisenescent) biosynthetic pathways. Cycloheximide inhibited ethylene production at high concentrations, but stimulated it in citrus peel at concentrations below 10 μM
(Riov *et al.*, 1969; Riov and Yang, 1982a; Baldwin and Biggs, 1983).
Exogenous ethylene stimulated phenylalanine ammonia-lyase (EC
4.3.1.5) activity in flavedo discs (Riov *et al.*, 1969).

Abscission chemicals

Abscission chemicals promoted wound ethylene which caused fruit
loosening at the abscission zone (Evensen *et al.*, 1981). It was hoped that
use of these chemicals would allow mechanical harvesting, but problems
arose such as immature fruit and leaf drop (Grierson and Wilson, 1983),
loosening failures, dependence on weather (Kossuth *et al.*, 1979), extreme
phytotoxicity, and erratic performance (Wilson, 1971, 1973). Some of the
ethylene-inducing or -releasing chemicals used included ethephon
(Samson, 1986), cycloheximide (Acti-aid) (Ismail, 1971), glyoxime (Pick
off), and chlorothalonil (Sweep) (Rasmussen, 1977). Thus far, use of
abscission chemicals together with mechanical harvesting has not proven
to be economically feasible for commercial use.

4.4   POSTHARVEST

4.4.1   Fresh fruit handling

Fruit destined for the fresh market are treated with 5–10 μl l^{-1} ethylene
gas if needed for degreening, washed, sanitized in water with hypo-
chlorous acid of sodium ortho-phenyl phenate (SOPP) or chlorine, sometimes coloured, and waxed (Salunkhe and Desai, 1986). Fruit packed in

cartons may be stored for several months with 90–95% relative humidity. With the exception of grapefruit, which requires storage temperatures of 10°C or above, most citrus is stored at 3° to 8°C (Samson, 1986).

4.4.2 Postharvest physiology

Respiration of orange was reported to be a function of O_2 tension in the range of 0–8% O_2 with anaerobic reactions occurring below 2.5% O_2 (Biale, 1961). Application of waxes to citrus can affect respiration and internal atmosphere composition. The problem with this procedure is that there is a trade-off between reduced water loss and the development of anaerobic conditions in the fruit, resulting in off-flavours due to increased internal ethanol, acetaldehyde, and CO_2 and decreased internal O_2 (Vines and Oberbacher, 1961; Cohen et al., 1990). Similar results, except for retardation of water loss, were observed when citrus fruit were subjected to low O_2 storage (Bruemmer and Roe, 1969; Ke and Kader, 1990; Shaw et al., 1990). Low O_2 storage also resulted in stimulation of pyruvic dehydrogenase (EC 1.2.4.1) and alcohol dehydrogenase (EC 1.1.1.1) accompanying the shift to anaerobic respiration. A decrease in the acetaldehyde-to-ethanol ratio was observed with decreasing O_2 and increasing CO_2 levels, and this effect was compounded by waxing of harvested fruits (Davis, 1970). Storage of fruit in air with acetaldehyde decreased acidity and accelerated degreening (Pesis and Avissar, 1989). Use of semi-permeable edible coatings on oranges resulted in decreased internal O_2 and increased internal CO_2, as well as increased levels of ethanol, methanol and some important flavour volatiles in the juice (Nisperos-Carriedo et al., 1990). Seal-packaging of individual fruit with a 10 μm film of high density polyethylene, however, doubled fruit storage life while maintaining normal flavour (Ben-Yehoshua, 1978).

Plant growth regulators have been used as postharvest treatments. Shelf life of fruit treated with 2,4-D, 2,4,5-trichlorophenoxyacetic acid (2,4,5-T), and NAA was increased by one to two months at several temperatures, while sugar increased and acidity and ascorbic acid content decreased (Kefford and Chandler, 1970). Increased shelf life was due both to delayed senescence and reduced fungal decay.

Chilling injury in grapefruit can be incurred during prolonged storage at temperatures below 10°C. However, temperatures of 0° to 1°C are required for quarantine treatments for Caribbean fruit fly disinfestation (Adsule et al., 1984) for shipment of grapefruit to Japan and California. To prevent chilling injury, techniques such as intermittent warming (Davis and Hofmann, 1973; Purvis, 1989), preconditioning at higher temperatures (McDonald, 1986; Miller et al., 1990), and applications of oils or emulsions (Aljuburi and Huff, 1984) and plastic wraps (Ben-Yehoshua et al., 1981; McDonald, 1986) have been employed. Preconditioned fruit, which become more chilling tolerant, had higher levels of putrescine, a stress-related antisenescent polyamine (McDonald, 1989), and squalene, a highly unsaturated C_{30} isoprene hydrocarbon component in the epicuticular fruit wax (Nordby and McDonald, 1990). Accumulation of carbohydrates and proline was associated with chilling injury avoidance in grapefruit (Purvis and Yelenosky, 1983). Reducing sugars increased while sucrose decreased in intermittently warmed grapefruit (Purvis, 1989). These reactions may directly or indirectly provide some measure of protection against chilling injury.

4.4.3   Postharvest pathology

Postharvest losses in the past have been as high as 15–30% (Salunkhe and Desai, 1986). Physiological and mechanical injury allows entrance of pathogens through wounds. There are two major forms of decay in citrus fruit: stem-end rot (*Diplodia*, *Phomopsis*, and *Alternaria*) and mould (*Penicillium glaucum* and *P.digitatum*, green moulds and *P. italicum*, blue mould) (Salunkhe and Desai, 1986; Droby *et al.*, 1989). Stem-end rots in harvested fruit arise from quiescent infections in the stem button (calyx + disk) once this tissue becomes senescent and begins to abscise (Salunkhe and Desai, 1986). Degreening of citrus fruit with 50 µl l^{-1} ethylene gas produces a substantial increase in stem-end rot during storage (Brown, 1986). Various fungicides are applied directly to the fruits by fumigation or by solution or suspension in water and wax formulations (Salunkhe and Desai, 1986). The yeast *Debaryomyces hansenii*, cus-7, later identified as *Candida quilliermondii*, isolated from the surface of lemon fruit, was found to provide biological control of several citrus fruit pathogens (Droby *et al.*, 1989). Application of 2,4-D to lemons before storage delayed senescence of the button, reducing susceptibility to attack by *Alternaria*. Oranges and grapefruits, when similarly treated, showed reduced stem-end rot due to *Alternaria* and *Diplodia* during transport. Other treatments with GA inhibited decay via increased host resistance (Salunkhe and Desai, 1986).

4.4.4   Processing

The juice of processed fruit is either sterilized as single strength canned juice, pasteurized for not-from-concentrate chilled produce, shipped fresh under refrigeration, or converted to frozen concentrate. Some of these products acquire a cooked taste due to thermally-altered flavour volatiles (Salunkhe and Desai, 1986). Popular cultivars grown for orange juice processing include 'Hamlin' and 'Pineapple' in the US, 'Valencia' in the US and many other citrus-producing regions, and 'Pera' and 'Natal' in Brazil (Cooper and Chapot, 1977; Fox, 1991).

4.5   GENERAL PHYSIOLOGY

4.5.1   Anatomy

Anatomically, citrus fruits are superior ovaries composed of 6 to 20 united carpels which form locules (Albrigo and Carter, 1977; Roth, 1977). The pericarp exterior to the locules is subdivided into the exocarp (flavedo or exterior peel), mesocarp (albedo or interior peel) and endocarp (locule or segment membrane). The juice vesicles, which are the edible portion of citrus fruit and therefore of economic value, arise from epidermal or subepidermal primordia on the surface of the endocarp and grow to fill the locular cavity (Roth, 1977) (Fig. 4.1)

Flavedo

The exocarp or flavedo, which is the coloured portion of the peel, contains pigments in chloroplasts or chromoplasts and oil glands formed by special cells that produce terpenes and oils. The oil gland cavity forms schizogenously and/or lysogenously by the separation and/or lysis of cell walls of the central cells (Thomson *et al.*, 1976; Bosabalidis and Tsekos, 1982). Flavedo epidermal cells produce cutin and waxes and contain actinocytic type stomates (Roth, 1977).

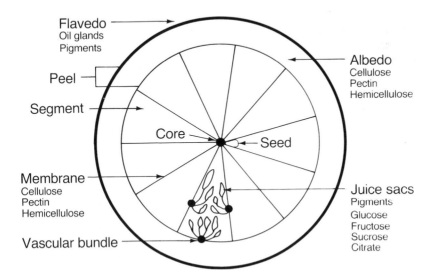

Fig. 4.1   Diagrammatic equatorial cross-section through a citrus fruit. (From Koch *et al.*, 1986; McCready, 1977.)

Albedo

The mesocarp or white albedo portion of the peel consists of colourless cells which are typically eight-armed, parenchymous, highly vacuolated, and tube-like (Albrigo and Carter, 1977). The tissue contains large air spaces, imparting a spongy nature. Where two arms of adjacent cells meet, the transverse wall remains thin and develops pits to enhance translocation of assimilates. A network of vascular tissue branches through the albedo and extends from the main bundles that run parallel to the fruit axis to the outside of the segments at three locations per segment from where juice vesicles are attached (Koch *et al.*, 1986). The albedo serves as a storage parenchyma and contains a higher flavanone content than juice vesicles, section membranes, or flavedo, and is also an important source of limonin (Albrigo and Carter, 1977). The albedo, although much more vulnerable to pathogen invasion than flavedo if exposed, was found to contain a polygalacturonase inhibitor (Barmore and Nguyen, 1985) believed to be a form of defence against pathogens such as *Diplodia natalensis*, which uses polygalacturonase to gain entrance into fruit tissue. The core of the fruit resembles albedo and contains vascular bundles and parenchymous tissue (Albrigo and Carter, 1977).

Peel

Albedo and flavedo together make up what is called the peel or rind, and contain more bitter principles and pectin than other parts of the fruit. The higher concentration of glycosides (flavonones), limonin, and water-soluble pectins are inversely correlated to processed juice flavour and these components are found in higher quantities in juice that contains more peel extract from hard squeezed fruit (Albrigo and Carter, 1977).

Juice vesicles

The endocarp portion of citrus fruit is the most complex, giving rise to juice sacs or vesicles (Roth, 1977). They are classified as both 'emergences' or multicellular hairs (Roth, 1977) and form elliptical-shaped single or branched body structures on a non-vascular stalk (Koch *et al.*, 1986). Juice sac cells are highly vacuolated and the narrow cytoplasm contains lipid droplets in plastids, leucoplasts and chromoplasts (Shomer *et al.*, 1975). Juice within the vacuole of these cells contains essentially all the titratable acids and other soluble materials such as amino acids and salts. Calcium oxalate, hesperidin and naringin crystals can occur in citrus fruits in the peripheral layers of the rind and juice sacs (Roth,

1977). Limonin and naringin contribute bitterness to juice flavour (Horowitz and Gentili, 1977; Maier *et al.*, 1977).

4.5.2 Colour and texture

The colour of citrus fruit is due to the presence of chlorophyll in green fruit and carotenoids in yellow to orange fruit. These pigments are located in plastids in the flavedo (peel) and juice vesicles (pulp). Citrus contains one of the most complex carotenoid patterns among fruits (Gross, 1977).

Orange pigments

In general, the citrus peel has 2 to 6 times the carotenogenic capacity of the endocarp (pulp); thus 70% of the fruit carotenoids are in the peel. The orange-yellow colour of orange peel and juice is due to the presence of carotenes and xanthophylls in chromoplasts, with the esterification of xanthophylls progressing during maturation (Eilati *et al.*, 1972; Gross, 1977). Specific pigment profiles differ in different citrus fruit for both peel and pulp. Violaxanthin predominates in the peel and luteoxanthin and antheraxanthin in the highly coloured pulp of 'Valencia' fruit and violaxanthin, cryptoxanthin, and antheraxanthin in pulp of Navel oranges. The juice vesicles of 'Blood' oranges have been reported to contain anthocyanins which, like the carotenoids, require cool weather for development (Erickson, 1986; Ting and Attaway, 1971). In general, orange juice has about 23 times more total carotenoids than lemon or grapefruit, and yet the carotenoid level is only around 0.001% (Maier, 1969).

Mandarin pigments

The more intense (redder) peel colour of tangerines and other citrus fruits such as 'Sinton' citrangequat is due to the peel pigments cryptoxanthin, β-citraurin, β-apo-8'-carotenal, violaxanthin, antheraxanthin, and zeaxanthin, with β-citraurin being the main pigment contributing to the red tint of these fruit (Yokoyama and White, 1966; Gross, 1977). Cryptoxanthin predominates in the juice pulp of tangerines (Gross, 1977).

Grapefruit pigments

Carotenoid levels in grapefruit peel are low, with the colourless carotenoids phytoene and phytofluene making up 70% of the total. As a result, the pulp of white grapefruit is generally colourless. Of the coloured carotenoids present in the peel, violaxanthin is predominant. Pink and red grapefruit peel carotenoids are 60% colourless pigments with 11% lycopene (giving red colour) while pulp carotenoids are 60% lycopene and 27% β-carotene (Gross, 1977).

Lemon and lime pigments

In lemons, 25% of peel and pulp carotenoids are colourless. Of the coloured pigments, ζ-carotene and cryptoxanthin diepoxides predominate in the peel and epoxides of β-carotene and cryptoxanthin are high in the pulp (Gross, 1977). 'Bronzing' of lemons was found to be associated with presence of the red carotenoids reticulataxanthin and 3-hydroxysintaxanthin, and β-carotenone (Maier, 1969). Phlobatannin (an oxidation product of polyhydroxyphlobian) in the cell sap contributes to the juice colour of limes (Ting and Attaway, 1971).

Texture

Citrus does not soften due to cell wall changes as do other fruit such as strawberry or tomato. Some apparent firming or softening of the fruit occurs, partially due to changes in turgor pressure and/or respiratory loss of dry matter during growth, development, and senescence. These changes can also be related to environmental conditions, irrigation regimes, water loss after harvest and aging of the fruit (Kefford and Chandler, 1970; Sasson and Monselise, 1977). Malonic acid levels were found to increase with aging of fruit tissues as determined by softening (Sasson and Monselise, 1977). Preharvest sprays of GA_3 delayed rind senescence of navel orange, including softening and colour (accumulation of carotenoid pigments) (Coggins, 1981).

4.5.3 Growth and development

Citrus fruit mature on the tree for six to twelve months (longer in the cooler subtropics). Growth and development of citrus fruit occurs in three stages (Bain, 1958; Goren and Monselise, 1964). From four to nine weeks after fruit set, fruit size and weight increase due to growth of the peel by cell division and enlargement. Then, in the cell enlargement period, the fruit increases in size by cell enlargement alone as the albedo tissue differentiates and expands. Finally, in the maturation period, juice sac acids decrease, while the peel turns from green to yellow or orange and increases slightly in thickness (Bain, 1958; Ting and Attaway, 1971).

Auxin, cytokinin and GA activities were found to be associated with fruit growth. Higher cytokinin and GA-like activities were found in the peel of 'Shamouti' oranges during periods of growth and cell division, especially in large fruit with an excessively thick, rough peel. This disorder occurs with 'Shamouti' oranges grown under marginal conditions and can be partially overcome by preharvest treatment with the growth retardants daminozide and chlormequat (Monselise and Goren, 1978). Corky (silvery) spots on 'Marsh' grapefruit are thought to be a result of the inability of external peel layers to keep pace with rapidly expanding endocarp. The application of the growth promoters, 2,4-D and GA_3 resulted in less splitting of the peel (Goren et al., 1976). Treatment of citrus with benzyladenine (BA) stimulated fruit growth and GA_3 sharply enhanced assimilate export from leaves to developing fruit by promoting sink intensity. Polyamine levels and enzymes of polyamine synthesis were associated with growth activity in 'Murcott' mandarin fruits during development (Nathan et al., 1984). On the other hand, NAA and ABA restricted leaf and fruit growth and inhibited transport of assimilates (Mauk et al., 1986). Higher levels of ABA were shown to be related to slow growth periods in acid limes including the time during which ripening and senescence took place (Murti, 1988).

4.5.4 Ripening

Ripening of oranges and grapefruit proceeds from the peripheral regions of the endocarp towards the core. The peripheral regions were found to be higher in soluble solids and lower in acidity and total nitrogen. Axial gradients were also observed with soluble solids and total nitrogen contents increasing from stem-end to stylar-end of the fruit, while ABA was more uniformly distributed (Ting, 1969). Another explanation for these metabolite gradients could be transport preferences rather than ripening pattern. In California fruit, the decrease in acids and limonin and increase in Brix, commonly observed during ripening of oranges, correlated with accumulation of maximum heat from the first four months after bloom (Kimball, 1984).

Plant growth regulators can affect ripening changes in citrus fruit. Preharvest treatment of grapefruit with GA decreased accumulation of nootkatone (an important flavour volatile), increased peel oil concentration (Wilson et al., 1990), and increased peel resistance to puncture (McDonald et al., 1987) and, therefore, to attack by Caribbean fruit fly. GA treatments also affected membrane fluidity (Fuh et al., 1988), induced delay of colouration and senescence of mandarins (Greenberg and Goldschmidt, 1989), and promoted peel thickening (Goldschmidt, 1983). The hormone, GA_3 is highly persistent in citrus peel. Ethylene treatment increased GA metabolism slightly, but not enough to account for ethylene's enhancement of senescence (Shechter et al., 1989). Treatment of mature citrus fruit with ethylene induced chlorophyll destruction, increased respiration, release of free amino acids, accumulation of reducing sugars, and the appearance of phenylalanine ammonia lyase (EC 4.3.1.5, PAL) activity. Application of GA$_3$ and BA opposed or modified the effects of ethylene on all the above except PAL activity (Goldschmidt *et al.*, 1977).

4.6   BIOCHEMICAL CHANGES DURING DEVELOPMENT, RIPENING, AND STORAGE

4.6.1   Colour

Colour break

At colour break there is apparent degeneration of the thylakoids in peel chloroplasts, appearance of loosely aggregated swollen grana, and eventually complete loss of membranous structure (Thomson, 1969; Huyskens *et al.*, 1985). Accompanying this is a decrease of carotenogenesis, degradation of chlorophyll and finally increased carotenoid biosynthesis (Gross, 1981; Gross *et al.*, 1983). Thus green fruit turn yellow to orange at maturity. Degreening requires light (Goldschmidt, 1988) and is promoted by low temperatures, high sucrose levels (Huff, 1984; Takagi *et al.*, 1989), and ethylene (Eaks, 1977), while high nitrogen levels inhibit this process (Huff, 1984; Takagi *et al.*, 1989). Cytokinin and GA enhance regreening (Coggins and Lewis, 1962). Chlorophyllase (Purvis and Barmore, 1981), magnesium-dechelatase, and decarbomethoxylase (Shimokawa *et al.*, 1990) activities increased in response to ethylene-induced senescence of the peel. These changes can be reversed by chromoplast's reconversion to chloroplasts in a process called regreening. This involves *de novo* synthesis of chlorophyll and protein components of the photosynthetic apparatus (Mayfield and Huff, 1986) as well as reassembly of thylakoid structures (Thomson *et al.*, 1967). Regreening can occur on the tree during extended periods of warm weather, in detached fruit (Saks *et al.*, 1988), and in cultured epicarp segments (Huff, 1983).

Oranges and mandarins

Changes in the carotenoid content of oranges and mandarins have been studied during their maturation and generally involve the disappearance of most chlorophyll and chloroplast-type carotenoids in the peel, and the synthesis of chromoplast-type carotenoids in the peel and juice. Analyses of juices of one tangerine and seven orange cultivars during ripening showed increases in phytofluene, α-carotene, α-cryptoxanthin, β-carotene, ζ-carotene, and β-cryptoxanthin, with the last three being the most

Table 4.2   Quantitative changes in flavedo pigments during ripening of 'Dancy' tangerine. (From Gross, 1981.)

	June	Sept.	Nov. (Colour break)	Dec.	Jan.
Fruit diameter (cm)	1.2	3.9	5.3	6.0	6.8
Chlorophyll a (µg/gFW)	280.0	197.0	86.0	51.0	–
Chlorophyll b (µg/gFW)	120.0	53.0	27.0	12.0	–
Total carotenoids (µg/gFW)	93.0	56.0	93.5	195.0	315.0
Carotenoid pattern	*% of total carotenoids*				
phytofluene	–	–	3.9	1.4	2.5
α-carotene	6.1	12.6	4.2	0.8	0.3
β-carotene	14.6	13.5	4.0	1.6	0.3
ζ-carotene	–	–	tr.	0.3	0.6
τ-carotene	–	–	–	tr.	tr.
β-Apo-8'-carotenal	–	–	0.2	0.3	1.1
OH-a-carotene	1.6	1.6	–	–	–
cryptoxanthin	2.5	–	4.5	7.8	13.8
cryptoxanthin 5,6-epoxide	–	–	–	0.5	0.8
cryptoflavin	–	–	0.8	0.7	1.2
cryptoxanthin 5',6'-epoxide	–	–	0.4	0.8	2.4
citraurin	–	–	3.0	4.2	7.8
Lutein	31.8	30.4	13.5	6.1	2.3
Lutein 5, 6-epoxide	3.0	2.1	2.0	–	–
Zeaxanthin	5.8	1.9	0.8	0.9	1.3
Mutatoxanthin	–	–	–	–	0.6
Antheraxanthin a	1.9	6.1	2.3	5.1	3.2
Antheraxanthin b	–	–	–	–	4.3
Luteoxanthin	1.2	–	3.6	2.6	4.0
Violaxanthin a	9.4	20.7	19.6	18.3	11.6
Violaxanthin b	1.6	–	29.1	44.4	39.8
Neoxanthin a	12.7	10.0	5.4	2.6	1.4
Neoxanthin b	4.5	1.1	2.7	1.6	0.8
Trollixanthin-like	3.9	–	–	–	–

dramatic (Stewart, 1977; Gross, 1982). Peel of ripening 'Dancy' tangerine, in contrast, showed decreases in chlorophyll a and b as well as in chloroplast carotenoids, α- and β-carotene, lutein, zeaxanthin, and neoxanthin, but not violaxanthin (Gross, 1981). The appearance of, or increases in phytofluene, ζ- and τ-carotene, β-apo-8'-carotenal, crypto-xanthin and its epoxides, cryptoflavin, citraurin, antheraxanthin a and b, luteoxanthin and most dramatically violaxanthin a and b were evident at or after colour break (Table 4.2). Similar changes were observed for 'Shamouti' orange peel (Eilati *et al.*, 1975).

Grapefruits, lemons, and pummelos

Carotenoid concentrations were reported to decrease in grapefruit and lemon along with chlorophyll as the fruit matured, although an active period of carotenogenesis commenced prior to the decrease in chlorophyll (Yokoyama and Vandercook, 1967; Yokoyama and White, 1967). In ripening pummelo, peel chlorophyll and chloroplast-type carotenoids decreased, while chromoplast-type carotenoids such as ζ-carotene, neurosporene, mutatochrome, cryptoxanthin and its epoxide appeared. Phytofluene, a colourless carotenoid, constituted 60% of total carotenoids in both peel and pulp. The pulp remained mostly colourless although small amounts of chlorophyll and coloured carotenoids remained constant throughout ripening (Gross *et al.*, 1983).

4.6.2 Acids

Development and ripening

As citrus fruit mature on the tree, soluble solids (Brix) increase rapidly at first and then at a slower rate, while total acidity (due mostly to citric acid) increases early in development, then steadily decreases (Samson, 1986). This was observed in oranges (Shaked and Hasdai, 1985), grapefruit (Erickson, 1968; Robertson and Nisperos, 1983), tangelos (Harding *et al.*, 1959), and 'Murcott' tangors (Long *et al.*, 1962). Lemons, on the other hand, increased in total acidity as they matured, mostly due to an increase in citrate (Bartholomew and Sinclair, 1951). The decrease in titratable acid concentration in most citrus fruits during ripening may be partly due to dilution with increased fruit size and water content (Kimball, 1984). The concurrent increase in soluble solids, however, would apparently take place in spite of this dilution effect. Oranges and mandarins have about the same levels of acidity, grapefruits are more acid (Rygg and Getty, 1955), and lemons, limes, and sour oranges have the highest acid content of all the citrus fruits.

Citric acid accounts for most of the acidity in citrus fruit juice and characteristic levels are shown in Table 4.3. In sweet lime, however, malic acid exceeds citric acid levels (0.2% compared to 0.08%) (Clements, 1964a; Yamaki, 1989). Dark fixation of CO_2 may be involved in citric acid synthesis in juice vesicles (Yen and Koch, 1990). Alternatively, accumulation of citric acid may be due to a block in aconitase (EC 4.2.1.3) activity, which interferes with the conversion of citrate to aconitic acid. In fact, both aconitase and citrate lyase activities were absent in mature

Table 4.3 Organic acids in citrus juice. (From Clements, 1964a; Shaked and Hasdai, 1985; Yamaki, 1989).

	Citric		Malic	
	g/100 ml	%Total acids	g/100 ml	%Total acids
Orange/Tangerine	0.5–1.5	75–88	0.06–0.20	9–23
Grapefruit	1.0–2.0	82–96	0.04–0.06	2–17
Lemon	4.0–5.0	87–88	0.17–0.26	11–12

orange juice sacs (Echeverria and Valich, 1988). The ratio of the reduced form of nicotinamide adenine dinucleotide to the non-reduced form (NADH/NAD) is reported to increase in citrus during maturation, and this change in the redox ratio may function as a regulator of citrate formation by decreasing synthesis of oxaloacetate (Bruemmer, 1969).

Malate is the next most abundant acid in citrus juice (Salunkhe and Desai, 1986). In the study of 42 *Citrus* species, lesser amounts of acetate, pyruvate, oxalate, glutarate, fumarate, formate, succinate and α-ketoglutarate made up 2% of total organic acids (Yamaki, 1989). Isocitric, aconitic and malonic acids have also been reported in orange juice (Sasson and Monselise, 1977; Vandercook, 1977). Levels of malic acid were found not to change during development (Clements, 1964b; Shaked and Hasdai, 1985).

Organic acids in the peel differ from those in the juice. Oxalic, malonic, succinic, malic and citric were found in salt form due to a peel pH of 5.0–5.5 (Vandercook, 1977). In a study with Navel oranges, oxalate decreased (Clements, 1964b) while malonate increased in both albedo and flavedo as the fruit matured. At maturity, malonate predominated in the flavedo while oxalate was the most abundant acid in the albedo and peel overall. Malate, oxalate, citrate and malonate made up 30–50% of total anions in peel fractions (Clements, 1964b). Sasson and Monselise (1977) reported that for 'Shamouti' oranges, malonic and malic acids predominated in both peel tissues with the former acid having the highest concentration.

Citrus juices are an important source of vitamin C, with the juice containing 44–59 mg ascorbic acid/100 g in orange and mandarin (Kefford and Chandler, 1970), 21–49 mg/100 g in grapefruit (Schmandke and Guerra, 1969; Robertson and Nisperos, 1983), 25.4 mg/100 g in lime (Schmandke and Guerra, 1969), and 42.3 mg/100 g in lemon (Wimalasiri and Wills, 1983). This compound has been reported to decrease with maturity (Rygg and Getty, 1955) or remain constant (Kefford, 1959). In grapefruit juice, ascorbic acid levels were observed to decrease during ripening (Harding and Fisher, 1945), or remain constant until late in the season, at which point the levels decreased (Robertson and Nisperos, 1983). Only one-quarter of the ascorbic acid present in the whole fruit is found in the juice, however, the rest being present in the peel and rag fractions, especially in orange flavedo (150–340 mg/100 g) and albedo (30–200 mg/100 g) ((Kefford, 1959).

Fruit maturation and sugar and acidity levels have been reported to be influenced by preharvest chemical and hormonal treatments. Accelerated maturity and reduced acidity were observed in grapefruit due to preharvest spraying of lead arsenate (Deszyck et al., 1952; Rice et al., 1985). This also resulted in increased non-reducing and total sugar and flavonoid contents of the juice. It is postulated that arsenate competes with phosphate and interferes with citric acid accumulation due to uncoupled phosphorylation (with citrate as substrate), but without affecting oxygen uptake (Vines and Oberbacher, 1965). Application of GA_3 as a preharvest spray delayed maturation of lemons as evidenced by a delay in peel chlorophyll breakdown and carotenoid synthesis, increased firmness of the peel, decreased juice and soluble solids content, and increased ascorbic acid (Coggins et al., 1960). However, GA treatment of Navel oranges, although resulting in a general delay in maturation and senescence of the peel, showed no effect on juice content or composition (Coggins and Hield, 1962). Spraying 'Pineapple' oranges with 2,4-D and 2,4,5-T to reduce fruit drop retarded maturation of fruit (accumulation of soluble solids and decrease in acidity) by three to four weeks (Phillips and Meagher, 1966). Application of NAA to 'Wilking' mandarins for thinning purposes, however, had no effect on juice content or composition (Hield et al., 1962).

Postharvest

In stored oranges, acids decreased faster than sugars so that the fruit was predicted to be slightly sweeter in holding (Samson, 1986). Citrus and especially malic acid were shown to decrease in the juice during storage of 'Shamouti' oranges, while succinic and malonic acids generally increased. In flavedo, malic acid increased continually as malonic acid peaked and then declined, while the reverse was observed in albedo (Sasson and Monselise, 1977).

The glycolytic machinery was found to be present despite high cellular citrate concentrations which inhibit adenosine triphosphate (ATP) phosphofructokinase (EC 3.2.1.26). It is possible that acids are used for energy production and alcoholic fermentation in harvested citrus, which would account for the decline in acid levels observed. An increase in acid metabolizing enzymes also supports this hypothesis (Echeverria and Valich, 1989). Ripening of citrus is characterized by accumulation of ethanol in juice sacs (Roe et al., 1984), decreased aerobic respiration, and increased alcoholic fermentation which continues during storage of harvested fruit. The necessary enzymes for postharvest use of organic acids for energy production and conversion of sugars through gluconeogenesis were found in the juice (Echeverria and Valich, 1989).

Acid levels in the fruit can also be affected by controlled atmosphere treatments. Acidity was reduced in stored grapefruit during a 20-hour anaerobic treatment with N_2 or CO_2 at 40°C. The Brix/acid ratio increased 10% along with a 20-fold increase in ethanol (Bruemmer and Roe, 1969).

4.6.3 Sugars

Development and ripening

Sugar levels in citrus juice range from 1–2.3% glucose, 1–2.8% fructose, and 2–6% sucrose for oranges and tangerines, 2–5% reducing sugars and 2–3% sucrose in grapefruit, and 0.8–0.9% glucose and fructose and 0.2–0.3% sucrose in lemons and limes (Ting and Attaway, 1971; McCready, 1977). As most citrus fruit ripen, sugars, especially sucrose, increase – as was observed in oranges, tangerines (Ting and Attaway, 1971), and grapefruit (Koch *et al*, 1986). Other studies, however, have shown that sucrose remained fairly constant in composition during ripening of grapefruit (McCready, 1977). Sugar accumulation in citrus was found to be affected by temperature and light intensity. In field studies, fruit, or even the fruit portion, exposed to more sunlight accumulated more sugar compared to unexposed fruit (Reuther, 1973; Kimball, 1984).

Compartmentalization of components in mature orange juice sacs was studied in prepared protoplasts and vacuoles (Echeverria, 1987). The cytoplasm contained activities for acid and neutral invertase (EC 3.2.1.26), uridine diphosphoglucose (UDPG) phosphorylase (EC 2.7.7.9), ATP and pyrophosphate (PPi) phosphofructokinase (EC 2.7.1.11). No activity was found for sucrose synthase (EC 2.4.1.13), hexokinase (EC 2.7.1.1), fructokinase (EC 2.7.1.4) or aconitase in the protoplast, cytosol, or vacuolar fractions (Echeverria and Valich, 1988). Subcellular localization of compounds in juice sac protoplasts revealed that vacuoles contained 70% of the recovered malic acid, 89% of the citric acid, 75% of the fructose and glucose, 100% of the sucrose, as well as 100% of the α-mannosidase (EC 3.2.1.24), 5% of the phosphohexoisomerase (EC 5.3.1.9) and 29% of the phosphoglucomutase (EC 2.7.5.1) activities. Sugars stored in vacuoles of juice sacs are the major carbon supply for mature oranges (Barmore and Biggs, 1972; Echeverria and Valich, 1988), implying the existence of a sugar translocator at the tonoplast (Echeverria and Valich, 1988, 1989).

Starch, α-(1,4)glucan, is found during citrus fruit development in all components, but is most abundant in the albedo. It is, however, only a transitory storage carbohydrate providing energy for growth and respiration of immature fruits and usually disappears upon ripening (McCready, 1977).

Acid-soluble nucleotides were reported in abundance in mature citrus fruit, including adenosine mono- and di-phosphate (AMP and ADP), cytidine diphosphate (CDP), guanosine monophosphate (GMP), uridine monophosphate (UMP), cytidine monophosphate (CMP)-NAD, cytidine triphosphate-guanosine diphosphate (CTP-GDP), and uridine diphosphate (UDP)-ATP. These compounds would be essential in interconversion and synthesis of sugars in juice vesicles. GMP and ADP also modified flavour in that these components enhanced detection of the flavour compound, octanol (Barmore and Biggs, 1972).

Postharvest

Soluble solids comprising the Brix are mostly sugars (70–85%) (Salunkhe and Desai, 1986) with a small amount of other material in the juice including lipids, nitrogen and phosphorus-containing compounds, and especially pectin (McCready, 1977). Brix changes have been observed during storage of citrus fruit, but in one study, corresponding changes in the simple sugar content in stored tangerines and sweet limes did not account for the increased Brix. It was suggested that solubilization of cell-wall constituents of galactosidases and glucosidases present in citrus fruit (Burns, 1990a) may have contributed to the increase Brix levels observed (Echeverria and Ismail, 1990).

Sucrose breakdown in stored 'Persian' limes occurred in vacuoles of juice vesicles in the absence of sucrose-degrading enzymes (Echeverria and Burns, 1989). Sucrose degradation was postulated to occur non-enzymatically by acid hydrolysis in a vacuolar pH of 2.0 to 2.5 (Echeverria, 1990), as would be expected *in vitro* at similar pH values (Echeverria, 1991). Ethylene degreening treatment decreased sugar content of citrus fruit by increasing respiration, requiring utilization of sugars as an energy source (Biale and Young, 1962).

4.6.4   Assimilate translocation

Photosynthate movement in citrus fruit occurs from source leaves to corresponding aligned peel and pulp areas of the fruit (Koch and Avigne, 1984). This can result in uneven distribution of total soluble solids among juice segments. Photosynthates exit the vascular system on the surface of segments in three distinct locations which serve as areas of phloem unloading and transport (vascular bundles and segment epidermis) (refer to Fig. 4.1). The destination of assimilates and point of storage are the juice sacs, which are separated from the site of phloem unloading by the non-vascular juice stalk (Koch *et al.*, 1986).

Sucrose-metabolizing enzymes in sink tissues may have a role in storage/utilization of sugars in sink cells and/or may be associated with phloem transport (Lowell *et al.*, 1989). Sucrose-metabolizing enzymes were measured during grapefruit development; acid invertase activity was found only in very young fruit in stage I of development. Sucrose synthase activity was greater in extracts of transport tissue (vascular bundles and segment epidermis) during stage II and III of juice sac growth. Alkaline invertase/and sucrose phosphate synthase (EC 2.4.1.14) were most active in sites of sucrose storage (juice sacs) (Lowell *et al.*, 1989). The molarity of sucrose (the major transport sugar) increased during grapefruit development (from 159 to 490 mM), while inflow of water decreased (Koch *et al.*, 1986). Sucrose synthase appeared, in some instances, to be associated with functioning of transport tissues in addition to playing a role in deposition and/or utilization of imported assimilates (Lowell *et al.*, 1989).

In grapefruit, leaf photosynthates of light or dark $^{14}CO_2$ fixation were transported primarily to juice tissues, while fruit (peel) photosynthates remained mostly in the peel. It was concluded that fruit photosynthates may play an important role in peel development. Products of dark CO_2 fixation by intact fruit were found in all tissues, but predominated in peel of young fruit and in juice tissues of more mature fruit (Bean and Todd, 1960; Yen and Koch, 1990). Dark CO_2 fixation is considered a possible source of organic acid (especially citric) in citrus (Bean and Todd, 1960).

4.6.5   Cell walls

The main structural carbohydrates in citrus peel were analysed in the pectin, hemicellulose and cellulose fractions. Arabinose, galactose and galacturonic acid were detected in the pectin fraction (McCready, 1977); in orange albedo, glucose (10.2%), xylose (11.5%), galactose (27.9)%, arabinose (48.1%), and galacturonic acid (2.3%) were present in the hemicellulose fraction; and xylose (45.2%), arabinose (15.3%), galactose (7.8%), glucose (27.7%), and galacturonic acid (4.0%) in the cellulose fraction (Braddock and Graumlich, 1981). Rhamnose has also been reported in the cellulose fraction of citrus peel (McCready, 1977). 'Pineapple' orange peel, juice sacs, and segment membranes contained 1.5–1.8 g/100 g fresh weight hemicellulose, 3.5–3.8 g cellulose, 3.7–4.8 g pectin and 0.3–1.0 g lignin. In comparison, 'Marsh' grapefruit peel contained 1.1–1.4 g hemicellulose, 1.8–2.6 g cellulose, 4.0–4.9 g pectin and 0.7–1.2 g lignin (Braddock and Graumlich, 1981). Orange and grapefruit peel alcohol insoluble solids contained 9% araban, 5% galactan, 2.5% xylan, 15–28% cellulose, and 23% galacturonan, but these components accounted for only 53–70% of the alcohol-insoluble solids in the peel (Kefford and Chandler, 1970).

Neutral sugars in water-soluble pectins from juice vesicles of 'Marsh Seedless' grapefruit consisted of 0.3% rhamnose, 0.4% fucose, 10.3% arabinose, 4.2% galactose, and trace amounts of xylose and mannose. Neutral sugars in the hemicellulose fraction consisted of 1.7% fucose, 10.2% arabinose, 59.9% xylose, 8.0% mannose, 7.5% galactose, 12.5% glucose and trace amounts of rhamnose (Hwang *et al.*, 1990).

Hemicelluloses remained fairly constant as fruit matured (McCready, 1977), although in one study more galactose was found in immature grapefruit peel compared to mature peel (Eaks and Sinclair, 1980). There have been inconsistent reports of pectin changes during citrus fruit ripening (McCready, 1977).

Pectin

Generally, pectin changes in meristematic and parenchymous tissues during maturation include a decrease in water-insoluble and highly methylated pectin and an increase, then decrease, in soluble pectins and pectinates (methyl-free pectin) as the fruit progressed into an overripe stage (Rouse, 1977). The rate of change from water-insoluble to water-soluble pectin was found to be greater in pulp than in albedo. However, levels of water-soluble pectins in oranges, and grapefruit were reported to fluctuate, increase, or decrease in flavedo, albedo, pulp, juice, and seeds during maturation (Sinclair and Jolliffe, 1961; Rouse, 1977), but tended to increase in peel and pulp of lemons, while decreasing in lemon seeds (Rouse, 1977). Total pectin levels of flavedo and albedo increased in early growth stages, decreased on a total solids basis during maturation, and increased again during storage. Ammonium oxalate-soluble pectin generally remained unchanged in the peel or increased slightly in pulp, juice, and seeds during maturation of oranges and grapefruit (Sinclair and Jolliffe, 1961; Rouse, 1977). In lemon, total pectin levels fluctuated over the season and ammonium oxalate-soluble pectin showed no clear trends (Rouse, 1977). The degree of methylation remained unchanged or decreased slightly as oranges matured (Rouse, 1977). In other studies, however, pectin methoxyl content and molecular weight, alcohol-insoluble pectin, and intrinsic viscosity, were reported to increase in grapefruit peel (Nisperos and Robertson, 1982a) and methylation increased to 80% in oranges (Sinclair and Jolliffe, 1961) as fruit ripened. Pectin extracted from fresh grapefruit peel had a degree of methylation of 73% (Fishman *et al.*, 1986) and 80% methylation has been reported for mature lime peel pectin (Aspinall and Cottrell, 1970). Some esterification of citrus pectin with acetic acid is also reported with acetyl levels reaching 0.23% (Nelson *et al.*, 1977).

Citrus pectin is produced commercially from albedo tissue whose primary cell walls contain large amounts of high molecular weight pectin (30–35% based on dry weight) (Nelson *et al.*, 1977; Rouse, 1977). Pectin polymers in citrus are present in a wide range of molecular weights (10kDa–400kDa). Molecular size and degree of methylation are important commercially, for both parameters have an effect on the ability of pectin to form certain gels. In this respect, some degree of methylation is necessary and the higher the polymerization of the pectin molecule, the greater its gel-forming ability and viscosity. The degree of methylation, however, was not found to be related to viscosity (Nelson *et al.*, 1977).

Cell-wall modifying enzymes

The cell-wall modifying enzymes identified in citrus fruit so far are pectinmethylesterase (EC 3.1.1.11) (PME) (Rouse, 1977), pectin acetylase (Williamson, 1991), α- and β-galactosidase (EC 3.2.1.22, EC 3.2.1.23), and α- and β-glucosidase (Burns, 1990a) and cellulase (EC 3.2.1.4) (Kuraoka *et al.*, 1975). Grapefruit PME activity increased over the season (Robertson, 1976) and was highest in juice sacs followed by seeds, segment membranes, and peel (Rouse, 1977). In lemon, PME activity was greater in peel than in pulp. Activity generally decreased with maturity in peel and juice, while in pulp it increased initially and then decreased (Rouse, 1977). Orange peel, membranes, and juice sacs all contained PME, with juice sacs being most active and showing an increase in activity during ripening (Monselise *et al.*, 1976; Rouse, 1977). PME activity was generally greatest when the Brix/acid ratio was highest (Kefford and Chandler, 1970; Rouse, 1977).

The cell wall-bound enzyme, PME (Jansen *et al.*, 1960), is of concern to the juice processing industry because it may precipitate the 'cloud' (which is composed mostly of cell wall materials) in citrus juice. In most processed citrus juices, loss of cloud or clarification is undesirable. Juice, therefore, must be heated to 194°C to inactivate all PME, including one heat-stable isoenzyme (Eagerman and Rouse, 1976; Versteeg *et al.*, 1978) of which there is enough to demethylate all the pectin in the peel in a matter of minutes. It is concluded, therefore, that PME is most likely compartmentalized in the cell wall primarily for the purpose of cell wall loosening during growth (Williamson, 1991). A pectin acetylase has also been recently purified from orange peel which de-esterified triacetin and pectin. This enzyme showed increased activity for pectin that was pretreated with PME. Activities of α- and β-galactosidase and α- and β-glucosidase were found in orange juice vesicles. However, only one of the α-galactosidase isoenzymes was able to digest isolated cell walls of juice vesicles. The activity of these enzymes was found to be higher in granulated juice vesicles (a physiological disorder) compared to healthy tissue (Burns, 1990b). During maturation of 'Satsuma' mandarin, cellulase activity decreased in the albedo and to a lesser extent in the flavedo as the fruit matured (Kuraoka *et al.*, 1975).

Section drying is a late season physiological disorder of citrus, and especially grapefruit, that results in granulation or collapse of juice vesicles, usually at the stylar or stem ends of the fruit. Vesicle collapse was associated with a decrease in molecular weight of water- and chelate-soluble pectins, but not of hemicellulose. Granulated vesicles contained twice the cell wall material of normal vesicles due to cell-wall thickening and deposition of lignin (Burns and Achor, 1989; Hwang *et al.*, 1990). The ratio of calcium to pectic material in the chelate-soluble fraction of fruit cell walls was higher in granulated juice sacs than in healthy tissue (Goto, 1989).

Creasing in oranges is a peel disorder that results in random grooves on the fruit surface due to depression of the flavedo tissue in areas where the albedo breaks and cracks (Erickson, 1968). The maximum development is attained in loose-skinned, mandarin-type fruits where the peel and pulp are completely separated causing puffiness. This condition is characteristic for certain genotypes, but is undesirable for oranges. The activity of PME was promoted and water-soluble pectins were increased in creased albedo tissue. Pectin degradation and loosening of the connection between cells was evident in electron micrographs of affected tissue. Preharvest applications of GA induced more compact tissues that were less prone to creasing (Monselise *et al.*, 1976). Nutrition also affects creasing of citrus fruit in that potassium applications partially reduce the occurrence of this disorder (Monselise and Goren, 1978).

4.6.6 Flavour volatiles and secondary metabolites

Volatile flavour compounds

The essence of citrus flavour is a complex mixture of volatile compounds of which some 200 have been identified. The most important volatiles are esters and aldehydes, followed by alcohols, ketones, and hydrocarbons (Bruemmer, 1975). Some components believed to contribute to citrus flavour are listed in Table 4.4.

Studies on the biosynthetic pathways of these flavour components in extracts from Valencia orange juice vesicles revealed that alcohol:NAD reductase oxidized alcohols to corresponding aldehydes in the homologous series from ethanol to octanol and reduced aldehydes to corresponding alcohols in the series from acetaldehyde to decanal. Alcohol dehydrogenase (EC 1.1.1.1) was active on C_2 to C_{10} aldehydes and C_2 to

Table 4.4   Components important to citrus juice flavour. (From Shaw, 1991).

Component	Oranges	Grapefruit	Tangerine	Lemon/lime
Aldehydes	geranial neral acetaldehyde decanal octanal sinensal (E)-2-pentenal	acetaldehyde decanal	acetaldehyde decanal octanal α-sinensal	geranial neral nonanal
Esters	ethyl acetate ethyl propionate methyl butanoate ethyl 2-methyl- butanoate ethyl-3-hydroxy- hexanoate	ethyl acetate methyl butanoate ethyl butanoate	dimethyl anthranilate	*methyl epijasmonate neryl acetate geranyl acetate
Alcohols and thiols	ethanol (E)-2-hexenol (Z)-3-hexenol linalool α-terpineol	1-p-menthene- 8-thiol	ethanol thymol	α-bisabolol geraniol
Ketones	1-carvone 1-penten-3-one	nootkatone		
Hydrocarbons	d-limonene myrcene α-pinene valencene	limonene	τ-terpinene β-pinene	τ-terpinene β-pinene bergamotene **caryophyllene
Ethers		ether		germacrene B *carvyl ethyl *8-p-cymenyl ethyl ether *fenchyl ethyl ether *myrcenyl ethyl ether *α-terpinyl ethyl ether
Bitter components	limonin	naringin		

* Lemon only
**Lime only

C_7 alcohols in reversible oxidation–reduction reactions (Breummer, 1975). Citrus has the coenzyme capability (NAD, NADP, NADH, NADPH) to support synthesis of oxygenated compounds which contribute to flavour (Bruemmer, 1969). Pyruvic decarboxylase from orange and grapefruit juice was most active with pyruvic acid as a substrate, but also decarboxylated 2-ketobutyric, 2-ketovaleric, and 2-ketocaproic acids to 3-, 4-, and 5-carbon aldehydes. Hexanal and homologues are probably breakdown products of fatty acid hydroperoxides formed from unsaturated fatty acids by lipoxygenase (EC 1.13.11.12). The terpenoid compounds are products of the mevalonic acid pathway. Orange juice vesicles contain enzymes that synthesize linalool, geraniol and nerol from mevalonic acid; linalool may be a precursor to limonene, the most abundant component (83–97%) in orange and tangerine fruit oil (Bruemmer, 1975). Information is lacking concerning development of flavour volatiles in citrus fruit during ripening.

Flavonoids and triterpene derivatives

Citrus contains may flavanone glycosides including narirutin, naringin, hesperidin, neohesperidin (Rouseff *et al.*, 1987), eriocitrin, and diosmin (Tisserat *et al.*, 1989c). Anthocyanin in blood oranges is also a flavonoid (Ting and Rouseff, 1986). Hesperidin, although tasteless, is very insoluble and an important component of the cloud in lemon and orange juice. Triterpene derivatives are also present in citrus, the most abundant of which is limonin. Limonin, nomilin, ichangin and nomilinate of the triterpene compounds and naringin and neohesperidin of the flavonoids, cause bitterness in orange and grapefruit juice, respectively (Kefford, 1959; Maier *et al.*, 1977; Ting and Rouseff, 1986; Hasegawa and Maier, 1990). These components can be removed from juice using polyvinylpyrrolidone (Nisperos and Robertson, 1982b), β-cyclodextrin polymers (Wilson *et al.*, 1989), or bacterial enzymes in bioreactors (Hasegawa and Maier, 1990). Levels of naringin and limonin, the major bitter compounds, must be below specific levels for acceptable juice quality and their concentrations are used as an index in quality control of juice products (Robertson and Nisperos, 1983). Quantitation of limonin was reported using HPLC (Rouseff and Fisher, 1980; Shaw and Wilson, 1984), an enzyme immunoassay with ^{125}I, tritium, or enzyme tracers (the latter utilizes the colorimetric reaction of alkaline phosphatase) (Mansell and Weiler, 1980), and anti-limonin antibody coated microtitration wells (Ram *et al.*, 1988). The most precise method to quantify individual flavanoids is by HPLC (Fisher and Wheaton, 1976).

Grapefruit naringin and total flavonoids have been reported to increase (Robertson and Nisperos, 1983), decrease (Kimball, 1984) or fluctuate during maturation (Albach *et al.*, 1981). Limonin levels decreased in citrus fruits with advancing maturity (Robertson and Nisperos, 1983) via dilution and degradation (Kimball, 1984). Hesperidin in lemon and naringin in grapefruit increased rapidly in early developmental stages, then tapered off in the cell enlargement growth stage (Jourdan *et al.*, 1985; Tisserat *et al.*, 1989c). The flavonoid, erocitrin, increased in lemons of 25–30 mm in diameter (after the increase in hesperidin had levelled off) until the fruit reached full size. Another flavonoid, diosmin, increased gradually throughout lemon development. These compounds decreased, however, on a fresh weight basis, due to increased water content in juice vesicles during maturation. The changes for lemon appeared to occur at a slightly faster rate in cultured vesicles compared to those from tree-grown fruit (Tisserat *et al.*, 1989c).

4.6.7   Respiration and metabolism

The photosynthetic flavedo was observed to have higher respiration rates compared to the albedo and juice vesicles indicating higher metabolic activity. Oxygen uptake was highest in flavedo and lowest in juice vesicles (Purvis, 1985). As citrus fruit matured, aerobic respiration decreased and the supplementary anaerobic pathway through alcohol dehydrogenase reactions increased (Bruemmer, 1989). Ethanol and acetaldehyde levels increased in citrus over the season (Davis, 1970).

Along with the increase in ethanol and acetaldehyde, a decrease in pyruvate and oxaloacetate was evident in Hamlin orange juice as fruit matured. The increased pyruvate decarboxylase activity observed (converting pyruvate to acetaldehyde) explained the decrease in pyruvate and increase in acetaldehyde. The acetaldehyde was subsequently reduced to ethanol. Increases in alcohol dehydrogenase and malic enzyme were also reported. Citrate synthase activity increased during fruit ripening while malate dehydrogenase (EC 1.1.1.37), isocitrate dehydrogenase (EC 1.1.1.42) and cytochrome oxidase (EC 1.9.3.1) activity remained constant (Roe *et al.*, 1984; Bruemmer and Roe, 1985). Similar changes were observed during maturation of Satsuma mandarin (Hirai and Ueno, 1977). It was concluded that the terminal oxidase was not a factor in the abrupt increase in anaerobic metabolism. The NADH/NAD ratio increased as 'Hamlin' oranges matured, implying that the oxidative capacity of the pathway was not able to maintain redox equilibrium and the availability of NAD for dehydrogenase reactions, necessitating operation of the anaerobic pathway. Increases in ethanol and pyruvate dehydrogenase activity implicated this enzyme as a regulatory enzyme in pyruvate metabolism. Levels of ATP and ADP decreased as fruit matured but the ratio was stable. The ubiquinol

(UQH)/ubiquinone ratio, which is an indicator of the redox state of the aerobic respiratory pathway, increased, suggesting a change in equilibrium between substrate availability and oxidative function (Bruemmer and Roe, 1985).

In mitochondrial fractions of ripening oranges, NADH oxidase became more sensitive to potassium cyanide (KCN) plus salicylhydroxamic acid (SHAM) during ripening. The SHAM-sensitive oxidase accounted for 24–30% of total respiration and the residual oxidase, 11–28% (Bruemmer, 1989). Respiration in citrus fruit increased with stress, e.g. pathogen infection, chilling, mechanical damage, and ethylene treatment (Vines *et al.*, 1965), and with physiological disorders such as section drying (granulation of juice sacs) (Burns, 1990b).

4.6.8   Nitrogen and lipid composition of citrus

Nitrogen

Total nitrogen was measured in Navel oranges during development. The juice increased in total and amino nitrogen with amino nitrogen making up 50–70% of the total, and protein nitrogen comprising only a small fraction. The peel decreased in total and protein nitrogen in stages I and II of growth, with protein nitrogen making up 60% of the total (Tadeo *et al.*, 1988). In 'Shamouti' oranges, total nitrogen and protein nitrogen decreased over the growing season when expressed as percentage of fruit dry weight, but increased on a per fruit basis (Goren and Monselise, 1964).

In juice, proline and total amino acids increased during ripening, with proline making up 50% of total amino acids and 2.67% of the solids in juice of 'Valencia' oranges (Clements and Leland, 1962). Proline increase occurred concurrent with temperature decreases (Purvis and Yelenosky, 1982; Tadeo *et al.*, 1988). High proline levels observed may play a role as reserve nitrogen (Moreno and Garcia-Martinez, 1984). Arginine increased (Moreno and Garcia-Martinez, 1984; Tadeo *et al.*, 1988), while asparagine and aspartic acid decreased with maturation. Of the total amino acids, serine and glutamic acid made up 10%, glutamine, alanine, and τ-aminobutyric acid, 2%, and the rest, less than 1%. In peel, the three main amino acids were asparagine, proline, and arginine. Total amino acids decreased in stages I and II, then increased in stage III. Proline and arginine increased while asparagine decreased (Moreno and Garcia-Martinez, 1984; Tadeo *et al.*, 1988). Total protein amino acids decreased with maturation, especially during stage I and II. This amino acid component contained 10% leucine and alanine. Hydrolysis of protein occurred during chloroplasts conversion to chromoplasts (Tadeo *et al.*, 1988).

Lipids

Lipid content of citrus fruit influences juice stability and modifies taste after pasteurization and storage. Degradation products derived from oxidation of unsaturated fatty acids may be contributors to the development of off-flavours in citrus juice (Nordby and Nagy, 1974). Fatty acid composition can be used to distinguish citrus species (Nagy and Nordby, 1971; Nordby and Nagy, 1974) and even cultivar (Nagy and Nordby, 1973).

Fatty acid composition was followed in juice sacs of ripening fruit of six orange cultivars. Palmitic and palmitoleic acids decreased during maturation in early ripening cultivars, while palmitic acid increased in late ripening varieties. Linoleic and linolenic acids were higher in overripe than ripe fruit. Palmitic, palmitoleic, oleic, linoleic, and linolenic acids (the most abundant) were found in linear and branch chain fatty acids of citrus fruit (Nagy and Nordby, 1971). Citrus fruit wax is composed of C_{20} to C_{25} alkanes and alkenes in immature fruit, while C_{27} to C_{33} alkanes and alkenes predominate in wax of mature fruit (Nordby and Nagy, 1977).

4.7   CITRUS BIOTECHNOLOGY

Citrus species and cultivars are polyembryonic, i.e. from one seed, two or three plants may germinate. Of those, one may be a sexual embryo as a result of fertilization and the others (frequently all) are nucellar embryos which are clones of the mother plant. The exceptions are *Citrus medica, C. grandis*, mandarin cvs. Temple and Clementine, and 'Meyer' lemon which are monoembryonic with heterogeneous seedlings. As a result of polyembryony and vegetative propagation, the majority of citrus cultivars are well established clones, but breeding work is more difficult. Mutations appear regularly, however, providing chances for improvement (Samson, 1986).

4.7.1   Tissue and organ culture

Underdeveloped ovules from mature polyembryonic citrus fruits initiated embryogenic cultures resulting in plantlets (Moore, 1985). Internodal seedling stem sections of *Citrus* rootstocks were also successfully cultured (Moore, 1986). These systems are useful for developing new techniques for protoplast fusion and genetic transformation of citrus. Meristem culture is currently being adapted commercially to produce virus-free budwood in the US (Florida) (Fisher, 1991).

Citrus juice vesicles from immature fruit were maintained in culture with little callus growth and developed similarly to vesicles in fruit grown on trees. This system offers the opportunity to study vesicle growth, nutrient transport, influence of various compounds on metabolism, and ripening physiology (Tisserat *et al.*, 1989b). Tissue proliferation was demonstrated via adventitious vesicle branching in vesicle culture which was enhanced by GA, NAA, and BA treatments (Tisserat *et al.*, 1989a).

4.7.2   Genetic transformation

Genetic transformation has recently been achieved via *Agrobacterium* in 'Washington' Navel and 'Trovita' oranges. Transformation frequencies of 0.5% per callus colony in suspension cells of these citrus species were reported and transgenic plants were achieved (Hidaka *et al.*, 1990). In another report, citrus protoplasts were genetically transformed by introduction of plasmid DNA via polyethylene glycol (PEG) treatments, resulting in transgenic plants (Vardi *et al.*, 1990). The requirement for embryogenic callus in these methods is a limitation for some citrus types that do not readily produce callus (Gmitter and Moore, 1986) and, in fact, embryogenic callus production is frequently difficult to obtain from amenable geneotypes. A recent report offers the alternative of *Agrobacterium*-mediated transformation of internodal stem segments followed by regeneration of shoots (Moore *et al.*, 1992). Transformation of plants in each of the above mentioned studies was confirmed by Southern analysis.

The development of plant transformation techniques, has resulted in experiments which have used antisense RNA technology to block specific enzyme activities in tomato fruits (Smith *et al.*, 1990). These techniques should be explored in citrus as a means to reduce certain enzyme activities, such as those resulting in production of bitter components such as naringin and limonin. Reduction in the activity of the heat stable PME isozyme that interferes with juice processing would also be beneficial. Insertion of genes that promoted cold hardiness, improved colour, increased sugar, and enhanced flavour would be other potential goals relating to fruit quality.

4.7.3   Protoplast fusion

Another recent technique that has been used to produce new genetic combinations in citrus is protoplast fusion. This is an additive process that combines complete nuclear genomes of both parents (Grosser and Gmitter, 1990). Protoplast fusion resulted in regenerated cybrid plants from three donor–recipient combinations: Poorman × *Poncirus trifoliata* with sour orange, sour orange with 'Villafranca' lemon, and Poorman × *P. trifoliata* with 'Villafranca' lemon. Gamma-irradiation was used to arrest nuclear division of donor protoplasts and the antimetabolite, iodoacetate, was added to produce transient metabolic inhibition (Vardi *et al.*, 1987). Somatic hybridization via protoplast fusion produced allotetraploid hybrids when *Poncirus trifoliata* leaf protoplasts were fused with embryogenic callus of *Citrus sinensis*, cv. Tronta (Ohgawara *et al.*, 1985) and when juvenile leaf tissue protoplasts from *P. trifoliata* cv. Flying Dragon were fused with *C. sinensis* cv. Hamlin embryogenic suspension culture protoplasts, providing plants of unique genetic combination. Fusion has also been achieved between protoplasts of sexually incompatible species, e.g. *C. sinensis* cv. Hamlin embryogenic suspension culture-derived protoplasts with seedling epicotyl callus-derived protoplasts of *Severinia disticha* (Blanco) Swing (Grosser and Gmitter, 1990). Capacity for somatic embryogenesis and plant regeneration was enhanced when a protoplast from nucellar tissue of 'Valencia' orange (*C. sinensis*) was fused with a protoplast from non-embryogenic tissue from 'Femminell' lemon (*C. limon*) (Tusa *et al.*, 1990). This technique accesses previously unavailable germplasm for breeding material with the potential for improvement of fruit quality.

REFERENCES

Adsule, P.G., Ismail, M.A. and Fellers, P.J. (1984) Quality of citrus fruit following cold treatment as a method of disinfestation against the Caribbean fruit fly. *Journal of the American Society of Horticultural Science* **109**, 851–854

Albach, R.F., Redman, G.H. and Cruse, R.R. (1981) Annual and seasonal changes in naringin concentration of Ruby Red grapefruit. *Journal of Agriculture and Food Chemistry* **29**, 808–811

Albrigo, L.G. and Carter, R.D. (1977) Structure of citrus fruits in relation to processing. In *Citrus Science and Technology*, Vol. I, (eds. S.Nagy, P.E. Shaw, and M.K. Veldhuis), AVI Publishing Co., Inc., Westport, CT, pp. 38–73

Aljuburi, H.J. and Huff, A. (1984) Reduction in chilling injury to stored grapefruit (*Citrus paradisi* Macf.) by vegetable oils. *Scientia Horticulturae*, **24**, 53–58

Aspinall, G.O. and Cottrell, I.W. (1970) Lemon-peel pectin. II. Isolation of homogenous pectins and examination of some associated polysaccharides. *Canadian Journal of Chemistry*, **48**, 1283–1289

Bain, J.M. (1958) Morphological, anatomical, and physiological changes in the developing fruit of the Valencia orange, *Citrus sinensis* (L.) Osbeck, *Australian Journal of Botany*, **6**, 1–28

Baldwin, E.A. and Biggs, R.H. (1983) Ethylene biosynthesis of citrus peel explants as influenced by cycloheximide. *Proceedings Florida State Horticultural Society*, **96**, 183–185

Baldwin, E.A. and Biggs, R.H. (1988) Cell-wall lysing enzymes and products of cell-wall digestion elicit ethylene in citrus. *Physiologia Plantarum*, **73**, 58–64

Barmore, C.R., and Biggs, R.H. (1972) Acid-soluble nucleotides of juice vesicles of citrus fruit. *Journal of Food Science*, **37**, 712–714

Barmore, C.R. and Nguyen, T.K. (1985) Polygalacturonase inhibition in rind of Valencia orange infected with *Diplodia natalensis*. *Phytopathology*, **75**, 446–449

Bartholomew, E.T. and Sinclair, W.B. (1951) *The Lemon Fruit: Its Composition, Physiology, and Products*, University of California Press, Berkeley

Basiouny, F.M. and Biggs, R.H. (1976) Pectin-polygalacturonase in *Citrus*. *Planta*, **128**, 271–273

Bean, R.C. and Todd, G.W. (1960) Photosynthesis and respiration in developing fruits. I. $^{14}CO_2$ uptake by young oranges in light and in dark. *Plant Physiology*, **35**, 425–429

Ben-Yehoshua, S. (1978) Delaying deterioration of individual citrus fruit by seal-packaging in film of high density polyethylene 1. General effects. *Proceedings International Society Citriculture*, **1**, 110–115

Ben-Yehoshua, S., Kobiler, I. and Shipiro, B. (1981) Effects of cooling versus seal-packaging with high-density polyethylene on keeping qualities of various citrus cultivars. *Journal of the American Society of Horticultural Science*, **106**, 536–540

Biale, J.B. (1961) Postharvest physiology and chemistry. In: *The Orange – Its Biochemistry and Physiology*, pp. 96–130. (ed W.B. Sinclair), Univ. of Calif. Press, Berkeley

Biale, J.B. and Young, R.E. (1962) The biochemistry of fruit maturation. *Endeavour*, **21**, 164–174

Biggs, R.H. (1971) Citrus abscission. *HortScience*, **6**, 388–392

Biggs, R.H. and Baldwin, E.A. (1981) Ethylene and citrus abscission. *Proceedings of the Plant Growth Regulation Society* **1**, 182–183

Bosabalidis, A. and Tsekos, I. (1982) Ultrastructural studies on the secretory cavities of *Citrus deliciosa* Ten. II. Development of the essential oil-accumulating central space of the gland and process of active secretion. *Protoplasma*, **112**, 63–70

Braddock, R.J. and Graumlich, T.R. (1981) Composition of fiber from citrus peel, membranes, juice vesicles and seeds. *Lebensmittel-Wissenschaft und Technologie*, **14**, 229–231

Brown, G.E. (1986) Diplodia stem-end rot, a decay of citrus fruit increased by ethylene degreening treatment and its control. *Proceedings Florida State Horticultural Society*, **99**, 105–108

Bruemmer, J.H. (1969) Redox state of nicotinamide-adenine dinucleotides in citrus fruit. *Journal of Agricultural Food Chemistry*, **17**, 1312–1315

Bruemmer, J.H. (1975) Aroma substances of citrus fruits and their biogenesis. In *Symposium on Fragrance and Flavour Substances*, (ed F. Drawart), Verlag Hans Carl, Nuremburg, Germany, pp. 167–176

Bruemmer, J.H. (1989) Terminal oxidase activity during ripening of Hamlin orange. *Phytochemistry*, **28**, 2901–2902

Bruemmer, J.H. and Roe, B. (1969) Post-harvest treatment of citrus fruit to increase Brix/acid ratio. *Proceedings Florida State Horticultural Society*, **82**, 212–215

Bruemmer, J.H. and Roe, B. (1985) Pyruvate dehydrogenase activity during ripening of Hamlin oranges. *Phytochemistry*, **24**, 2105–2106

Burns, J.K. (1990a) α- and β-galactosidase activities in juice vesicles of stored Valencia oranges. *Phytochemistry*, **29**(8), 2425–2429

Burns, J.K. (1990b) Respiratory rates and glycosidase activities of juice vesicles associated with section-drying in citrus. *HortScience*, **25**, 544–546

Burns, J.K. and Achor, D.S. (1989) Cell wall changes in juice vesicles associated with 'section drying' in stored late-harvested grapefruit. *Journal of the American Society of Horticultural Science*, **114**, 283–287

Clements, R.L. (1964a) Organic acids in citrus fruits. I. Varietal differences. *Journal of Food Science*, **29**, 276–280

Clements, R.L. (1964b) Organic acids in citrus fruits. II. Seasonal changes in the orange. *Journal of Food Science*, **29**, 281–286

Clements, R.L. and Leland, H.V. (1962) An ion-exchange study of the free amino acids in the juices of six varieties of citrus. *Journal of Food Science*, **27**, 20–25

Coggins, C.W., Jr. (1981) The influence of exogenous growth regulators on rind quality and internal quality of citrus fruits. *Proceedings of the International Society for Citriculture*, **1**, 214–216

Coggins, C.W., Jr. and Hield, H.Z. (1962) Navel orange fruit response to potassium gibberellate. *Proceedings of the American Society of Horticultural Science*, **81**, 227–230

Coggins, C.W., Jr., Hield, H.Z. and Boswell, S.B. (1960) The influence of potassium gibberellate on Lisbon lemon trees and fruit. *Proceedings of the American Society of Horticultural Science*, **76**, 199–207

Coggins, C.W., Jr. and Lewis, L.N. (1962) Regreening of Valencia orange as influenced by potassium gibberellate. *Plant Physiology*, **37**, 625–627

Cohen, E., Shalom, Y., Rosenberger, I. (1990) Postharvest ethanol buildup and off-flavour in 'Murcott' tangerine fruits. *Journal of the American Society of Horticultural Science*, **115**, 775–778

Cooper, W.C. and Chapot, H. (1977) Fruit production with special emphasis on fruit for processing. In: *Citrus Science and Technology*, (eds S. Nagy, P.E. Shaw and M.K. Veldhuis), Vol. 2, AVI Publishing Co., Inc., Westport, CT, pp. 1–127

Davis, P.L. (1970) Relation of ethanol content of citrus fruits to maturity and to storage conditions. *Proceedings Florida State Horticultural Science*, **83**, 294–298

Davis, P.L. and Hofmann, R.C. (1973) Reduction of chilling injury of citrus fruits in cold storage by intermittent warming. *Journal of Food Science*, **38**, 871–873

Deszyck, E.J., Reitz, H.J., and Sites, J.W. (1952) Effect of copper and lead arsenate sprays on total acid and maturity of 'Duncan' Grapefruit. *Proceedings Florida State Horticural Society*, **65**, 38–42

Droby, S., Chalutz, E., Wilson, C.L. and Wisniewski, M. (1989) Characterization of the biocontrol activity of *Debaryomyces bansenii* in the control of *Penicillium digitatum* on grapefruit. *Canadian Journal of Microbiology*, **35**, 794–800

Eagerman, B.A. and Rouse, A.H. (1976) Heat inactivation temperature-time relationships for pectinesterase inactivation in citrus juices. *Journal of Food Science*, **41**, 1396–1397

Eaks, I.L. (1977), Physiology of degreening – summary and discussion of related topics. *Proceedings of the International Society for Citriculture*, **1**, 223–226

Eaks, I.L. and Sinclair, W.B. (1980) Cellulose-hemicellulose fractions in the alcohol-insoluble solids of Valencia orange peel. *Journal of Food Science*, **45**, 985–988

Echeverria, E. (1987), Preparation and characterization of protoplasts from citrus juice vesicles. *Journal of the American Society of Horticultural Science*, **112**, 393–396

Echeverria, E. (1990) Developmental transition from enzymatic to acid hydrolysis of sucrose in acid limes (*Citrus aurantifolia*). *Plant Physiology*, **92**, 168–171

Echeverria, E. (1991) Regulation of acid hydrolysis of sucrose in acid lime juice sac cells. *Physiologia Plantarum*, (in press)

Echeverria, E. and Burns, J.K. (1989) Vacuolar acid hydrolysis as a physiological mechanism for sucrose breakdown. *Plant Physiology*, **90**, 530–533

Echeverria, E. and Ismail, M. (1990) Sugars unrelated to Brix changes in stored citrus fruits. *HortScience*, **25**, 710

Echeverria, E. and Valich, J. (1988) Carbohydrate and enzyme distribution in protoplasts from Valencia orange juice sacs. *Phytochemistry*, **27**, 73–76

Echeverria, E. and Valich, J. (1989) Enzymes of sugar and acid metabolism in stored 'Valencia' oranges. *Journal of the American Society of Horticultural Science*, **114**, 445–449

Eilati, S.K., Budowski, P. and Monselise, S.P. (1972) Xanthophyll esterification in the flavedo of citrus fruit. *Plant and Cell Physiology*, **13**, 741–746

Eilati, S.K., Budowski, P. and Monselise, S.P. (1975) Carotenoid changes in the 'Shamouti' orange peel during chloroplast – chromoplast transformation on and off the tree. *Journal Experimental Botany*, **26**, 624–632

Erickson, L.C. (1968) The general physiology of citrus. In *Citrus Industry: Anatomy, Physiology, Genetics, and Reproduction*, (eds W. Reuther, L.D. Batchelor and H.J. Webber), Vol. II, University of California Press, Riverside, CA, pp. 86–126

Erickson, L.C. and Brannaman, B.L. (1950) Some effects on fruit growth and quality of a 2,4-D spray applied to Bearss lime trees. *Journal of the American Society of Horticultural Science*, **56**, 79–82

Erickson, L.C. and Brannaman, B.L. (1960) Abscission of reproductive structures and leaves of orange trees. *Journal of the American Society of Horticultural Science*, **75**, 222–229

Even-Chen, Z., Mattoo, A.K. and Goren, R. (1982) Inhibition of ethylene biosynthesis by aminoethoxyvinylglycine and by polyamines shunts label from 3,4-[^{14}C]methionine into spermidine in aged orange peel discs. *Plant Physiology*, **69**, 385–388

Evensen, K., Bausher, M.G. and Biggs, R.H. (1981) Wound-induced ethylene production in peel explants of 'Valencia' orange fruit. *HortScience*, **16**, 43–44

Feinstein, B., Monselise, S.P. and Goren, R. (1975) Studies on the reduction of seed number in mandarins. *HortScience*, **10**, 385–386

Fisher, J. (1991) Lab makes blight-free citrus trees. *Florida Grower and Rancher*, **84** (4), 12–13

Fisher, J.F. and Wheaton, T.A. (1976) A high pressure liquid chromatographic method for the resolution and quantitation of naringin and naringenin rutinoside in grapefruit juice. *Journal of Agricuture and Food Chemistry*, **24**, 898–899

Fishman, M.L., Pepper, L., Damert, W.C., Phillips, J.G. and Barford, R.A. (1986) A critical reexamination of molecular weight and dimensions of citrus pectin. In *Chemistry and Function of Pectins*, (eds M. Fishman and J. Jen), ACS

Symposium Series 310, American Chemical Society, Washington, DC, pp. 22–38

Fox, K (1991) Status update of the worldwide citrus industry. *1991 Citrus Engineering Conference*, Fla. Sec. Amer. Soc. Mech. Eng. 1991

Fuh, B.S., Ichii, T., Kawai, Y. and Nakanishi, T. (1988) Changes in lipid composition in the flavedo tissues of Naruto (*Citrus medioglobosa*) during storage, and the effects of growth regulators and storage temperature. *Journal of Japanese Society of Horticultural Science*, **57**, 529–537

Gmitter, F.G. and Moore, G.A. (1986) Plant regeneration from underdeveloped ovules and embryogenic calli of *Citrus* embryo production, germination, and plant survival. *Plant Cell Tissue and Organ Culture*, **6**, 139–147

Goldschmidt, E.E. (1983) Asymmetric growth of citrus fruit peel induced by localized application of gibberellins in lanolin pastes. *Scientia Horticulturae*, **21**, 29–35

Goldschmidt, E.E. (1988) Regulatory aspects of chloro-chromoplast interconversions in senescing *Citrus* fruit peel. *Israel Journal of Botany*, **37**, 123–130

Goldschmidt, E.E., Aharoni, Y., Eilati, S.K., Riov, J.W. and Monselise, S.P. (1977) Differential counteraction of ethylene effects by gibberellin A_3 and N_6-benzyladenine in senescing citrus peel. *Plant Physiology*, **59**, 193–195

Goren, R. and Monselise, S.P. (1964) Morphological features and changes in nitrogen content in developing Shamouti orange fruits. *Israel Journal of Agricultural Research*, **14**, 65–74

Goren, R., Monselise, S.P. and Ben Moshe, A. (1976) Control of corky (silvery) spots of grapefruit by growth regulators. *HortScience*, **11**, 421–422

Goto, A. (1989) Relationship between pectic substances and calcium in healthy, gelated, and granulated juice sacs of Sanbokan (*Citrus sulcata* hort. ex Takanashi) fruit. *Plant Cell Physiology*, **30**, 801–806

Greenberg, J. and Goldschmidt, E.E. (1989) Acidifying agents, uptake, and physiological activity of gibberellin A_3 in *Citrus*. *HortScience*, **24**, 791–793

Grierson, W. and Wilson, W.C. (1983) Influence of mechanical harvesting on citrus quality: Cannery vs. fresh fruit crops. *HortScience*, **18**, 407–409

Gross, J. (1977), Carotenoid pigments in citrus. In *Citrus Science and Technology*, (eds S. Nagy, P.E. Shaw and M.K. Veldhuis), vol. I, AVI Publishing Co., Inc., Westport, CT, pp. 302–354

Gross, J. (1981) Pigment changes in the flavedo of Dancy tangerine (*Citrus reticulata*) during ripening. *Z. Pflanzenphysiol. Bd.*, **103**, 451–457

Gross, J. (1982) Carotenoid changes in the juice of the ripening Dancy tangerine (*Citrus reticulata*). *Lebensm.-Wiss. u.-Technol.*, **15**, 36–38

Gross, J., Trimberg, R. and Graef, M. (1983) Pigment and ultrastructural changes in the developing pummelo *Citrus grandis* 'Goliath'. *Botanical Gazelte*, **144**, 401–406

Grosser, J.W. and Gmitter, F.G., Jr. (1990) Somatic hybridization of *Citrus* with wild relatives for germplasm enhancement and cultivar development. *HortScience*, **25**, 147–151

Harding, P.L. and Fisher, D.F. (1945) Seasonal changes in Florida grapefruit. *US. Department of Agricutlure Technical Bulletin 886*

Harding, P.L., Sunday, M.B. and Davis P.L. (1959) Seasonal changes in Florida tangelos. *US. Department Agriculture Technical Bulletin 1205*

Hasegawa, S. and Maier, V.P. (1990) Biochemistry of limonoid citrus juice bitter principles and biochemical debittering processes. In *Bitterness in Foods and Beverages*, (ed. R.L. Rouseff), Elsevier Science Publishing Co., New York

Hidaka, T., Omura, M., Ugaki, M., Tomiyama, M., Kato, A., Ohshima, M. and Motoyoshi, F. (1990) *Agrobacterium*-mediated transformation and regeneration of *Citrus* spp. from suspension cells. *Japan J. Breed.*, **40**, 199–207

Hield, H.Z., Burns, R.M. and Coggins, C.W., Jr. (1962) Some fruit thinning effects of naphthaleneacetic acid on Wilking mandarin. *Journal of the American Society of Horticultural Science*, **81**, 218–222

Hirai, M. and Ueno. I. (1977) Development of citrus fruits: Fruit development and enzymatic changes in juice vesicle tissue. *Plant and Cell Physiology*, **18**, 791–799

Horowitz, R.M. and Gentili, B. (1977) Flavonoid constituents of citrus. In *Citrus Science and Technology*, (eds S. Nagy, P.E. Shaw and M.K. Veldhuis), Vol. I, AVI Publishing Co. Inc., Westport, CT, pp. 397–426

Huff, A. (1983) Nutritional control of regreening and degreening in citrus peel segments. *Plant Physiology*, **73**, 243–249

Huff, A. (1984) Sugar regulation of plastid interconversions in epicarp of citrus fruit. *Plant Physiology*, **76**, 307–312

Huyskens, S., Timberg, R. and Gross, J. (1985) Pigment and plastid ultrastructural changes in kumquat (*Fortunella margarita*) 'Nagami' during ripening. *Journal of Plant Physiology*, **118**, 61–72

Hwang, Y.S., Huber, D.J. and Albrigo, L.G. (1990) Comparison of cell wall components in normal and disordered juice vesicles of grapefruit. *Journal of the American Society of Horticultural Science*, **115**, 281–287

Hyodo, H. (1977) Ethylene production by albedo tissue of Satsuma mandarin (*Citrus unshiu* Marc.) fruit. *Plant Physiology*, **59**, 111–113

Ismail, M.A. (1971) Seasonal variation in bonding force and abscission of citrus fruit in response to ethylene, ethephon, and cycloheximide. *Proceedings of Florida State Horticultural Society*, **84**, 77–81

Jansen, E.F., Jang, R. and Bonner, J. (1960) Orange pectinesterase binding activity. *Food Research*, **25**, 64–72

Jourdan, P.S., McIntosh, C.A. and Mansell, R.L. (1985) Naringin levels in citrus tissues. *Plant Physiology*, **77**, 903–908

Ke, D. and Kader, A.A. (1990) Tolerance of 'Valencia' oranges to controlled atmospheres as determined by physiological responses and quality attributes. *Journal of the American Society of Horticultural Science*, **115**, 779–783

Kefford, J.F. (1959) The chemical constituents of citrus fruits. *Advances in Food Research*, **9**, 285–372

Kefford, J.F. and Chandler, B.V. (1970) General composition of citrus fruits. In *The Chemical Constituents of Citrus Fruits*, (eds C.O. Chichester, E.M. Mrak and G.F. Stewart), Academic Press, New York, pp. 5–22

Kimball, D.A. (1984) Factors affecting the rate of maturation of citrus fruits. *Proceedings Florida State Horticultural Society*, **97**, 40–44

Koch, K.E. and Avigne, W.T. (1984) Localized photosynthate deposition in citrus fruit segments relative to source-leaf position. *Plant and Cell Physiology*, **25**, 859–866

Koch, K.E., Lowell, C.A. and Avigne, W.T. (1986) Assimilate transfer through citrus juice vesicle stalks: A nonvascular portion of the transport path. In *Phloem Transport*, Alan R. Liss, Inc., New York, NY, pp. 247–258

Kossuth, S.V., Biggs, R.H. and Martin, F.G. (1979) Effect of physiological age of fruit, temperature, relative humidity, and formulations on absorption of ^{14}C-release by 'Valencia' oranges. *Journal of the American Society of Horticultural Science*, **104**, 323–327

Kuraoka, T., Iwasaki, K. and Tsuji, H. (1975) Studies on the peel puffing of the Satusma mandarin. II. Changes of cell morphology and cellulase activity during the development of the fruit rind. *Journal Japanese Society of Horticultural Science*, **44**, 7–14

Liu, Y., Hoffman, N.E. and Yang S.F. (1985) Ethylene-promoted malonylation of 1-aminocyclopropane-1-carboxylic acid participates in autoinhibition of ethylene synthesis in grapefruit flavedo discs. *Planta*, **164**, 565–568

Long, W.G., Sunday, M.B. and Harding, P.H. (1962) Seasonal changes in Florida Murcott Honey oranges. *U.S. Dept. Agr. Tech. Bull. 1271*

Lowell, C.A., Tomlinson, P.T., Koch, K.E. (1989) Sucrose-metabolizing enzymes in transport tissues and adjacent sink structures in developing citrus fruit. *Plant Physiology*, **90**, 1394–1402

Mabesa, L.B. (1990) Calamansi or Calamondin. In *Fruits of Tropical and Subtropical Origin*, eds S. Nagy, P.E. Shaw, and W F. Wardowski, Florida Science Source, Inc., Lake Alfred, Fl, pp. 348–372

Maier, V.P. (1969) Compositional studies of citrus: Significance in processing, identification, and flavor. *Proceedings First International Citrus Symposium*, **1**, 235–243

Maier, V.P., Bennett, R.D. and Hasegawa, S. (1977) Limonin and other limonoids In *Citrus Science and Technology*, (eds S. Nagy, P.E. Shaw and M.K. Veldhuis), vol. I, AVI Publishing, Westport, CT, pp. 355–396

Mansell, R.L. and Weiler, E.W. (1980) Immunological tests for the evaluation of citrus quality. In *Citrus Nutrition and Quality*, (eds S. Nagy and J.A. Attaway), The American Chemical Society, Washington, DC

Mauk, C.S., Bausher, M.G. and Yelenosky, G. (1986) Influence of growth regulator treatments on dry matter production, fruit abscission, and ^{14}C-assimilate partitioning in citrus. *Journal of Plant Growth Regulation*, **5**, 111–120

Mayfield, S.P. and Huff, A. (1986) Accumulation of chlorophyll, chloroplastic proteins and thylakoid membranes during reversion of chromoplasts to chloroplasts in *Citrus sinensis* epicarp. *Plant Physiology*, **81**, 30–35

McCready, R.M. (1977) Carbohydrates: Composition, distribution, significance. In *Citrus Science and Technology*, (eds S. Nagy, P.E. Shaw and M.K. Veldhuis), vol. I, AVI Publishing, Inc., Westport, CT, pp. 74–109

McDonald, R.E. (1986) Effects of vegetable oils, CO_2 and film wrapping on chilling injury and decay of lemons. *HortScience*, **21**, 476–477

McDonald, R.E. (1989) Temperature-conditioning affects polyamines of lemon fruits stored at chilling temperatures. *HortScience*, **24**, 475–477

McDonald, R.E., Shaw, P.E., Greany, P.D., Hatton, T.T. and Wilson, C.W. (1987) Effect of gibberellic acid on certain physical and chemical properties of grapefruit. *Tropical Science*, **27**, 17–22

Miller, W.R., Chun, D., Risse, L.A., Hatton, T.T. and Hinsch, R.T. (1990) Conditioning of Florida grapefruit to reduce peel stress during low-temperature storage. *HortScience*, **25**, 209–211

Monselise, S.P. (1986) Citrus. In *CRC Handbook of Fruit Set and Development*, (ed. S.P. Monselise), CRC Press, Inc., Boca Raton, FL, pp. 87–108

Monselise, S.P. and Goren, R. (1978) The role of internal factors and exogenous control in flowering, peel growth, and abscission in citrus. *HortScience*, **13**, 134–139

Monselise, S.P., Weiser, Shafir, N., Goren, R. and Goldschmidt, E.E. (1976) Creasing of Orange Peel – Physiology and Control. *Journal of Horticultural Science*, **51**, 341–351

Moore, G.A. (1985), Factors affecting in vitro embryogenesis from undeveloped ovules of mature *Citrus* fruit. *Journal of the American Society of Horticultural Science*, **110**, 66–70

Moore, G.A. (1986), *In vitro* propagation of *Citrus* rootstocks. *HortScience* **21**, 300–301.

Moore, G.A., Jacano, C.C., Neidigh, J.L., Lawrence, S.D. and Cline, K. (1992) *Agrobacterium*-mediated transformation of *Citrus* stem segments and regeneration of transgenic plants. *Plant Cell Reports*, (in press)

Moreno, J. and Garcia-Martinez, J.L. (1984) Nitrogen accumulation and mobilization in *Citrus* leaves throughout the annual cycle. *Physiologia Plantarum*, **61**, 429–434

Murti, G.S.R. (1988) Changes in abscisic acid content during fruit development in acid lime (*Citrus aurantifolia*, Swingle). *Plant Physiology and Biochemistry*, **15**, 138–143

Nagy, S. and Nordby, H.E. (1971), Distribution of free and conjugated sterols in orange and tangor juice sacs. *Lipids*, **6**, 826–830

Nagy, S. and Nordby, H.E. (1973) Saturated and mono-unsaturated long-chain hydrocarbon profiles of sweet oranges. *Phytochemistry*, **12**, 801–805

Nagy, S., Shaw, P.E. and Veldhuis, M.K. (1977) (eds) *Citrus Science and Technology*, vol. I and II, AVI Publishing, Inc., Westport, CT

Nathan, R., Altman, A. and Monselise, S.P. (1984) Changes in activity of polyamine biosynthetic enzymes and in polyamine contents in developing fruit tissues of 'Murcott' mandarin. *Science and Horticulture*, **22**, 359–364

Nelson, D.B., Smit, C.J.B. and Wiles, R.R. (1977) Commercially important small pectic substances. In *Food Colloids*, (ed. H.D. Grahm), AVI Publishing Co., Westport, CT

Nisperos, M.O. and Robertson, G.L. (1982a) Extraction and characterization of pectin from New Zealand grapefruit peel. *Philippine Agriculture*, **65**, 259–268

Nisperos, M.O. and Robertson, G.L. (1982b) Removal of naringin and limonin from grapefruit juice using polyvinylpyrrolidone. *Philippine Agriculture*, **65**, 275–282

Nisperos-Carriedo, M.O., Shaw, P.E. and Baldwin, E.A. (1990) Changes in volatile flavor components of Pineapple orange juice as influenced by the application of lipid and composite films. *Journal of Agricultural Food Chemistry* **38**, 1382–1387

Nordby, H.E. and McDonald, R.E. (1990) Squalene, a possible natural protectant from chilling injury. *Lipids*, **25**, 807–810

Nordby, H.E. and Nagy, S. (1974) Fatty acids of triglycerides and sterol esters from Duncan grapefruit, Dancy mandarin, and their tangelo hybrids. *Phytochemistry*, **13**, 2215–2218

Nordby, H.E. and Nagy, S. (1977) Relationship of alkane and alkene long-chain hydrocarbon profiles to maturity of sweet oranges. *Journal of Agricultural Food Chemistry*, **25**, 224–228

Ohgawara, T., Kobayashi, S., Ohgawara, H., Uchimaya, H. and Ishii, S. (1985) Somatic hybrid plants obtained by protoplast fusion between *Citrus sinensis* and *Poncirus trifoliata*. *Theoretical and Applied Genetics*, **71**, 1–4

Pesis, E. and Avissar, I. (1989) The post-harvest quality of orange fruits as affected by pre-storage treatments with acetaldehyde vapour or anaerobic conditions. *Journal of Horticultural Science*, **64**, 107–113

Phillips, R.L. and Meagher, W.R. (1966) Physiological effects and chemical residues resulting from 2,4-D and 2,4,5-T sprays used for control of preharvest fruit drop in 'Pineapple' oranges. *Proceedings Florida State Horticultural Society*, **79**, 75–79

Purvis, A.C. (1985) Low temperature induced azide-insensitive oxygen uptake in grapefruit flavedo tissue. *Journal of the American Society of Horticultural Science*, **110**, 782–785

Purvis, A.C. (1989) Soluble sugars and respiration of flavedo tissue of grapefruit stored at low temperatures. *HortScience*, **24**, 320–322

Purvis, A.C. and Barmore, C.R. (1981) Involvement of ethylene in chlorophyll degradation in peel of citrus fruits. *Plant Physiology*, **68**, 854–856

Purvis, A.C. and Yelenosky, G. (1982) Sugar and proline accumulation in grapefruit flavedo and leaves during cold hardening of young trees. *Journal of the American Society of Agricultural Science*, **107**, 222–226

Purvis, A.C. and Yelenosky, G. (1983) Translocation of carbohydrates and proline in young grapefruit trees at low temperatures. *Plant Physiology*, **73**, 877–880

Ram, B.P., Jang, L., Martins, L. and Singh, P. (1988) An improved enzyme immunoassay for limonin. *Journal of Food Science*, **53**, 311–312

Rasmussen, G.K. (1975) Cellulase activity, endogenous abscisic acid and ethylene in four citrus cultivars during maturation. *Plant Physiology*, **56**, 765–767

Rasmussen, G.K. (1977) Loosening oranges with Pik-off, Release, Acti-acid, and Sweep combinations. *Proceedings Florida State Horticultural Society*, **90**, 4–6

Reuther, W. (1973) Climate and citrus behavior. In The Citrus Industry, ed. Ruether, W., University of California Press, Berkeley, CA

Rice, J.D., Nikdel, S. and Purvis, A.C. (1985) Maturity spray residue determination and early season acid accumulation in grapefruit. *Proccedings Florida State Horticultural Society*, **98**, 224–228

Riov, J. (1974) A polygalacturonase from citrus leaf explants. *Plant Physiology*, **53**, 312–316

Riov, J. and Yang, S.F. (1982a) Effect of cycloheximide on ethylene production in intact and excised citrus fruit tissues. *Journal of Plant Growth Regulation*, **1**, 95–104

Riov, J. and Yang, S.F. (1982b) Autoinhibition of ethylene production in citrus peel discs: Supression of 1-aminocyclopropane-1-carboxylic acid synthesis. *Plant Physiology*, **69**, 687–690

Riov, J., Monselise, S.P. and Kahan, R.S. (1969) Ethylene-controlled induction of phenylalanine ammonialyase in citrus fruit peel. *Plant Physiology*, **44**, 631–635

Robertson, G.L. (1976) Pectinesterase in New Zealand grapefruit juice. *Journal of Food & Agriculture Science*, **27**, 261–265

Robertson, G.L. and Nisperos, M.O. (1983) Changes in the chemical constituents of New Zealand grapefruit during maturation. *Food Chemistry*, **11**, 167–174

Roe, B. and Davis, P.L. and Bruemmer, J.H. (1984) Pyruvate metabolism during maturation of Hamlin oranges. *Phytochemistry*, **23**, 713–717

Roth, I. (1977) Species of citrus (type: hesperidium). In *Fruits of Angiosperms*, Gebruder Borntraeger, Berlin, pp. 494–642

Rouse, A.H. (1977) Pectin: Distribution and significance. In *Citrus Science and Technology*, (eds S. Nagy, P.E. Shaw, and M.K. Veldhuis), Vol. 1, AVI Publishing Co., Inc., Westport, CT, pp. 110–207

Rouseff, R.L. and Fisher, J.G. (1980) Determination of limonin and related limonoids in citrus juices by high performance liquid chromatography. *Analytical Chemistry*, **52**, 1228–1233

Rouseff, R.L., Martin, S.F. and Youtsey, C.O. (1987) Quantitative survey of narirutin, naringin, hesperidin, and neohesperidin in *Citrus*. *Journal of Agriculture and Food Chemistry*, **35**, 1027–1030

Rygg, G.L. and Getty, M.R. (1955) Seasonal changes in Arizona and California grapefruit. *U.S. Dept. Agr. Tech. Bul. No. 1130*

Saks, Y., Weiss, B., Chalutz, E., Livne, A. and Gepstein, S. (1988) Regreening of stored pummelo fruit. *Proceedings of the International Society for Citriculture*, **3**, 1401–1406

Salunkhe, D.K. and Desai, B.B. (1986) Citrus. In *Postharvest Biotechnology of Fruits*, Vol. 1, CRC Press, Inc., Boca Raton, Fl, pp. 59–75

Samson, J.A. (1986) Citrus. In *Tropical Fruits*, Second edition, Longman Group UK. Ltd., Essex, UK pp. 73–138

Sasson, A. and Monselise, S.P. (1977) Organic acid composition of 'Shamouti' oranges at harvest and during prolonged postharvest storage. *Journal of the American Society of Horticultural Science*, **102**, 331–336

Schmandke, H. and Guerra, O.O. (1969) Uber den Carotin, L-ascorbinasure-, degydroascorbinasure- und tocopherolgehalt pflanzlicher produkte Kubas. *Nahrung*, **13**, 523–530

Shaked, A. and Hasdai, D. (1985), Organic acids in the juice of developing nucellar and old-line clone Shamouti oranges. *Journal of Horticultural Science*, **60**, 563–568

Shaw, P.E. (1991) Fruits II. In *Volatile Compounds in Foods and Beverages*, (ed. H.Maarse), Marcel Dekker, Inc., New York, N Y, pp. 305–327

Shaw, P.E. and Wilson, C.W. (1984) A rapid method for determination of limonin in citrus juices by HPLC. *Journal of Food Science*, **49**, 1216–1218

Shaw, P.E., Carter, R.D., Moshonas, M.G. and Sadler, G. (1990) Controlled atmosphere storage of oranges to enhance aqueous essence and essence oil. *Journal of Food Science*, **55**, 1617–1619

Shechter, S., Goldschmidt, E.E and Galili, D. (1989) Persistence of [^{14}C] gibberellin A_3 and [^{3}H] gibberellin A_1 in senescing, ethylene treated citrus and tomato fruit. *Plant Growth Regulation*, **8**, 243–253

Shimokawa, K., Hashizume, A. and Shioi, Y. (1990) Pyropheophorbide *a*, a catabolite of ethylene-induced chlorophyll *a* degradation. *Phytochemistry*, **29**, 2105–2106

Shomer, I., Ben-Gera, I. and Fahn, A. (1975) Epicuticular wax on the juice sacs of citrus fruits: A possible adhesive in the fruit segments. *Journal of Food Science*, **40**, 925–930

Sinclair, W.B. and Jolliffe, V.A. (1961) Pectic substances of Valencia oranges at different stages of maturity. *Journal of Food Science*, **26**, 125–130

Smith, C.J.S., Watson, C.F., Morris, P.C., Bird, C.R., Seymour, G.B., Grey, J.E., Arnold, C., Tucker, G.A., Schuch, W., Harding, S. and Grierson, D. (1990) Inheritance and effect on ripening of antisense polygalacturonase genes in transgenic tomatoes. *Plant Molecular Biology*, **14**, 369–379

Stewart, I. (1977) Provitamin A and carotenoid content of citrus juices. *Journal of Agriculture and Food Chemistry*, **25**, 1132–1137

Swingle, W.T. and Reece, P.C. (1967) The botany of citrus and its wild relatives. In *The Citrus Industry*, (eds. W. Reuthers, H.J. Webber, and L.D. Batchelor), Vol. 1, University of California Press, Berkeley, CA, pp. 190–430

Tadeo, J.L., Ortiz, J.M., Martin, B. and Estelles, A. (1988) Changes in nitrogen content and amino acid composition of navel oranges during ripening. *Journal of Science & Food Agriculture*, **43**, 201–209

Takagi, T., Masuda, Y., Ohnishi, T. and Suzuki, T. (1989) Effects of sugar and nitrogen content in peel on colour development in Satsuma mandarin fruits. *Journal of Japanese Society of Horticultural Science*, **58**, 575–580

Talon, M., Zacarias, L. and Primo-Millo, E. (1990) Hormonal changes associated with fruit set and development in mandarins differing in their parthenocarpic ability. *Physiologia Plantarum*, **79**, 400–406

Thomson, W.W. (1969) Ultrastructural studies on the epicarp of ripening oranges. *Proceedings of First International Citrus Symposium*, **3**, 1163–1169

Thomson, W.W., Lewis, L.N. and Coggins, C.W. (1967) The reversion of chromoplasts to chloroplasts in Valencia oranges. *Cytologia*, **32**, 117–124

Thomson, W.W., Platt-Aloia, K.A. and Endress, A.G. (1976) Ultrastructure of oil gland development in the leaf of *Citrus sinensis*, L. *Botanical Gazette*, **137**, 330–340

Ting, S.V. (1969) Distribution of soluble components and quality factors in the edible portion of citrus fruits. *Journal of the American Society of Horticultural Science*, **94**, 515–519

Ting, S.V. and Attaway, J.A. (1971) Citrus fruits. In *The Biochemistry of Fruits and their Products*. Vol. 2, (ed. A.C. Hulme), Academic Press, London, pp. 107–169.

Ting, S.V. and Rouseff, R.L. (1986) *Citrus Fruits and their Products: Analysis and Technology*. Marcel Dekker, Inc., New York, NY

Tisserat, B., Galletta, P.D. and Jones, D. (1989a) Induction of adventitious branches from cultured citrus juice vesicles – a potential means of proliferation. *American Journal of Botany*, **76**, 1650–1758

Tisserat, B., Jones, D. and Galletta, P.D. (1989b) Growth responses from whole fruit and fruit halves of lemon cultured *in vitro*. *American Journal of Botany*, **76**, 238–246

Tisserat, B., Vandercook, C.E and Berhow, M. (1989c) Citrus juice vesicle culture: A potential research tool for improving juice yield and quality. *Food Technology*, **43**, 95–100

Tusa, N., Grosser, J.W. and Gmitter, F.G., Jr. (1990) Plant regeneration of 'Valencia' sweet orange, 'Femminello' lemon, and the interspecific somatic hybrid following protoplast fusion. *Journal of the American Society of Horticultural Science*, **115**, 1043–1046

Vandercook, C.E. (1977) Organic acids. In *Citrus Science and Technology.* (eds. S. Nagy, P.E. Shaw and M.K. Veldhuis), Vol. 1. AVI Publishing Co., Inc., Westport CT, pp. 208–228

Vardi, A., Bleichman, S. and Aviv, D. (1990) Genetic transformation of *Citrus* protoplasts and regeneration of transgenic plants. *Plant Science,* **69**, 199–206

Vardi, A., Breiman, A. and Galun, E. (1987) *Citrus* cybrids: production by donor-recipient protoplast-fusion and verification by mitrochondrial-DNA restriction profiles. *Theoretical and Applied Genetics,* **75**, 51–58

Versteeg, C., Rombouts, F.M. and Pilnik, W. (1978) Purification and some characteristics of two pectinesterase isoenzymes from orange. *Lebensm.-Wiss. u.-Technol.,* **11**, 267–274

Vines, H.M. and Oberbacher, M.F. (1961) Changes in carbon dioxide concentrations within fruit and containers during storage. *Proceedings of Florida State Horticultural Society,* **74**, 243–247

Vines, H.M. and Oberbacher, M.F. (1965) Response of oxidation and phosphorylation in citrus mitochondria to arsenate. *Nature,* **206**, 319–320

Vines, H.M., Edwards, G.J. and Grierson, W. (1965) Citrus fruit respiration. *Proceedings Florida State Horticultural Society,* **78**, 198–202

Ward, R.W. and Kilmer, R.L. (1989) *The Citrus Industry. A Domestic and International Economic Perspective.* Iowa State Univ. Press, Ames, Iowa

Williamson, G. (1991) Purification and characterization of pectin acetylesterase from orange peel. *Phytochemistry,* **30**, 445–449

Wilson, C.W. III, Wagner, C.J. Jr. and Shaw, P.E. (1989) Reduction of bitter components in grapefruit and navel orange juices with β-cyclodextrin polymers of XAD Resins in a fluidized bed process. *Journal of Agricultural and Food Chemistry,* **37**, 14–18

Wilson, C.W. III, Shaw, P.E., McDonald, R.E., Greany, P.D. and Yokoyama, H. (1990) Effect of gibberellic acid and 2-(3,4-Dichlorophenoxy) triethylamine on nootkatone in grapefruit peel oil and total peel oil content. *Journal of Agricultural Food Chemistry,* **38**, 656–659

Wilson, W.C. (1971) Field testing of cycloheximide for abscission of oranges grown in the Indian River area. *Proceedings Florida State Horticultural Society,* **84**, 67–70

Wilson, W.C. (1973) A comparison of cycloheximide with a new abscission chemical. *Proceedings Florida State Horticultural Society,* **86**, 56–60

Wimalasiri, P. and Wills, R.B.H. (1983) Simultaneous analysis of ascorbic acid and dehydroascorbic acid in fruit and vegetables by high-performance liquid chromatography. *Journal of Chromatography,* **256**, 368–371

Yamaki, Y.T. (1989) Organic acids in the juice of citrus fruits. *Journal Japan Horticultural Science,* **58**, 587–594

Yen, C. and Koch, K.E. (1990) Developmental changes in translocation and localization of ^{14}C-labelled assimilates in grapefruit: light and dark CO_2 fixation by leaves and fruit. *Journal of the American Society of Horticultural Science,* **115**, 815–819

Yokoyama, H. and Vandercook, C.E. (1967) Citrus carotenoids. I. Comparison of carotenoids of mature-green and yellow lemons. *Journal of Food Science,* **32**, 42–48

Yokoyama, H. and White, M.J. (1966) Citrus carotenoids, VI. Carotenoid pigments in the flavedo of Sinton citrangequat. *Phytochemistry*, **5**, 1159–1173

Yokoyama, H. and White, M.J. (1967) Carotenoids in the flavedo of Marsh seedless grapefruit. *Journal of Agricultural Food Chemistry*, **15**, 693–696

Yomaki, Y.T. (1989) Organic acids in the juice of citrus fruits. *Journal of Japanese Society of Horticultural Science*, **58**, 587–594

Zacarias, L., Tudela, D. and Primo-Millo, E. (1990) Autoinhibition of ethylene biosynthesis in isolated citrus petals. *Plant Physiology and Biochemistry*, **28**, 561–565

Chapter 5

Exotics

J.E. Taylor

5.1 INTRODUCTION

This chapter will discuss a selected range of so-called 'exotic' fruit. These are typically tropical in origin and at present represent only a very small proportion of the fruit imported into Western Europe and America. However, demand for many of these fruit is increasing within western societies and a progressive increase in trade will probably occur in the future. Some of the tropical fruit now entering the market place, their sites of origin and areas of production are listed in Table 5.1. Photographs of a representative range of exotics are shown in Plates 1–5. A major problem preventing the full commercial exploitation of these fruit is a lack of suitable postharvest handling techniques to ensure a relatively cheap and reliable supply of fruit. Several postharvest ripening control methods have been investigated for these fruit. A full description of these is outside the scope of this chapter, but some methods of interest have been summarized in Table 5.2. The postharvest handling of these fruit would benefit significantly from detailed information on their ripening physiology and biochemistry. However, these fruits have, in general, been poorly studied in this respect. In some cases, postharvest problems, for instance astringency in persimmon and cracking in lychee, have initiated quite intense investigation while other aspects of the biochemistry of ripening in these fruit have been largely ignored. This chapter will attempt to bring together the biochemical knowledge available on many of the fruit listed in Table 5.1. In most cases this is limited largely to a determination of the biochemical constituents of the fruit and how these alter during ripening or postharvest storage.

Biochemistry of Fruit Ripening. Edited by G. Seymour, J. Taylor and G. Tucker. Published in 1993 by Chapman & Hall, London. ISBN 0 412 40830 9

Table 5.1 Some exotic species: their sites of origin and areas of production.

Common name	Species	Site of origin	Area of production
Feijoa	Acca sellowiana Berg	Southern Brazil	New Zealand, Florida, Brazil
Guava	Psidium guajava L.	New World Tropics	Hawaii, India, Canaries, Florida, and most tropical regions
Lychee	Litchi chinensis Sonn.	China, South-east Asia	China, South Africa, Australia, Hawaii, Brazil, Tropical America
Passion Fruit	Passiflora edulis Sims.	South America	Australia, Hawaii, Brazil, Tropical America
Persimmon	Diospyros kaki L.	China	Florida, California, Japan, South-east Asia
Rambutan	Nephelium lappaceum L.	South-east Asia	South-east Asia
Soursop	Annona muricata L.	Tropical America	Venezuela, Colombia, Dominican Republic
Star Fruit	Averrhoa carambola L.	South-east Asia	Florida, Malaysia, Canaries
Tamarillo	Cyphomandra betacea (Cav.) Sendt		South America, New Zealand, Australia.
Mangosteen	Garcinia mangostana L.	Malaysia	Malaysia, Thailand, Indonesia

5.2 PERSIMMON

Diospyros kaki, persimmon or kaki, is an important crop in Japan and China and is becoming prominent in Mediterranean countries (Samson, 1986). The Japanese persimmon has at least 1000 varieties (Itoo, 1980) and may be divided into two distinct types, namely those which show no change in the colouration of the flesh under the influence of pollination, and those in which the fruit is light coloured when seedless, and dark coloured when seeded as a result of pollination. They can be further subdivided into astringent and non-astringent varieties (Table 5.3).

Astringency has a major impact on those varieties cultivated with, in Japan, non-astringent (NA) fruit accounting for 60–65% of the total yield

Table 5.2   Postharvest storage techniques for some exotic fruit.

Fruit	Technique	Reference
Persimmon	Controlled atmosphere 8% CO_2, 3–5% O_2, 1°C	Itoo (1980)
	Modified atmosphere (polyethylene bags)	Young Koo Son et al. (1981)
		Zuthi and Ben-Arie (1989)
Lychee	Perforated polyethylene bags/wax emulsion	Bhullar et al. (1983)
Guava	Low temperature	Vazquez-Ochoa and Colinas-Leon (1990) Wills et al. (1983)
Cherimoya	Controlled atmosphere 10% CO_2, 2% O_2, 9°C	Calvo et al. (1983)
Carambola	Low temperature	Kenney and Hull (1987) Campbell (1989a)
Passion Fruit	Low temperature	Collazos et al. (1984)
Mangosteen	Low temperature	Srivasta et al. (1962)
Feijoa	Low temperature	Thorp and Klein (1987)
Rambutan	Low temperature + 7% CO_2	Mohamed and Othman (1988)

Table 5.3   Classification of Persimmon varieties. (After Itoo, 1980.)

Pollination Constant and Non-Astringent (PCNA)	Varieties – Fuyu,Jiro,Gosho,Suruga. These have dark tannin spots in the flesh.
Pollination Variant and Non-Astringent (PVNA)	Varieties – Zenjimaru, Shogatsu, Mizushina, Amahyakume. These have dark tannin spots and may become astringent when seedless.
Pollination Constant and Astringent (PCA)	Varieties – Yokono, Yotsumizo, Gionbo. These do not have tannin spots.
Pollination Variant and Astringent (PVA)	Varieties – Aizumishirazu, Emon, Koshuhyakume. These fruit are astringent when pollinated, and have dark tannin spots around the seeds.

(Itoo, 1980). Such fruit are eaten fresh, whereas astringent (AS) varieties require removal of astringency before eating, or consumption as the dried product.

Takata (1983) examined respiration, ethylene production and ripening of Japanese persimmon fruit harvested at various stages of development. Rates of respiration were found to vary considerably among individual fruit of the cultivar Fuyu. When fruits were harvested in June and July and held at 25°C they had a climacteric-like rise in respiration and then subsequently softened and became yellow. However, when fruits were harvested in August they softened and changed colour while showing an increase in respiration without a peak. Consequently Japanese persimmon fruit were considered to be of a type which differs from both climacteric and non-climacteric fruits. In contrast Akamine and Goo (1981) suggested that persimmon fruit did show typical climacteric respiration as indicated by CO_2 production patterns at two stages of maturity, namely mature-green and 5% red-coloured.

Takata (1983) also found a close correlation between respiration and ethylene production in fruits harvested in July. However, fruits harvested in August still showed an increase in ethylene production with a peak characteristic of a climacteric fruit, while fruits harvested later produced only small quantities of ethylene. Such results suggested that fruit ability to synthesize ethylene decreases rapidly during maturation.

When Japanese persimmon fruit were treated with 10 μl.l^{-1} ethylene for up to three days following harvest, respiration was stimulated and ethylene production and fruit softening increased, irrespective of the developmental stage (Takata, 1982); however, treatments have progressively less effect as the fruit mature. The rate of ethylene evolution at the induced climacteric also decreased as the fruit matured.

Fruit ripening also responds to the application of other plant growth regulators and in some cases this could be beneficial for storage. It has been reported that prior to harvest the application of gibberellic acid in the form of GA$_3$ to trees (and hence to fruit) may have a delaying effect on fruit ripening. When sprayed on to trees of the cultivar Triumph, at a concentration of 10^{-3}–10^{-4} M, fruit ripening was delayed, as shown by a retardation in colour development and growth rate of the fruits, extending the harvest season by three weeks (Ben-Arie *et al.*, 1986). The treatment also led to an increase in fruit firmness at harvest, delayed softening both during and following cold storage, with a consequent extension to storage life of approximately one month. A similar delay in colour development following GA treatment was attributed to an inhibition of chlorophyll degradation and of carotenoid biosynthesis (Gross *et al.*, 1984). In contrast, Sive and Resnizky (1987) found that GA treatment prior to harvest had no effect on storage life.

Storage life may also be increased by simply maintaining high quality fruit while still on the tree itself, by harvesting when the fruits reach full colour development, and by filling the controlled atmosphere (CA) store within a period of five days (Sive and Resnizky, 1987).

5.2.1 Composition

Fructose, glucose and sucrose are the major sugars in the fruit pulp of persimmon (Hirai and Yamazaki, 1984), the total sugar content varying from 10.2–19.6% in a range of cultivars of astringent fruit, and from 10.1–16.7% in sweet cultivars. These sugars were present in all cultivars studied, from mature green to the fully ripe stages of fruit development (Senter *et al.*, 1991). Sorbitol and inositol are also present in minor quantities, the levels varying significantly with cultivar and stage of maturity (Senter *et al.*, 1991). Total sugars rise from approximately 10% to 14% during the change from immature to mature fruits (Sugiura *et al.*, 1979, 1983). Sucrose content is also influenced by time of harvesting and by astringency removal by processing (Hirai and Yamazaki, 1984). Glucose and fructose levels also vary significantly with both cultivar and maturity (Senter *et al.*, 1991). Zheng and Sugiura (1990) related seasonal changes in the accumulation of sucrose and reducing sugar content with changes in invertase (EC 3.2.1.26) activity in six cultivars which they examined.

Cell wall polysaccharides have also been studied. Fishman (1986) studied 18 varieties of persimmon, finding the highest pectin content in 'Costata' at 11.69%. During ripening both protopectin and pectin are hydrolysed, softening appearing to follow the onset of this protopectin degradation. Fishman and Oganesyan (1984) extracted and measured the pectolytic enzymes from fruit at different stages of development; pectolytic activity decreased from 7956 units.g^{-1} to 295 units.g^{-1} approximately two weeks later, and exopolygalacturonase (EC 3.2.1.67) activity decreased from 0.32 to 0.21 units.g^{-1}, whereas pectinesterase (EC 3.1.1.11) activity increased from 0.07 to 0.18 units.g^{-1}. It has been suggested, however, that changes in the composition of cell wall polysaccharides during softening are mainly due to the decomposition of hemicellulose rather than of the pectic component (Itamura *et al.*, 1989).

The major non-volatile organic acids found in persimmon fruit include succinic, malic, citric and quinic acid (Senter *et al.*, 1991); malic acid appears to increase with maturity, whereas citric acid declines (Senter *et al.*, 1991).

Vitamins form a major nutritive component of these fruit. When examined in a range of cultivars, provitamin A activity ranged from 17 retinol equivalents (RE) 100 g^{-1} in 'Aizumi Shiraza', to 120 RE 100 g^{-1} in 'Hana Gosho' (Homnava *et al.*, 1990). β-carotene was found to be the predominant provitamin A isomer in 11 out of 15 cultivars studied, with β-cryptoxanthin predominating in the remainder. Total ascorbic acid (vitamin C) levels vary considerably; for example, the ascorbic acid level in Hachiya was found to be only 35ng.100 g^{-1}, but 218 ng.100 g^{-1} in Fuyu (Homnava *et al.*, 1990). Evidence suggests that long-term storage of persimmons is possible only with fruit showing low pectin and ascorbic acid fluctuations (Tovkach, 1979).

The predominant unsaturated fatty acids in the cultivars Khachia and Khiakume are linolenic, oleic, linoleic and palmitoleic (Kolesnik *et al.*, 1987); Suzuki *et al.* (1982) found a similar distribution in the cultivars Fuyu, Zenjimaru and Koshuhyakume, i.e. linolenic (20.2–31.3%), oleic (20.4–28.2%), palmitic (16.8–23.6%) and palmitoleic (12.1–18.3%). These findings were the result of an examination of the neutral lipids, glycolipids and phospholipids in the pulp and peel of these cultivars.

GCMS analysis of volatile fractions from both astringent and non-astringent persimmon fruit have identified the major components to be bornyl acetate and (E)-2-hexenal for all the cultivars studied (Horvat *et al.*, 1991), while minor constituents included (E)-2-hexanol, phenylacetaldehyde, borneol, benzothiozole, neryl acetate, and palmitic acid (Horvat *et al.*, 1991).

5.2.2 Astringency

Astringency is a major quality problem in persimmon and one which has occupied a major proportion of research input on this fruit. An immature persimmon fruit is markedly astringent due to water-soluble tannins present in tannin cells. The soluble tannins, which spread easily over cut surfaces, have a strong protein binding capacity and have been widely used as deproteinizing agents in the brewing of saké. Coagulated or polymerized tannins do not show astringency, primarily because they are insoluble in water.

As the fruit ripens, the tannin cells increase in size and number, their size and density differing markedly in different varieties. Non-astringent (NA) and PCNA varieties have particularly small tannin cells. The variety Yotsumizo (PCA) has several times as many tannin cells as the varieties Jiro or Fuyu (PCNA). In mature fruit of astringent varieties the amount of soluble tannin is 0.8–1.94% (av. 1.41%) of the fresh weight.

Experiments by Matsuo and Itoo (1978) suggested that kaki-tannin consists of polymers of catechin, catechin-3-gallate, gallocatechin and gallocatechin-3-gallate in the ratio 1:1:2:2, and belongs to the proanthocyaninidin B group with a carbon–carbon interflavan linkage, from C_4 of one unit to C_6 or C_8 of another unit. Nakabayashi (1971) examined the difference between the polyphenolic components of 10 varieties, and between astringent and non-astringent immature fruits, with the polyphenolic compounds, β-D-glucogallin (β-1-*o*-galloyl-D-glucose) being isolated and identified only from astringent immature fruits. Although β-D-glucogallin does not have an astringent taste, it was thought that the presence of β-D-glucogallin might be used to differentiate between astringent and non-astringent varieties.

Astringency and tannin content

The reduction of astringency during growth and maturation of astringent varieties and the disappearance of astringency from non-astringent varieties is reflected in the tannin content of the fruit. Kato (1984a) found a relatively high correlation between the degree in astringency and tannin concentration, with fruits containing approximately 0.25% tannin being slightly astringent, while those containing <0.1% were almost non-astringent.

It has been suggested that the capacity for natural removal of astringency in fruits of cultivars of the PVA, PVNA and PCA types depends upon the relative amounts of volatile compounds such as ethanol and acetaldehyde produced by seeds during fruit development (Sugiura *et al.*, 1979; Taira *et al.*, 1986). The production of ethanol by the seeds is probably triggered by anaerobic conditions and high CO_2 concentrations early in fruit development. It has been suggested that two different mechanisms may be involved in the loss of astringency. One is associated with PVNA, PVA, and PCA types and is dependent upon the production and accumulation of ethanol and presumably acetaldehyde; the second is associated with PCNA types which do not produce these volatile substances (Sugiura *et al.*, 1979).

Yonemori (1986) examined four varieties. In PVNA, the fruit on the tree showed a rise in ethanol and acetaldehyde concentrations and a decrease in soluble tannins. In PCNA, neither ethanol nor acetaldehyde could be detected. In PCNA, fruit tannins of low molecular weight predominate and these are chemically less reactive than those of the other fruit types (in which tannins of high molecular weight predominate), and also the final size of the tannin cell is smaller.

Reduction of astringency

Many studies have focused on methods of treating fruit to reduce astringency. These include spraying with ethanol (Kato, 1984b; Kawashima *et al.*, 1984; Dauriach 1986), or by storing with highly absorbent polymer sheets soaked in 30% ethanol (Kawashima *et al.*, 1984). Manabe (1982) found that when astringency was removed by ethanol treatment, treated fruit contained 13 times more insoluble tannic substances and twice as much alcohol-insoluble N, Ca^{2+}, Mg^{2+}, P and K^+ than untreated fruits.

Kato (1984b) harvested fruits of the cultivars Aizumishirazu and Hiratenenashi which were sprayed with 0–50% ethanol for up to four days, resulting in fruit ethanol concentrations of 0.01–1.4% fresh weight. Evidence indicated that higher ethanol concentrations reduced the lag time before the decrease in tannin concentration, and increased the rate of tannin loss once initiated. Removal of astringency occurs more readily in younger fruit than in more mature fruit. Taira *et al.* (1990) suggested that this may be due to a more active conversion of the applied ethanol to acetaldehyde.

An alternative method used to increase ethanol accumulation, and hence remove astringency, has been to wrap fruit in polyethylene (PE) bags, and using evacuated N_2 or CO_2 atmospheres (Pesis *et al.*, 1986). It was found that acetaldehyde accumulated most rapidly in fruits under CO_2 atmosphere, where the deastringency process was fastest, but the fruits were susceptible to flesh browning. Fruits stored under vacuum or N_2 atmospheres accumulated ethanol and acetaldehyde and maintained high quality and firmness for two weeks at 20°C and three months at –1°C. Flesh appearance and taste after 14 weeks' storage at –1°C was best under N_2 atmospheres, although this treatment produced the lowest levels of ethanol and acetaldehyde.

Pesis *et al.* (1986) however were able to demonstrate that the removal of astringency was achieved within 3–4 days at 20°C by vacuum-packing fruits (cv. Triumph) in 0.08 mm PE bags. This was the result of the accumulation of CO_2, acetaldehyde and alcohol within the bags. Following the removal of astringency, the fruits maintained a high quality and firmness during cold storage at –1°C, followed by stimulated marketing at 20°C. Good flavour was maintained for approximately 10 days at 20°C, but thereafter an off-flavour developed due to the accumulation of volatile compounds.

Moderate to severe discoloration may occur in fruits containing more than 0.1% ethanol when treated under excessive moisture conditions (Kato, 1984b; Kawashima *et al.*, 1984), and in all fruits containing more than 0.3% ethanol (Kato, 1984b). For adequate stringency removal and fruit quality, the optimal ethanol concentration in the fruit would appear to be 0.1–0.2% at 10–20°C, and about 0.2% at 30°C.

In addition to examining the effects of packaging and gaseous treatments on the level of astringency, more detailed experiments have been performed on the tannin cells themselves. Yonemori and Matsushima (1985) studied the development of tannin cells in PCNA fruits using an image analysis technique. Tannin cells in this type of cultivar stopped enlarging at the end of June, whereas those in PVNA, PVA, and PCA type cultivars continued enlarging until late July. The total number of tannin cells was similar in all four types, but the cells were much smaller in the PCNA type. Tannins in the PCNA fruits had not coagulated by early August, whereas those in the PVNA fruits had formed insoluble complexes. The authors suggested that the decrease in astringency of PCNA fruits is related to the smaller size of tannin cells, rather than to the degree of coagulation of the tannins.

Yonemori and Matsushima (1987a, b) also made a more detailed study of tannin cells using fluorescent microscopy (FM) and scanning electron microscopy (SEM) of cells from all four types of fruit. FM examination of highly astringent fruit indicated that tannin cells of all fruit possessed some discontinuous portions in their walls. At one month after the initial observations, these pores in the tannin cell walls of Fuyu (PCNA) and Chokenji (PVNA) had disappeared, and the surface of coagulated internal contents had become smooth. At the same stage, Hiratanenashi (PVA) and Kuramitsu (PCA) showed little change in the structure from that observed previously. Pore occlusion occurred in Hiratanenashi fruits that were treated on the tree with ethanol fumes to remove astringency, indicating that loss of astringency induces structural changes in tannin cell walls. Subsequently, Yonemori and Matsushima (1987b) demonstrated that occlusion coincided with cessation of tannin accumulation, as determined by soluble tannin content and fruit fresh weight.

5.3 LYCHEE

The lychee is indigenous to Southern China and possibly Northern Vietnam. It is grown in Burma, India, Australia, West Indies, Hawaii, South Africa and Florida. An evergreen, subtropical crop, the lychee requires a period of low temperature or a dry period to induce flowering and hence fruiting (Samson, 1986). The fruit is consumed both in its fresh and preserved forms. When mature, the fruit pericarp is thin, hard and may be pale green, bright red or rose coloured depending upon the variety. The fruit are round to oval in shape, measure 2.5–4.0 cm, and weigh around 22g. The edible portion of the fruit is a white-to-cream-coloured translucent pulp which is grape-like in texture, very succulent and aromatic, and is characterized by a sweet, acid taste. The aril composes 75% of the weight, the skin 14% and the seed 11%. Death of the embryo results in 'chicken-tongue' seeds, with bigger and tastier fruits (Cavaletto, 1980). Menzel and Simpson (1991) have recently described the important cultivars in the major producing countries, and Nip (1988) has reviewed the literature on the handling and processing of lychee.

One major problem during handling is that of desiccation which may result in the development of browning, rendering the fruit unsuitable for market. Percentage fruit weight loss appears to increase with storage duration (Ajay Singh and Abidi, 1986) and as such is a feature which most storage strategies try to overcome. Bain et al. (1983) found that water loss, which is responsible for browning, was prevented by packaging the fruit in small punnets and overwrapping with a cling plastic film. Other treatments include treating precooled lychee fruits with 0.5% lecithin and 2.5% sodium bicarbonate (Zhang et al., 1986) then packing in 0.04 mm polyethylene (PE) bags and storing at room temperature. Under such conditions after eight days' storage, the fruit retained its fresh-red colour, with a weight loss of 2–3%, while controls (not packed in bags) became totally brown and shrunken, with a weight loss of 18–20%.

5.3.1 Fruit growth and physiology

Lychee fruit have a sigmoidal pattern of growth (Jaiswal *et al.*, 1982; Huang and Xu, 1983; Zhong and Wu, 1983; Paull *et al.*, 1984; Singh and Yadav, 1988; Ghosh *et al.*, 1989). Many cultivars have been examined, including Nowici, Groff, Shui Dong, Gui Wei, Mei, Shahi, China, CHES-80-1, Late Bedana, Bombai and Elachi.

There appears to be three relatively distinct phases of growth which vary in length, depending upon the cultivar. The first stage involves the formation of the embryo, which is then followed by the rapid growth of the pericarp, which in turn is followed by a rapid growth of the flesh or aril (Zhong and Wu, 1983). The increase in fruit weight amounts to approximately 80% of the total fresh weight. The development of the seedcoat and pericarp may last for 25–40 days and aril development may last for between 35–41 days, depending upon the cultivar (Singh and Yadav, 1988).

Classification of lychee fruit as either climacteric or non-climacteric is difficult. Prasad and Jha (1979) demonstrated that O_2 uptake showed a climacteric peak in partially ripe fruits and then subsequently declined. However, Chen *et al.* (1986) in studies with fruits of several cultivars stored at room temperature, found that the rate of ethylene release rose during the first 3–5 days of storage and then declined, but with no obvious climacteric peak. Fruits that were stored at low temperature (8°C) had a low respiration rate and rate of ethylene production. Fruits stored at 30°C and treated with thiabenzadazole at 1000 $\mu l.l^{-1}$ had a reduced rate of ethylene release, extended storage life and improved fruit quality.

The use of growth regulators has been examined as a possible aid to early ripening of lychee fruit. Ethephon at 400 $\mu l.l^{-1}$, sprayed on the panicles just before colour break, significantly increased the fresh weight and total soluble solids of the fruits and hastened ripening by eight days. Other growth regulators that have been examined include kinetin, gibberellic acids (GA) and nappthaleneacetic acid (NAA). Jaiswal *et al.* (1987) were able to show that treatment of partially ripe fruit with kinetin caused chlorophyll retention and a lower level of total anthocyanins which later increased during senescence. Using the cultivar Kasba, Jaiswal *et al.* (1986) had also demonstrated that the application of kinetin at 0.5–10 $\mu g.ml^{-1}$ increased total phenolics at the fully ripe stage, particularly in the rind, but the loss of phenolics increased with advancing fruit senescence.

The spraying of lychee with NAA appears to have a major positive effect on the quality of the fruit produced. Jain *et al.* (1985) found that the highest juice percentage (47%) and fruit ascorbic acid content (41.1 mg.100 ml^{-1}) was obtained when panicles were sprayed with NAA at 10 μl.l^{-1}. In the same study, it was found that the highest TSS (18.6 mg.ml^{-1}) occurred when the fruit were sprayed with 2,4,5-T at 25 μl.l^{-1}. Similarly, Sharma and Dhillon (1984) found that the maximum fruit weight, the highest aril percentage and the lowest skin and seed percentages were obtained with NAA at 25 μl.l^{-1}.

5.3.2 Composition

Sugars, water-soluble pectin and total pectin have all been shown to increase during fruit ripening, whereas acid-soluble pectin declines in the four cultivars, Early Large Red, Calcuttia, Muzaffarpur and Bedana (Ajay Singh and Abidi, 1986). The cultivars Groff, Shui Dong, Gui Wei and Mei have been shown to have slightly different ratios of sucrose, glucose and fructose. However, in all cases total sugars increased from about 3% to 16% during the period of rapid growth associated with aril development (Paull *et al.*, 1984). Prasad and Jha (1979) have also reported an increase in fructose throughout maturation.

Ascorbic acid content has been shown to increase during ripening (Ajay Singh and Abidi, 1986). Ghosh *et al.* (1989) found that the ascorbic acid of the cultivars Bombai and Elachi increased during early stages of fruit development, but declined on maturity. Paull and Chen (1987a) studied postharvest changes in the cultivars Hei Ye and Chen Zi during storage; ascorbic acid content fell during storage from 1 to 0.4 mg.g^{-1} in four days, irrespective of storage method or temperature. Ascorbic acid changes in Chen Zi were more marked than in Hei Ye, the content declining from 1.2 to 0.2 mg.g^{-1} at 22°C after eight days.

Paull *et al.* (1984) suggested that titratable acidity was due mainly to succinic and malic acids, the acidity decreasing over two weeks in the middle of the fruit growth period from 65 to 15 meq.100g^{-1}. Succinic acid decreased in the same period from about 350 to 0.4 meq.100g^{-1} and remained low for the remaining 40 days of development. Malic acid declined from 75 to 12 meq.100g^{-1}, and citric acid remained constant at around 3 meq.100g^{-1}. A major portion of the decline in succinic acid was due to dilution.

TSS (°Brix) was found not to be suitable as a maturity indicator for lychee, unlike a considerable number of other fruits, whereas titratable acidity (TA) and TSS/TA ratios were both good predictors of taste (Batten, 1989). Batten (1989) suggested that a TA of 4.4 cmolH$^+$.kg^{-1} juice or a TSS/TA ratio of 4.3 would be suitable maturity indices.

Skin chlorophyll begins to decrease logarithmically soon after the start of fruit growth (Paull *et al.*, 1984), and this coincides with the active synthesis of flavanoids, particularly anthocyanins, the concentration of which increases up to a fully ripe stage (Jaiswal *et al.*, 1987). Anthocyanins in the skin of lychee fruit have been isolated, cyanidin-3-rutinoside being the major component (Lee and Wicker, 1991a). Cyanidin-3-glucoside and malvidin-3-acetyl glucoside have also been identified (Lee and Wicker, 1991a).

Fruit colour can change significantly during storage. During a 48-day storage period, fruits of the variety Brewster showed a decline in total anthocyanin content from 1.77 to 0.73 mg.g^{-1} fresh weight (Lee and Wicker, 1991b). Thus, during postharvest storage of lychee fruits there may be a loss of red-colouration and a darkening of the fruit. Coincident with the onset of fruit discolouration is a decline in peroxidase (EC 1.11.1.7) activity and an increase in the activity of polyphenoloxidase (EC 1.10.3.1; PPO) (Huang *et al.*, 1990). It has been demonstrated that fruit colour may be preserved by SO_2 fumigation, followed by a dip in dilute HCl, and is the result of an inhibition of PPO activity (Zauberman *et al.*, 1991). It has been suggested that if PPO activity is inhibited soon after harvest, the red colour of the lychee fruit may be preserved as long as the rind remains acidic (Zauberman *et al.*, 1991).

Jaiswal *et al.* (1986) were able to demonstrate that during ripening both rind and aril of the cultivars Deshi and Kasba were characterized by large increases in the concentration of total phenolics which rapidly declined at senescence, with the rind being richer than the aril. In Kasba, application of kinetin at 0.5–10 μg.ml^{-1} increased total phenolics at the fully ripe stage, particularly in the rind, but the loss of phenolics increased with advancing fruit senescence. Treatment with cyclo-heximide at 1.0–100 μg.ml^{-1} resulted in lower phenolic levels in the rind at all stages of senescence in Kasba. Among the phenolics, tannic, caffeic, vanillic and salicylic acids predominated, whereas gentisic, ferulic and *p*-hydrobenzoic acids were minor components.

5.3.3   Fruit cracking in lychee

Fruit cracking is a major problem during the postharvest storage of lychee fruit. Measurements of the fruit growth pattern in the cultivars Dehra Dun (highly susceptible to cracking), Calcuttia (moderately susceptible) and Hong Kong (resistant) suggested that diametrical growth had a greater effect on cracking than longitudinal growth (Kanwar and Nijar, 1984). A higher diametrical fruit growth rate from mid-May to early June, when the temperature is high and the relative humidity low, renders Dehra Dun more susceptible to cracking than the other two cultivars in which diametrical growth rates are much lower during this period. Sharma and Dhillon (1986) measured the internal level of gibberellins in relation to fruit cracking in the cultivar Dehra Dun, in the skin, aril, and seed. In the two years of study, seed gibberellin levels near final harvest time were significantly higher in cracked than in normal fruit. Sharma and Dhillon (1987) also found that fruit cracking in the cultivar Dehra Dun was significantly reduced when sprayed with 25 μl.l^{-1} NAA. Several other compounds have been used to inhibit fruit cracking including $ZnSO_4$ and GA_3 (Sharma and Dhillon, 1987); Cu or B (Misra and Khan, 1981); ethephon sprays (Shrestha, 1981) which when applied at 10 μl.l^{-1} reduced percentage fruit cracking in the cultivar Early Large Red from 12% to 6%.

5.4   GUAVA

Guava is a small tree of the Myrtaceae family that attains heights of up to 10 m and grows well in most tropical and subtropical soils. The tree is native to tropical America but by the 17th century had been distributed to Asia, and is now present in every tropical and subtropical country (Knight, 1980). Although popular for fresh consumption in many places, fruit are widely used in processed form. These features, together with the ease of culture and high nutritional value, mean that this fruit is important in international trade as well as in the domestic economies of most tropical countries (Samson, 1986).

The fruit is a berry, topped by calyx lobes with many seeds embedded in a white or reddish pulp. The fruit may take on one or several shapes; for example fruit of the cultivar Supreme oval are pyriform in shape, may be 10 cm long and average 285 g in weight; Red Indian produces a globose fruit that may reach 7 cm in diameter and weigh on average 170 g; Ruby bears an ovate fruit, approximately 7.5 cm long with an average weight close to 200 g (Knight, 1980). The varieties Pyriferum and Pomiferum are often referred to as pear and apple guavas, respectively (Wilson, 1980). The fruit has many small hard seeds, from 153 to 664 per fruit, which are located in the centre of the flesh. Seedless (triploid) varieties have been developed, though these produce small amounts of misshapen fruit. The numerous store cells (sclereids) in the fleshy portion of the fruit impart a gritty texture to the flesh. The fruit colour varies from white to deep pink and salmon red. Ripe fruits emit a sweet aroma and have a pleasant sour sweet taste, but in a later stage produce a penetrating scent (Wilson, 1980).

Firm, yellow mature guavas do not keep well and are usually transported rapidly to processing plants. Guavas are usually shipped in small wooden boxes, rather than in orange crates because they are easily crushed or bruised, and damaged fruit deteriorate rapidly. Fruit may be graded by specific-gravity into one of three groups, i.e. group I (<1.0), group II (1.0–1.02), and III (>1.02). Fruits of group I are the largest, highest in vitamin C content, and have a higher acceptability score and a shelf life of six days. Fruits of group II have a shelf life of up to eight days. Group III fruits are the smallest, are relatively immature compared with the other two groups and hence poor in colour development and inferior in nutrient quality, especially with respect to vitamin C content (Tandon et al., 1989).

Weight increase and reduction in hardness of Vietnamese-type fruit both change in a sigmoidal manner during development (Salmah and Mohamed, 1987). The maximum weight increase occurs between weeks 10–12, and the greatest reduction in firmness between weeks 12–14. Fruit density shows an initial rapid increase, which then remains constant between weeks 4–12, and subsequently decreases. The moisture content also increases rapidly at first, followed by a more gradual increase, and remains constant after week 14. Harvesting after week 16 has been recommended (Salmah and Mohamed, 1987). Guava can be stored for several weeks at temperatures of between 3.5° and 7°C (Vazquez-Ochoa and Colinas-Leon, 1990) but are damaged at 0°C (Wills et al., 1983). Postharvest life can be influenced by various treatments; thus morphactin chloroflurecol, when applied to the cultivar Allahabad Safeda as a postharvest treatment, at a concentration of 100 μl.l^{-1} was found to maintain the market value of the fruit by retarding various harmful textural, physiological and biochemical changes (Gupta and Mukherjee, 1980).

Various other postharvest treatments have been examined, including fruit cooling and dipping in Waxol (Kartar Singh and Chauhan, 1984); treatment with metabisulphite (Ahlawat et al., 1980); dipping fruit in 1% calcium nitrate (Singh et al., 1981); treating with NAA at 150 μl.l^{-1} and packing in PE bags (Dhoot et al., 1984); and dipping in 1000 μl.l^{-1} cyclocel for five minutes (Tandon et al., 1984).

Studies of postharvest changes indicate that mature fruit show greater colour and texture changes than less mature fruit during storage (Brown and Wills, 1983). Hastening of colour development and also softening may be achieved by dipping light green fruit in ethephon (Singh et al., 1979).

5.4.2   Composition

Fructose is the major sugar present in the ripe fruits of both Florida (Wilson et al., 1982a) and Vietnamese type cultivars (Salmah and Mohamed, 1987). Fructose comprises 55.93% and 52.85% of the sugar in white and pink cultivars respectively (Mowlah and Itoo, 1982). Glucose is also present and changes in a sigmoidal pattern, the maximum glucose increase occurring between weeks 12–14 (Salmah and Mohamed, 1987).

Increase in fruit diameter and fructose content show a more linear change. Sucrose and inositol, also present in small quantities, increase during ripening and occur in greatest amounts in fully ripe fruits (Mowlah and Itoo, 1982).

A study of the pectin component of the fruits of seven cultivars at different stages of development revealed that the pectin content in Allahabad Safeda, Banaras and Red Flesh was highest in ripe fruits, whereas that in the other cultivars was highest in immature fruits (Pal and Selvaraj, 1979). Similarly, pectin content appears to be affected by the cropping season (rainy or winter) and stage of fruit maturity (green mature or half ripe) (Dhingra et al., 1983). In fruits of the Vietnamese-type the change in total pectin content was shown to display a sigmoid curve (Salmah and Mohamed, 1987); both soluble and insoluble pectin increased during development but insoluble pectin was converted to soluble pectin during maturity (after week 14). Some workers have found reductions in both total and soluble pectin during ripening (Dhillon et al., 1987; Esteves et al., 1984b), whereas others have found an increase (Mowlah and Itoo, 1983), though these variations may be a consequence of studying different varieties. During ripening there is a reduction in alcohol-insoluble solids, total and soluble pectin, cellulose and hemicellulose with a consequent start in fruit softening about 93 days after full bloom (Esteves et al., 1984b).

Mowlah and Itoo (1983) found that polygalacturonase activity was very low until the fruit were fully ripe, and increased thereafter. Pectinesterase (EC 3.1.1.11) and cellulase (EC 3.2.1.4) activities also increased between maturation and ripening. Pal and Selvaraj (1979) found that pectinesterase activity in the seven cultivars examined, was highest in over-ripe and immature fruit. These enzymes may be responsible for the changes in the wall polysaccharides during ripening.

Vitamin C content increases during ripening of guava fruit and decreases during senescence (Esteves et al., 1984a). The absolute contents have been found to be similar in red and white cultivars, and to vary from 215 to 372 mg.100g^{-1} at 100 to 114 days after full bloom. Similar results were found by Mowlah and Itoo (1983), ascorbic acid content in the two cultivars examined increasing throughout maturation, ripening and storage, reaching over 270 mg.100g^{-1} in both cultivars. Itoo et al. (1980) reported that guava fruits have ten times more ascorbic acid than oranges; however, ascorbic acid levels have been shown to vary from 0.04% to 0.44% depending upon the cultivar (Wilson et al., 1982a). Similarly, ascorbic acid content of the cultivar Allahabad Safeda has been shown to be more than three times higher in the winter than in the rainy season (329 vs. 93 mg.100g^{-1}), and may differ by a factor of two for these seasons in other varieties (Chauhan et al., 1986). Most ascorbic acid is present in the peel, with less in the pulp and still less in the core (Itoo et *al.*, 1980). The change in ascorbic acid level shows a sigmoid curve (Salmah and Mohamed, 1987).

Wilson *et al.* (1982a) analysed the acid content of the ripe fruits of five Florida cultivars. While citric acid was the major acid in all the cultivars, glycollic, malic and ascorbic acid were also present, sometimes in substantial amounts; traces of fumaric acid were also detected.

Sharaf and El-Saadany (1986) found that total free amino acids increased with ripening, whereas total organic acids decreased. The content of stearic and palmitic acids increased with ripening, whereas the linolenic acid content diminished. In ripe fruits the content of linoleic acid was twice that of green-mature fruits, but fell slightly with over-ripening.

Volatile aroma compounds have been determined from both white and pink-fleshed fruit using GCMS. Askar *et al.* (1986) identified over 80 volatiles, mycrene, *cis-* and *trans*-ocimene and β-caryophyllene being the most important. The characteristic guava aroma is mainly due to cinnamyl derivatives, β-caryophyllene and C_6-derivatives. The volatile constituents of a commercially produced guava natural flavour concentrate have also been analysed (Pino *et al.*, 1990); of the 29 compounds identified 10 had not previously been described in flavour concentrates.

There are quantitative and qualitative differences in the volatile constituents between the inner flesh (including seeds) and the outer flesh–peel of guava, with the outer flesh having the greatest number and highest content of total volatile compounds (Chyau and Wu, 1989). Volatile production also changes during maturation of the fruit; immature fruit contain high levels of isobutanol, butanol and sesquiterpenes, which decrease on maturation (Askar *et al.*, 1986). In mature and ripe fruit levels of ethyl acetate, ethyl caproate, ethyl caprylate and *cis*-hexenyl acetate are particularly high.

The polyphenol content of guava fruit changes during maturation of the fruit. Young guava fruit contain 620 mg.100g^{-1} fresh weight, 65% of which are condensed tannins, but this level decreases as the fruit matures. This loss occurs particularly in the high molecular weight polyphenols (Itoo *et al.*, 1987). A large percentage of the polyphenols in guava fruit are flavans, especially proanthocyanidin heteropolymers composed of (+)-catechin and (+)-gallocatechin (Itoo *et al.*, 1987). Associated with this loss is a marked increase in polyphenoloxidase (EC 1.10.3.1) activity during ripening (Mowlah and Itoo, 1982).

5.5   *ANNONA* SPECIES

The *Annonas* are native to tropical America, but may grow in most tropical, or frost-free subtropical countries. The *Annonas* are small trees

Table 5.4 Annona species and their common names. (After Samson, 1986.)

Latin name	Common name
Annona squamosa	Sweetsop, Sugar Apple, Custard Apple
Annona reticulata	Bullock's Heart, Custard Apple
Annona cherimola	Cherimoya
Annona diversifolia	Ilama
Annona muricata	Soursop
A. cherimola × *A. squamosa*	Atemoya

which grow between 5 and 8 m tall. They are self-incompatible and therefore require cross pollination (Samson, 1986). There are several species within the genus which bear edible fruit (Table 5.4). The fruit is formed as a result of the fusion of single-seeded fruitlets, which contain a white pulp, with the receptacle.

Cherimoya and Sweetsop are best for eating fresh, Sweetsop yielding 60–70 fruits per tree, or about 20 tonnes.ha^{-1}. The fruit is harvested at a stage when it is still firm but when the skin between segments is beginning to turn yellow. The segments will begin to split apart within one week at this stage, but if the fruit is harvested at a more immature stage, quality is lost (Samson, 1986).

Soursop fruit are irregular in shape, but usually round or heart-shaped. The skin is matt-green in mature fruit and bears short curved spines, while at the immature stage fruit have a shine to the skin surface (Bueso, 1980).

The Soursop was suggested by Biale and Barcus (1970) to have a diffuse type of climacteric, having two apparent maxima. They explained this pattern of respiration as the result of a compound fruit which consists of tissues of variable physiological age. The first peak of respiration, two days after harvest consumed about 70 ml O_2.kg^{-1}.h^{-1}, while the second peak observed between two and four days later, consumed 120 ml O_2.kg^{-1}.h^{-1}. Bruinsma and Paull (1984) found a similar increase in respiration as measured by an increase in CO_2 production two days after harvest, followed by an increase which accompanied autocatalytic ethylene production and other ripening-related phenomena. However, the latter authors concluded, from evidence obtained using fruit discs, that the initial rise in respiration is an inherent characteristic of fruit development and is not a feature of differences between the developmental stages on the constituent ovaries of the fruit.

Ethylene production by Soursop fruit increases 24–48h after the climacteric has been initiated, reaching a peak in production of

290 μl.kg^{-1}.h^{-1} at about the same time that the respiration rate reaches a plateau of 108 μl.kg^{-1}.h^{-1}, six days after harvest (Paull, 1982). Similarly, the onset of rapid ethylene production has been shown to occur following the onset of the respiratory climacteric in *A. cherimola*, *A. squamosa* and *A. atemoya* (Brown *et al.*, 1988).

5.5.2 Composition

Sucrose, fructose and glucose have all been identified as sugar components of Soursop fruit. Sucrose concentration reaches a peak three days after harvest, which then declines to 40% of the peak value, whereas fructose and glucose increase slowly to a peak five days after harvest; the ratio of sucrose, fructose and glucose attained at the edible ripe stage is 4.4:3.0:3.2 (Paull *et al.*, 1983). Such changes may contribute to the increase in the TSS in the five days following harvest from 10° to 16° Brix (Paull, 1982). During the same time period fruit pulp pH declines from 5.8 to 3.6 and titratable acidity increases from 1.0 to 10.0 meq.100g^{-1} of tissue. Associated with these changes is an increase in the content of particular acids, for example an 11-fold increase in ascorbic acid (Paull, 1982), a 7-fold increase in malic acid and a 3-fold increase in citric acid (Paull *et al.*, 1983). Starch breakdown leading to sugar and organic acid production was found to occur before any rise in ethylene production, and was suggested to be an important initiating event in Soursop fruit ripening (Paull *et al.*, 1983). Increases in amylase activity during ripening (Paull *et al.*, 1983) may be involved in this starch breakdown. Changes in cell wall polysaccharides are poorly documented; however, Paull *et al.* (1983) have demonstrated increases in the activities of polygalacturonase and cellulase during ripening.

Polyphenoloxidase, peroxidase (EC 1.11.1.7), catalase (EC 1.11.1.6) and acid phosphatase (EC 3.1.3.2) activities have all been identified in cherimoya fruit (Sanchez de Medina *et al.*, 1987; Martinez-Cayuela *et al.*, 1988). During measurements made up to five weeks after harvest it was found that polyphenoloxidase activity increased with a peak 10–16 days after harvest. Peroxidase rose in both mesocarp and epicarp with advancing ripening, whereas catalase activity increased at the beginning of ripening in the mesocarp, but not in the epicarp.

Volatile production

Volatile production has been examined in several of the Annona species, principally by GCMS and infra-red analysis. Volatile production by Soursop fruit paralleled ethylene production, reaching a peak five days after harvest which coincided with the peaks in total sugars, organic acids and the attainment of an edible ripe stage. After the peak there is a drop over the next few days in the production of the major esters, with an associated increase in two unknown volatiles which may be responsible for the off-odour imparted by over-ripe fruit (Paull *et al.*, 1983). Approximately 80% of the Soursop aroma components have been found to be a chemically closely related series of esters (Macleod and Pieris, 1981), methyl hexanoate (31%) and methyl hex-2-enoate (27%) being the most abundant, and together amounting to 0.7 mg.kg^{-1} fruit. Similarly, in Custard Fruit, 94% of the volatile components identified were esters, the major components being methyl and ethylbutanoate and methyl hexanoate (Bartley, 1987). All of the compounds comprising the volatile flavour fraction of fresh atemoya fruits have been identified as sesquiterpenes, with α- and β-pinene, germacrene D, and bicyclo-germacrene constituting the major components (Wyllie *et al.*, 1987). The major terpene components identified in Custard Apple are α- and β-pinene, E-ocinene and germacrene-D (Bartley, 1987). In this case there appears to be no marked change in the composition of the volatiles during ripening (Wyllie *et al.*, 1987). In cherimoya, 47 volatile acids have been identified, the major components being hexanoic (3mg.kg^{-1}) and octanoic (1 mg.kg^{-1}) acids (Idstein *et al.*, 1985). However, quantitatively, alcohols such as 1-butanol, 3-methyl-1-butanol, 1-hexanol and linalool, and a series of butanoates and 3-methylbutyl esters have been found to comprise the major part of the volatiles (Idstein *et al.*, 1984).

5.6   CARAMBOLA

Averrhoa carambola is grown in Florida (Campbell, 1989b), Malaysia and Taiwan (Campbell, 1989c), Surinam (Lewis and Groeizaam, 1989) and was introduced to the Canary Islands in 1985 (Galán-Saúco *et al.*, 1989). It is possible to grow two crops in one year, in late summer and in winter (Campbell, 1989b).

Several cultivars are grown; Arkin is the predominant cultivar in Florida (Campbell, 1989b), other cultivars include Golden Star, Fwang Tung, Maha, B–10, Newcomb, Thayer (Galán-Saúco *et al.*, 1989).

Carambola is also known as the Star Apple because of its shiny five-sided shape, and typical length of between 5 and 8 cm. Fruit size may however be highly variable, for example for the varieties Arkin and Golden Star the fruit may be between 51–103 mm long, so that length is not a reliable indicator of maturity (Campbell and Koch, 1989). The fruit are a translucent pink to yellow colour, with a thin skin and firm, juicy pulp of tart flavour. Both sweet and acid forms exist (Samson, 1986; Lewis and Groeizaam, 1989; Campbell, 1989a).

Fruit development takes 60–70 days, and results in fruit weighing between 80 and more than 200g (Campbell, 1989c). Choice of cultivar for commercial growing has centred around those varieties which are resistant to physical damage (Ritter, 1989) and provide the best quality fruit (Knight, 1989). The main storage problem is stem-end browning (Campbell, 1989a). General commercial practice is to store this fruit at 10°C or higher in order to avoid chilling damage (Campbell et al., 1989).

As a large percentage of fruit may not meet fresh fruit standards, studies have been carried out to determine if such fruit might be utilized as refrigerated slices. Of the various treatments undertaken, dipping in citric acid and then vacuum packing provides fruit that retain satisfactory colour, texture and flavour for six weeks (Matthews et al., 1989).

Carambola fruits have ethylene, carbon dioxide and ripening patterns which are typical of the non-climacteric fruit (Lam and Wan 1983; Oslund and Davenport, 1983). Fruits of the cultivar B–10 show a rise in respiration rate during the period when the fruits change from green (unripe) to an over-ripe stage, but this is not sudden or marked (Lam and Wan, 1983). In addition, fruit do not display autocatalytic production when treated with ethylene (Lam and Wan, 1987).

5.6.1 Composition

The sweet and sour types of carambola are a reflection of their total acid content, of which oxalic acid is a major component. In Florida about 97% of the trees grown are of the sweet type, mainly the cultivar Arkin, while 3% are of the tart or acid type, predominantly Golden Star (Crane et al., 1989). Examination by HPLC of 15 such Florida-grown cultivars, demonstrated that oxalic acid levels between the two types can vary almost 10-fold, ranging from 0.08g to 0.73 g.100 g^{-1} fruit (Wilson et al., 1982b). Joseph and Mendonca (1989a) have also indicated that oxalic acid levels are higher in sour cultivars, from 0.5–1 g.100 g^{-1} fruit, when compared with those of the sweet type, 0.07–0.17 g.100 g^{-1} fresh weight; more specifically in Arkin the level is 1 mg.g^{-1} fresh weight and in Golden Star, 7 mg.g^{-1} fresh weight (Campbell and Koch, 1989). The pH of sweet and sour types are between 3.8–4.1 and between 2.2–2.6 respectively, while the acid:sugar ratio is 0.01–0.02 for sweet types and 0.1 for sour types (Joseph and Mendonca, 1989b). Although Total Titratable Acidity (TTA) and TTA:TS (Total Sugars) decreases, the pH to TTA ratio increases during fruit ripening. Joseph and Mendonca (1989b) identified two types of sweet carambola; fruits of type 1 have a high sugar content, with an average TS of 14 g.100 g^{-1}, compared with those of type 2 which have a sugar content of 6.25 g.100 g^{-1}. At harvest, sucrose makes up 15–20% of the soluble sugars, the remainder comprising glucose and fructose (Campbell and Koch, 1989).

The carotenoid pattern of star fruit is unusual, the main pigments being phytofluene (17%), ζ-carotene (25%), β-cryptoflavin (34%) and mutatooxanthin (14%). Also present, but in small amounts, are β-carotene, β-apo-8'-carotenal, cryptoxanthin, cryptochrome and lutein. The total carotenoid content of ripe fruit has been found to be 22 μg.g^{-1} fresh weight (Gross *et al.*, 1983).

Esters and lactones are significant components of the aroma, together with several compounds which are derived from carotenoid precursors; for example, the four isomeric megastigma-4,6,8-trienes, and megastigma-5,8(E) and (2)-dien-4-one (MacLeod and Ames, 1990).

5.7 TAMARILLO

The tamarillo is a fast growing, early fruiting tree which is grown in subtropical climates (Morton, 1983). The fruit may be eaten raw or stewed (Samson, 1986), or processed (Patterson, 1980) and may potentially be used in juices and jams (Rotundo *et al.*, 1983). Tamarillo fruits grow rapidly, reaching full size approximately 16 weeks after anthesis (Heatherbell *et al.*, 1982) and maturity at 27 weeks after anthesis. Tamarillos are harvested when purple, such that they continue to ripen after harvest, becoming more soft and juicier (El-Zeftawi *et al.*, 1988).

5.7.1 Composition

Starch content attains a maximum of 14% of fruit fresh weight during development, and falls to less than 1% at maturity (Heatherbell *et al.*, 1982). Total sugars (mainly sucrose, fructose and glucose) reach a peak of 5% of fruit fresh weight approximately 18 weeks post-anthesis.

The content of anthocyanins and organic acids in fruit flesh decrease with maturity, whereas peel anthocyanin content increases. Pectin content is also reduced from 1 to 0.75% of fruit fresh weight at maturity (Heatherbell *et al.*, 1982). Such changes in colour, firmness, juice content and °Brix levels have been identified as indicators of maturity (El-Zeftawi *et al.*, 1988). The carotenoids that have been identified in both the pulp and peel of tamarillos are β-carotene, β-cryptoxanthin, S-carotene, 5,6-monoepoxy-β-carotene, lutein and zeaxanthin (Rodriguez-Amaya *et al.*, 1983). A high vitamin A value (2475 IU.100 g^{-1} edible portion) is due to β-carotene and β-cryptoxanthin, which are present in the highest amounts and possess provitamin A activity.

5.8 PASSION FRUIT

Passiflora is a large genus of perennial climbing plants, many with edible fruits. *Passiflora edulis* var. edulis is probably the best known, i.e. the purple passion fruit. This thrives in the sub-tropics and tropical high-lands (Samson, 1986). *Passiflora edulis* var. flavicarpa is the yellow form which has an aromatic and rather acid juice, and is better adapted to the tropical lowlands (Samson, 1986). Passion fruit attain their maximum size and full maturity at 24 and 88 days post-anthesis respectively (Gachanja and Gurnah, 1981). Potential yield may be reduced by either low temperatures which restrict vegetative growth, or by high temperatures which prevent flower production (Menzel *et al.*, 1987).

The fruit contains between 100–150 seeds. Passion fruit juice contains 1.2% protein, 18% sugars, a lot of vitamin A and substantial amounts of vitamins B_1 and C (Samson, 1986).

There are two crops annually in the sub-tropics, but in the tropics cropping may be continuous. Purple passion fruit are picked for the factory when the purple colour is well defined (Samson, 1986); yellow passion fruit is simply gathered every other day.

Not only are passion fruit eaten as the fresh product, the fruit juice is used as a mixture with other fruit juices, and the waste products of seeds and peel are used as a source of pectin and oil (Lopez, 1980; Prasad, 1980).

5.8.1 Composition

A considerable amount of the research carried out on passion fruit has centred on the nature of their volatile constituents. Casimir *et al.* (1981) have reviewed the volatile flavour compounds of purple and yellow passion fruit, and the effect that the cultivar and stage of maturity have on the flavour of the processed product. An analysis of *P. edulis* f. flavicarpa fruit pulp has identified a series of monoterpene hydro-carbons, alcohols and oxides. Linalool, nerol, geraniol and α-terpeniol were shown to be present in the glycosidic, rather than the free form (Engel and Tressl, 1983). The components of fruit juice of *P. edulis* and the form *flavicarpa* which are responsible for flavour are ionone-related compounds, their quantity and quality, differing markedly between the two types of fruit (Whitfield, 1982). Flavour comparisons between purple and yellow Passion Fruit, and their commercial hybrid, have indicated that the hybrid would be best for commercial juice production (Chen *et al.*, 1982). More recently, further examination of passion fruit juice has revealed several other C_{13} norterpenoid aglycons (Winterhalter, 1990; Herderich and Winterhalter, 1991).

Potentially toxic levels of HCN have been found to be present in passion fruit, especially if the fruit is picked from the vine (Spencer and Siegler, 1983). Levels of HCN in immature yellow fruit have been found to be as high as 59.4 mg HCN.100 g^{-1} fresh weight, decreasing to 14–17 mg.100 g^{-1} at maturity and 6.5 mg.100 g^{-1} after abscission. Contents of HCN in immature and mature purple passion fruit were found to be 10–13.3 mg.100 g^{-1} (Spencer and Siegler, 1983).

5.9   MANGOSTEEN

The mangosteen (*Garcinia mangostana* L.) is grown extensively in the areas to which it is a native, namely Indonesia and Malaysia (Samson, 1986). The fruit is small and contains 4–8 segments of differing widths (Martin, 1980). Unlike the citrus family, the segments have no limiting membrane, are white, and are contained within a thick, bitter-tasting cortex, which attains a brown-to-purple colouration when the fruit is ripe (Martin, 1980). The maximum physical growth of the fruit is reached 103 days from full bloom, when the skin has red patches. During ripening on the tree the skin achieves a purple colour at 114 days after full bloom (Sosrodiharjo, 1980).

Optimal commercial maturity is achieved at 103 days after full bloom (Sosrodiharjo, 1980), however fruits may be picked at the mature green stage (90 days post full bloom) and will ripen normally at ambient temperature (Sosrodiharjo, 1980). Mangosteens are very juicy fruit, and so techniques have been developed for the production of juice, jellies, syrup and canning fruit segments. Processing however appears to lead to a loss of aroma (Martin, 1980).

5.9.1   Composition

Pulp-soluble solids increase with increasing days from full bloom until the skin becomes purple (Sosrodiharjo, 1980), but during storage soluble solids, total acidity and ascorbic acid content decrease with a concomitant rise in reducing sugars (Martin, 1980). During storage, fruits undergo a climacteric peak in respiration approximately 10 days after the beginning of storage (Nagy and Shaw, 1980).

The major components of the aroma volatiles of ripe fruit have been identified as hexyl acetate, *cis*-hex-3-enyl acetate (both of which contribute to the characteristic mangosteen aroma) and *cis*-hex-3-en-l-ol (MacLeod and Pieris, 1982).

Several xanthones have been isolated from the hulls of fruit (Sen *et al.*, 1980, 1981) and xanthone content has been shown to increase in quantity and in range during ripening (Martin, 1980).

5.10   FEIJOA

Acca sellowiana, feijoa or pineapple guava, combines in its fruit the flavours of pineapple and strawberry. It is native to South Brazil and is now cultivated on a fairly large scale in New Zealand for local and export fresh fruit markets (Samson, 1986; Thorp and Klein, 1987). It is also grown in Azerbaijan (Babaev, 1986), Western Georgia (Barbakadze and Gogatadze, 1989), and the USA (Samson, 1986).

The bush is 2–4 metres high, evergreen, with large axillary flowers and fruit that may weigh between 25 and 60 grams each. Feijoa is susceptible to frost damage, but is salt and drought tolerant (Samson, 1986). The fruit contain many small seeds and have an aromatic taste. The sweetness of the fruit may vary considerably with cultivar; Thorp and Klein (1987) described the cultivar Gemini as a tart fruit, Triumph as a medium sweet fruit, and Apollo as a sweet fruit.

There are many well known cultivars, including Triumph and Mammoth in New Zealand, Coolidge in California, Choiceana and Superba in Florida (Samson, 1986). Coolidge is self-compatible, but the other cultivars have to be planted in mixed stands for pollination.

The pattern of fruit growth of the cultivar Mammoth has been shown to follow a double sigmoid curve, with characteristic rapid final growth phase that continues until the point at which the fruit abscinds (Harman, 1987).

The fruit are easily damaged (Thorp, 1988), although some cultivars are more resistant to bruising than others (Thorp and Klein, 1987). The fruits attain optimum harvest maturity just prior to the natural abscission of the crop. In order to restrict damage to the fruit they may be touch picked, that is, the fruit is tilted gently while still on the tree and harvested at a time when it comes away easily from the parent plant. However, various studies have suggested that suspended catching nets should be used (Thorp and Klein, 1987). Not only does this practice reduce bruising, it is also quicker and provides fruit which are fully mature (Thorp, 1988). The cultivars Apollo, Gemini, Marion and Triumph have been shown to have a maximum storage life of four weeks at 4°C and a shelf life of five days at 20°C (Thorp and Klein, 1987). At temperatures less than 0°C fruit frequently demonstrated signs of chilling injury (Thorp and Klein, 1987).

A biochemical study of fruit of the cultivar Mammoth indicated that the respiration rate of young fruits was high at 120 mg $CO_2.kg^{-1}.h^{-1}$, but declined to 20–30 mg $CO_2.kg^{-1}.h^{-1}$ at 80–100 days after anthesis, and then increased again as the fruit matured and ripened (Harman, 1987).

5.10.2 Composition

Carbohydrate and organic acid contents of the fruit increased markedly from about 90 days after anthesis, the main sugars being sucrose, fructose and glucose, and the main acids malic and citric. Sugar and organic acid content of fruit harvested less than 80 days after anthesis was low.

The analysis of volatiles from mature Californian-grown fruit resulted in the identification of 85 different compounds, most of which were hydrocarbons or alcohols (Binder and Flath, 1989). Sesquiterpenes constituted 53% of the volatiles, with germacrene D, bicyclogermacrene and β-caryophyllene predominating. Other major constituents were methyl benzoate, (2)-3-hexenyl benzoate, linalool, humalene and 3-octanone (Binder and Flath, 1989). A similar study has been carried out on the cultivar Mammoth in which methyl benzoate constituted 82% of the extract (Shaw *et al.*, 1990); Shaw *et al* (1983) also suggested that ethyl benzoate may be important in determining the optimum ripeness of Feijoa.

5.11   RAMBUTAN

The rambutan (*Nephelium lappaceum* L.) requires a humid equatorial climate for growth (Samson, 1986). The tree produces bunches of fruit which change their colour from green to red or orange on ripening. The word 'rambut' means hair in Indonesia, and the fruit was so-named because of the soft, fleshy spines which surround the fruit, providing it with a large gooseberry-like appearance. The fruit is oblong to round in shape and may measure between 5–6 cm in length (Samson, 1986). The rind surrounds a translucent and fleshy aril, which contains a single seed. A two-part review details the botanical and taxanomic character-istics of this fruit, its growth and development, and agronomic and economic aspects (Delabarre, 1989a,b).

After harvesting, rambutans undergo a rapid darkening of the skin which is associated with dehydration. This occurs in 72 hours or less when stored at ambient temperature (Mohamed and Othman, 1988). Attempts have been made to avoid this deterioration by the use of reduced temperature storage and wrapping in low permeability pack-aging. When these measures are combined with the introduction of CO_2 of up to 7%, respiration is suppressed and the shelf life may be extended to one month (Mohamed and Othman, 1988). The use of dipping and SO_2 treatments have also been studied, but have been found to be less effective than cold-temperature storage (Mohamed *et al.*, 1988)

5.11.1   Composition

The change in composition of rambutan fruit has been studied during storage at different temperatures (Paull and Chen, 1987b). Sucrose content increased in the aril from 51 to 76 mg.g^{-1} when the fruit was stored at 12°C; however, there was no change in sucrose content following storage at 22°C. Titratable acidity was found to initially decrease and then increase under all storage conditions. Succinic acid was found only at harvest, declining to undetectable levels during storage, whereas high concentrations of citric acid (approximately

Table 5.5 The nutritive value of tropical fruits* (After Parkinson et al., 1989.)

Fruit	Calories	Protein (g)	Fat (g)	Carbohydrate (g)	Calcium (mg)	Iron (mg)	Vit A (mg)	Vit C (mg)	Thiamin (mg)
Carambola	28	0.3	0.4	7	8	1	160	38	0.05
Custard Apple	76	1.5	0.3	19	27	0.5	20	21	0.11
Durian	124	2.5	1.6	28.3	20	0.9	trace	37	0.27
Guava	58	1	0.4	13	15	1	60	200	0.05
Jack Fruit	94	1.7	0.3	23.7	27	0.6	235	9	0.09
Lychee	65	0.8	0.4	16.3	10	0.3	0	50	0.05
Mangosteen	57	0.5	0.3	14.7	10	0.5	0	4	0.03
Passion Fruit	92	2.3	2	16	10	1	6	20	NA
Rambutan	64	1.0	0.1	16.5	20	1.9	0	23	trace
Sapodilla	76	0.4	0.7	19.1	27	0.6	25	13	trace
Soursop	93	1	NA	22	25	0.5	NA	30	0.10

*All values expressed per 100 g prepared raw food.
NA = not applicable

175 meq.g^{-1}) were maintained throughout storage. Ascorbic acid levels decreased during storage, whereas total phenols showed little change. Paull and Chen (1987b), concluded that there were only minor compositional changes in rambutan when the fruit was stored at 12°C for as long as 20 days.

SUMMARY

This chapter covers only a small selection of the more exotic fruits becoming available to the consumer. Other fruit such as Sapodilla or Jackfruit have not been dealt with here, but this in no way reflects on their relative importance.

Exotic fruits are now gaining an increased share of the market, especially as public demand for more variety in the diet increases. These trends are illustrated by the increases in consumption of avocado and kiwifruit. However, unlike avocado and kiwifruit, our knowledge of the physiology and biochemistry of many of these 'exotic' fruits is very limited. What little information is available is concerned primarily with the composition of the fruits (Table 5.5), and it is apparent that many of these exotic fruits are a rich source of vitamins and minerals. However, a greater understanding of the biochemical processes associated with the ripening of these fruits should lead to improved methods of their storage and handling, thereby not only reducing their cost to the consumer, but also increasing their availability.

REFERENCES

Ahlawat, V.P., Yamdagni, R. and Jindal, P.C. (1980) Studies on the effect of post-harvest treatments on storage behaviour of guava (*Psidium guanjava* L.) cv. Sardar (L49). *Haryana Agricultural University Journal of Research* **10**, 242–247

Ajay Singh and Abidi, A.B (1986) Level of carbohydrate fractions and ascorbic acid during ripening and storage of litchi (*Litchi chinensis* Sonn.) cultivars. *Indian Journal of Agricultural Chemistry*, **19**, 197–202

Akamine, E.K. and Goo, T. (1981) Carbon dioxide and ethylene production in *Diospyros discolor* Willd. *Hort Science*, **16**, 519

Askar, A., El-Nemr, S.E. and Bassionny, S.S. (1986) Aroma constituents in white and pink guava fruits. *Alimenta*, **25**, 162–167

Babaev, M.M. (1986) Effectiveness of *Feijoa* irrigation and mulching in the Lenkoran zone of Azerbaijan. *Subtropicheskie Kul'tury*, No. 5, 123–127

Bain, J.M., Chaplin, G.R., Scott, K.J., Brown B.I. and Willcox, M.E. (1983) Control of wastage in litchi fruit during marketing. *Australian Horticultural Research Newsletter*, No. 55

Barbakadze, T.P. and Gogatadze, N.K. (1989) Feijoa cultivation in Western Georgia. *Subtropicheskie Kul'tury*, No. 6, 26–31

Bartley, J.P. (1987) Volatile constituents of custard apple. *Chromatographia* **23**, 129–131

Batten, D.J. (1989) Maturity criteria for litchis (lychees). *Food Quality and Preference*, **1**, 149–155

Ben-Arie, R., Bazak, H. and Blumenfeld, A. (1986) Gibberellin delays harvest and prolongs storage life of persimmon fruits. *Acta Horticultura*, **179**, 807–813

Bhullar, J.S., Dhillon, B.S. and Randhawa, J.S. (1983) Extending the post-harvest life of litchi cultivar Seedless Late. *Journal of Research, Punjab Agricultural University*, **20**, 467–470

Biale, J.B. and Barcus, D.E. (1970) Respiratory patterns in tropical fruits of the Amazon basin. *Tropical Science*, **12**, 93–104

Binder, R.G. and Flath, R.A. (1989) Volatile components of pineapple guava (feijoa). *Journal of Agricultural Food Chemistry*, **37**, 734–736

Brown, B.I. and Wills, R.B.H. (1983) Post-harvest changes in guava fruit of different maturity. *Scientia Horticulturae*, **19**, 237–243

Brown, B.I., Wong, L.S., George, A.P. and Nissen, R.J. (1988) Comparative studies on the postharvest physiology of fruit from different species of *Annona* (custard apple). *Journal of Horticultural Science*, **63**, 521–528

Bruinsma, J. and Paull, R.E. (1984) Respiration during postharvest development of soursop fruit, *Annona muricata* L. *Plant Physiology*, **76**, 131–138

Bueso, C.E. (1980) Soursop, Tamarind and Chironja. In *Tropical and Subtropical Fruits, Composition, Properties and Uses*, (eds S. Nagy, and P.E. Shaw), Chapter 10, AVI Publishing Co., Inc., Westport, CT, pp. 375–406

Calvo, M.L., Plaza, J.L. de la., Alique, R. and Rodrigo, M.E. (1983) Influence of pretreatment with CO_2 on cherimoya storage in normal and controlled atmospheres. Almeria, Spain; Sociedad Española de Ciencias Horticolas Primer Congreso Nacional II, 1005–1009

Campbell, C.A. (1989a) Storage and handling of Florida carambola. *Proceedings of the Interamerican Society for Tropical Horticulture*, **33**, 79–82

Campbell, C.W. (1989b) Carambola production in the United States. *Proceedings of the Interamerican Society for Tropical Horticulture*, **33**, 47–54

Campbell, C.W. (1989c) Propagation and production systems for carambola. *Proceedings of the Interamerican Society for Tropical Horticulture*, **33**, 66–71

Campbell, C.A., Huber, D.J. and Koch, K.E. (1989) Postharvest changes in sugars, acids and colour of carambola fruit at various temperatures. *Hort Science*, **24**, 472–475

Campbell, C.A. and Koch, K.E. (1989) Sugar/acid composition and development of sweet and tart carambola fruit. *Journal of the American Society of Horticultural Science*, **114**, 455–457

Casimir, D.J., Kefford, J.F., and Whitfield, F.B. (1981) Technology and flavour chemistry of passion fruit juices and concentrates. *Advances in Food Research*, **27**, 243–295

Cavaletto, C. (1980) Lychee. In *Tropical and Subtropical Fruits, Composition, Properties and Uses*. (eds S. Nagy and P.E. Shaw) Chapter 14, AVI Publishing, Inc., Westport, CT, pp. 469–478

Chauhan, R., Kapoor, A.C. and Gupta, O.P. (1986) Note on the effect of cultivar and season on the chemical composition of guava fruits. *Haryana Journal of Horticultural Sciences*, **15**, 228–230

Chen, C.-C., Kuo, M.-C., Hwang, L.S. Wu, J.S.-B. and Wu, C.-M. (1982) Headspace components of passion fruit juice. *Journal of Agricultural and Food Chemistry* **30**, 1211–1215

Chen, F., Li, Y. and Chen, M. (1986) Production of ethylene by litchi fruits during storage and its control. *Acta Horticulturae Sinica*, **13**, 151–156

Chyau, C.C. and Wu, C.M. (1989) Differences in volatile constituents between inner and outer flesh-peel of guava (*Psidium guajava*, Linn.) fruit. *Lebensmittel-Wissenschaft und Technologie*, **22**, 104–106

Collazos, E.O., Bautista, G.A., Millán, M.B. and Mapura, M.B. (1984) Effect of polyethylene bags on passion fruit (*Passiflora edulis* var. *flavicarpa* Degener), tasco (*P. mollissima* HBK Bailey) and tomato (*Lycopersicon esculentum* Miller) storage. *Acta Agronómica*, **34**, 53–59

Crane, J.H., Campbell, C.W. and Olszack, R. (1989) Current statistics for commercial carambola groves in South Florida. *Proceedings of the Interamerican Society for Tropical Horticulture*, **33**, 94–99

Dauriach, J. (1986) Persimmon, a question of astringency. *Arboriculture Frutiere*, **33**, 27–29

Delabarre Y. (1989a) Bibliographic summary on rambutan or hairy litchi (*Nephelium lappaceum* L.) *Fruits (Paris)*, **44**, 33–44

Delabarre, Y. (1989b) Bibliographic summary on rambutan or hairy litchi (*Nephelium lappaceum* L.) *Fruits (Paris)*, **44**, 91–98

Dhillon, B.S., Singh, S.N., Kundal, G.S. and Minhas, P.P.S. (1987) Studies on the developmental physiology of guava fruit (*Psidium guajava* L.) II. Biochemical characters. *Punjab Horticultural Journal*, **27**, 212–221

Dhingra, M.K., Gupta, O.P. and Chundawat, B.S. (1983) Studies on pectin yield and quality of some guava cultivars in relation to cropping season and fruit maturity. *Journal of Food Science and Technology, India*, **20**, 10–13

Dhoot, L.R., Desai, U.T. and Rane, D.A. (1984) Studies on the shelf-life of guava fruits with polythene packaging and chemical treatments. *Journal of Maharashtra Agricultural Universities*, **9**, 185–188

El-Zeftawi, B.M., Brohier, L., Dooley, L., Goubran, F.H., Holmes R. and Scott, B. (1988) Some maturity indices for tamarillo and pepino fruits. *Journal of Horticultural Science* **63**, 163–169

Engel, K.-H. and Tressl, R. (1983) Formation of aroma components from non-volatile precursors in passion fruit. *Journal of Agricultural Food Chemistry*, 998–1002

Esteves, M.T. da C., Carvalho, V.D. de., Chitarra, M.I.F., Chitarra, A.B. and Paula, M.B. de. (1984a) Characteristics of fruits of six guava (*Psidium guajava* L.) cultivars during ripening. II. Vitamin C and tannin contents. In *Anais do VII Congresso Brasileiro de Fruiticultura. Florianopolis, Brazil; Empresa Catarinense de Pesiquisa Agropecuária S.A.*, Vol. 2, 490–500

Esteves, M.T.da C., Chitarra, M.I.F., Carvalho, V.D. de., Chitarra, A.B. and Paula, M.B. de. (1984b) Characteristics of fruits of six guava (*Psidium guajava* L.) cultivars during ripening. III. Pectin, cellulose and hemicellulose. In *Anais do*

VII Congresso Brasileiro de Fruiticultura. Florianópolis, Brazil; Empresa Catarinense de Pesiquisa Agropecuária S.A., Vol. 2, 510–513

Fishman, G.M. (1986) Pectins in some processable species of subtropical crops. *Subtropicheskie Kul'tury*, No. 4, 127–131

Fishman, G.M. and Oganesyan, K.V. (1984) Studies on the fruit enzyme system of the subtropical persimmon cv. Kachia. *Subtropicheskie Kul'tury*, No. 3, 119–122

Gachanja, S.P. and Gurnah, A.M. (1981) Flowering and fruiting of purple passion fruit at Thika. *East African Agricultural and Forestry Journal*, **44**, 47–51

Galán-Saúco, V., Hernandes-Delgado, P.M. and Fernández-Galván, D. (1989) Preliminary observations on carambola in the Canary Islands. *Proceedings of the Interamerican Society for Tropical Horticulture*, **33**, 55–58

Ghosh, S.K., Dhua, R.S., Sen, S.K. and Maiti, S.C. (1989) Fruit growth and development of two important litchi cvs. Bombai and Elachi. *Horticultural Journal*, **2**, 26–32

Gross, J., Bazak, H., Blumenfeld, A. and Ben-Arie, R. (1984) Changes in chlorophyll and carotenoid pigments in the peel of "Triumph" persimmon (*Diospyros kaki* L.) induced by pre-harvest gibberellin (GA$_3$) treatment. *Scientia Horticulturae*, **24**, 305–314

Gross, J., Ikan, R. and Eckhardt, G. (1983) Carotenoids of the fruit of *Averrhoa carambola*. *Phytochemistry* **22**, 1479–1481

Gupta, V.K. and Mukherjee, D. (1980) Effect of morphactin on the storage behaviour of guava fruits. *Journal of the American Society of Horticultural Science*, **105**, 115–119

Harman, J.E. (1987) Feijoa fruit: growth and chemical composition during development. *New Zealand Journal of Experimental Agriculture*, **15**, 209–215

Heatherbell, D.A., Reid, M.S. and Wrolstad, R.E. (1982) The tamarillo; chemical composition during growth and maturation. *New Zealand Journal of Science*, **25**, 239–243

Herderich, M. and Winterhalter, P. (1991), 3-Hydroxy-*retro*-α-ionol: a natural edulans in purple passionfruit (*Passifloa edulis* Sims.). *Journal of Agricultural Food Chemistry*, **39**, 1270–1274

Hirai, S. and Yamazaki, K. (1984) Studies on sugar components of sweet and astringent persimmon by gas chromatography. *Journal of Japanese Society of Food Science Technology*, **31**, 24–30

Homnava, A., Payne, J., Koehler, P. and Eitenmiller R. (1990) Provitamin A (α-carotene, β-carotene and β-cryptoxanthin) and ascorbic acid content of Japanese and American persimmons. *Journal of Food Quality*, **13**, 85–95

Horvat, R.J., Senter, S.D., Chapman, G.W.Jr. and Payne, J.A. (1991) Volatile compounds from mesocarp of persimmons. *Journal of Food Science*, **56**, 262–263

Huang, H. and Xu, J. (1983) The developmental patterns of fruit tissues and their correlative relationships in *Litchi chinnensis* Sonn. *Scientia Horticulturae*, **19**, 335–342

Huang, S., Hart, H., Lee, H. and Wicker, L. (1990) Enzymic and colour changes during post-harvest storage of lychee fruit. *Journal of Food Science*, **55**, 1762–1763

Idstein, H., Bauer, C. and Schreirer, P. (1985) Volatile acids from tropical fruits: Cherimoya (*Annona cherimola* Mill.), guava (*Psidium guajava* L.), mango (*Mangifera indica* L. var. Alphonso), pawpaw (*Carica papaya* L.). *Zeitschrift für Lebensmittel-Untersuchung und-Forschung*, **180**, 394–397

Idstein, H., Herres, W. and Schreier, P. (1984) High-resolution gas chromatography–mass spectrometry and Fourier transform infrared analysis of cherimoya (*Annona cherimola* Mill.) volatiles. *Journal of Agricultural Food Chemistry*, **32**, 383–389

Itamura, H., Fukushima, T. and Kitamura, T. (1989) Changes in cell wall polysaccharide composition during fruit softening of Japanese persimmon (*Diospyros kaki* Thunb. var. Hiratanenashi). *Journal of Japanese Society of Food Science and Technology* **36**, 647–650

Itoo, S. (1980) Persimmon. In *Tropical and Subtropical Fruits. Composition, Properties and Uses*. (eds S. Nagy, and P.E. Shaw), Chapter 13 AVI Publishing Company, Inc., Westport, CT, pp. 442–468

Itoo, S., Matsuo, T., Ibushi, Y. and Tamari, N. (1987) Seasonal changes in the levels of polyphenols in guava fruit and leaves and some of their properties. *Journal of Japanese Society of Horticultural Science*, **56**, 107–113

Itoo, S., Yamaguchi, T., Oohata, J.T. and Ishihata, K. (1980) Studies on qualities of subtropical fruits. II. Ascorbic acid and stone cells of guava fruits (*Psidium guajava* L.). *Bulletin of the Faculty of Agriculture, Kagoshima University*, No. 30, 47–54

Jain, B.P., Das, S.R. and Verma, S.K. (1985) Effect of growth substances and minor elements on the synthesis of major chemical constituents of litchi (*Litchi chinensis* Sonn.). *Haryana Journal of Horticultural Science* **14**, 1–3

Jaiswal, B.P., Jha, A.K., Sah, N.L. and Prasad, U.S. (1982) Characteristics of fruit growth and development in litchi cultivars. *Indian Journal of Plant Physiology*, **25**, 411–414

Jaiswal, B.P., Sah, N.L. and Prasad, U.S. (1986) Studies on phenolics of rind and aril during ripening and senescence of litchi fruits. *Plant Physiology and Biochemistry* **13**, 40–45

Jaiswal, B.P., Sah, N.L. and Prasad, U.S. (1987), Regulation of colour break during litchi (*Litchi chinensis* Sonn.) ripening. *Indian Journal of Experimental Biology*, **25**, 66–72

Joseph, J. and Mendonca, G. (1989a) Oxalic acid content of carambola (*Averrhoa carambola*, L.) and bilimbi (*Averrhoa bilimbi* L.). *Proceedings of the Interamerican Society for Tropical Horticulture*, **33**, 117–120

Joseph, J. and Mendonca, G. (1989b) Chemical characteristics of *Averrhoa carambola* L. *Proceedings of the Interamerican Society for Tropical Horticulture*, **33**, 111–116

Kanwar, J.S. and Nijar, G.S. (1984) Comparative evaluation of fruit-growth in relation to cracking of fruits in some litchi cultivars. *Punjab Horticultural Journal*, **24**, 79–82

Kartar Singh and Chauhan, K.S. (1984) Effect of Waxol and KMnO$_4$ with pre-cooling on storage life of cv. L-49 of guava. *Haryana Journal of Horticultural Sciences*, **11**, 192–198

Kato, K. (1984a) Conditions for tanning and sugar extraction, the relationship of tannin concentration to astringency and the behaviour of ethanol during the removal of astringency by ethanol in persimmon fruits. *Journal of Japanese Society of Horticultural Science*, **53**, 127–134

Kato, K. (1984b) Astringency removal and ripening as related to ethanol concentration in persimmon fruits. *Journal of Japanese Society of Horticultural Science*, **53**, 278–289

Kawashima, K., Kamihisa, Y. and Katabe, K. (1984) The control of pericarp blackening during ethanol treatment for the removal of astringency in Japanese persimmon. *Journal of Japanese Society of Horticultural Science*, **53**, 290–297

Kenney, P and Hull, L. (1987) Effects of storage condition on carambola quality. *Proceedings of Florida State Horticultural Society*, **99**, 222–224

Knight, R. Jr.(1980) Origin and World Importance of Tropical and Subtropical Fruit Crops. In *Tropical and Subtropical Fruits. Composition, Properties and Uses.* (eds S. Nagy, and P.E. Shaw), Chapter 1, AVI Publishing, Inc., Westport, CT, pp. 1–120

Knight, R.J. Jr. (1989) Carambola cultivars and improvement programs. *Proceedings of the Interamerican Society for Tropical Horticulture*, **33**, 72–78

Kolesnik, A.A., Golubev, V.N., Kostinskaya, L.I. and Khalilov, M.A. (1987) Lipids in *Diospyros kaki* fruits. *Khimiya Prirodnykh Soedinii*, no. 4, 501–505

Lam, P.F. and Wan, C.K. (1983) Climacteric nature of the carambola (*Averrhoa carambola* L.) fruit. *Pertanika*, **6**, 44–47

Lam, P.F. and Wan, C.K. (1987) Ethylene and carbon dioxide production of starfruits (*Averroha carambola* L.) stored at various temperatures and in different gas and relative humidity atmospheres.*Tropical Agriculture* **64**, 181–184

Lee, H.S. and Wicker, L. (1991a) Anthocyanin pigments in the skin of lychee fruit. *Journal of Food Science*, **56**, 466–468

Lee, H.S. and Wicker, L. (1991b) Quantitative changes in anthocyanin pigments of lychee fruit during refrigerated storage. *Food Chemistry* **40**, 263–270

Lewis, D. and Groeizaam, M. (1989) The cultivation and utilization of carambola in Surinam. *Proceedings of the Interamerican Society for Tropical Horticulture*, **33**, 59

Lopez, A.S. (1980) Lipids from the seeds of passion fruit. *Revista Theobroma*, **10**, 47–50

MacLeod, G. and Ames, J.M. (1990) Volatile components of starfruit. *Phytochemistry* **29**, 165–172

MacLeod, A.J. and Pieris, N.M. (1981) Volatile flavour components of soursop (*Annona muricata* L.). *Journal of Agricultural Food Chemistry*, **29**, 488–490

MacLeod, A.J. and Pieris, N. (1982) Volatile flavour components of mangosteen, *Garcinia mangostana*. *Phytochemistry*, **21**, 117–119

Manabe, T. (1982) Changes of insoluble nitrogen compounds during the process of removing astringency in the Japanese persimmon cultivar Saijyo. *Journal of Japanese Society of Food Science and Technology*, **29**, 677–679

Martin, F.W. (1980) Durian and Mangosteen. In *Tropical and Subtropical Fruits. Composition, Properties and Uses.* (eds. S. Nagy, and P.E. Shaw), Chapter 11, AVI Publishing, Inc., Westport, CT, pp. 407–414

Martinez-Cayuela, M., Sanchez de Medina, L., Faus, M.J. and Gil, A. (1988) Cherimoya (*Annona cherimola* Mill.) polyphenoloxidase : monophenolase and dihydroxyphenolase activities. *Journal of Food Science* **53**, 1191–1194

Matsuo, T. and Itoo, S. (1978) The chemical structure of *Kaki*-tannin from immature fruit of the persimmon. *Agricultural and Biological Chemistry*, **42**, 1637–1643

Matthews, R.F., Lindsey, J.A., West, P.F. and Leinart, A. (1989) Refrigerated vacuum packaging of carambola slices. *Proceedings of Florida State Horticultural Society*, **102**, 166–169

Menzel, C.M. and Simpson, D.R. (1991) A description of lychee cultivars. *Fruit Varieties Journal*, **45**, 45–46

Menzel, C.M., Simpson, D.R. and Winks, C.W. (1987) Effect of temperature on growth, flowering and nutrient uptake of three passion fruit cultivars under low irradiance. *Scientia Horticulturae*, **31**, 259–268

Misra, R.S. and Khan, I. (1981) Effect of 2, 4, 5-trichlorophenoxyacetic acid and micronutrients on fruit-size, cracking, maturity and quality of litchi cv. Rose Scented. *Progress in Horticulture*, **13**, 87–90

Mohamed, S. and Othman, E. (1988) Effect of packaging and modified atmosphere on the shelf life of rambutan (*Nephelium lappaceum*). *Pertanika*, **11**, 217–228

Mohamed, S., Othman, E. and Abdullah, F. (1988) Effect of chemical treatments on the shelf life of rambutans (*Nephelium lappaceum*). II. *Pertanika*, **11**, 407–417

Morton, J.F. (1983) The tree tomato, or 'tamarillo', a fast-growing early-fruiting small tree for subtropical climates. *Proceedings of Florida State Horticultural Society*, **95**, 81–85

Mowlah, G. and Itoo, S. (1982) Guava (*Psidium guajava* L.) sugar components and related enzymes at stages of fruit development and ripening. *Journal of Japanese Society of Food Science and Technology*, **29**, 472–476

Mowlah, G. and Itoo, S. (1983) Changes in pectic components, ascorbic acid, pectic enzymes and cellulase activity in ripening and stored guava (*Psidium guajava* L.). *Journal of Japanese Society of Food Science and Technology*, **30**, 454–461

Nagy, S. and Shaw, P.E. (1980) *Tropical and Subtropical fruits*. AVI Publishing Inc., Westport, C.T

Nakabayashi, T. (1971) Studies on tannin of fruits and vegetables. Part 7. Difference of the components of tannin between astringent and non-astringent persimmon fruits. *Journal of Japanese Society of Food Science and Technology*, **18**, 33–37

Nip, W.K. (1988) Handling and preservation of lychee (*Litchi chinensis*, Sonn.) with emphasis on colour retention. *Tropical Sciences*, **28**, 5–11

Oslund, C.R. and Davenport, T.L. (1983) Ethylene and carbon dioxide in ripening fruit of *Averrhoa carambola*. *Hort Science*, **18**, 229–230

Pal, D.K. and Selvaraj, Y. (1979) Changes in pectin and pectinesterase in developing guava fruits. *Journal of Food Science and Technology*, **16**, 115–116

Parkinson, S., Stacy, P. and Mattinson, A. (1989) In *Taste of the Tropics*, David Bateman Ltd., Australia and New Zealand, pp. 50

Patterson, K.J. (1980) A new cultivar. Yellow tamarillos for canning. *Fruit and Produce*, **24**, 26

Paull, R.E. (1982) Postharvest variation in composition of soursop (*Annona muricata* L.) fruit in relation to respiration and ethylene production. *Journal of the American Society of Horticultural Science*, **107**, 582–585

Paull, R.E. and Chen, N.J. (1987a) Effect of storage temperature and wrapping on quality characteristics of litchi fruit. *Scientia Horticulturae*, **33**, 223–226

Paull, R.E. and Chen, N.J. (1987b) Changes in longan and rambutan during postharvest storage. *Hort Science*, **22**, 1303–1304

Paull, R.E., Chen, N.J., Deputy, J., Huibai Huang, Guiwen Cheng and Feifei Gao (1984) Litchi growth and compositional changes during fruit development. *Journal of the American Society of Horticultural Science*, **109**, 817–821

Paull, R.E., Deputy, J. and Chen, N.J. (1983) Changes in organic acids, sugars, and headspace volatiles during fruit ripening of soursop (*Annona muricata* L.). *Journal of the American Society of Horticultural Science*, **108**, 931–934

Pesis, E., Levi, A. and Ben-Arie, R. (1986) Deastringency of persimmon fruit by creating a modified atmosphere in polyethylene bags. *Journal of Food Science*, **51**, 1014–1016

Pino, J., Gutierrez, S. and Rosada, A. (1990) Volatile constituents from a guava (*Psidium guajava* L.) natural flavour concentrate. *Nahrung*, **34**, 279–282

Prasad, J. (1980) Pectin and oil from passion fruit waste. *Fiji Agricultural Journal* **42**, 45–48

Prasad, U.S. and Jha, O.P. (1979) Variations in the levels of oxygen uptake, fructose and pyruvate in fruit aril of *Litchi chinensis* Sonn. during maturation and ripening. *Indian Journal of Plant Physiology*, **22**, 163–165

Ritter, S. (1989) Dooryard carambola survey. *Proceedings of the Interamerican Society for Tropical Horticulture*, **33**, 100–107

Rodriguez-Amaya, D.B., Bobbio, P.A. and Bobbio, F.O. (1983) Carotenoid composition and vitamin A value of the Brasilian fruit *Cyphomandra betacea*. *Food Chemistry*, **12**, 61–65

Rotundo, A., Rotundo, S. and Gherardi, S. (1983) The tamarillo. *Rivista di Frutticoltura e di Ortofloricoltura*, **45**, 37–40

Salmah, Y. and Mohamed, S. (1987) Physico-chemical changes in guava (*Psidium guajava* L.) during development and maturation. *Journal of the Science of Food and Agriculture*, **38**, 31–39

Samson, J.A. (1986) *Tropical Fruits*, Second Edition. Longman Scientific and Technical, Longman Group UK. Ltd

Sanchez de Medina, L., Plata, M.C., Martinez-Cayuela, M., Faus, M.J. and Gil, A. (1987) Changes in polyphenoloxidase, peroxidase, catalase and acid phosphatase activities for cherimoya fruit (*Annonaa cherimola*) during ripening in controlled temperature and relative humidity. *Revista de Agroquimica y Tecnologia de Alimentos*, **26**, 529–538

Sen, A.K., Sartar, K.K., Majumder, P.C. and Banerji, N. (1981) Minor xanthones of *Garcinia mangostana* (fruit hulls). *Phytochemistry*, **20**, 183–185

Sen, A.K., Sarkar, K.K., Mazumder, P.C., Banerji, N., Uusvuori, R. and Hase, T.A. (1980) A xanthone from *Garcinia mangostana*. *Phytochemistry*, **19**, 2223–2225

Senter, S.D., Chapman, G.W., Forbus, W.R. Jr. and Payne J.A. (1991) Sugar and non-volatile acid composition of persimmons during maturation. *Journal of Food Science*, **56**, 989–991

Sharaf, A. and El Saadany, S.S. (1986) Biochemical studies on guava fruits during different maturity stages. *Annals of Agricultural Science, Moshtohor*, **24**, 975–984

Sharma, S.B. and Dhillon, B.S. (1984) Effect of zinc sulphate and growth regulators on the growth of litchi fruit. *Progressive Horticulture*, **16**, 19–22

Sharma, S.B. and Dhillon, B.S. (1986) Endogenous levels of gibberellins in relation to fruit cracking in litchi (*Litchi chinensis* Sonn.). *Journal of Research, Punjab Agricultural University*, **23**, 432–434

Sharma, S.B. and Dhillon, B.S. (1987) Effect of zinc sulphate and growth regulators on fruit cracking in litchi. *Indian Journal of Horticulture*, **44**, 57–59

Shaw, G.J., Allen, J.M., Yates, M.K. and Ranich, R.A. (1990) Volatile flavour constituents of feijoa (*Feijoa sellowiana*)-analysis of fruit flesh. *Journal of the Science of Food and Agriculture*, **50**, 357–361

Shaw, G.J. Ellingham, P.J. and Birch, E.J. (1983) Volatile constituents of feijoa – headspace analysis of intact fruit. *Journal of the Science of Food and Agriculture*, **34**, 743–747

Shrestha, G.K. (1981) Effects of ethephon on fruit cracking of lychee (*Litchi chinensis* Sonn.). *Hort Science* **16**, 498

Singh, B.P., Singh, H.K. and Chauhan, K.S. (1981) Effect of post-harvest calcium treatments on the storage life of guava fruits. *Indian Journal of Agricultural Sciences*, **51**, 44–47

Singh, I.S., Singh, H.K. and Gupta, A.K. (1979) Effect of post-harvest application of ethephon on quality of guava (*Psidium guajava* L.) cv. Lucknow-49. *Haryana Journal of Horticultural Sciences* **8**, 12–16

Singh, H.P. and Yadav, I.S. (1988) Physico-chemical changes during fruit development in litchi cultivars. *Indian Journal of Horticulture*, **45**, 212–218

Sive, A. and Resnizky, D. (1987) Experiments on the storage of persimmons cv. Triumph under CA conditions: 1985-6 season. *Alon Hanotea*, **41**, 633–642

Sosrodiharjo, S. (1980) Picking maturity of mangosteen. *Buletin Perelitian Horticultura*, **8**, 11–17

Spencer, K.C. and Siegler, D.S. (1983) Cyanogenesis in *Passiflora edulis*. *Journal of Agriculture and Food Chemistry*, **31**, 794–796

Srivasta, H.C., Singh, K.K. and Mathur, P.B. (1962) Refrigerated storage of mangosteen (*Garcinia mangostana*). *Food Science (Mysore)*, **11**, 226–228

Sugiura, A., Kataoka, I. and Tomana T. (1983) Use of refractometer to determine soluble solids of astringent fruits of Japanese persimmon (*Diospyros kaki* L.). *Journal of Horticultural Science*, **58**, 241–246

Sugiura, A., Yonemori, K., Harada, H. and Tomana, T. (1979) Changes in ethanol and acetaldehyde contents of Japanese persimmon fruits in relation to natural loss of astringency. *Studies from the Institute of Horticulture, Kyoto University*, **9**, 41–47

Suzuki, K., Itoh, S. and Tsuyuki, H. (1982) Studies on lipids in persimmon. Part III. Total neutral lipids in persimmon fruits. *Journal of Japanese Society of Food Science and Technology*, **29**, 484–489

Taira, S., Itamura, H., Abe, K., Ooi, K. and Watanabe, S. (1990) Effect of harvest maturity on removal of astringency in Japanese persimmon (*Diospyros kaki* Thunb.), Hiratanenashi fruits. *Journal of Japanese Society of Horticultural Science*, **58**, 813–818

Taira, S., Sugiura, A. and Tomana, T. (1986) Seasonal changes of internal gas composition in fruits of Japanese persimmon (*Diospyros kaki*, Thunb.) and effects of different gas environments on production of ethanol by their seeds. *Journal of Japanese Society of Horticultural Science*, **55**, 228–234

Takata, M. (1982) Effects of ethylene on respiration, ethylene production and ripening of Japanese persimmon fruits harvested at different stages of development. *Journal of Japanese Society of Horticultural Science*, **51**, 203–209

Takata, M. (1983) Respiration, ethylene production and ripening of Japanese persimmon fruit harvested at various stages of development. *Journal of Japanese Society of Horticultural Science*, **52**, 78–84

Tandon, D.K., Adulse, P.G.and Kalra, S.K. (1984) Effect of certain post-harvest treatments on the shelf life of guava fruits. *Indian Journal of Horticulture*, **41**, 88–92

Tandon, D.K., Singh, B.K. and Kalra, S.K. (1989) Storage behaviour of specific-gravity-graded guava fruit. *Scientia Horticulturae*, **41**, 35–41

Thorp, G. (1988) Post-harvest handling of export feijoas and the use of "catching nets". *Orchardist of New Zealand*, **61**, 40

Thorp, T.G. and Klein, J.D. (1987) Export feijoas: post-harvest handling and storage techniques to maintain optimum fruit quality. *Orchardist of New Zealand*, **60**, 164–166

Tovkach, S.P. (1979) Changes in the biochemical indices of persimmon fruit during ripening and storage. In *Subtrop. Kul'tury Uzbekistana. Tashkent, Uzbek SSR.*, No. 4, 52–54

Vazquez-Ochoa, R.I. and Colinas-Leon, M.T. (1990) Changes in guavas at three maturity stages in response to temperature and relative humidity. *Horticultural Science*, **25**, 86–87

Whitfield, F.B. (1982) The chemistry of food acceptance. *CSIRO Food Research Quarterly*, **42**, 52–57

Wills, R.B., Mulholland, E.E. and Brown, B.I. (1983) Storage of two new cultivars of guava fruit for processing. *Tropical Agriculture*, **60**, 175–178

Wilson, C.W.III (1980) Guava. In *Tropical and Subtropical Fruits. Composition, Properties and Uses.* (eds S. Nagy, and P.E. Shaw), Chapter 6, AVI Publishing, Inc., Westport, CT. pp. 279–299

Wilson, C.W., Shaw, P.E. and Campbell, C.W. (1982a) Determination of organic acids and sugars in guava (*Psidium guajava* L.) cultivars by high-performance liquid chromatography. *Journal of the Science of Food and Agriculture*, **33**, 777–780

Wilson, C.W. III, Shaw, P.E. and Knight, R.J. Jr. (1982b) analysis of oxalic acid in carambola (*Averrhoa carambola* L.) and spinach by high-performance liquid chromatography. *Journal of Agriculture and Food Chemistry*, **30**, 1106–1108

Winterhalter, P. (1990) Bound terpenoids in the juice of the purple passion fruit (*Passiflora edulis* Sims). *Journal of Agriculture and Food Chemistry*, **38**, 452–455

Wyllie, S.G., Cook, D., Brophy, J.J. and Richter, K.M. (1987) Volatile flavour components of *Annona atemoya* (custard apple). *Journal of Agriculture and Food Chemistry*, **35**, 768–770

Yonemori, K. (1986) Studies on the removal of astringency in Japanese persimmon fruits. Natural loss of astringency. *Bulletin of the Faculty of Agriculture, Mie University*, No. 72, 1–62

Yonemori, K. and Matsushima, J. (1985) Development of tannin cells in non-astringent Japanese persimmon fruits (*Diospyros kaki*) and its relationship to natural loss of astringency. *Journal of Japanese Society of Horticultural Science*, **54**, 201–208

Yonemori, K. and Matsushima, J. (1987a) Morphological characteristics of tannin cells in Japanese persimmon fruit. *Journal of American Society of Horticultural Science*, **112**, 812–817

Yonemori, K. and Matsushima, J. (1987b) Changes in tannin cell morphology with growth and development of Japanese persimmon fruit. *Journal of American Society of Horticultural Science*, **112**, 818–821

Young Koo Son, In Wha Yoon and Pan Joo Han. (1981) Studies on the storage of astringent persimmons in polyethylene film bags. *Research Reports of the Office of Rural Development, Agricultural Engineering, Farm Products Utilization and Farm Management, Suwan* **23**, 95–102

Zauberman, G., Ronen, R., Akerman, M., Weksler, A., Rot, I. and Fuchs, Y. (1991) Post-harvest retention of the red-colour of litchi fruit pericarp. *Scientia Horticulturae*, **47**, 89–97

Zhang, Q.C., Huang, J.W., Tan, P.F., and Ye, W. (1986) A brief report on the study of fresh litchi preservation. *Plant Physiology Communications*, No. 1, 35–36

Zheng, G.H. and Sugiura, A. (1990) Changes in sugar composition in relation to invertase activity during growth and ripening of persimmon (*Diospyros kaki*) fruits. *Journal of Japanese Society of Horticultural Science*, **59**, 281–287

Zhong, Y.W. and Wu, S.X. (1983) A study on the growth pattern of fruits of *Litchi chinensis* cultivar Nomici. *Guangdong Agricultural Science*, No. 4, 1–5

Zuthi, J. and Ben-Arie, R. (1989) Modified atmosphere storage of Fuyu persimmons. *Hassadeh*, **70**, 410–414

Chapter 6

Grape

A. K. Kanellis and
K. A. Roubelakis – Angelakis

6.1 ORIGIN AND DISTRIBUTION

Grapevine belongs to the genus *Vitis*, which was created in 1700 by Tourneford and was one of the first genera studied by Linnaeus in 1735. The word *Vitis* is derived from the Latin word *viere* which means 'to attach'; it was used by Virgil to describe the climbing habit of vines.

The *Vitis* are shrubs of the northern hemisphere, especially the temperate zones of Asia, North America, and Central America. Cultivars are found in all the temperate zones of Europe, in northern and southern Asia, in Australia, in northern and southern America and New Zealand. The European grape *Vitis vinifera* is believed to have originated in the Middle East, perhaps in Iran, Afghanistan or the southern part of the former Soviet Union, and to have been carried from there to the Mediterranean countries before the Christian era.

The species of *V. vinifera*, with its numerous varieties bearing both black and green or pale grapes, is the only species of European origin and commercial use. *V. amurensis* of Asian origin has no commercial application, while the American species, *V. riparia*, *V. rupestris*, *V. berlandieri*, *V. rotundifolia*, *V. champini*, and interspecific hybrids and others are generally used as phylloxera-resistant rootstocks for *V. vinifera*. It should be mentioned, however, that some accidental interspecific crosses of *V. vinifera* with native American species have produced some useful cultivars, Concord being the main commercial representative. Cultivars of the American *V. labrusca* are planted in areas with high temperature and relative humidity during the summer, such as the eastern United States.

Biochemistry of Fruit Ripening. Edited by G. Seymour, J. Taylor and G. Tucker. Published in 1993 by Chapman & Hall, London. ISBN 0 412 40830 9

The demand for new varieties which exhibit greater disease resistance and cold tolerance led to interspecific crosses between American and *vinifera* varieties. Such new hybrids, the 'hybrid direct producers' (HDP), are interspecific hybrid cultivars that resist parasites and produce satisfactory fruit; they have gained importance in the wine industry in the Eastern United States. The first HDP came from the USA and were for the most part hybrids of *V. labrusca*, *V. aestivalis*, *V. riparia* and *V. vinifera*; among them are Isabella, Clinton, Noah, Othello, Iacquez and Herbemond. These hybrids give fruit of low quality but have now tended to disappear, especially in Europe. French breeders continue to improve the early American hybrids and/or to develop new HDP. Generally, these are less resistant to diseases than the first American hybrids of the pure *vinifera* cultivars; examples are Villard, Plantet, Baco 22A, Baco noir, Garronet, Seyval blanc, Colobel, Marechal and Foch.

Since 1955, wine and table grape cultivars have been divided into three categories: 'recommended', 'authorized', and 'tolerated'. The hybrids may belong only to the last two categories, with the exception of Baco 22A, which remains 'recommended' for the Armagnac brandy appellation. Later, the 'tolerated' category was replaced by 'temporarily authorized'.

Data on total viticultural acreage and production in the world are given in Table 6.1. The distribution of vineyards among the continents is Europe 70%, Asia 15%, America 9%, Africa 5% and Oceania 1%.

The grape products can be classified according to their use. Table grapes are for fresh consumption; the largest percentage of the world's grape production is used for wine making, while a portion is used for production of raisins and grape juice.

6.2   GRAPE BERRY MORPHOLOGY

The grape fruit belongs to the large group of fleshy fruits (Coombe, 1976) and is classified as a berry (Pratt, 1971; Coombe, 1976), which develops after fertilization of the ovary (Pratt, 1971). The capacity for inflorescence

Table 6.1   Global viticultural land and production.*

Viticultural land	8 812 000 ha
Grape production	59 517 000 tonnes
Wine production	290 000 000 hl
Table grapes	7 719 900 tonnes
Raisins	917 600 tonnes

*Bulletin de l' Office International du Vin, O.I.V., 1990: 717–718 and 938–974.

initiation is a quantitative genetic character; the degree of its expression is highly affected by nutritional and environmental parameters. The berries are born on clusters and are attached through short pedicels containing vascular bundles (Coombe, 1987), through which the berries are supplied with solutes, water and nutrients (Pratt, 1971; Coombe, 1973).

The flesh of a berry consists of parenchyma cells which retain their protoplasts through ripening (Pratt, 1971). Generally, each berry consists of a pericarp and seeds. The pericarp is divided into exocarp (skin), mesocarp (flesh) and endocarp (Fig. 6.1). The skin is composed of the epidermis (thickness 6.5–10 μm) (Alleweldt *et al.*, 1981), which has smaller cells than other dermal layers (Considine, 1981) and the hypodermis (thickness 107–246 μm) (Alleweldt *et al.*, 1981) with some flesh cells (Pratt, 1971). Mesocarp tissue consists of the pulp of about 25–30

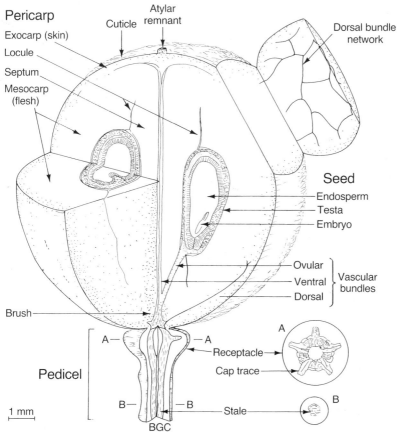

Fig. 6.1   Grape berry structure. (From Coombe, 1987.)

layers of cells, where most of the constituents are stored during ripening. The mesocarp and endocarp is divided topographically into that either outside or inside of the peripheral vascular bundles. The seeds, which number up to 4, originate from the ovules of the ovary.

6.2.1 Skin cells

The skin cells have distinct and active metabolism involving many physiological and biochemical changes, occurring during development and ripening. It has been proposed that this tissue has an endocrinal function, regulating the development of the other pericarp tissues (Coombe, 1973) with its portion accounting for 5–12% of the total berry weight (Lavee and Nir, 1986). The number of layers in the skin of grape berries and their size and volume are cultivar specific. The outer epidermis is covered by non-living layers, namely cuticle, lenticels, wax and collenchymatous hypodermal cells (Pratt, 1971; Winkler et al., 1974). The thickness of epidermal cell walls is the only parameter showing positive correlation with resistance to stress (Considine, 1981). The cuticle and epicuticular wax layer of table grape berries contributes to their quality by scattering light and imparting a frosted appearance which is desirable to consumers. In addition, berry waxes are the primary source of waxes in wines and may contribute to colloidal turbidity of wines (Rosenquist and Morrison, 1988).

The genesis of cuticle begins about three weeks before anthesis as highly organized, tightly oppressed cuticular ridges, which become progressively more disorganized during post anthesis berry growth. Its thickness ranges from 1.6–3.8 μm and decreases slightly during ripening. Epicuticular wax formation begins with the appearance of small, simple wax platelets, which increase in size, number and complexity as the berries mature (Rosenquist and Morrison, 1988). Cuticular wax consists mainly of oleanolic acid (79%), in addition to long-chain alcohols, fatty acids and traces of esters, aldehydes and paraffins (Winkler et al., 1974).

Contrary to earlier reports (Pratt, 1971), skin of grape berries contain stomata whose occurrence and frequency vary depending on cultivar (Swift et al., 1973; Nakagawa et al., 1980; Blanke and Leyhe, 1987). Stomata activity of the grape berry decreases with fruit ontogeny; transpiration through the cuticle is the main avenue of water loss in grape berries and increases with fruit development (Blanke and Leyhe, 1987). Therefore, the thickness and toughness of the skin are factors which contribute to resistance of table grapes to handling injury during harvest, packing, transport and storage. Furthermore, the epidermal and subepidermal cell layers contain most of the colour, aroma, and flavour constituents of the berries and are richer in vitamin C than the pulp (Winkler et al., 1974).

6.2.2 Pulp and seed

The cells of the pulp have large vacuoles containing the cell sap, which is the main constituent of the juice found in the berry at maturity and accounts for 64–90% of the berry weight (Lavee and Nir, 1986).

The seeds constitute up to 10% of the weight of the fruit. They are rich in phenolic compounds, which form a large proportion of wine tannins (5–8%). They also contain large amounts of oil (10–20%) and lesser amounts of resinous materials (Winkler *et al.*, 1974). In some seedless varieties, fertilization does not occur and the berries are truly parthenocarpic (Pratt, 1971; Coombe, 1973), whereas in others the development of the embryo ceases at various stages of development resulting in stenospermocarpic seeds (Sultanina). Generally the size and composition of the berries is a function of seed development: the greater the number of seeds, the heavier the berry, with relatively lower sugar and nitrogen concentrations, but higher level of acidity (Peynaud and Ribéreau-Gayon, 1971).

Seeds are generally a rich source of hormones. Seeded grape berries contain higher amounts of gibberelin-like compounds and abscisic acid than seedless berries. In this context, it is suggested that seeds contribute to the hormonal regulation of the berry development and ripening.

6.3   FRUIT GROWTH AND DEVELOPMENT

Berry growth follows a double sigmoid curve, characteristic of all berry fruits (Coombe, 1976). The exact form of the growth curve can vary considerably and divisions into two, three, or four phases have been proposed (Coombe, 1973, 1976, 1980; Alleweldt *et al.*, 1975; Alleweldt and Koch, 1977; Coombe and Bishop, 1980; Alleweldt *et al.*, 1984). Three phases have been designated by using the parameters of accumulative berry diameter, length, and volume or weight of seeded berries. In this context, the three phases are the following: (I) period of rapid growth, (II) period of slow growth and (III) period of second and final increase in size (Fig. 6.2). The duration and manifestation of each growth period varies according to cultivar and environmental conditions (Coombe, 1973; Hale and Buttrose, 1974; Farmahan and Pandey, 1976). Growth curves of seedless cultivars exhibit generally less distinct phases, because of a less enunciated phase II (Iwahori *et al.*, 1968; Pratt, 1971; Farmahan and Pandey, 1976).

The division of berry growth into four phases adds a period designated as (0), which lasts from anthesis up to 10 days thereafter, and is characterized by a small increase in dry and fresh weight (Nitsch *et al.*, 1960). Some other workers also divide the berry development into four phases by defining the early cell division period as phase I. Thus, phase

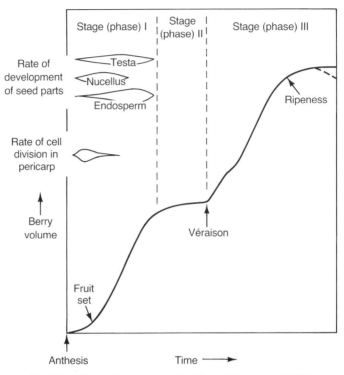

Fig. 6.2 Grape berry growth. (From Coombe, 1973.)

II would include the first rapid growth and cell enlargement period, phase III would be the slow-growing (lag) phase and phase IV the second rapid growth and ripening phase (Coombe, 1960; Alleweldt *et al.*, 1975, 1984; Alleweldt and Koch, 1977).

Recently, the division of grape berry growth into two phases was proposed without transition points of slow growth either at the beginning of the first growth period or between the first and second rapid growth periods (Staudt *et al.*, 1986). In the present review, a three-phase curve has been adopted for simplicity and comparison reasons, since most of the work on fruit development and compositional changes refers to the three-stage division of grape berry growth.

6.3.1 Stages of berry growth

Stage I, which varies from 45 to 60 days, is characterized by a period of about 2 to 3 weeks with a very rapid cell division rate followed by a marked cell enlargement (Pratt, 1971; Coombe, 1976; Lavee and Nir, 1986). However, cell divisions continue until well into the first growth period (Coombe, 1976; Staudt *et al.*, 1986). Most of the cell divisions in the pericarp take place 5 to 10 days post anthesis (Coombe, 1973; Pratt, 1971). Cell division ceases first in the placenta and inner pericarp 7 to 11 days post anthesis, later in the outer pericarp and finally in the hypodermis and epidermis 32 to 40 days after bloom (Considine and Knox, 1981; Pratt, 1971; Staudt *et al.*, 1986).

By the end of stage I, the seeds attain nearly their full size (Pratt, 1971), whereas seed dry weight shows a biphasic pattern during the whole fruit growth. The growth of the embryo shows no relation to the double-sigmoidal growth of the pericarp and final size is reached 70–75 days after anthesis (Staudt *et al.*, 1986). During stage I, chlorophyll is the predominant pigment and berries display active metabolism, with high rates of respiration and rapid acid accumulation. Acid content is comparable to that of sugar content (20 g kg^{-1}) (Peynaud and Ribéreau-Gayon, 1971; Winkler *et al.*, 1974).

Phase (stage) II starts 35 to 80 days after anthesis (Winkler *et al.*, 1974). The duration of slow growth phase or lag phase (stage II) depends on the cultivar (early or late, seedless or seeded), timing of flowering (early versus late, primary versus secondary clusters), competition between clusters and the vine's environment (Pratt, 1971; Coombe, 1973, 1976; Alleweldt, 1977; Lavee and Nir, 1986). For example, late maturing or late flowering berries of seeded varieties show a distinct and prolonged lag phase, while the lag phase of early and parthenocarpic cultivars is barely discernible (Coombe, 1980).

During the lag phase, the embryos develop rapidly and generally reach their maximum size towards the end of this period, the berries lose chlorophyll and soften. In addition, acidity reaches its highest level. The end of the lag phase marks the beginning of stage III.

The inception of stage III is called *véraison* and is characterized by a rapid change in appearance and constitution of the berries. The events that take place, as stage III begins, are acceleration of growth, softening of the berry and increase in deformability, increase in the content of glucose, fructose, total and free amino acids (especially arginine and proline), total proteins and total nitrogen (Peynaud and Ribéreau-Gayon, 1971; Coombe, 1973; Lavee and Nir, 1986). Also, decreases in the concentration of organic acids (mainly malic acid) and ammonia, loss of chlorophyll from the skin, accumulation of anthocyanins, a decrease in respiration rates, and finally an increase in activity of some enzymes, e.g. sucrose phosphate synthetase, sucrose synthetase and hexokinase are found during stage III. This period lasts 5–8 weeks (Winkler *et al.*, 1974).

6.4 GROWTH REGULATION–PHYTOHORMONES

6.4.1 Auxins

The contribution of phytohormones to the growth and development of grape berry has been demonstrated. During the initial stages of berry development (stage I), when active cell divisions take place, the concentration of indoleacetic acid (IAA) increases, reaches its maximum at the end of the lag phase (stage II) and decreases thereafter (Coombe, 1960; Nitsch *et al.*, 1960; Alleweldt and Hifny, 1972; Coombe and Hale, 1973; Alleweldt *et al.*, 1975). The effect of exogenously applied synthetic auxins to control grape berry development greatly depends on the type of applied auxin, the rate and time of application and the cultivar's characteristics. A monophenoxy auxin, 4-chloro-phenoxyacetic acid (4-CPA) applied after fruit set to seedless cultivars enhances berry growth and delays maturation (Weaver, 1953). Dipping of clusters in a low concentration of 2,4 dichlorophenoxyacetic acid (2,4-D) at full bloom slightly reduces the number of seeds and enhances sugar accumulation. In coloured cultivars a late spray delays ripening (Weaver *et al.*, 1961).

6.4.2   Cytokinins

Cytokinins (CK) play an important role in the regulation of flower initiation (Shrinivasan and Mullins, 1979) and berry development (Weaver *et al.*, 1968; Pool, 1975). Their mode of action seems to be closely related to RNA and protein synthesis. A CK-binding protein, which consists of at least two glycoprotein components with high activity has been isolated from grape berries (Harada, 1980). Cytokinin content in developing grape berries is low until the onset of the lag phase (Alleweldt *et al.*, 1975). During the lag phase (stage II) the cytokinin content increases sharply and decreases with the new active growth phase (stage III). Cytokinin application at full bloom to grapevines grown at low light regimes was not effective in increasing the translocation rate of photosynthates to inflorescences, thus increasing the fruit set (Roubelakis and Kliewer, 1976), whereas under normal light regimes it increases the berry size (Weaver and van Overbeek, 1963; Weaver, 1965; Roubelakis and Kliewer, 1976).

6.4.3   Gibberellin

Gibberellin (GA)-like substances have been found in grape berries (Coombe, 1960; Weaver and Pool, 1965; Scienza *et al.*, 1978). In seedless and seeded berries, GA activity is mainly in the acidic ethyl acetate fraction. In seeded berries there is high activity at the early fruit-set stage, persisting for about three weeks, decreasing to very low levels thereafter and increasing again to a second peak two weeks later. Thereafter, it decreases and remains at very low levels throughout fruit ripening (Scienza *et al.*, 1978) (Fig. 6.3). The GA-like substances content is correlated to the number of seeds. Three-seeded berries have a GA-like substances content which, especially at the second peak, is over five-fold higher compared to single-seeded berries (Scienza *et al.*, 1978) (Fig. 6.3).

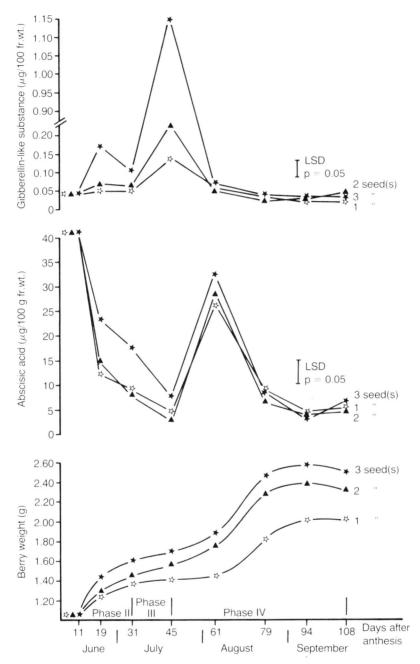

Fig. 6.3 Gibberellin-like substances and abscisic acid in developing berries with different number of seeds (phases II, III and IV correspond to stages of grape berry development I, II and III, respectively in the text). (From Scienza *et al.*, 1978.)

The same pattern is found in seedless berries, except that the decrease occurs earlier and the second peak also appears considerably earlier (Iwahori *et al.*, 1968).

Girdling, which is an old practice applied to seedless cultivars such as the parthenocarpic Black Corinth, causes a marked increase in GA in berries and an increase in berry size (Weaver and Pool, 1965). In addition, exogenously applied GA could compensate for missing seeds and enlarge the berries of seeded cultivar with only a partial number of seeds to their genetically determined full size (Lavee, 1960). Thus, GA has been considered as a significant regulator of growth and development of grape berries (Lavee and Nir, 1986). Application of GA to seedless cultivars is a commercial practice for controlling bunch shape and berry size. Post bloom applications greatly increase berry size, whereas the full bloom application causes thinning of bunches (Lynn and Jensen, 1966). Defoliation (50%) of grapevines causes a reduction of GA in grape berries. Exogenous application of GA to defoliated vines results in an increase in grape GA equal to that of foliated vines (Sihamed and Kliewer, 1980).

6.4.4 Abscisic acid

The level of abscisic acid (ABA) is high during anthesis (Scienza *et al.*, 1978); is low in berries at an early stage and rises during the commencement of the second active growth phase (Coombe, 1973; Coombe and Hale, 1973; Düring, 1973, 1974; Downton and Loveys, 1978; Scienza *et al.*, 1978) and decreases thereafter (Cawthon and Morris, 1982; Düring, 1974) (Fig. 6.3). The highest levels of ABA are found in the exocarp of the berries (Coombe, 1973), which may facilitate the accumulation of carbohydrates in the cells (Alleweldt et al., 1975). Exogenous application of ABA during the lag phase of berry growth results in an increase in endogenous ABA level and enhances sugar accumulation (Coombe, 1973). Also, application of ABA increases the activity of gluconeogenic enzymes, glucose-6-phosphatase, fructose-1,6-bisphosphatase (EC 3.1.3.11) and malate dehydrogenase (EC 1.1.1.37); this effect is nullified by cycloheximide, indicating the *de novo* synthesis of these enzymes (Palejwala *et al.*, 1985).

6.4.5 Ethylene

The involvement of ethylene in grape berry ripening is discussed in section 6.5. However, the effect of exogenous ethylene or ethylene-releasing compounds under the generic name ethephon [(2-chloroethyl) phosphoric acid] on berry development and ripening, fruit set and abscission has been extensively studied in the last 10–15 years (Szyjewicz *et al.*, 1984; Lavee and Nir, 1986). In general, spray applications of ethephon enhance colour development in pigmented cultivars, increase soluble solids, reduce acidity, induce abscission of berries (thus facilitating mechanical harvesting), and reduce firmness in some varieties (Szyjewicz *et al.*, 1984). The extent and the efficacy of ethephon applications are affected by factors such as ambient conditions which may change the spray solution, cultivar, concentration, pH, timing, method of application and grapevine water status (Szyjewicz *et al.*, 1984).

6.5   POSTHARVEST PHYSIOLOGY

Grape is a non-climacteric fruit with a relatively low rate of physiological activity (Harris *et al.*, 1971; Alleweldt and Koch, 1977; Pandey and Farmahan, 1977; Koch and Alleweldt, 1978; Weaver and Singh, 1978). The classification of grape as a non-climacteric was based on the very small amounts of ethylene produced during development. A peak of ethylene production at bloom followed by a continuous decrease until harvest has been reported (Coombe and Hale, 1973; Weaver and Singh, 1978). However, ethylene evolution at or after *véraison* slightly increases and may coincide with a peak of internal ethylene concentration mentioned by Alleweldt and Koch (1977) during the same period. Respiration expressed on a per fresh weight basis or a per protein basis shows a continuous decrease during development and ripening (Harris *et al.*, 1971; Pandey and Farmahan, 1977; Koch and Alleweldt, 1978). When respiration is expressed as ml O_2/berry (Alleweldt and Koch, 1977; Koch and Alleweldt, 1978) a situation resembling the peak in internal ethylene is observed.

Although there appears to be a minute increase in internal ethylene (up to 0.4 ml/l) (Alleweldt and Koch, 1977) or ethylene production (up to 0.5 ml/kg day^{-1}) (Weaver and Singh, 1978), this increase is not associated with a climacteric respiratory or ethylene rise in the generally accepted sense. On the other hand, the dramatic increase in the respiratory quotient values after *véraison*, coupled with the synchronous acid decrease, massive sugar accumulation and colour change, suggest an alteration in the physiological and biochemical characteristics of the grape berry, in which ethylene, even in minute quantities, alone or in combination with other plant hormones, may play a role.

Levels of ABA and ethylene produced in the fruit in response to ethylene and ethylene-releasing compounds during development and ripening show that endogenous ABA levels in the flesh may mediate the response to exogenous or endogenous ethylene (Coombe and Hale, 1973). ABA must accumulate to a certain threshold level value in order to act synergistically with ethylene in promoting ripening. As a consequence, berry ripening is delayed, if ABA is below this threshold level and exogenously applied or induced increases in ethylene maintain ABA at low levels. This implication of ABA in grape berry ripening may involve changes in cell wall permeability at *véraison*, allowing carbo- hydrates and water to enter the berry cells more readily (Alleweldt *et al.*, 1975). Alternatively, ABA appears to enhance ripening by accelerating gluconeogenesis (Palejwala *et al.*, 1985). This involvement of ABA implies also the requirement of *de novo* protein synthesis (Palejwala *et al.*, 1985). It is also interesting to mention that increasing ethephon levels stimulate peroxidase activity, with a concomitant new synthesis of two peroxidase isoenzymes (Kochhar *et al.*, 1979).

It is apparent that the evidence for the hormonal regulation of grape berry ripening is fragmentary at best. In view of the recent developments in hormone detection techniques, in biochemical and molecular biology methods which have been applied to elucidate other plant processes (i.e. ripening of climacteric fruit, refer to Chapters 2 and 14), it is imperative that more attention should be paid towards understanding the ripening of the grape and other non-climacteric fruits.

6.6   HANDLING AND STORAGE

Grapes, like other non-climacteric fruits, do not ripen after harvest; they are allowed to attain the optimum stage of acceptability in appearance, flavour and texture while still on the vine.

Harvest maturity is based upon the total soluble solids content (or °Brix) of the berries. Titratable acidity and sugar-to-acid ratio are also used as maturity indices (Guelfat-Reich and Safran, 1971). The minimum requirements vary with variety, growing region and use of grapes. Appearance is determined mainly by colour, especially in pigmented varieties in which minimum colour requirements exist that vary with variety and grade standard. The grade standards designate the per- centage of berries on the cluster, which must show a certain minimum colour intensity and coverage (Nelson, 1979). The 'bloom' or wax on the grape berry's surface is a very important quality factor in table grapes; rough handling (rubbing) destroys this bloom, making the skin shiny rather than the desirable lustre effect of bloom.

Grapes are subjected to serious water loss after harvest, which can cause stem drying and browning, berry shatter, and even wilting and shrivelling of berries (Nelson, 1979). Thus, grapes should be cooled as soon as possible after harvest. Forced-air-cooling is the preferred method of precooling, since grapes do not tolerate wetting, which is the pre- requisite of hydrocooling (Nelson, 1979). In this context, 'sweating' of fruit, as a result of drastic and rapid changes in temperature during storage, causes serious problems associated with *Botrytis* rot infection. The most important postharvest decay organism is *Botrytis cinerea* Pers, which along with water loss represents the most serious problems determining the postharvest life and preservation of table grapes (Nelson, 1979; Lavee and Nir, 1986). Sulphur dioxide (SO_2) fumigation immediately after packing is a standard procedure for the phyto-sanitation of grapes. This treatment combined with proper temperature handling and periodic SO_2 treatments at a lower base during storage and transportation reduce decay markedly (Nelson, 1979). Most grapevine cultivars store best at $-1°C$ and 90–95% relative humidity (Nelson, 1979).

The best source of information regarding the proper postharvest handling procedures such as maturity, harvest, precooling, proper SO_2 fumigation, storage and transportation conditions continues to be Nelson's handbook (1979).

6.7   COMPOSITIONAL CHANGES

6.7.1   Nitrogenous compounds and enzymes

Nitrate and ammonium are the most common forms of nitrogen available to grapevine in the soil. The relative importance and magnitude of each of these ions depends on the genetic, developmental and physiological status of each plant, as well as on soil properties, such as texture, structure, water content and pH. Nitrate absorbed by the roots of grapevine is partially reduced in the roots, while ammonium ions produced are either assimilated *in situ* or transported to aerial parts of the grapevine including the reproductive parts. In flower clusters, nitrate concentration is 2 mM and reaches a peak of 85–139 mM at flowering. The berries show a maximum nitrate concentration before *véraison* of about 1.6 mM, with highest concentrations in the skins (Schaller *et al.*, 1985).

Active amino acid and protein synthesis takes place in developing grape berries. The amino acid fraction (amino acids and low molecular weight peptides) accounts for 50% to 90% of the total nitrogen in the juice and the non-amino acid N-fraction accounts for 10% to 56%. Eight free amino acids (arginine, proline, alanine, γ-aminobutyric acid, aspartic acid, glutamic acid, serine, and threonine) account for 29% to 85% of the total nitrogen in the juice of grapes and 50% to 95% of the total free amino acids present. Cultivar, degree of maturity, rootstock, temperature, mineral nutrition, crop level, trellising system and diseases may influence the nitrogenous composition of grapes.

There are two phases of intense N incorporation in grape berries; the first occurs about two weeks before the 'pea size' stage of the berries and

the second starts at the beginning of ripening and lasts about two weeks
(Lohnertz, 1988). Towards the end of fruit ripening, concentrations of
soluble and total nitrogen increase again (Fig. 6.4) while at harvest, half
of the N present in annual structures of grapevines is located in the
reproductive parts (Alexander, 1957; Nassar and Kliewer, 1966; Kliewer,
1969; Wermelinger and Koblet, 1990; Lohnertz, 1991). Total nitrogen
concentration in grape juice ranges from 36.7–213.0 mM and the amino
acid fraction from 2.0–13.0 mM eq. leucine.

   In grape berries during maturation, the organic nitrogen in the fruit
steadily increases, including total amino acids and proteins, while
ammonia decreases (Peynaud, 1947; Peynaud and Maurie, 1953). In
immature fruit ammonium ions account for more than half the total
nitrogen. Peynaud (1947) found concentrations of 1.1 to 8.0 mM for
ammonia and 8.9 to 48.9 mM for total nitrogen in grapes grown in
Bordeaux. In California, ammonia concentration in wine cultivars ranged
from 1.5 to 18.8 mM, with an average value of 7.7 mM (Ough, 1969).
Synthesis of amino acids, peptides, and proteins occurs mainly during
the last six to eight weeks of berry ripening and during this period
ammonia decreased sharply. Total free amino acids in grape juice

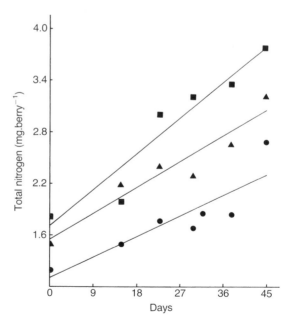

Fig. 6.4   Total nitrogen in ripening grape berries from vines fertilized with 112
(●), 448 (triangle) and 672 (square) kg N.ha⁻¹. (From Roubelakis-Angelakis,
unpublished data.)

increased two- to five-fold during ripening and ranged from 2 to 8 g.l^{-1} juice (as leucine equivalents) (Kliewer, 1968).

Concentrations of arginine and proline differ by as much as ten- to twelve-fold among cultivars and from two- to six-fold between early- and late-harvested fruits of the same cultivar. The level of proline in grapes ranged from 304 to 4600 mg.l^{-1} and the 'Cabernet' group of cultivars ('Cabernet Saugvinon' and 'Merlot') were found to be particularly high in proline (Ough, 1968). Proline is usually present in larger amounts in grapes during warm rather than cool years (Flanzy and Poux, 1965), whereas other amino acids are largely unaffected by temperature. Proline levels increase 25–30-fold during berry ripening (Juhasz, 1985). It has been suggested that this increase is due to proline biosynthesis from other amino acids, and that the intensive carbohydrate synthesis during the late season hinders its synthesis. Arginine accounts for 6–44% of total nitrogen in grape juice.

The assimilation of ammonium ions into amino acids is mediated by the enzymes glutamine synthetase/glutamate synthase (GS/GOGAT) and probably by glutamate dehydrogenase (GDH) (EC 1.4.1.3). In grape berries, GDH activity shows two maxima, the first at one week following anthesis and the second at four weeks after *véraison*; also GS shows three maxima of activity, at anthesis, at one week before véraison and at two weeks thereafter (Ghisi *et al.*, 1984). These findings indicate that both pathways could be active in ammonium assimilation in grape berries. Recently, it has been shown that GDH in grapevine tissues exhibits a seven-isoenzyme pattern (Loulakakis and Roubelakis-Angelakis, 1990a, b). The more anodal isoenzymes show high NADH-GDH activity, whereas the more cathodal forms show high NAD-GDH activity. Two subunits, α and β, with molecular weights of 43.0 and 42.5 kDa respectively, and each with a different charge, participate in the structure of the seven isoGDHs in an ordered ratio (Loulakakis and Roubelakis-Angelakis, 1991). These isoenzymes with a higher proportion of α subunits (the more anodal forms) show high NADH-GDH activity, while more cathodal isoenzymes with a higher proportion of β-subunits show higher NAD-GDH activity. Thus, it has been proposed that the two subunits, when assembled to form isoforms of GDH, exhibit anabolic and catabolic activities, respectively (Loulakakis and Roubelakis-Angelakis, 1991). Furthermore, *de novo* synthesis of the α-subunit and thus of the anodal isoenzymes is induced by increased concentration of ammonium (Loulakakis and Roubelakis-Angelakis, 1990c, 1992; Roubelakis-Angelakis *et al.*, 1991) (Fig. 6.5). It can be assumed, therefore, that in grapes GDH exhibits anabolic activity towards amination of α-ketoglutarate to form glutamate when the intracellular concentration of ammonium exceeds the minimum level required for the expression of the α-GDH subunit gene. Further study is needed on the critical *in vivo* concentration of ammonium for induction of NADH-GDH activity and the molecular regulation of GDH, as well as for the relative contribution of GS/GOGAT and GDH in ammonium assimilation in grapes.

As mentioned earlier, arginine is one of the most abundant amino acids in grapes, and plays a key role in the biosynthesis of polyamines and guanidines. The enzymes mediating the reactions of its synthesis, namely ornithine transcarbamoylase (OTC), arginosuccinate synthetase and lyase, and arginase have been studied in grapevine tissues (Roubelakis and Kliewer 1978a,b,c). In developing berries, OTC and arginase activities are higher at *véraison* (Roubelakis-Angelakis and Kliewer, 1981). Also, aspartase (EC 4.3.1.1), which catalyses the reversible deamination of L-aspartic acid to yield fumaric acid and ammonia is present in grape berries. An activation effect of malate and an inhibitory effect of several electrolytes are observed suggesting that this enzyme plays a key role in the regulation of the organic acid pool in developing berries (Robin *et al.*, 1987).

Several soluble proteins are present in the juice of grapes, though these differ somewhat between cultivars (Koch and Sajak, 1959). Whether grapes for wine are harvested at the peak of protein content or later may affect the clouding of wine. Greater amounts of soluble proteins are formed during warm seasons than during cool seasons. Reported soluble protein content in juice differ significantly, ranging from 1.5 to over 100 mg.l^{-1} (Hsu and Heatherbell, 1987; Yokotsuka *et al.*, 1988). SDS-PAGE separation of the soluble protein fraction of grape juice has revealed between 13 and 41 protein bands with molecular weights ranging from 11.2 to 100 kDa (Nakanishi *et al.*, 1986; Murphey *et al.*, 1989). Proteins with high molecular weight are found in the insoluble protein fraction. In developing berries, the highest protein concentration is found at stage I; it decreases at stage II, increases again early at stage III and

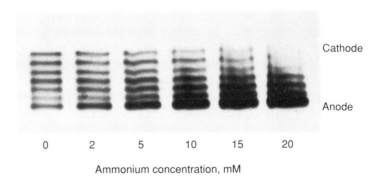

Cathode

Anode

0   2   5   10   15   20

Ammonium concentration, mM

Fig. 6.5 Isoenzymic pattern of glutamate dehydrogenase from grapevine callus grown on varying ammonium concentrations. (From Roubelakis-Angelakis *et al.*, 1991.)

decreases thereafter (Koch and Sajak, 1959; Ghisi *et al.*, 1984). Further work is required to elucidate the mechanism(s) of protein synthesis and regulation in grape berries during grape ontogeny, and to identify the structure of specific genes encoding enzyme proteins, which mediate reactions, the products of which contribute to the quality of grapes.

6.7.2 Sugars

Glucose and fructose account for 99% or more of the carbohydrates in the grape juice and from 12% to 27% or more of the fresh weight of the mature berry, constituting a large proportion of the total soluble solids (Winkler *et al.*, 1974; Hofacher *et al.*, 1976). In green berries, during the early stages of berry development, glucose accounts for 85% of the sugar content; during *véraison*, glucose predominates over fructose, in ripe berries the ratio of glucose to fructose content approaches unity, while in over-ripe grapes fructose generally exceeds glucose (Peynaud and Ribéreau-Gayon, 1971; Winkler *et al.*, 1974; Possner and Kliewer, 1985; Coombe, 1987). Varieties containing more fructose at maturity may be picked and consumed earlier than varieties containing more glucose, as fructose tastes sweeter than glucose. Warm seasons tend to result in a lower glucose to fructose ratio (Lavee and Nir, 1986).

In addition to glucose and fructose, several other sugars are present in small amounts in grapes. Among them, sucrose (less than 0.1% in ripe berries), raffinose, stachyose, melibiose, maltose and galactose have been found in different varieties (Winkler *et al.*, 1974). It seems that *V. vinifera* varieties contain less sucrose (0.019–0.6%) than *V. rotundifolia* or *V. labrusca* (0.2–5%). Further, sucrose concentration in the flesh is lower than in the skin and brush (Coombe, 1987). Pentoses, mainly arabinose and traces of xylose are present in small amounts (0.3–1.0 g/l of juice) in ripe grapes (Winkler *et al.*, 1974).

Glucose and fructose are more or less evenly distributed throughout the berry (Possner and Kliewer, 1985; Coombe, 1987). However, before the onset of ripening, the skin and the core zones are characterized by significantly higher hexose concentrations than the tissue in between (Possner and Kliewer, 1985). These results correlate well with the finding that ^{14}C-labelled photosynthates enter the berry via the peripheral and central vascular systems (Kriedemann, 1969).

Sugars begin to accumulate at a rapid rate at the onset of ripening (Saito and Kasai, 1978; Hrazdina *et al.*, 1984; Possner and Kliewer, 1985) coinciding with the beginning of softening (Coombe and Phillips, 1982). This dramatic increase in sugar content during phase III could not be attributed to an increase in photosynthetic activity (Peynaud and Ribéreau-Gayon, 1971). Source–sink relationships, that is competition between fruit growth and other developing organs of the grapevine, are critical for fruit development. Berries exert a stronger sink capacity after *véraison* (Alleweldt *et al.*, 1975; Lavee and Nir, 1986).

Most of the sugars in the berries are synthesized in the leaves (Winkler *et al.*, 1974) and move through the phloem to the berries, mainly as sucrose (Lavee and Nir, 1986). In the berries the sucrose is hydrolysed to glucose and fructose (Hardy, 1968; Saito and Kasai, 1978). This accumulation of hexose sugars is correlated with high levels of invertase (EC 3.2.1.26) activity (Düring and Alleweldt, 1984). However, other studies suggest that invertase activity remains fairly constant before and after *véraison* (Hawker, 1969; Coombe, 1989). Carbohydrate translocation to, and accumulation in, the fruit involves a sequence of reactions: phosphorylation of glucose and fructose, synthesis of sucrose phosphate, hydrolysis of sucrose phosphate to sucrose in the leaves, transport and hydrolysis of sucrose in the berry (Hawker, 1969).

The mechanisms by which sugar moves into a grape berry are not well understood. Two possible mechanisms of sugar transport have been tested in the past (Brown and Coombe, 1982, 1984; Coombe 1992) and a third has been proposed recently (Lang and Düring, 1991). The first mechanism envisages that a group translocator may operate at the tonoplast, giving metabolic channelling of cytoplasmic hexoses, followed by vectorial release of sucrose phosphate (Brown and Coombe, 1982, 1984), though recent data indicate that sugar transport involves sugar porters (Niemietz and Hawker, 1988). The second mechanism proposes the existence of pumps which transport sugars out of phloem cells into the berry (Coombe *et al.*, 1987; Coombe, 1992). The third mechanism suggests that sugar flow is caused by 'leakiness' of plasma membranes in pericarp cells (Lang and Düring, 1991; Coombe, 1992). The later two mechanisms of sugar transport into the grape berry, along with the elucidation of the mechanism which triggers the sugar accumulation into the grape berries during and after *véraison*, need further exploitation and verification.

6.7.3 Organic acids

The acid fraction of grape consists mainly of tartaric and malic acids, accounting for 90% or more of the total acidity (Winkler *et al.*, 1974). Other organic acids found in variable, but always low concentrations, are citric (5–10% of total acidity), succinic, fumaric, acetic, glycolic, lactic, aconitic, quinic, shikimic and mandelic acids. Since these are all inter-mediates of glycolysis, the Krebs cycle, glyoxylic acid cycle and shikimic acid pathway, the inference is that all these metabolic pathways are operating in grapes (Winkler *et al.*, 1974; Ruffner, 1982a). Grape acid content is an important quality factor (Winkler *et al.*, 1974; Coombe *et al.*, 1980; Lavee and Nir, 1986), since high acid (and in some instances extremely low concentrations) not only affect the palatability of table grapes, but also influence the suitability of wine grapes for vinification (Ruffner, 1982a). Excessive tartness of the fruit normally correlates with low sugar concentrations, resulting in poor wine quality, whereas low acid levels at harvest can be accompanied either by low or high sugar content, depending on the preceding climatic conditions; in both cases unbalanced and 'flat' wines are produced.

Acid content of ripening grapes is affected by exogenous factors, mainly temperature (Kliewer, 1973; Hale and Buttrose, 1974; Winkler *et al.*, 1974; Klenert *et al.*, 1978; Ruffner, 1982a). Continuous warm conditions result in enhanced degradation of malic acid during ripening, resulting in a lower acid content at maturity (Buttrose *et al.*, 1971; Ruffner *et al.*, 1976; Ruffner, 1982a).

The ratio between tartrate and malate contents varies considerably depending on the grape variety. Varieties having higher tartrate-to-malate ratio are more suitable for wine making, especially in warm climates (Buttrose *et al.*, 1971; Lavee and Nir, 1986). Despite their close chemical similarity, tartaric and malic acids exhibit distinctly different accumulation patterns during grape berry ontogeny and are linked to different biosynthetic and metabolic pathways. Malic and tartaric acids are synthesized mainly in the fruit, most certainly from carbohydrate precursors (Ruffner, 1982a).

Tartaric acid

Grape seem to be the only commercially cultivated fruit to accumulate tartaric acid in appreciable amounts (Saito and Kasai, 1968; Ruffner, 1982a). Tartaric acid occurs in Vitaceae as the optically active L-(+)-stereoisomer (Wagner *et al.*, 1975); it accumulates in grape berries before *véraison* and decreases thereafter. However, tartrate levels appear to remain remarkably constant when calculated on a per berry basis, which takes into account the considerable increase in berry volume occurring during ripening (Ruffner, 1982a; Iland and Coombe, 1988). Due to the temporal coincidence between cell divisions and intensive tartrate synthesis at the beginning of berry development, a correlation between the two processes has been proposed (Winkler *et al.*, 1974). However, there is no evidence for direct participation of tartrate in cell division.

Tartaric acid is stored predominantly as the salt form. Due to the abundance of potassium in grapes, it was assumed that potassium was the natural counter-cation, and that the crystalline precipitates found in grape tissue were potassium hydrogen tartrates (Winkler *et al.*, 1974). However, this hypothesis has been disputed, firstly by circumstantial evidence (Hale, 1977) showing the obvious lack of correlation between potassium and tartrate contents in grapes at any stage of development and secondly, by experimental evidence (Ruffner, 1982a), indicating that the insoluble tartrates occur as highly ordered crystalline bundles of calcium salts. In grape material, calcium tartrate is sequestered in huge, specialized cells, the idioblasts (Zindler-Frank, 1974; Ruffner, 1982a), while a portion of tartaric acid is also stored in the vacuoles in the form of the free acid (Moskowitz and Hrazdina, 1981).

Tartaric acid is believed to originate exclusively from the carbohydrate pool (Ruffner, 1982a). Various compounds have been identified as precursors of tartaric acid. The first to be suggested, based on labelling experiments, was glucose (Peynaud and Ribéreau-Gayon, 1971). Both [1–^{14}C] glucose and [6–^{14}C] glucose serve as precursors of tartaric acid, with transformation rates of [6–^{14}C] glucose under photosynthetic conditions amounting to 50–60% of the value observed with [1–^{14}C] glucose (Ruffner and Rast, 1974; Ruffner, 1982a). Based on the evidence that [6–^{14}C] glucose does not relay radioactive carbon to tartrate in the dark, it has been suggested that in *Vitis*, two different pathways of tartaric acid synthesis from glucose are effective: one retains the original carbon sequence of glucose with C_1 of the sugar entering C_1 of tartaric acid, while the other includes a formal inversion of the hexose skeleton (Ruffner and Rast, 1974; Ruffner, 1982a). Redistribution of label between terminal carbons of hexose due to glycolysis prior to tartrate biosynthesis has been offered as an alternative hypothesis to account for tartrate labelling from [6–^{14}C] glucose (Saito and Kasai, 1978).

Another compound which has been identified as a precursor of tartaric acid is glycolate (Maroc, 1967), based on the high incorporation of [1–^{14}C] glycolate and [2–^{14}C] glycolate and the resulting labelling pattern of tartrate (Ruffner and Rast, 1974); a tail-to-tail condensation of two C_2 molecules, presumably glycolaldehyde, is proposed (Ruffner, 1982a).

[^{14}C] Ascorbic acid is another extremely effective precursor of tartaric acid in grape berries, with transformation rates of more than 70% within 24 hours (Saito and Kasai, 1969). Despite the difficulties and conflicting results in identifying the intermediates between ascorbic acid and tartaric acid (Ruffner and Rast, 1974; Wagner and Loewus, 1974; Helsper and Loewus, 1985; Wagner *et al.*, 1975; Saito and Kasai, 1978; Williams and Loewus, 1978; Saito and Loewus, 1979; Williams *et al.*, 1979), recent reports demonstrate that L-ascorbic acid serves as a true physiological intermediate in L-tartaric acid biosynthesis in grape berries (Saito and Kasai, 1982, 1984; Malipiero *et al.*, 1987). The pathway of tartaric acid biosynthesis from ascorbic acid seems to involve the following reactions: ascorbic acid → 2-keto-L-idonic acid → L-idonic acid → 5-keto-D-gluconic acid → tartaric acid (Saito and Kasai, 1984).

There is no conclusive evidence concerning the biochemical nature of the dissimilatory pathway of tartrate; respiration is one route of its degradation (Ruffner, 1982a).

Malic acid

Malic acid is a very active intermediate in grape metabolism (Peynaud and Ribéreau-Gayon, 1971; Winkler *et al.*, 1974), and plays a significant role in anabolic reactions, such as dark fixation of carbon dioxide, and also in the acid catabolizing processes of fruit ripening (Ruffner, 1982b). Malic acid accumulates in berries up to 15 mg.g^{-1} fresh weight, or 95 µmol/berry during phase I of rapid growth (Fig. 6.6). At *véraison*, a rapid decrease in malate content, ranging from 2 to 3 mg.g^{-1} fresh weight is observed (Ruffner and Hawker, 1977; Ruffner, 1982b).

Grapevine and young green berries assimilate carbon dioxide by the C$_3$-mechanism (Ruffner, 1982b; Ruffner *et al.*, 1983a,b). Thus, malate synthesis in leaves and very young berries involves the assimilation of CO$_2$ mediated by ribulose-1,5-bisphosphate carboxylase to form phosphoglyceric acid as the primary product, followed by β-carboxylation of phosphoenol pyruvate (PEP) (Brem *et al.*, 1981; Ruffner *et al.*, 1983b) (Fig. 6.7). The last reaction is catalysed by PEP-carboxylase (PEPC) (Hawker, 1969; Lakso and Kliewer, 1975, 1978; Ruffner and Kliewer, 1975; Ruffner, 1982b; Ruffner *et al.*, 1983b). The activity of PEPC is inhibited by the presence of malic acid (Lakso and Kliewer, 1975), probably as a feedback mechanism preventing excessive malate accumulation in the cytoplasm (Lakso and Kliewer, 1978).

The high malic enzyme activity observed even during the acid accumulating phase of berry development is indicative of alternative biosynthesis of malic acid via the carboxylation of pyruvate (Peynaud

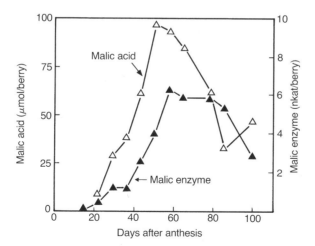

Fig. 6.6 Developmental changes in malic acid content and malic enzyme activity in grape berries. (From Ruffner *et al.*, 1984.)

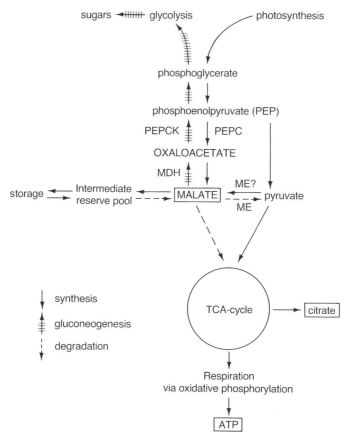

Fig. 6.7 Biosynthesis, dissipation and regulation of malic acid in grape berries. (Modified from Ruffner, 1982b; Ruffner *et al.*, 1983a.)

and Ribéreau-Gayon, 1971; Winkler *et al.*, 1974) (Figs 6.6 and 6.7). Also, the *in vitro* reversibility of the reaction supports the possibility that malate may be formed under conditions facilitating carboxylation of pyruvate, notably high CO_2/HCO_3^- and NADPH/NADP ratios (Ruffner *et al.*, 1984). However, a very limited incorporation of ^{14}C into malate and the uniform labelling pattern of the dicarboxylic acid after administration of [U-14C]alanine to grape berries, before and after the onset of ripening, indicate that the 'reverse' reaction does not contribute significantly to grape malate synthesis (Possner *et al.*, 1981; Ruffner *et al.*, 1984).

Sucrose is the main precursor of malic acid in grape berries, via the common glucose utilizing pathways, glycolysis and oxidative pentose phosphate pathway, and via β-carboxylation (Ruffner *et al.*, 1976; Ruffner and Hawker, 1977). Ripening of grape berries is accompanied by a marked alteration in malate metabolism (Kriedemann, 1968; Possner *et al.*, 1983). The acid decrease during ripening involves mainly malate consumption (Ruffner and Hawker, 1977; Saito and Kasai, 1978) most likely by NADP-dependent malic enzyme (Hawker, 1969; Lakso and Kliewer, 1975, 1978), by respiration of the formed pyruvate, and to a lesser extent, by phosphoenolpyruvate carboxykinase (Ruffner and Kliewer, 1975). The respiratory quotient has values close to unity in the green berry and this increases to about two in post *véraison* fruit (Kriedemann, 1969; Koch and Alleweldt, 1978) and the higher $^{14}CO_2$-evolution from ripening berries, after administering ^{14}C-malate, compared to green berries (Hardy, 1968; Kriedemann, 1968; Steffan *et al.*, 1975), supports the theory that respiration contributes significantly to malate degradation.

Malic enzyme (ME) has been purified from grape berries. It has a pI of 5.8; it is NADP-specific and requires divalent cations for activity (Spettoli *et al.*, 1980; Possner *et al.*, 1981). Substitution of Mg^{2+} by Mn^{2+} increases maximal turnover rates and enzyme affinity for malate. In the malate decarboxylating direction at pH 7.4, grape ME displays positive cooperativity towards its substrate, the curve approaching normal Michaelis–Menten kinetics at pH 7.0 (Possner *et al.*, 1981). In addition, grape ME is characterized by a very pronounced specificity for the naturally occurring L(–) form of malic acid (Ruffner *et al.*, 1984). These features are considered indicative of the regulatory properties of the enzyme.

The close correlation between acid decrease and massive sugar accumulation in the berry at *véraison* led to the hypothesis that a metabolic inter-relationship exists between these two processes. The distribution of ^{14}C-activity within glucose upon administration of [2,3-^{14}C] malate for relatively short metabolic periods (6 hours), shows that the hexose is mainly (80%) labelled in the peripheral carbons 1,2 and 5,6 (Ruffner *et al.*, 1975). The reassimilation of $^{14}CO_2$ after decarboxylation would result in ^{14}C incorporation into C-atoms 3 and 4 of glucose. The labelling pattern of glucose is indicative of the gluconeogenic metabolism of malic acid: oxidation of malate to oxaloacetic acid, decarboxylation and subsequent formal reversion of glycolysis to yield hexose.

The oxidation of malate to oxaloacetate is catalysed by malate dehydrogenase (MDH) (EC 1.1.1.37) and the decarboxylation reaction by PEP-carboxykinase (Hawker, 1969; Dal Belin Peruffo and Pallavicini, 1975; Ruffner and Kliewer, 1975; Palejwala *et al.*, 1985). Changes in activity of MDH throughout maturation are contradictory (Hawker, 1969; Dal Belin Peruffo and Pallavicini, 1975); a shift in the isoenzyme pattern (Dal Belin Peruffo and Pallavicini, 1975) may indicate, as was pointed out by Ruffner (1982b) and reported by Palejwala *et al.* (1985), that MDH is synthesized *de novo* after *véraison* in some compartments and, although this change may not affect the overall MDH activity, the enzyme could be locally more active than before.

Phosphoenol pyruvate-carboxykinase activity in grape berries is highest before *véraison* (Ruffner and Kliewer, 1975). However, gluconeogenic activity is nearly 5-fold higher at *véraison* than in immature or ripe berries (Hardy, 1968; Ruffner *et al.*, 1975). The contribution of gluconeogenic transformation of malate to sugar does not exceed 5%, assuming that no fresh malate is formed during this period. Since malate is further metabolized predominantly via respiration (Harris *et al.*, 1971; Alleweldt and Koch, 1977), gluconeogenesis must be considered as a mechanism mediating the flow of carbon between the two groups of compounds rather, than a quantitatively important pathway of hexose accumulation (Ruffner and Hawker, 1977).

Malate content at maturity is negatively correlated with temperature (Buttrose *et al.*, 1971; Kliewer *et al.*, 1972; Kliewer, 1973; Hale and Buttrose, 1974; Winkler *et al.*, 1974; Klenert *et al.*, 1978). This is due to higher respiratory rates (Rapp *et al.*, 1971; Steffan *et al.*, 1975; Ruffner, 1982a), and to switching from carbohydrate degradation to acid break-down as the main line of supply of respiratory substrates. This is supported by the accompanying increase in respiratory quotient (RQ) values (Koch and Alleweldt, 1978), the higher ME activity (Ruffner *et al.*, 1976, 1984), the failure of green berries, in contrast to ripening fruit, to metabolize exogenous malic acid (Kriedemann, 1968), the considerably lower rates of incorporation and respiration of [^{14}C] sucrose after *véraison* (Possner *et al.*, 1983) and the low [^{14}C] malate levels after metabolism of labelled sucrose in ripening berries (Possner *et al.*, 1983).

Therefore, the decrease in malic acid content during maturation is the result of a reduced malate synthesis combined with accelerated breakdown. The decrease in the synthesis of malic acid, is a consequence of an inhibition of glycolysis which results in an accumulation of imported sugars (Ruffner *et al.*, 1983a). Upon inhibition of this pathway, the intermediates of glycolysis and the Krebs cycle are consumed and replenished from the storage compartment (Fig. 6.7). Thus, the mechanism which controls the rate of malic acid remetabolism during ripening, may play a role in controlling the malic acid concentration during development and ripening of grape berry.

6.8 PHENOLIC COMPOUNDS

Flavonoids and other phenolic compounds cover a large variety of substances containing one or more hydroxylated aromatic rings in their molecules. In grapes, they are of special significance because they contribute significantly to the colour, taste and flavour of the fresh fruit and also to the colour, taste and body of wines and other products. The chemical structure of flavonoids is based on a C_{15} skeleton with a chromane ring bearing a second aromatic ring B in position 2,3, or 4. Various groups of flavonoids are classified according to the oxidation and substitution patterns of ring C. Both the oxidation state of the heterocyclic ring and the position of ring B are important in the classification (Hahlbrock, 1981). The major phenols present in grape constituents are anthocyanins, benzoic acids, cinnamic acids, flavonols and tannins (Singleton and Esau, 1969; Peynaud and Ribéreau-Gayon, 1971)

6.8.1 Anthocyanins

Anthocyanins are the major pigments of grape cultivars. Skin colour of grape berries varies and cultivars are classified accordingly into white, red and black. The structure of anthocyanins is discussed in Chapter 1.

The anthocyanidin pelargonidin, is fairly common in many fruits and berries but is very rare, if not absent, in grapes (Webb, 1970). Anthocyanidins in grapes may be considered primitive pigments (cyanidin and delphinidin), stable pigments (peonidin and malvidin) and petunidin, which is in the middle of the chain of transformation from cyanidin to malvidin (Roggero *et al.*, 1986).

Anthocyanins in grapes are the anthocyanidins modified by attachment of a molecule of glucose at the 3-OH position, or by attachment of glucose at both the 3 and the 5 positions. The diglucosides are characteristic of the American *Vitis* species *rupestris, riparia, labrusca*, etc., whereas only monoglucosides are present in *V. vinifera*. Further complication can result from esterification of the glucose hydroxyls with organic acids. The acids acylating anthocyanins in grapes for which the evidence is convincing are *p*-coumaric, caffeic and acetic (Webb, 1970). The anthocyanins are water and lower-alcohol soluble, while the coloured benzopyrilium ion form requires a low pH medium for stability so that the extracting solvent is usually an acidified alcohol solution.

Anthocyanin accumulation in berries starts at or shortly after the increased accumulation of sugars (stage III, Fig. 6.8a) and is mostly localized in the first three to six subepidermal layers (Kataoka *et al.*, 1983; Hrazdina *et al.*, 1984). Anthocyanins are at equilibrium between the red-coloured flavylium salt, the purple-coloured anhydrous base, and the colourless carbinol base forms (Hrazdina and Moskowitz, 1980). Cyanidin and delphinidin contents increase sharply during the first weeks of ripening but decrease thereafter, indicating that they are transformed into stable pigments.

Endogenous sugars were considered to be the causal agents for synthesis of anthocyanins and other phenolic compounds (Pirie and Mullins, 1976, 1977), but other results have shown that treatments, which

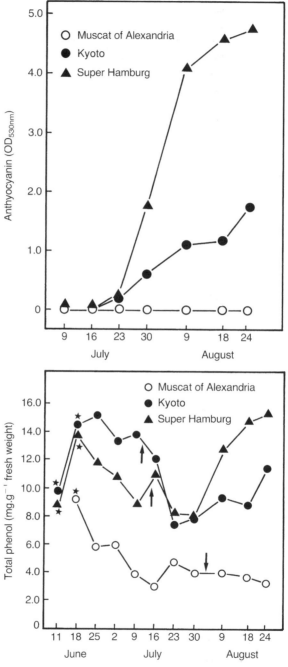

Fig. 6.8 Total phenolics and anthocyanins in ripening grape berries. Asterisks indicate that whole berries were used for extraction and arrows denote the beginning of *véraison*. (From Kataoka *et al.*, 1983.)

increase the synthetic rate of anthocyanins, such as light and ethylene, do not affect the levels of sugars in grape skins (Wicks and Kliewer, 1983). However, it has not been determined whether these treatments affect the redistribution of soluble carbohydrates within the epidermal and sub-epidermal cell layers and between the sub-cellular organelles and compounds without affecting the total soluble skin carbohydrates.

Several exogenous factors affect synthesis and accumulation of anthocyanins in berries. Among these are light, temperature, crop size and leaf area (Buttrose et al., 1971; Kliewer and Weaver, 1971; Ribéreau-Gayon, 1971, 1972; Kliewer and Schultz, 1973; Flora, 1978; Roubelakis-Angelakis and Kliewer, 1986). Also, plant growth regulators and in particular ethylene and abscisic acid cause an increase in anthocyanin accumulation (Jensen et al., 1975; Lee et al., 1979b; Tomana et al., 1979; Kataoka et al., 1982).

The astringent properties of tannins depend on their total content, on the structure of the elementary units and on their degree of polymerization. In white grapes, trans-caffeoyl tartaric acid, cis-coumaroyl tartaric acid, and trans-coumaroyl tartaric acid are the major compounds of acidic phenols and catechin, epicatechin, and two unidentified compounds are the major constituents in the neutral phenols. The two unidentified compounds have been tentatively identified as catechin-gallate and isomers of catechin-catechin-gallate (Ong and Nagel, 1978; Lee and Jaworski, 1987). Also, the hydroxycinnamic acid tartrates, flavan-3-ols and oligomeric procyanidins are present in high concentrations early in the growing season and decrease to their lowest concentrations at harvest (Lee and Jaworski, 1989). There exist spatial, seasonal and cultivar variations in the amount and the kind of phenolics in grape berries. Berry skins contain more hydroxycinnamic tartrates than the flesh, while larger quantities of flavan-3-ols and procyanidins are found in the berry flesh (Lee and Jaworski, 1989); these findings partially explain the differences in browning potential of grapes in relation to cultivar and seasonal variations.

6.8.2   Flavonols

Flavonols are localized in the solid parts of the cluster. In red cultivars, flavonols are present in much smaller quantities than anthocyanins; co-pigmentation of anthocyanins with flavonols has been shown to influence the colour of other plants and this may also occur in grapes. Using paper chromatography, the following flavonols have been identified in red cultivars: the 3-glucosides of kaempferol, quercetin and myricetin and quercetin-3 glucuronide; myricetin derivatives are absent in white cultivars (Ribéreau-Gayon, 1964; Cheynier and Rigaud, 1986). High performance liquid chromatographic separation has verified the above flavonols and has revealed the presence of kaempferol-3-galactoside, isorhamnetin-3-glucoside and probably kaempferol and myricetin-3-glucuronides and three diglycosides in the grape berry skins (Roggero *et al.*, 1986).

The grape seeds contain high amounts of phenolic compounds, among which are gallic acid, monomeric flavanols (such as catechin, epicatechin), polymeric flavans (such as procyanidins B_1, B_2, B_3, B_4), and epicatechin gallate and epigallocatechin (Singleton *et al.*, 1966; Su and Singleton, 1969; Singleton, 1980; Oszmianski *et al.*, 1986). These compounds contribute significantly to the oxidative browning of grapes.

During ripening, the concentration of total phenols in grape berries generally rises for a short time early in the developmental process and then shows a steady decline (Dumazert *et al.*, 1973; Kataoka *et al.*, 1983; Singleton, 1966) (Fig. 6.8). However, total phenols per berry increase until rather late in maturation, showing that there is synthesis but this is insufficient to match berry enlargement (Singleton, 1966).

The enzymic reactions leading to the formation of flavonoids from compounds of intermediary metabolism have been summarized in Chapter 1. These arise from general phenylpropanoid metabolism via the conversion of L-phenylalanine to 4-coumaroyl-coenzyme A or via the conversion of acetyl-CoA to malonyl-CoA (Hahlbrock, 1981). Phenylalanine ammonia-lyase (PAL) (EC 4.3.1.5) is the enzyme mediating the elimination of NH_3 from L-phenylalanine to give *trans*-cinnamic acid, and is the first enzyme in the pathway that channels phenylalanine away from the synthesis of proteins towards that of phenylpropanoid and flavonoid compounds. PAL is located mainly in the epidermal cell layers of the berry and the *in vitro* determination of its activity can be masked by several factors (Roubelakis-Angelakis and Kliewer, 1985). Also, PAL activity in berry skins is high during the early stages of berry development and this is followed by a rapid decline toward *véraison*. In coloured cultivars, PAL activity increases again with the beginning of berry colouration and there is a correlation between activity and the colour intensity (Kataoka *et al.*, 1983; Hrazdina *et al.*, 1984) (Fig. 6.9).

The activity of chalcone synthase, the first enzyme in the specific flavonoid pathway, increases rapidly during the third phase of berry development and declines sharply thereafter (Hrazdina *et al.*, 1984). Cinnamate 4-monooxygenase, hydroxycinnamate: CoA-lyase, chalcone isomerase (EC 5.5.1.8) and UDP- glucose flavonoid glucosyltransferase are also present in developing grape berries (Hrazdina *et al.*, 1984).

Polyphenol oxidases (PPO) (EC 1.10.3.1) comprise a large group of enzymes which can utilize molecular oxygen for the oxidation of phenolic substances. They are divided into two main groups: the catechol oxidases and the laccases (EC 1.10.3.2), which have been classified under

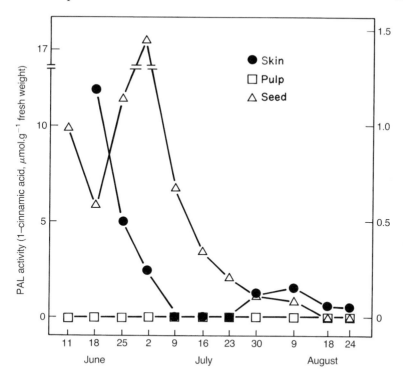

Fig. 6.9 Phenylalanine ammonia-lyase activity in the grape during ripening. (From Kataoka *et al.*, 1983.)

the general name monophenol monooxygenase (EC 1.14.18.1) (monophenol, dihydroxy-L-phenylalanine: oxygen oxidoreductase). The activity of PPO in grape berries is considerably affected by several factors, such as cultivar, developmental stage, and environmental conditions (Mayer and Harel, 1979; Wisseman and Lee, 1980; Sapis *et al.*, 1983a,b). Grapes generally contain high PPO activity (Harel and Mayer, 1971; Raddler and Totokfalvy, 1973; Dubernet and Ribéreau-Gayon, 1974; Kidron *et al.*, 1978; Wisseman and Lee, 1980; Nakamura *et al.*, 1983; Sapis *et al.*, 1983a,b; Interesse *et al.*, 1984). Multiple forms of grape PPO have been distinguished by polyacrylamide gel electrophoresis (Harel and Mayer, 1971; Harel *et al.*, 1973; Dubernet and Ribéreau-Gayon, 1974; Kidron *et al.*, 1978; Interesse *et al.*, 1984) and in some cases PPO has been purified and characterized (Harel and Mayer, 1971; Wisseman and Lee, 1980; Nakamura *et al.*, 1983; Interesse *et al.*, 1984). In grape musts PPO activity increases as the yield increases (Yokotsuka *et al.*, 1988). The insoluble precipitate contains 74.0% of the protein and 42.2% of the PPO activity; this PPO activity is solubilized at pH 10 and exhibits optimum activity at pH 6. The molecular weight of the purified latent PPO from grape berries cv Monastrell is 38 to 39 kDa (Sanchez-Ferrer *et al.*, 1989), whereas the molecular weight of a PPO purified from Koshu grapes was 39 to 41 kDa (Nakamura *et al.*, 1983). Polyphenol oxidase activity varies throughout the ripening period (Valero *et al.*, 1989), with activity very low during the first week before *véraison* and increasing thereafter (Wisseman and Lee, 1980; Valero *et al.*, 1989).

Although appreciable work has been performed on grape flavonoids, there is still much to be done to further elucidate their biosynthetic pathways, transformations, and the molecular regulation of the enzymes mediating these reactions.

6.9 AROMA

Aroma is a significant and complex character of grape quality. The characteristic aroma of grapes and wines is a natural blend of several hundred chemically different compounds (Gunata *et al.*, 1985a,b; Strauss *et al.*, 1986) which are synthesized during ripening. These aromatic substances are mainly localized in the skin of the berries (Winkler *et al.*, 1974). Although the precursor compounds of these substances are synthesized in the leaves (Gunata *et al.*, 1986) the synthesis and evolution of the aroma takes place in the berries (Winkler *et al.*, 1974).

Aroma is a distinctive varietal character. Generally, varieties of *labrusca* and *rotundifolia* have a distinct and pronounced odour, whereas those of *vinifera* (with the exception of Muscat group) have an aroma more delicate and subdued. The foxy aroma of *V. labrusca* is due to the compound methyl anthranilate (Winkler *et al.*, 1974). Muscat varieties have an aroma almost as distinguished as that of the *labrusca* varieties and hybrids; on the other hand, varieties like Thompson Seedless are only mildly vinous.

The most aromatic fraction of Muscat odour is mainly composed of monoterpenes, especially linalool and genaniol. In addition, certain alcohols such as benzyl and 2-phenylethyl alcohols, as well as ethers, aldehydes, hydrocarbons and polyfunctional derivatives are among the compounds responsible for grape aroma (Gunata *et al.*, 1985b; Ribéreau-Gayon *et al.*, 1975; Strauss *et al.*, 1986; Williams *et al.*, 1982a). Genaniol and nerol are localized in the skin, whereas linalool is distributed almost equally between the juice and the solid parts of the berry. The terpenes are found either in the free or glycosidically-bound forms (Cordonnier and Bayonore, 1974; Williams *et al.*, 1982a, b). These bound forms occur as disaccharide glycosides, i.e. β-rutinosides (6-*O*-α-L-rhamno-pyranosyl-β-D-glucopyranosides and 6-*O*-α-L-arabinofuranosyl-β-D-glucopyranosides) (Williams *et al.*, 1982a).

The levels of free and bound monoterpenes increase during development and maturation of the berry. The bound forms are abundant in the green-berry stage and exhibit higher levels than the free forms throughout maturation (Gunata *et al.*, 1985a,b; Wilson *et al.*, 1984). The bound fraction continues to increase even after over-ripeness, whereas accumulation of free forms stops or even declines (Gunata *et al.*, 1985a,b). The release of the glycosidic forms may contribute to enhancement of the aroma of grape juice; this may occur via either acid or enzymic hydrolysis (Williams *et al.*, 1982b; Aryan *et al.*, 1987; Gunata *et al.*, 1988, 1989, 1990).

Enzymic hydrolysis of grape monoterpenyl disaccharide glycosides in grape berries is a sequential reaction requiring three glycosidases; firstly, α-L-arabinofuranosidase or α-L-rhamnopyranosidase and secondly, β-D-glucopyranosidase (Gunata *et al.*, 1988). These enzymes are present in grape berries of both Muscat and non-Muscat varieties and increase during ripening (Aryan *et al.*, 1987; Gunata *et al.*, 1989). Surprisingly, β-glycosidase and β-galactosidase activities are mostly present in the pulp/juice rather than in the skins, while bound and free glycosides are mostly localized in the skin (Gunata *et al.*, 1985; Aryan *et al.*, 1987).

Lipids may contribute adversely to the aroma of grapes (van Wyk *et al.*, 1967) via the actions of lipoxygenase (EC 1.13.11.12) (Caryel *et al.*, 1983) and the subsequent actions of cleaving enzyme and alcohol dehydrogenase (EC 1.1.1.1). The products of these reactions are six-carbon aldehydes and alcohols (hexanal, 2-hexanal, etc.) (Drawert *et al.*, 1966). This requires the crushing of berries so that aroma precursors previously kept separate come into contact with the corresponding enzymic systems. In grapes, the occurrence of lipoxygenase and alcohol dehydrogenase activities have been reported (Caryel *et al.*, 1983; Molina *et al.*, 1986).

It must be emphasized, however, that despite the considerable work directed towards identifying the aromatic compounds in grape fruit and wine, elucidation of their enzymatic synthesis and regulation is still lacking.

6.10   CELL WALL METABOLISM

Softening in grapes is due largely to flaccidity from water loss (Nelson, 1979). However, recent studies indicate that changes occur in cell walls of grape berries during ripening and these may contribute to berry softening. Despite their considerable importance, information on the cell wall constituents of grapes and changes in their composition during development and ripening is scant. Recently, more attention has been paid to elucidating the structure of the cell wall of grape pulp and especially the structure of pectic substances, which play an important role in wine-making. Soluble pectic polymers from the mesocarp occurring in musts may play a detrimental role in white wine-making by restricting juice extraction (Robertson *et al.*, 1980) and delaying must clarification (Saulnier and Brillouet, 1988; Saulnier *et al.*, 1988; Brillouet *et al.*, 1990).

The composition and structure of the fruit cell wall is covered in detail in Chapter 1. In grapes, cellulose accounts for about 17% of cell wall material, hemicellulose 25% and pectin substances almost 50% (Jona *et al.*, 1983). The pectin content of ripe fruit ranges from 0.02–0.6%. Galacturonic acid comprises 92% of the total pectins, with rhamnose, galactose, arabinose and xylose as minor components (Yokotsuka *et al.*, 1982). The neutral, and acidic, soluble pectins from the pulp of grape berries have very complex structures. They are mainly pectic polysaccharides made of galactose, arabinose and galacturonic acid incorporated into type II arabinogalactans, arabinans and rhamno-galacturonans. Their concentrations and the neutral-acidic sugar ratio vary greatly depending upon the cultivar (Brillouet, 1987). Conversely, water-insoluble pectin substances, extracted by dilute acid and endo-pectin-lyase (EC 4.2.2.10), show arabinan-like structures associated with minor proportions of arabinogalactans of both types I and II. The free pectin fractions contain minor amounts of proteins rich in hydroxy-proline, serine and glycine (Saulnier and Brillouet, 1988, 1989). The pectic polysaccharides contain homogalacturonan with regions degraded by endopolygalacturonase and endopectin-lyase and resistant regions rich in neutral side-chains (Saulnier and Thibault, 1987a, b).

Total pectins in grapes are present in lower amounts, but are more methylated than in other fruits (Jona and Foa, 1979). There is a con-siderable variation in the degree of esterification of grape pectin, varying from 43.9% to 65.3% (Ishii and Yokotsuka, 1973) or even 70% (Yokotsuka *et al.*, 1982). This variation is possibly due to the different cultivars examined, which contain varying amount of pectin-methylesterase (PME) (EC 3.1.1.11). Indeed, PME activity is higher in red than in white varieties, with activity in *vinifera* lower than in *labruscana* (Lee *et al.*, 1979a). PME exhibits maximum activity within a pH range of between 7.1 and 7.5 (Datunashvili *et al.*, 1976; Lee *et al.*, 1979a) and is largely found in the skin; it shows an increase of more than 100% in Concord and Riesling grapes during ripening (Lee *et al.*, 1979a). On the other hand, grape polygalacturonase (EC 3.2.1.15) has optimum activity within the pH range 4.5 to 4.7 and has a pI of 5.3 (Tyrina, 1977). The presence of these two cell wall hydrolysing enzymes in grape berry pulp may account for the continuous decrease in total pectin substances during the ripening of grapes (Robertson *et al.*, 1980).

PROSPECTS

Despite the considerable contribution of grapes to the world produce and processing industries (table grape, raisins and wines), the elucidation of metabolic processes taking place during development and ripening and their regulation at the biochemical and molecular level are still poorly understood. Reasons for this lack of knowledge may be attributed to the inherent nature of this fruit. Due to its 'non-climacteric' character, all physiological and biochemical changes take place very slowly; therefore, grape is not an attractive candidate in which to study fruit ripening. Furthermore, its high levels of carbohydrates, acids, flavonoids, etc., is in most cases a barrier to biochemical and molecular studies. Increasingly, immunological and molecular biological methods should be applied to grape ripening; recent reports on methods describing one- and two-dimensional protein electrophoresis (Loulakakis and Roubelakis-Angelakis, 1990a,b,c, 1991, 1992; Tesniere and Robin, 1992), analysis of berry RNA (Tesniere and Vayda, 1991), production of grape berry skin callus, grape berry and grape berry skin RNA (Loulakakis, Roubelakis-Angelakis and Kanellis, unpublished data) and DNA isolation (Harding and Roubelakis-Angelakis, unpublished data) together with transformation of grapevine should ignite further developments and knowledge on grape berry ripening. Areas of research should include: gene expression, protein synthesis and regulation of enzymes catalysing sugar formation and translocation, organic acid, and especially malic acid synthesis, degradation and compartmentation, anthocyanin synthesis, hormonal regulation of ripening, aroma formation and cell wall metabolism.

Acknowledgements

We thank Drs B.G. Coombe and V.L. Singleton for reading the manuscript and for their suggestions, and Dr H.P. Ruffner for his constructive comments on the manuscript. We also thank Mrs Mairyannakis-Toutexis for typing assistance.

REFERENCES

Alexander, D. Mc. E. (1957) Seasonal fluctuations in the nitrogen content of the Sultana vine. *Australian Journal of Agricultural Research*, **8**, 162–178

Alleweldt, G. (1977) Growth and ripening of grape berry. In O.I.V. Int. Symp. on the Quality of the Vintage, Cape Town, South Africa, pp. 129–136

Alleweldt, G. and Hifny, H.A.A. (1972) Zur stiellahme der reben. II. Kausalanalytische untersuchungen. *Vitis*, **11**, 10–28

Alleweldt, G. and Koch, R. (1977) Der athylengehalt reifender weinbeeren. *Vitis*, **9**, 263–273

Alleweldt, G., Düring, H. and Waitz, G. (1975) Utersuchungen zum mechanismus der zuckereinlagerung in die qachsenden weinbeeren. *Angew. Bot.*, **49**, 65–73

Alleweldt, G., Düring, H. and Jung, K.H. (1984) Zum einfluss des klimas auf beerenentwicklung, ertrag und Qualitat bei Reben: Ergebnisse einer siebenjahrigen Factorenanalyse. *Vitis*, **23**, 127–142

Alleweldt, G., Engel, M. and Gebbing, H. (1981) Histologische Untersuchungen an weinbeeren. *Vitis*, **20**, 1–7

Aryan, A.P., Wilson, B., Strauss, C.B. and Williams, P.J. (1987) The proportion of glycosides of *Vitis vinifera* and a comparison of their β-glycosidase activity with that of exogenous enzymes. As assessment of possible applications in enology. *American Journal of Enology and Viticulture*, **38**, 182–188

Blanke, M.M. and Leyhe, A. (1987) Stomatal activity of the grape berry cv. Riesling, Muller-Thurgan and Ehrenfelser. *Journal of Plant Physiology*, **127**, 451–460

Brem, S., Ruffner, H.P. and Rast, D.M. (1981) Aspects of malate formation in grape leaves. In *Photosynthesis VI*, (ed. G. Akoyunoglou), Balaham Inst. Sci. Services, Philadelphia, pp. 181–187

Brillouet, J.M. (1987) A study of pectic polysaccharides in must from various mature grapes grown in the Pech Rouge experimental vineyards. *Biochemie*, **69**, 713–721

Brillouet, J.M., Bosco, C. and Miutaunet, M. (1990) Isolation, purification and characterization of an arabinogalactan from a red wine. *American Journal of Enology and Viticulture*, **41**, 29–36

Brown, S.C. and Coombe, B.G. (1982) Sugar transport by an enzyme complex at the tonoplast of grape pericarp cells? *Naturwissenschaften*, **69**, 43–45

Brown, S.C. and Coombe, B.G. (1984) Solute accumulation by grape pericarp cells. II. Studies with protoplasts and isolated vacuoles. *Biochem. Physiol. Pflanzen.*, **179**, 157–171

Buttrose, M.S., Hale C.R. and Kliewer, W.M. (1971) Effect of temperature on the composition of "Cabernet Sauvignon" berries. *American Journal of Enology and Viticulture*, **22**, 71–75

Caryel, A., Crouzet, J. and Chan, H.W. (1983) Evidence for the occurence of lipoxygenase activity in grapes (variety Carignane). *American Journal of Enology and Viticulture*, **34**, 77–82

Cawthon, D.L. and Morris, J.R. (1982) Relationship of number and maturity to berry development, fruit maturation, hormonal changes, and uneven ripening of "Concord" (*Vitis labrusca* L.) grapes. *Journal of American Society of Horticultural Science*, **107**, 1097–1104

Cheynier, V. and Rigaud, J. (1986) HPLC separation and characterization of flavonols in the skins of *Vitis vinifera* var. Cinsault. *American Journal of Enology and Viticulture*, **37**, 248–251

Considine, J.A. (1981) Stereological analysis of the dermal system of fruit of the grape *Vitis vinifera* L. *Australian Journal of Botany*, **29**, 463–474

Considine, J.A. and Knox, R.B. (1979) Development and histochemistry of the cells, cell walls and cuticle of the dermal system of fruit of the grape, *Vitis vinifera* L. *Protoplasma*, **99**, 347–365

Considine, J.A. and Knox, R.B. (1981) Tissue organs, cell lineages and patterns of cell division in the developing dermal system of the fruit of *Vitis vinifera* L. *Planta*, **151**, 403–412

Coombe, B.G. (1960) Relationship of growth and development to changes in sugars, auxins, and gibberellins in fruit of seeded and seedless varieties of *Vitis vinifera*. *Plant Physiology*, **35**, 241–250

Coombe, B.G. (1973) Regulation of set and development of the grape berry. *Acta Horticulture*, **34**, 261–269

Coombe, B.G. (1976) The development of fleshy fruits. *Annual Review of Plant Physiology*, **27**, 207–228

Coombe, B.G. (1980) Development of the grape berry. I. Effects of time of flowering and competition. *Australian Journal of Agricultural Research*, **31**, 125–131

Coombe, B.G. (1987) Distribution of solutes within the developing grape berry in relation to its morphology. *American Journal of Enology and Viticulture*, **38**, 129–127

Coombe, B.G. (1989) The grape berry as a sink. *Acta Horticulturae*, **239**, 149–158

Coombe, B.G. (1992) Research on development and ripening of the grape berry. *American Journal of Enology and Viticulture*, (in press)

Coombe, B.G. and Bishop, G.R. (1980) Development of the grape berry. II. Changes in diameter and deformability during *véraison*. *Australian Journal of Agricultural Research*, **31**, 499–509

Coombe, B.G. and Hale, C.R. (1973) The hormone content of ripening grape berries and the effect of growth substances treatments. *Plant Physiology*, **51**, 629–634

Coombe, B.G. and Phillips, P.E. (1982) Development of the grape berry II. Compositional changes during *véraison* measured by sequential hypodermic sampling. *Proceedings of International Symposium on Grapes and Wine*, Davis, Calif. June, 1980, pp. 132–136

Coombe, B.G., Bovio, M. and Schneider, A. (1987) Solute accumulation by grape pericarp cells. *Journal of Experimental Botany*, **38**, 1789–1798

Cordonnier, R. and Bayonore, C. (1974) Mise en evidence dans la baie de raisin, varieté muscat d'Alexandrie, de monoterpenes lies revelables par une ou plusieurs enzymes du fruit. *CR Acad. Sci. Paris*, **D 278**, 3387–3390

Dal Belin Peruffo, A. and Pallavicini, C. (1975) Enzymatic changes associated with ripening of grape berries. *Journal of the Science of Food and Agriculture*, **26**, 559–566

Datunashvili, E.N., Tyrina, S.S. and Kardash, N.K. (1976) Some properties of grape pectin methyl esterase. *Applied Biochemistry and Microbiology*, **12**, 36–40

Downton, W.J.S. and Loveys, B.R. (1978) Compositional changes during grape berry development in relation to abscisic acid and salinity. *Australian Journal of Plant Physiology*, **5**, 415–423

Drawert, F., Heimann, W., Emberger, R. and Tobback, P. (1966) Uber die biogenese von armastoffen bei pflanzen und fruchen. II. Enzymatisch produktion von 2-hexanol, hexanal und deren vorsterffen. *Ann. Chem.*, **694**, 200–208

Dubernet, M. and Ribéreau-Gayon, P. (1974) Isoelectric point changes in *Vitis vinifera* catechol oxidase. *Phytochemistry*, **13**, 1085–1087

Dumazert, G., Margulis, H. and Montreau, F.R. (1973) Evolution des composes phenoliques au cours de la maturation d'un *Vitis vinifera* blanc: le Mauzac. *Annals Technology Agriculture*, **22**, 137–151

Düring, H. (1973) Abscisinsaure in *Vitis vinifera* fruchten wahrend der reife. *Natuzwissenschaften*, **60**, 301–302

Düring, H. (1974) Abscisinsaure in *Vitis vinifera* fruchten wahrend der reife. *Vitis*, **13**, 112–119

Düring, H. and Alleweldt, G. (1984) Zue moglichen Bedeuting der Avscisinsaure bei der zuckereinlagerung in die Weinbeere. *Ber. Deutsch. Bot. Ges.*, **97**, 101–113

Farmahan, H.L. and Pandey, P.M. (1976) Hormonal regulation of the lag phase in seeded and seedless grapes (*Vitis vinifera* L.). *Vitis*, **15**, 227–235

Flanzy, G. and Poux, C. (1965) Sur la teneur en acides amine dans le raisin et la mout en relation aux conditions climatiques. *Ann. Technol. Agric.*, **14**, 87–91

Flora, L.F. (1978) Influence of heat cultivar and maturity on the anthocyanidin-3,5-diglucosides of muscadine grapes. *Journal of Food Science*, **43**, 1819–1821

Ghisi, R., Jannini, B. and Passera, C. (1984) Changes in the activities of enzymes involved in nitrogen and sulphur assimilation during leaf and berry development of *Vitis vinifera*. *Vitis*, **23**, 257–267

Guelfat-Reich, S. and Safran, B. (1971) Indices of maturity of table grapes as determined by variety. *American Journal of Enology and Viticulture*, **22**, 13–18

Gunata, Y.Z., Bayonove, C.L., Baumes, R.L. and Cordonnier, R.E. (1985a) The aroma of grapes. 1. Extraction and determination of free and glycosidically bound fractions of some grape aroma components. *Journal of Chromatography*, **331**, 83–90

Gunata, Y.Z., Bayonove, C.L., Baumes, R.L. and Cordonnier, R.E. (1985b) The aroma of grapes. Localization and evolution of free and bound fractions of some grape aroma components cv Muscat during first development and maturation. *Journal of the Science of Food and Agriculture*, **36**, 857–862

Gunata, Y.Z., Bayonove, C.L., Baumes, R.L. and Cordonnier, R.E. (1986) Changes in free and bound fractions of aromatic components in vine leaves during development of Muscat grapes. *Phytochemistry*, **25**, 944–946

Gunata, Y.Z., Bayonove, C.L., Cordonnier, R.E., Arnaud, A. and Galzy, P. (1990) Hydrolysis of grape monoterpenyl glycosides by *Candida molischiana* and *Candida wickerhamii* β-glycosides. *Journal of the Science of Food and Agriculture*, **50**, 499–506

Gunata, Y.Z., Biron, C., Sapis, J.C. and Bayonove, C.L. (1989) Glycosidase activities in sound and rotten grapes in relation to hydrolysis of grape monoterpenyl glycosides. *Vitis*, **28**, 191–197

Gunata, Y.Z., Bitteur, S., Brillouet, J.M., Bayonove, C. and Cordonnier, R.E. (1988) Sequential enzymic hydrolysis of potentially aromatic glycosides from grape. *Carbohydrate Research*, **184**, 139–149

Hahlbrock, K. (1981) Flavonoids. In *The Biochemistry of Plants*, (eds P.K. Stumpf and E.E. Conn), Vol. 7, pp. 425–456

Hale, C.R. (1977) Relation between potassium and the malate and tartrate contents of grape berries. *Vitis*, **16**, 9–19

Hale, C.R. and Buttrose, M.S. (1974) Effect of temperature on ontogeny of berries of *Vitis vinifera* L. *Journal of the American Society of Horticultural Science*, **99**, 390–394

Harada, H. (1980) Cytokinin-binding protein in grape berries. *Vitis*, **19**, 216–225

Hardy, P.J. (1968) Metabolism of sugars and organic acids in immature grape berries. *Plant Physiology*, **43**, 224–228

Harel, E. and Mayer, A.M. (1971) Partial purification and properties of catechol oxidases in grapes. *Phytochemistry*, **10**, 17–22

Harel, E., Mayer, A.M. and Lehman, E. (1973) Multiple forms of *Vitis vinifera* catechol oxidase. *Phytochemistry*, **12**, 2649–2654

Harris, J.M., Kriedemann, P.E. and Possingham, J.V. (1971) Grape berry respiration: effects of metabolic inhibitors. *Vitis*, **9**, 291–298

Hawker, J.S. (1969) Changes in the activities of malic enzyme, malate dehydrogenase, phosphopyruvate carboxylase during the development of a non-climacteric fruit (the grape). *Phytochemistry*, **8**, 19–23

Helsper, J.P.F.G. and Loewus, F.A. (1985) Studies on L-ascorbic acid biosynthesis and metabolism in *Parthenocissus quinquefolia* L. (Vitaceae). *Plant Science*, **40**, 105–109

Hofacher, W., Alleweldt, G. and Khader, S. (1976) Einfluss von umweltfaktoren aut Beenrenwachstum und must qualitat bei der rebe. *Vitis*, **15**, 96–112

Hrazdina, G. and Moskowitz, H. (1980) Subcellular status of anthocyanins in grape skins. *Proceedings of International Symposium on Grape and Wine*, Davis Ca., pp. 245–253

Hrazdina, G., Parsons, G.F. and Mattick, L.R. (1984) Physiological and biochemical events during development and maturation of grape berries, *American Journal of Enology and Viticulture*, **35**, 220–227

Hsu, J.C. and Heatherbell, D.A. (1987) Isolation and characterization of soluble proteins in grapes, grape juice and wine. *American Journal of Enology and Viticulture*, **36**, 6–10

Iland, P.G. and Coombe, B.G. (1988) Malate, tartrate, potassium, and sodium in flesh and skin of Shiraz grapes during ripening: concentration and compartmentation. *American Journal of Enology and Viticulture*, **39**, 71–76

Interesse, E.S., Alloggio, Y., Lamparelli, F. and D'Avella, G. (1984) Characterization of the oxidative enzymatic system of the phenolic compounds from Muscat grapes. *Lebensm. Wiss. Technol.*, **17**, 5–10

Ishii, S. and Yokotsuka, T. (1973) Susceptibility of fruit juice to enzymatic clarification by pectin lyase and its relation to pectin in fruit juice. *Journal of Agricultural Food Chemistry*, **21**, 269–272

Iwahori, S., Weaver, R.J. and Pool, R.M. (1968) Gibberellin-like activity of berries and seedless Tokay grapes. *Plant Physiology*, **43**, 333–337

Jensen, F.L., Kissler, J.J., Peacock, W.L. and Leavitt, G.M. (1975) Effect of ethephon on color and fruit characteristics of "Tokay" and "Emperor" table grapes. *American Journal of Enology and Viticulture*, **26**, 79–81

Jona, R. and Foa, E. (1979) Histochemical survey of cell-wall polysaccharides of selected fruits. *Scientia Horticulturae*, **10**, 141–147

Jona, R., Vallania, R. and Rosa, C. (1983) Cell wall development in the berries of two grapevines. *Scientia Horticulturae*, **20**, 169–178

Juhasz, O. (1985) Changes in the free proline content of grape berries during ripening (in Hungarian). *Acta Agron. Acad. Sci. Hung.*, **34**, 243–248

Kataoka, I., Suglura, A., Utsunomiya, N. and Tomana, I. (1982) Effect of abscisic acid and defoliation on anthocyanin accumulation in Kyoto grapes (*Vitis vinifera* L. × *V. labruscana* Bailey). *Vitis*, **21**, 325–332

Kataoka, I., Yasutaka, K., Sugiura, A. and Tomana, T. (1983) Changes in L-phenylalanine ammonia-lyase activity and anthocyanin synthesis during berry ripening of three grape cultivars. *Journal of Japanese Society of Horticultural Science*, **52**, 273–279

Kidron, M., Harel, S. and Mayer, A.M. (1978) Catechol oxidase activity in grapes and wine. *American Journal of Enology and Viticulture*, **29**, 30–35

Klenert, M., Rapp, A. and Alleweldt, G. (1978) Einfluss der traubentemperatur aut Beerenwachstum und Beerenreipe der Rebsorte Silvaner. *Vitis*, **17**, 350–360

Kliewer, W.M. (1968) Changes in the concentration of free amino acids in grape acids in grape berries during maturation. *American Journal of Enology and Viticulture*, **19**, 166–174

Kliewer, W.M. (1969) Free amino acids and other nitrogenous substances of table grape varieties. *Food Science*, **34**, 274–278

Kliewer, W.M. (1973) Berry composition of *Vitis vinifera* cultivars as influenced by photo- and nycto-temperature during maturation. *Journal of American Society of Horticultural Science*, **98**, 153–159

Kliewer, W.M. and Schultz, H.B. (1973) Effect of sprinkler cooling of grapevines on fruit growth and composition. *American Journal of Enology and Viticulture*, **24**, 17–26

Kliewer, W.M. and Weaver, R.J. (1971) Effect of crop level and leaf area on growth, composition, and coloration of "Tokay" grapes. *American Journal of Enology and Viticulture*, **22**, 172–177

Kliewer, W.M., Lider, L.A. and Ferrari, N. (1972) Effects of controlled temperature and light intensity on growth and carbohydrate levels of "Thompson Seedless" grapevines. *Journal of American Society of Horticultural Science*, **97**, 185–188

Koch, R. and Alleweldt, G. (1978) Der gaswechsel reinfender weinbeeren. *Vitis*, **17**, 30–44

Koch, J. and E. Sajak (1959) A review and some studies on grape protein. *American Journal of Enology and Viticulture*, **10**, 114–123

Kochhar, S., Kochhar, V.K. and Khanduja, S.D. (1979) Changes in the pattern of isoperoxidases during maturation of grape berries cv Gulabi as affected by ethephon (2-chloroethyl) phosphonic acid. *American Journal of Enology and Viticulture*, **30**, 275–277

Kriedemann, P.E. (1968) Observations on gas exchange in the developing Sultana berry. *Australian Journal of Biological Sciences*, **21**, 907–916

Kriedemann, P.E. (1969) Sugar uptake by the grape berry. A note on the absorption pathway. *Planta*, **85**, 111–117

Lakso, A.N. and Kliewer, W.M. (1975) The influence of temperature on malic acid metabolism in grape berries. *Plant Physiology*, **56**, 370–372

Lakso, A.N. and Kliewer, W.M. (1978) The influence of temperature on malic acid metabolism in grape berries. II. Temperature responses of net dark CO_2

fixation and malic acid pools. *American Journal of Enology and Viticulture*, **29**, 145–149

Lang, A. and Düring, H. (1991) Partitioning control by water potential gradient: evidence for compartmentation breakdown in grape berries. *Journal of Experimental Botany*, **42**, 1117–1122

Lavee, S. (1960) Effect of gibberellic acid on seeded grapes. *Nature (London)*, **185**, 395

Lavee, S. and Nir, G. (1986) Grape, In *CRC Handbook of fruit set and development*, (ed S.P. Monselise), CRC Press, Boca Raton, FL, pp. 167–191

Lee, C.Y. and Jaworski, A. (1987) Phenolic compounds in white grapes grown in New York. *American Journal of Enology and Viticulture*, **38**, 277–281

Lee, C.Y. and Jaworski, A. (1989) Major phenolic compounds in ripening white grapes. *American Journal of Enology and Viticulture*, **40**, 43–46

Lee, C.Y., Smith, N.L. and Nelson, P.P. (1979a) Relationship between pectin methylesterase activity and the formation of methanol in Concord grape juice and wine. *Food Chemistry*, **4**, 143–148

Lee, J.C., Tomana, T., Utsunomiya, N. and Kataoka, I. (1979b) Physiological studies on the anthocyanin development on grape. I. Effect of fruit temperature on the anthocyanin development in Kyoto grape. *Journal of Korean Society of Horticultural Science*, **20**, 55–65

Lohnertz, O. (1988) Untersuchungen zum zeitlichen veraluf der nahrstoffaujnahme bei *Vitis vinifera* (cv Riesling) Thesis, University of Giessen, Geisenheimer Berichte, p. 228

Lohnertz, O. (1991) Soil nitrogen and the uptake of nitrogen in grapevines. *Proceedings International Symposium on Nitrogen in Grape and Wine*, Seattle, Wa., pp.1–11

Loulakakis, K.A. and Roubelakis-Angelakis, K.A. (1990a) Intra-cellular localization and properties of NADH-glutamate dehydrogenase from *Vitis vinifera* L.: Purification and characterization of the major isoenzyme. *Journal of Experimental Botany*, **41**, 1223–1230

Loulakakis, K.A. and Roubelakis-Angelakis, K.A. (1990b) Immunocharacterization of NADH-glutamate dehydrogenase from *Vitis vinifera* L. *Plant Physiology*, **94**, 109–113

Loulakakis, K.A. and Roubelakis-Angelakis, K.A. (1990c) Effect of trophic conditions on glutamate dehydrogenase regulation in grapevine callus. *Physiologia Plantarum*, **79**, 745

Loulakakis, K.A. and Roubelakis-Angelakis, K.A. (1991) Plant NAD(H) glutamate dehydrogenase consists of two polypeptide subunits and their participation in the seven isoenzymes occurs in an ordered ratio. *Plant Physiology*, **97**, 109–111

Loulakakis, K.A. and Roubelakis-Angelakis, K.A. (1992) Ammonia-induced increase in NAD(H)-glutamate dehydrogenase activity is caused by *de novo* synthesis of α-subunit. *Planta*, **187**, 322–327

Lynn, C.D. and Jensen, F.L. (1966) Thinning effects of bloomtime gibberellin spray on Thompson Seedless grapes. *American Journal of Enology and Viticulture*, **17**, 286–289

Malipiero, U., Ruffner, H.P. and Rast D.M. (1987) Ascorbic tartaric acid conversion in grapevines. *Journal of Plant Physiology*, **129**, 33–40

Maroc, J. (1967) La conversion du glucolate en glucose et ses relations avec la biogenese de l'acide tartarique. *Physiologie Vegetale*, **5**, 37–46

Mayer, A.M. and Harel, E. (1979) Polyphenoloxidases in plants. A review. *Phytochemistry*, **18**, 193–215

Molina, I., Nicolas, M. and Croujet, J. (1986) Grape alcohol dehydrogenase. I. Isolation and characterization. *American Journal of Enology and Viticulture*, **37**, 169–173

Moskowitz, A. and Hrazdina, G. (1981) Vacuolar contents of fruit epidermal cells from *Vitis* species. *Plant Physiology*, **68**, 686–692

Murphey, J.M., Spayd, S.E. and Powers, J.R. (1989) Effect of grape maturation on soluble protein characteristics of Gewürztraminer and White Riesling juice and wine. *American Journal of Enology and Viticulture*, **40**, 199–207

Nakagawa, S.H., Komatsu, H. and Yuda, Y. (1980) A study of micromorphology of grape berry surface during their development with special reference to stomata. *Journal of Japanese Society of Horticultural Science*, **49**, 1–7

Nakamura, K., Amano, Y. and Kagami, M. (1983) Purification and some properties of a polyphenoloxidase from Koshu grapes. *American Journal of Enology and Viticulture*, **34**, 122–127

Nakanishi, K., Uesugi, T., Sato, T. and Yokotsuta, K. (1986) Isolation and characterization of soluble and insoluble proteins in Koshu grape juice. *Journal of the Institute of Enology and Viticulture*, **24**, 7–14

Nassar, A.R. and Kliewer, W.M. (1966) Free amino acids in various parts of *Vitis vinifera* at different stages of development. *Proceedings of American Society of Horticultural Science*, **89**, 281–224

Nelson, K.E. (1979) Harvesting and handling California table grapes for market. University of California, Publication 4095

Niemietz, C. and Hawker, J.S. (1988) Sucrose accumulation in red beet vacuoles by UDP glucose-dependent group translocation – fact or artefact? *Australian Journal of Plant Physiology*, **15**, 359–366

Nitsch, J.P., Pratt, C., Nitsch, C. and Shaulis, N.J. (1960) Natural growth substances in Concord and Concord seedless grapes in relation to berry development. *American Journal of Botany*, **47**, 566–576

Ong, B.Y. and Nagel, C.W. (1978) Hydroxycinnamic acid tartaric acid ester content in mature grapes during maturation of White Riesling grapes. *American Journal of Enology and Viticulture*, **29**, 277–281

Oszmianski, J., Romeyer, F.M., Sapis, J.C. and Macheix, J.J. (1986) Grape seed phenolics extraction as affected by some conditions occurring during wine processing. *American Journal of Enology and Viticulture*, **37**, 7–12

Ough, C.S. (1968) Proline content of California grapes. *American Journal of Enology and Viticulture*, **7**, 321–331

Ough, C.S. (1969) Ammonia content of California grapes. *American Journal of Enology and Viticulture*, **20**, 213–220

Palejwala, V.A., Parikh, H.R. and Modi, V.V. (1985) The role of abscisic acid in the ripening of grapes. *Physiologia Plantarum*, **65**, 498–502

Pandey, R.M. and Farmahan, H.L. (1977) Changes in the rate of photosynthesis and respiration in leaves and berries of *Vitis vinifera* grapevines at various stages of berry development. *Vitis*, **16**, 106–111

Peynaud, E. (1947) Contribution a l'étude biochemique de la maturation du raisin et de la composition des vins. *Bulletin O.I.V.*, **20**, 34–51

Peynaud, E. and Maurie, A. (1953) Sur l'évolution de azore dans les differences partes du raisin au cours de la maturation. *Ann. Technol. Agr.*, **2**, 12–35

Peynaud, E. and Ribéreau-Gayon, G.P. (1971) The grape. In *The Biochemistry of Fruits and Their Products*, Vol. 2 (ed. A.C. Hulme), Academic Press, London. pp. 179–205

Pirie, A. and Mullins, M.G. (1976) Changes in anthocyanin and phenolics content of grapevine leaf and fruit tissues treated with sucrose, nitrate and abscisic acid. *Plant Physiology*, **58**, 468–472

Pirie, A. and Mullins, M.G. (1977) Interrelationships of sugars, anthocyanins, total phenols and dry weight in the skin of grape berries during ripening. *American Journal of Enology and Viticulture*, **28**, 204–209

Pool, R. (1975) Effect of cytokinins on *in vitro* development of Concord flowers. *American Journal of Enology and Viticulture*, **26**, 43–46

Possner, D. and Kliewer, W.M. (1985) The localisation of acids, sugars, potassium and calcium in developing grape berries. *Vitis*, **24**, 229–240

Possner, D., Ruffner, H.P. and Rast, D.M. (1981) Isolation and biochemical characterization of grape malic enzyme. *Planta*, **151**, 549–554

Possner, D., Ruffner, H.P. and Rast, D.M. (1983) Regulation of malic acid metabolism in berries of *Vitis vinifera*. *Acta Horticulturae*, **139**, 117–122

Pratt, C. (1971) Reproductive anatomy in cultivated grapes – a review. *American Journal of Enology and Viticulture*, **22**, 92–106

Raddler, F. and Totokfalvy, E. (1973) The affinity for oxygen of polyphenol-oxidase in grapes. *Z. Lebensm. Unters. Forsch.*, **152**, 38–41

Rapp, A., Staffan, H., Kupter, G. and Ullemeyer, H. (1971) Uder der saurestoffwechsel in weihbeeren. *Angew. Chem.*, **22**, 925

Ribéreau-Gayon, G.P. (1964) The composes phenoliques du raisin et du vin. II. Les flavonosides et les anthocyanosides. *Annals de Physiologie Vegetale*, **6**, 211–242

Ribéreau-Gayon, G.P. (1971) Evolution des composes phenoliques au cours de la maturation du raisin. I. Experimentation 1969. *Connaissance de la Vigne et du Vin*, **2**, 247–261

Ribéreau-Gayon, G.P. (1972) Evolution des composes phenoliques au cours de la maturation du raisin. II. Discussion des resultats obtenus en 1969, 1970 et 1971. *Connaissance de la Vigne et du Vin*, **2**, 161–175

Ribéreau-Gayon, G.P., Boidron, J.N. and Terrier, A. (1975) Aroma of muscat grape varieties. *Journal of Agricultural Food Chemistry*, **21**, 1042–7

Robertson, G.L., Eschenbruch, R. and Cresswell, K.S. (1980) Seasonal changes in the pectin substances of grapes and their implication in juice extraction. *American Journal of Enology and Viticulture*, **31**, 162–164

Robin, J.P., Romieu, C., Sauvage, F.X., Nicol, M.Z. and Flanzy, C. (1987) Evidence of an aspartase activity in *Vitis vinifera* berries. *Plant Physiology and Biochemistry*, **25**, 797–804

Roggero, J.P., Coen, S. and Ragonnet, B. (1986) High performance liquid chromatography. Survey on changes in pigment content in ripening grapes of Syrah. An approach to anthocyanin metabolism. *American Journal of Enology and Viticulture*, **137**, 77–83

Rosenquist, J.K. and Morrison, J.C. (1988) The development of the cuticle and epicutical wax of the grape berry. *Vitis*, **27**, 63–70

Roubelakis, K.A. and Kliewer, W.M. (1976) Influence of light intensity and growth regulators on fruit-set and ovule fertilization in grape cultivars under low temperature conditions. *American Journal of Enology and Viticulture*, **27**, 163–167

Roubelakis, K.A. and Kliewer, W.M. (1978a) Enzymes of Krebs-Henseleit Cycle in *Vitis vinifera* L. I. Ornithine carbamoyl-transferase: isolation and some properties. *Plant Physiology*, **62**, 337–339

Roubelakis, K.A. and Kliewer, W.M. (1978b) Enzymes of Krebs-Henseleit Cycle in *Vitis vinifera* L. II. Arginosuccinate synthetase and lyase. *Plant Physiology*, **62**, 340–344

Roubelakis, K.A. and Kliewer, W.M. (1978c) Enzymes of Krebs-Henseleit Cycle in *Vitis vinifera* L. III. *In vivo* and *in vitro* studies of arginase. *Plant Physiology*, **62**, 344–347

Roubelakis-Angelakis, K.A. and Kliewer W.M. (1981) Influence of nitrogen fertilization on activities of ornithine transcarbamoylase and arginase in Chenin blanc berries at different stages of development. *Vitis*, **20**, 130–135

Roubelakis-Angelakis, K.A. and Kliewer, W.M. (1985) Phenylalanine ammonia-lyase in berries of *Vitis vinifera* L.: Extraction and possible sources of error during assay. *American Journal of Enology and Viticulture*, **36**, 314–315

Roubelakis-Angelakis, K.A. and Kliewer, W.M. (1986) Effects of exogenous factors on phenylalanine ammonia-lyase activity and accumulation of anthocyanins and total phenolics in grape berries. *American Journal of Enology and Viticulture*, **37**, 275–280

Roubelakis-Angelakis, K.A., Loulakakis, K.A. and Kanellis, A.K. (1991) Synthesis and regulation of glutamate dehydrogenase in grapevine callus. *Proceedings of International Meeting on Nitrogen in Grape and Wine*, Seattle, Wa., June 1991, pp. 306–311

Ruffner, H.P. (1982a) Metabolism of tartaric and malic acids in *Vitis*: a review – Part A. *Vitis*, **21**, 247–259

Ruffner, H.P. (1982b) Metabolism of tartaric and malic acids in *Vitis*: a review – Part B. *Vitis*, **21**, 346–358

Ruffner, H.P. and Hawker, J.S. (1977) Control of glycolysis in ripening berries of *Vitis vinifera*. *Phytochemistry*, **16**, 1171–1175

Ruffner, H.P. and Kliewer, W.M. (1975) Phosphoenolpyruvate carboxykinase activity in grape berries. *Plant Physiology*, **56**, 67–71

Ruffner, H.P. and Rast, D.M. (1974) Die Biogenese von tartrat in der Weinrebe. *Z. Pflanzephysiol.*, **73**, 45–55

Ruffner, H.P., Brem, S. and Malipiero, U. (1983a) The physiology of acid metabolism in grape berry ripening. *Acta Horticulturae*, **139**, 123–128

Ruffner, H.P., Brem, S. and Rast, D.M. (1983b) Pathway of photosynthetic malate formation in *Vitis vinifera*, a C_3 plant. *Plant Physiology*, **73**, 582–585

Ruffner, H.P., Hawker, J.C. and Hale, C.R. (1976) Temperature and enzymic control of malate metabolism in berries of *Vitis vinifera*. *Phytochemistry*, **15**, 1877–1880

Ruffner, H.P., Koblet, W. and Rast, D.M. (1975) Gluconeogenese in reifenden Beeren von *Vitis vinifera*. *Vitis*, **13**, 319–328

Ruffner, H.P., Possner, D., Brem, S. and Rast, D.M. (1984) The physiological role of malic enzyme in grape ripening. *Planta*, **160**, 144–448

Saito, K. and Kasai, Z. (1968) Accumulation of tartaric acid in the ripening process of grapes. *Plant Cell Physiology*, **9**, 529–537

Saito, K. and Kasai, Z. (1969) Tartaric acid synthesis from L-ascorbic acid-1 ^{14}C in grape berries. *Phytochemistry*, **8**, 2177–2182

Saito, K. and Kasai, Z. (1978) Conversion of labelled substrated to sugars, cell wall polysaccharides, and tartaric acid in grape berries. *Plant Physiology*, **62**, 215–219

Saito, K. and Kasai, Z. (1982) Conversion of L-ascorbic acid to L-idonic acid, L-idono-γ-lactone and 2-keto-L-idonic acid in slices of immature grapes. *Plant Cell Physiology*, **23**, 499–507

Saito, K. and Kasai, Z. (1984) Synthesis of L(+)-tartaric acid from ascorbic acid via 5-keto-D-gluconic acid in grapes. *Plant Physiology*, **76**, 170–174

Saito, K. and Loewus, F.A. (1979) The metabolism of L-6[14]C ascorbic acid in detached leaves. *Plant Cell Physiology*, **20**, 1481–1488

Sanchez-Ferrer, A., Bru, R. and Garcia-Carmena, F. (1989) Novel procedure for extraction of a latent grape polyphenoloxidase using temperature-induced phase separation in Triton X-114. *Plant Physiology*, **91**, 1481–1487

Sapis, J.C., Macheix, J.C. and Cordonnier, R.E. (1983a) The browning capacity of grapes. I. Changes in polyphenol oxidase activities during development and maturation of the fruit. *Journal of Agriculture and Food Chemistry*, **31**, 342–345

Sapis, J.C., Macheix, J.C. and Cordonnier, R.E. (1983b) The browning capacity of grapes. II. Browning potential and polyphenol oxidase activities in different mature grape varieties. *American Journal of Enology and Viticulture*, **34**, 157–162

Saulnier, L. and Brillouet, J.M. (1988) Structural studies of peptic substances from the pulp of grape berries. *Carbohydrate Research*, **182**, 63–78

Saulnier, L. and Brillouet, J.M. (1989) An arabinogalactan-protein from the pulp of grape berries. *Carbohydrate Research*, **188**, 137–144

Saulnier, L., Brillouet, J.M. and Moutounet, M. (1988) Nouvelles aquisitions structurales sur les substances pectiques de la pulpe de raisin. *Connaissance de la Vigne et du Vin*, **22**, 135–158

Saulnier, L. and Thibault, J.F. (1987a) Extraction and characterization of pectic substances from pulp of grape berries. *Carbohydrate Polymers*, **7**, 329–343

Saulnier, L. and Thibault, J.F. (1987b) Enzymic degradation of isolated peptic substances and cell wall from pulp of grape berries. *Carbohydrate Polymers*, **7**, 343–345

Schaller, K., Lohnertz, O., Oswald, D. and Sprengart, B. (1985) Nitratanreicherung in reben. 3. Mitteilung: Nitratdynamik in rappen und beeren wahrend einer vegetationsperiode in verschiedenen rebsorten. *Wein Wissen*, **40**, 147–159

Scienza, A., Mieavalle, R., Visai, C. and Fregoni, M. (1978) Relationships between seed number, gibberellin and abscisic acid levels and ripening in Cabernet Sauvignon grape berries. *Vitis*, **17**, 361–368

Shrinivasan, C. and Mullins, M.G. (1979) Flowering in *Vitis*: Conversion of tendrils into inflorescences and bunches of grapes. *Planta*, **145**, 187–192

Sihamed, O.A. and Kliewer, W.M. (1980) Effects of defoliation, gibberellic acid and 4-chlorophenoxyacetic acid on growth and composition of Thompson

Seedless grape berries. *American Journal of Enology and Viticulture*, **31**, 149–153

Singleton, V.L. (1966) The total phenolic content of grape berries during the maturation of several varieties. *American Journal of Enology and Viticulture*, **25**, 107–117

Singleton, V.L. (1980) Grape and wine phenolics: background and prospects. *Cent. Davis Univ. Proc.*, pp. 215–227

Singleton, V.L., Draper, D.E. and Rossi, J.A. (1966) Paper chromatography of phenolic compounds from grapes, particularly seeds and some variety ripeness relationships. *American Journal of Enology and Viticulture*, **17**, 206–217

Singleton, V.L. and Esau, P. (1969) Phenolic substances in grapes and wine and their significance. *Advances in Food Research*, **1**, 1–282

Spettoli, P., Bottacin, A. and Zamorani, A. (1980) Purificazione per chromatografia de affinita dell' enzima malico estratto da uva. *Vitis*, **19**, 4–12

Staudt, G., Schneider, A. and Leidel, J. (1986) Phases of berry growth in *Vitis vinifera*. *Annals of Botany*, **58**, 789–800

Steffan, H., Rapp, A., Ullemeyer, H. and Kupper, G. (1975) Uder den reifeabhangigen saure- zucker- stoffwechsel bei beeren von *Vitis-vinifera* – sorten, untersucht mit ^{14}C-verbindungen. *Vitis*, **14**, 181–189

Strauss, C.R., Wilson, B., Gooley, P.R. and Williams, P.J. (1986) The role of monoterpenes in grape and wine flavor – a review. In *Biogeneration of Aroma Compounds* (eds T.H. Parliment and R.B. Crotean), ASC Symposium Series 317, American Chemical Society, Washington, DC, pp. 222–242

Su, C.T. and Singleton, V.L. (1969) Identification of three flavan-3-ols from grapes. *Phytochemistry*, **8**, 1553–1558

Swift, J.S., Buttrose, M.S. and Possingham, J.C. (1973) Stomata and starch in grape berries. *Vitis*, **12**, 38–45

Szyjewicz, E., Rosner, N. and Kliewer, W.M. (1984) Ethephon (2-chloroethylphosphonic acid, Ethrel, CEPA) in viticulture – a review. *American Journal of Enology and Viticulture*, **35**, 117–123

Tesniere, C. and Robin, J.R. (1992) Two-dimensional electrophoresis of total polypeptides in ripe red grape berries. *Electrophoresis*, (in press)

Tesniere, C. and Vayda, M.E. (1991) Method for the isolation of high-quality RNA from grape berry tissues without contaminating tannins or carbohydrates. *Plant Molecular Biology Reporter*, **9**, 242–251

Tomana, I., Utsunomiya, N. and Kataoka, I. (1979) The effect of environmental temperatures on fruit ripening on the tree. II. The effect of temperatures around whole vines and clusters on coloration of Kyoto grapes. *Journal of Korean Society of Horticultural Science*, **48**, 261–266

Tyrina, S.S. (1977) Properties of grape polygalacturonase. *Applied Biochemistry and Microbiology*, **13**, 142–145

Valero, E., Ferrez, A.S., Varon, R. and Carmona, F.G. (1989) Evolution of polyphenol oxidase activity and phenolic content during maturation and vinification. *Vitis*, **28**, 85–95

Van Wyk, C.J., Webb, A.D. and Kepner, R.E. (1967) Some volatile compounds of *Vitis vinifera*, variety White Riesling. 1. Grape juice. *Journal of Food Science*, **32**, 660–664

Wagner, G. and Loewus, F. (1974) L-Ascorbic acid metabolism in *Vitaceae, Plant Physiology*, **54**, 784–787

Wagner, G., Yang, J.C. and Loewus, F.A. (1975) Stereoisomeric characterization of tartaric acid produced during L-ascorbic acid metabolism in plants. *Plant Physiology*, **55**, 1071–1073

Weaver, R.J. (1953) Further studies on the effects of 4-chlorophenoxyacetic acid on development of Thompson Seedless and Black Corinth grapes. *Proceedings of American Society of Horticultural Science*, **61**, 135–143

Weaver, R.J. (1965) Induction of fruit set in *Vitis vinifera* L. by a kinin. *Nature (London)*, **206**, 952–953

Weaver, R.J., Leonard, O.A. and McCune, S.B. (1961) Response of clusters of *Vitis vinifera* grapes to 2,4-diclorophenoxyacetic acid and related compounds. *Hilgardia*, **31**, 113–125

Weaver, R.J. and Pool, R.M. (1965) Relation of seededness and ringing to gibberellin-like activity in berries of *Vitis vinifera*. *Plant Physiology*, **40**, 770–776

Weaver, R.J. and Singh, I.S. (1978) Occurence of endogenous ethylene and effect of plant regulators on ethylene production in the grapevine. *American Journal of Enology and Viticulture*, **29**, 282–285

Weaver, R.J. and van Overbeek, J. (1963) Kinins stimulate grape growth. *Californian Agriculture*, **17**, 12

Weaver, R.J., van Overbeek, J. and Pool, R.M. (1968) Effect of kinins on fruit set and development in *Vitis vinifera*. *Hilgardia*, **37**, 181–201

Webb, A.D. (1970) Anthocyanin pigments in grapes and wines. *Suomen kemistilehti*, **A43**, 67–74

Wermelinger, B. and Koblet, W. (1990) Seasonal variation and nitrogen distribution in grapevine leaves, shoots and grapes. *Vitis*, **29**, 15–26

Wicks, A.S. and Kliewer, W.M. (1983) Further investigations into relationships between anthocyanins, phenolics and soluble carbohydrates in grape berry skins. *American Journal of Enology and Viticulture*, **34**, 114–116

Williams, M. and Loewus, F.A. (1978) Biosynthesis of (+)tartaric acid from L-4-^{14}C ascorbic acid in grape and geranium. *Plant Physiology*, **61**, 672–674

Williams, G., Saito, K. and Loewus, F.A. (1979) Ascorbic acid metabolism in geranium and grape. *Phytochemistry*, **18**, 953–956

Williams, P.J., Strauss, C.R., Wilson, B. and Massy-Westropp, R.A. (1982a) Novel monoterpene disaccharide glycosides of *Vitis vinifera* grapes and wines. *Phytochemistry*, **21**, 2013–2020

Williams, P.J., Strauss, C.R., Wilson, B. and Massy-Westropp, R.A. (1982b) Studies on the hydrolysis of *Vitis vinifera* monoterpene precursor compounds and model monoterpene β-D-glucosides rationalizing the monoterpene composition of grapes. *Journal of Agriculture and Food Chemistry*, **10**, 1219–1223

Wilson, B., Strauss, C.R. and Williams, P.J. (1984) Changes in free and glycosidally bound monoterpenes in developing Muscat grapes. *Journal of Agriculture and Food Chemistry*, **32**, 919–924

Winkler, A., Cook, J., Lider, J.A. and Kliewer, W.M. (1974) *General Viticulture*. University of California Press, Berkeley

Wisseman, K.W. and Lee, C.Y. (1980) Polyphenol oxidase activity during grape maturation and wine production. *American Journal of Enology and Viticulture*, **31**, 206–211

Yokotsuka, K., Ito, K., Nozaki, K. and Kushida, T. (1982) Koshu grape pectins: Isolation, chemical composition and precipitation. *Journal of Institute of Enology and Viticulture, Yamanashi Univ.*, **17**, 59–63

Yokotsuka, K., Makino, S. and Singleton, V.L. (1988) Polyphenol oxidase from grapes: precipitation re-solubilization and characterization. *American Journal of Enology and Viticulture*, **39**, 293–302

Zindler-Frank, E. (1974) Die differezierung von kristallidioblasten im Dunkeln und bei Hemmung der glykolsaureoxidase. *Z. Pflanzenphysiol.* **73**, 313–325

Chapter 7

Kiwifruit

The late N. K. Given

7.1.   INTRODUCTION

But this beauty serves merely as a guide to birds and beasts in order that the fruit may be devoured and the manured seeds disseminated. (Darwin, 1859)

The genus *Actinidia*, to which the kiwifruit (*Actinidia deliciosa* (A. Chev.) C.F. Liang et A.R. Ferguson var *deliciosa*) belongs, has its origins in South Western China. Although the genus contains several other species which produce edible fruit (for example, *Actinidia chinensis* and *Actinidia arguta*), *Actinidia deliciosa* is the only member which has been developed commercially on a large scale (Ferguson, 1990a). The kiwifruit plant is a deciduous, perennial climber and is dioecious. Fruit belonging to several *Actinidia* species are collected in the wild, the total quantity being about the same as total current production of kiwifruit elsewhere in the world (Ferguson, 1990b).

The fruit is ovoid in shape, with a brown skin covered in short hairs. The flesh is bright green with a large number (up to 1500) of black seeds surrounding the white core. Several cultivars of the kiwifruit have been selected, which vary in shape as well as other characteristics. Fruit of the cultivar Hayward, which is the most important cultivar commercially, are usually 55–70 mm long, 40–50 mm wide and weigh 80–120 g when mature (Beever and Hopkirk, 1990).

Although it is a native of China, the kiwifruit takes its name from the kiwi (*Apteryx australis*), the national bird of New Zealand. This unusual situation has arisen because although kiwifruit plants were introduced into Europe and the USA at the beginning of this century, it was in New Zealand that the kiwifruit was developed commercially. The first large

Biochemistry of Fruit Ripening. Edited by G. Seymour, J. Taylor and G. Tucker. Published in 1993 by Chapman & Hall, London. ISBN 0 412 40830 9

scale plantings were made in the 1930s, but not until the 1970s did the development of kiwifruit as an export crop lead to rapid expansion of the industry (Ferguson and Bollard, 1990). The export of fruit from New Zealand to Europe and the USA resulted in the belated recognition of the potential for the crop in these countries. In some cases commercial development in the countries also depended on plant material introduced from New Zealand. The first commercial plantings in California were cropped in 1965. A similar pattern occurred in Europe, with commercial plantings in France, Italy, Spain and Greece in the 1970s. Expansion of these industries has been so rapid that in 1990, for the first time, New Zealand production was expected to be less than 50% of the total world crop (Ferguson and Bollard, 1990).

New Zealand is approximately 10 000 and 20 000 kilometres respectively from important markets in the USA and Europe. The success of the kiwifruit as an export crop depends on the ability to store the fruit for extended periods and ship it over these long distances. It is not surprising therefore that much of the New Zealand research related to kiwifruit maturity and ripening has been dedicated to providing the consumer with a product which is both appealing to the eye and tasty to eat, despite the long voyage. In particular, the identification of new cultivars which store well, and the development of procedures for harvesting, handling and storing the fruit have received high priority.

In the southern hemisphere, kiwifruit normally flower in November and are ready to harvest in April–May. In many fruit (the tomato, strawberry and some apple cultivars for example) the colour of the skin changes as the fruit ripens. The kiwifruit is unusual in that the outward appearance of the fruit changes very little as it ripens: judging the maturity of kiwifruit on the vine is therefore difficult. Kiwifruit which are harvested immature do not store well and never develop the characteristic attractive flavour of ripe fruit (Harman, 1981). The urgent need by the industry for a reliable maturity test to determine when the fruit is ready to harvest provided the incentive for research on kiwifruit ripening during the 1970s and early 1980s (Beever and Hopkirk, 1990).

Although the majority of the crop is exported as fresh fruit, considerable research effort has been directed at developing processing alternatives. Techniques have been developed to produce canned, pulped, dried, and candied kiwifruit. Kiwifruit juice, nectar and wine are also produced commercially (Lodge and Robertson, 1990).

To understand in full the physiology of this unusual fruit we must consider it from a biological, as well as a commercial, perspective. One possible biological function of a fleshy fruit is to aid in the spread of seed. The changes in colour, flavour and texture which characterize the ripening of many fruit may make these fruit attractive to animals. As a result of the animals eating the fruit, the seed is passed through their digestive systems and spread away from the parent plant. If one function of a fleshy fruit is to attract birds and animals to spread seed, then we might expect the development and maturation of the seed to be closely co-ordinated with the growth and ripening of the fruit.

Because much of the research on kiwifruit ripening has been driven from a commercial perspective and has been reviewed recently (Hopping, 1986; Beever and Hopkirk, 1990), this review will not refer to all literature relating to kiwifruit ripening, but will instead attempt to build a coherent picture of the changes in the physiology and bio-chemistry which underly the ripening of the kiwifruit. In particular, the review will focus on those aspects which might relate to the biological function of the fruit from the point of view of the kiwifruit plant.

7.2   PHYSIOLOGY

As shown in Chapter 1, the fleshy fruits of commerce are derived from many different parts of the fertilized flower. Many are therefore not true fruit in the botanical sense. Although an excellent review of this subject was published several years ago (Coombe, 1976) the significance of the anatomical origin of fleshy fruit on their subsequent development and ripening is very rarely discussed.

7.2.1   Fruit size

The kiwifruit is a berry which develops from a superior multicarpellate ovary borne on a 3 to 4 cm pedicel (Hopping, 1986). A developing kiwifruit increases in size over about 160 days from a pollinated flower to a fruit weighing approximately 120 g (Beever and Hopkirk, 1990). This increase in size involves both cell division and cell expansion, and does not occur at a constant rate. There are conflicting reports in the literature on the overall pattern of fruit development, which probably varies between cultivars. Pratt and Reid (1974) reported a triple sigmoid curve of volume growth for the cultivar 'Bruno', while Hopping (1976a) suggested that for the cultivar 'Monty' the growth curve was double sigmoid, and could be divided into three stages (Table 7.1).

Table 7.1   Growth stages of kiwifruit. (After Hopping, 1976a.)

Growth stage	Days after flowering	Development
I	0–58	Rapid growth and weight gain due to cell division initially, followed by cell enlargement.
II	58–76	Reduced growth due to slow cell enlargement.
III	76–160	A second period of growth due to enlargement of the cells of the inner pericarp.

Cell expansion and fruit growth can continue until well after the fruit is mature and therefore ready to harvest (Beever and Hopkirk, 1990).

An individual kiwifruit may contain up to 1500 seeds. There is good circumstantial evidence that the seeds play a major role in regulating fruit expansion. For fruit of the cultivar Hayward which are large enough for export, there is a linear relationship between seed number and fruit size (Hopping and Hacking, 1983; Grant and Ryugo, 1984; Pyke and Alspach, 1986). It is possible that a substance or substances produced by the seeds is involved in the regulation of expansion and possibly division of the fruit cells. The size of fruit that contain few seeds is increased by the application of exogenous plant growth regulators; application of mixtures of auxin, gibberellin, cytokinin, and zeatin after flowering is effective in increasing fruit size (Hopping, 1976b). N-(2-chloro-4-pyridyl)-N'-phenylurea (CPPU) is a diphenylurea derivative with cytokinin activity. Application of CPPU on or before anthesis causes parthenocarpic development of kiwifruit; in contrast, application of CPPU after anthesis stimulates fruit growth. (Iwahori *et al.*, 1988). In one of the few studies on endogenous hormones in kiwifruit, Young (1977) identified the cytokinins zeatin and zeatin ribotide in extracts of seeds and flesh. In 20-day-old and mature fruit, both zeatin and zeatin ribotide were present, whereas only zeatin was present in fruit sampled at 40 and 60 days after flowering (Okuse and Ryugo, 1981).

In summary, the large number of seeds in the kiwifruit and the linear relationship between fruit size and seed number suggest that the seeds may produce a substance which regulates the size and number of the kiwifruit cells. The observation that application of cytokinin markedly stimulated fruit growth and that the cytokinins, zeatin and zeatin ribotide, are present in both kiwifruit seed and the developing fruit suggest that cytokinin may be fulfilling this role. The response of kiwifruit to exogenous hormones (including CPPU) is most marked when seed number exceeds a threshold value, indicating that cytokinin alone does not cause fruit expansion.

7.2.2 Fruit development

The development of any fleshy fruit can be divided into two stages, maturation and ripening. The distinction between them is that physiologically mature fruit will continue to ripen after harvest to a stage where they are optimum for eating. In contrast, fruit which are picked when immature will never develop the full flavour, texture and colour which we associate with ripe fruit. The kiwifruit is unusual in that there are no obvious visible changes in the colour of the skin or the flesh during development. The identification of the optimum stage to harvest is therefore more difficult than in some fruit. This was particularly important during the development of the kiwifruit industry in New Zealand for two reasons:

1. Because the industry is based on exports, it was important that the fruit maintained high eating quality after prolonged storage.
2. Kiwifruit normally reached physiological maturity in April during the southern hemisphere autumn. Fruit left on the vine longer than necessary are likely to be damaged by frost or storms. The identification of a maturity index which could be reliably used to identify the harvest date for kiwifruit, and which would produce fruit of high quality after storage was therefore a priority for the New Zealand industry (Beever and Hopkirk, 1990).

Two characteristics which do alter as kiwifruit mature are firmness and carbohydrate composition. Firmness was rejected as a maturity index because it does not correlate with eating quality after storage (Harman, 1981). Starch accumulates during kiwifruit development and is subsequently hydrolysed into soluble sugars. In kiwifruit juice there is a close relationship between soluble solids content and sugar content, such that the refractive index of the juice reflects the soluble sugar content of the fruit. Trials reported by Ford (1971), and Harman (1981), showed that for the cultivar Hayward soluble solids content at harvest accurately predicted eating quality after storage. Fruit harvested at a low soluble solids content do not soften normally, and do not develop a soluble solids content as high as that of fruit which are harvested at maturity; nor do these fruit develop a good kiwifruit flavour after storage. In contrast, fruit harvested at a high soluble solids concentration have a normal internal appearance and ripen to an acceptable flavour (Harman, 1981). Therefore in New Zealand, soluble solids content was adopted as a standard maturity index, with harvesting of each crop starting when a sample of the fruit has reached a soluble solids content of 6.25% or greater. The date at which this soluble solids content is reached varies between different districts and different seasons (Beever and Hopkirk, 1990).

Although soluble solids content is also used as a maturity index for kiwifruit outside New Zealand, there is a growing recognition that it may not be the best predictor of fruit quality in other countries. In Australia total solids content and the soluble solids content of fruit which had been ripened by exposure to ethylene were better predictors of flavour quality than soluble solids content at harvest (Scott et al., 1986).

7.3 BIOCHEMISTRY

Ripening is associated with changes in the colour, flavour and texture of many fruit. The aim of this section is to describe the biochemical events underlying these and other ripening changes in the kiwifruit. The flavour of a fruit is the result of two different components – first, the balance between sugar and organic acid content, and second, the characteristic aroma of the fruit, which is due to the release of volatile compounds (chiefly aldehydes, alcohols and esters). The events associated with ripening in the kiwifruit will be described in turn, but the reader should be aware that these occur simultaneously in the fruit, and are co-ordinated to reach a climax in the characteristic colour, texture and flavour which we associate with a ripe kiwifruit.

7.3.1 Colour

The dark green colour of kiwifruit is due to the presence of chlorophyll in plastids in the pericarp cells of the flesh. Compared to leaf tissue the chlorophyll concentration of the flesh of immature fruit is low (about 2–3 mg.100 g^{-1} fresh weight) and declines further to about 1.2 mg.100g^{-1} at harvest (Fuke et al., 1985). This fall is mostly due to the loss of chlorophyll a. In some cultivars the chlorophyll content decreases during storage (Ben-Aire et al., 1982; Fuke et al., 1985).

7.3.2 Carbohydrate composition

Early in the development of a kiwifruit, photosynthate is translocated from the leaves to the developing fruit, where it accumulates as starch. There is conflicting evidence for the exact pattern of starch accumulation. Working in California, Okuse and Ryugo (1981) and Grant and Ryugo (1984) reported a steady accumulation of starch from mid-June until mid-September (20 weeks after anthesis), whereas in New Zealand, immature fruit (less than 10 weeks after anthesis) initially had a high starch content, expressed on a fresh weight basis, which declined until about 12 weeks after anthesis and then rose to a maximum level about 20 weeks after anthesis (Reid et al., 1982) (Fig. 7.1). The grains of starch are laid down initially in the cells immediately under the epidermis and then sequentially in the layers of cells closer to the core (Okuse and Ryugo, 1981; Fuke and Matsuoka, 1984). The average particle size of the starch granules increases during the season from 3–4 μm in immature fruit to reach a maximum of 20–30 μm at about 20 weeks after anthesis (Sugimoto et al., 1988).

As the fruit matures, this starch is converted to sugar. The concentration of glucose, fructose and sucrose remain low (less than 1% fresh weight) until 20 weeks after anthesis (Reid et al., 1982); the sugar alcohol, inositol, is present at a low level throughout the development of the fruit (Okuse and Ryugo, 1981).

At about 20 weeks after anthesis there is a major switch in carbohydrate metabolism with the rapid hydrolysis of the accumulated starch and a decrease in the size of the starch grains (Sugimoto et al., 1988). This

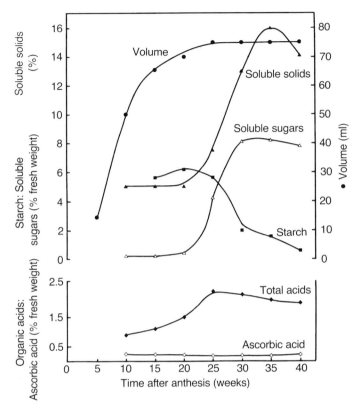

Fig. 7.1 Stylized diagram to illustrate changes in the size and composition of kiwifruit during growth and maturation on the vine. (Data from Pratt and Reid, 1974; Okuse and Ryugo, 1981; Reid *et al.*, 1982.)

starch hydrolysis is accompanied by a marked rise in the total sugar concentration of the fruit due to a major increase in the concentrations of glucose and fructose and a smaller increase in sucrose concentration (Okuse and Ryugo, 1981; Reid *et al.*, 1982; Grant and Ryugo, 1984) (Fig. 7.1).

If fruit are allowed to remain on the vine after the normal harvest date, the sugar content can increase to reach almost 10% of the fresh weight by about 35 weeks after anthesis (Fig. 7.1). Starch hydrolysis accounts for only about half of this increase, indicating that photosynthate is being translocated from the rest of the plant even at this late stage of development (Reid *et al.*, 1982). This suggestion has been confirmed by recent ^{14}C labelling studies (MacRae and Redgwell, 1990). The leaves of the kiwifruit vine start to abscise about 30 weeks after anthesis, and this supply of photosynthate is lost. The sugar content of fruit left on the vine after this stage starts to decline, presumably as a result of respiration (Reid *et al.*, 1982).

Soluble solids

The soluble solids concentration of kiwifruit juice accurately reflects the sugar concentration of the fruit (Fig. 7.1). Because soluble solids can be measured easily in the field using a hand-held refractometer, this parameter is used in preference to sugar content as a measure of maturity. Kiwifruit are normally harvested about 22 weeks after flowering when they reach a soluble solids content of greater than 6.2% (Harman, 1981). At this stage the starch content has decreased slightly from the maximum value and the sugar content has reached about 8% fresh weight (Reid *et al.*, 1982). The conversion of starch into sugar which occurs in fruit left on the vine, also continues in fruit harvested at this stage of maturity. In fruit held at 0°C, the starch content of the fruit declines from 40% dry weight at harvest to about 2% after 20 weeks storage. The decline in starch content is mirrored by an increase in soluble solids content from 6.1% to 13% (Hopkirk *et al.*, 1986). In fruit ripened at 20°C, and in the presence of ethylene, these changes occur much more quickly, the starch content declining rapidly with an increase in the concentrations of fructose, glucose and sucrose (Matsumoto *et al.*, 1983).

In summary, the early stages of kiwifruit development are characterized by the conversion of photosynthate to starch which accumulates as grains in the cells of the flesh from the sub-epidermis to the core. At about 20 weeks after anthesis, the concentration of starch reaches its maximum value. After this stage of development the starch is hydrolysed, resulting in the rapid accumulation of soluble sugars, particularly glucose, fructose and sucrose. This starch hydrolysis continues in fruit harvested at normal maturity, and stored at either 0° or 20°C.

7.3.3 Organic acids

Organic acids accumulate during the development of many fruit and contribute significantly to their flavour. Titratable acidity is a measure of the total organic acid content. The titratable acidity of kiwifruit increases steadily from 0.4% soon after harvest to 1.9% fresh weight at 19 weeks (Scienza *et al.*, 1983) (Fig. 7.1). In kiwifruit the predominant organic acids are malic, citric and quinic (Okuse and Ryugo, 1981; Reid *et al.*, 1982).

Studies on changes in the concentration of these compounds during fruit development both in New Zealand (Reid *et al.*, 1982) and California (Okuse and Ryugo, 1981) showed a small gradual increase in malic acid concentration during fruit development with a sharper rise in the concentration of citric acid, which reaches its maximum value at about the same time as the maximum starch concentration (about 20 weeks after anthesis) (Fig. 7.1). There is a major difference between the two studies in the amount of quinic acid in the immature fruit. In the Californian study, immature fruit had a very high concentration of quinic acid (up to 30% dry weight) which declined as the fruit matured, to a minimum of about 2% dry weight at 27 weeks after anthesis (Okuse and Ryugo, 1981). In contrast, the quinic acid concentration in the New Zealand fruit remained relatively constant during fruit development, at about 0.6% fresh weight (Reid *et al.*, 1982). The reasons for this difference in the pattern of quinic acid accumulation have not been identified, though different cultivars, Bruno (Reid *et al.*, 1982) and Hayward (Okuse and Ryugo, 1981) and slightly different analytical techniques were used in each case. Differences in the growing conditions in the two countries may also contribute to different patterns of fruit development in these studies. There is also a marked decline in the organic acid content of kiwifruit ripened at 20°C, in the presence of ethylene (Ben-Aire *et al.*, 1982; Matsumoto *et al.*, 1983)

Kiwifruit contain a significant concentration of ascorbic acid, although there is disagreement in the literature both on changes in the pattern of ascorbic acid accumulation during development and the final concentration reached in ripe fruit. In the cultivar Bruno grown in New Zealand, the concentration of ascorbic acid remained at about 0.2% fresh weight (Reid *et al.*, 1982) (Fig. 7.1), whereas in Hayward grown in California the fruit showed a rise in ascorbic acid concentration about 20 weeks after anthesis (Okuse and Ryugo, 1981). The concentration of ascorbic acid varies in different parts of the fruit, and is higher in the green carpellary tissue than in the white core (Okuse and Ryugo, 1981).

The cells of kiwifruit accumulate numerous raphides of calcium oxalate and this may contribute to the stability of ascorbic acid in the fruit (Okuse and Ryugo, 1981).

7.3.4   Aroma

The characteristic aroma of ripe kiwifruit is due to the combination of at least 26 volatile components, of which ethyl butanoate, hexanal and *trans*-hex-2-enal are important components of the aroma profile (Young *et al.*, 1983).

In unripe fruit certain aldehydes, particularly *trans*-hex-2-enal, give the aroma of the fruit a 'green' character, perhaps like freshly-mown grass. As the fruit ripens (i.e. firmness declines below 10 N), the concentration of these aldehydes declines, while the total concentration of volatile compounds, and the relative proportion of esters in the volatile fraction increase. Fruit which are harvested immature (soluble solids less than

6.2%) retain their high aldehyde content when they soften and never develop the aroma associated with ripe fruit (Young and Patterson, 1985).

7.3.5 Texture

Consumers judge the ripeness of many fruit, including kiwifruit, by their texture. Designing an easily quantifiable measurement of texture which accurately reproduces consumer assessments has proved difficult. One commonly-used method is to assess the firmness of the fruit by the force required to penetrate the flesh with a probe of a defined diameter. The results of such penetrometer measurements are usually expressed in units of Newton (N).

The flesh of immature kiwifruit remains very firm until starch hydrolysis starts and the concentration of soluble sugars begins to rise about 20 weeks after anthesis (Fig. 7.1). If the fruit are left on the vine their firmness declines steadily until at least 30 weeks after anthesis (Pratt and Reid, 1974). At the normal harvest date, at about 20–22 weeks after anthesis, the fruit usually have a firmness of 60–90 N, measured using a 7.9 mm diameter penetrometer probe (Beever and Hopkirk, 1990). The physiological basis for the decrease in the firmness of fruit left on the vine after this harvest date is unknown.

The pattern of ripening of harvested fruit is rather different from that of fruit left on the vine. In fruit stored in a packed tray at 20°C, firmness declines gradually from over 80 N to about 70 N during the first seven to ten days during storage and then declines sharply to reach 10 N about ten days later. Eating ripeness is reached at a firmness of 5–8 N after about three weeks' storage at 20°C (Beever and Hopkirk. 1990).

Storage

The success of the New Zealand kiwifruit industry is based on the ability to transport kiwifruit to distant markets in USA, Europe and Japan by sea, rather than by air. This depends on the ability to store fruit for prolonged periods at low temperature. The long storage life of fruit of the cultivar Hayward is one reason that this cultivar was chosen to form the basis of the industry in New Zealand.

Under optimum conditions, particularly in the absence of ethylene, kiwifruit can be stored for 4–6 months at 0°C, during which time the firmness of the fruit declines rapidly from 80 N to about 30 N in four to six weeks, but then much more slowly to about 10 N (Arpaia *et al.*, 1987; Hopkirk *et al.*, 1989).

At harvest, the composition of the walls of cells in different parts of the fruit – outer pericarp, inner pericarp, locule wall, and core are similar. The cell wall material contains 40–50% pectic galactans and 15–25%

hemicelluloses (Redgwell *et al.*, 1988). The sharp decline in the firmness of the fruit during storage, even at 0°C, is accompanied by an increase in the proportion of uronic acids present in the water-soluble fraction, and the release of galactose, arabinose and rhamnose, sugars which are normally associated with pectic polymers. These results suggest that the pectic polymers in the cell wall may undergo degradation during storage (Arpaia *et al.*, 1987). No consistent changes were found in the cellulose fraction or neutral sugars associated with hemicelluloses. The rate of softening of the fruit reaches its lowest value after about eight weeks' storage. From this point, the relative proportions of uronic acids, arabinose and rhamnose in the ethanol-insoluble (polymeric) and ethanol-soluble fractions remain constant.

After extended storage at 0°C, kiwifruit normally reach a firmness below 10 N. When the fruit is removed to 20°C, the fruit soften rapidly and reach an eating ripeness at about 5 N after one to two weeks. This final phase of fruit softening is associated with a decrease in the proportion of high molecular weight pectic polymers and hemicelluloses and an increase in the proportion of the equivalent low molecular weight compounds (Soda *et al.*, 1987a,b)

7.3.6 Respiration

Kiwifruit are generally considered to be climacteric. The respiration rate of fruit measured immediately after picking depends on the maturity of that fruit. In immature fruit the respiration rate is about 40 mg $CO_2.kg^{-1}.h^{-1}$ at harvest and declines to less than 20 mg $CO_2.kg^{-1}.h^{-1}$ after two weeks' storage at 20°C (Wright and Heatherbell, 1967; Pratt and Reid, 1974). The respiration rate measured immediately after harvest is lower in more mature fruit. Fruit harvested from 25 to 47 weeks after anthesis have a respiration rate of less than 20 mg $CO_2.kg^{-1}.h^{-1}$ (Pratt and Reid, 1974). These studies have recently been confirmed by measurement of the respiration rate of developing fruit which are still attached to the vine (Walton and de Jong, 1990). Understandably, the respiration rate of kiwifruit is sensitive to temperature, the rate at 2°C being about one-fifth of that at 20°C (Wright and Heatherbell, 1967). The rate of ripening and senescence of the fruit is delayed by low temperature, and this is the basis for the successful storage of kiwifruit for prolonged periods under refrigeration. The respiration rate is also sensitive to ethylene, the magnitude and nature of this effect depending on the maturity of the fruit. When immature fruit, picked nine weeks after anthesis, were exposed for several days to ethylene at high concentrations (at least 100 $\mu l.l^{-1}$) the respiration rate of the fruit rose to a plateau which was sustained for the duration of the exposure to ethylene. In fruit picked from 11 to 25 weeks after anthesis, respiration rate increased rapidly after the start of the ethylene treatment and then fell gradually. Mature fruit exposed to ethylene showed a response which was similar to the climacteric of fruit ripened normally – there was a rise in respiration rate accompanied by softening, development of aroma and synthesis of ethylene by the fruit (Pratt and Reid, 1974).

7.3.7 Ethylene

The production of ethylene by kiwifruit only begins to rise late in the ripening of the fruit. During ripening at 20°C, softening of the fruit below 10 N is associated with a sharp rise in ethylene production from trace levels and an increase in respiration rate of the fruit (Pratt and Reid, 1974). Peak ethylene concentrations reach 60–80 $\mu l.kg^{-1}.h^{-1}$. Reid *et al.* (1982) have suggested the kiwifruit should be classified as a climacteric fruit because of the simultaneous increase in respiration rate and ethylene production.

7.4   PATHWAYS AND ENZYMES

There is little published literature on the biochemical events which account for the physiological changes described above.

7.4.1   Polygalacturonase (EC 3.2.1.15)

Soda *et al.* (1986) measured changes in polygalacturonase activity in kiwifruit which had been ripened in the presence of ethylene. Two different assay techniques were used, viscometric and colorimetric. Polygalacturonase was not detected by either technique in unripe fruit or fruit ripened without ethylene, but was present in fruit ripened in the presence of ethylene.

7.4.2   Ethylene-Forming Enzyme

Membrane vesicles from ripe fruit are able to convert 1-amino-2-ethyl cyclopropane-1-carboxylic acid to ethylene. This ethylene-forming enzyme (EFE) activity also showed a high affinity for the natural substrate 1-aminocyclopropane-1-carboxylic acid (ACC) (Mitchell *et al.*, 1988). These authors concluded that the stereospecificity of the enzyme *in vitro* indicates that it represents authentic activity *in vivo*. Kiwifruit stored individually at 21°C began to produce ethylene exponentially after reaching a threshold value of about 0.1 $\mu l.kg^{-1}.h^{-1}$. EFE activity and concentrations of both ACC and malonyl-ACC rose with increasing ethylene production (Hyodo *et al.*, 1987).

7.4.3   Actinidin (EC 3.4.22.14)

The juice of raw kiwifruit tenderizes meat and prevents jellies based on gelatin from setting. This is due to the presence in the fruit of actinidin, a proteolytic enzyme (Arcus, 1959). The enzyme, a thiol protease, has been purified in crystalline form (McDowall, 1970) and the amino acid sequence (Carne and Moore, 1978), and three-dimensional structure determined, the latter by X-ray crystallography (Baker 1973, 1977, 1980; Baker *et al.*, 1980).

The amino acid sequence, three-dimensional structure and reaction mechanism are similar to those of papain (EC 3.4.22.2), a thiol protease from papaya. More recently, cDNA clones have been isolated for actinidin (Praekelt *et al.*, 1988; Podivinsky *et al.*, 1989). There appear to be two classes of actinidin, an acidic class (pI = 4.4, corresponding to the known protein sequence) and a basic class (pI = 8.8). The gene for the acidic actinidin is highly expressed, while the gene for the basic protein is expressed at a lower level. Expression of both genes is first detected when the fruit is about half-size and reaches maximum level when the fruit is ripe. The function of actinidin in the fruit is not known, although based on similarities with plant pathogenesis-related proteins it has been suggested that it plays a role in the protection of kiwifruit from attack by pathogens (Podivinsky *et al.*, 1991).

7.5   REGULATION OF RIPENING

7.5.1   Ripening patterns

The time taken by individual kiwifruit to ripen is variable, and this is true whether they are ripened on or off the vine. Some fruit left on the vine may ripen as early as 20 weeks after anthesis while others may stay attached to the vine for up to six months. Ripe fruit do not normally abscise and soft fruit will remain on the vine for several weeks after ripening (Pratt and Reid, 1974; Beever and Hopkirk, 1990). There is a ripening gradient within an individual fruit, with the soluble solids content of the blossom end of the fruit being higher than the stem end (Hopkirk *et al.*, 1986).

Individual fruit kept in a well ventilated atmosphere at 20°C sometimes soften to below 10 N in less than six weeks, whereas other fruit may take over six months (Pratt and Reid, 1974). Normally, of course, fruit are not kept separate but are packed in boxes under a plastic liner, under which conditions there is a much more uniform rate of ripening, which takes about three weeks. The accumulation of ethylene within the polyliner probably plays an important part in this process.

7.5.2   The role of ethylene

After harvest, kiwifruit are extremely sensitive to added ethylene. Fruit treated with as little as 5 µl.l^{-1} of ethylene soften rapidly. As the fruit ripen there are significant changes in the carbohydrate composition of the fruit, with concentrations of fructose, glucose and sucrose rising within five days of ethylene treatment (Matsumoto *et al.*, 1983). These changes are reflected in an increase in the soluble solids content of the fruit, and a decrease in the titratable acidity, starch and amylose content. The chlorophyll content of the fruit is unaffected. Treatment of fruit on the vine with ethephon, which releases ethylene, reduces the accumulation of starch in the core of the fruit, and later increases degradation of starch at the blossom end (Bowen *et al.*, 1988). Even during storage at 0°C, fruit will soften rapidly in the presence of ethylene. Because of this the concentration of ethylene in commercial cool stores is kept as low as possible, with a maximum of 0.03 µl.l^{-1} being accepted (Harris, 1981).

Most of the studies on the effect of ethylene on the ripening of kiwifruit have used softening as the measure of ripening. It is unclear whether all of the physiological processes associated with ripening are equally affected (Beever and Hopkirk, 1990), although the respiration rate of the fruit increases as a result of ethylene treatment (Wright and Heatherbell, 1967; Pratt and Reid, 1974) and there are also changes in the metabolism of carbohydrates and organic acids (Matsumoto *et al.*, 1983). The effects of ethylene described in the literature are primarily the result of the application of ethylene to the fruit, rather than of endogenous ethylene production. Clearly they show that physiologically mature fruit, at least, are very sensitive to this plant hormone. The role of ethylene in the normal development and ripening of the fruit has proved difficult to establish.

The major change in the physiology of the fruit occurs before harvest when the accumulation of starch ceases, and the sugar content and soluble solids content of the fruit begin to rise. Although the kiwifruit has been described as climacteric, because of the pattern of respiration during storage, it is unusual in that the rise in ethylene and respiration occur late in the ripening process. In other climacteric fruit eating ripeness occurs either at the same time as the climacteric rise (as in the avocado and banana) or considerably later (apple and tomato) (Beever and Hopkirk, 1990).

7.5.3   The seed

If one function of a fleshy fruit is to spread seed, then it would be expected that seed development and maturation would be closely co-ordinated with the development and ripening of the fruit. Although very little experimental evidence exists on this subject, it is worth noting that the seed is fully expanded and beginning to change colour at the same time as starch accumulation ceases and hydrolysis begins (Pratt and Reid, 1974; Hopping, 1976a).

7.6   FUTURE PROSPECTS

The development of a fertilized flower into a fleshy fruit with its characteristic colour, texture, flavour and aroma is, at least in my biased view, a good example of the beauty of the complexity of biology. There is beauty both in the way that expansion of the cells of the fruit is coordinated so that each different cultivar develops into its characteristic shape and in the regulation of the array of biochemical processes associated with ripening so that they reach their climax when the fruit is ready to eat.

We are only just beginning to discover how these processes are controlled at a genetic and biochemical level in the kiwifruit. The establishment of the kiwifruit industry in New Zealand on a commercial scale depended on scientists providing urgent answers to critical questions, in particular, which cultivars to grow in which parts of the country, when to harvest the fruit, and how to store and ship them. Many of these questions have been answered, at least for the moment. In the past research has been driven, appropriately, from a problem-solving approach rather than from the point of view of the biology of the fruit. Future progress will depend firstly on the development of new cultivars, not only through conventional breeding and biotechnology, but also on understanding the basic biology of the kiwifruit, and in particular how the development and ripening of the fruit are regulated.

The increase in the sugar content is used to determine the harvest date of the crop in New Zealand, yet surprisingly little has been published on the biochemical basis for the change in carbohydrate metabolism which occurs as the fruit reaches physiological maturity. Research is currently in progress on the identification of enzymes involved in starch degradation in kiwifruit and the isolation of genomic clones coding for amylases (Wegrzyn and McHale, 1991).

cDNA clones for several genes coding for enzymes involved in starch biosynthesis have been isolated from a number of different sources (Battacharyya *et al.*, 1990). It would be interesting and useful to isolate similar genes from kiwifruit and study changes in gene expression and enzyme activity during fruit development and ripening.

The spectacular increase in the sugar content of the kiwifruit which occurs during ripening would be expected to lead to an increase in the osmotic pressure inside the affected cells. The influx of water expected from this increase in osmotic pressure could provide the driving force for the increase in cell volume which occurs during the later stages in fruit development; this must be a closely regulated process to account for the development of fruit of uniform shape and texture. However, very little research appears to have been done on the control of cell size during fruit development. In the future, understanding how these processes are regulated may be critical in producing fruit of uniform shape, size and quality.

Although kiwifruit are very sensitive to applied ethylene, the role of ethylene in the development and ripening of the fruit is still unclear. In other fruit, studies with inhibitors of ethylene synthesis and action have proved valuable in understanding the role of ethylene in these fruit (Davies *et al.*, 1988), and the investigation of changes in the pattern of gene expression are providing a fundamental understanding of the regulation of ripening. The kiwifruit is unusual in that the respiratory climacteric and burst of ethylene production occur very late in ripening. Studies in the kiwifruit similar to those described above would prove very valuable in understanding the role of ethylene in ripening in this unusual climacteric fruit.

The quotation from Darwin emphasized one biological role of the fruit, to spread seed. One might expect that the development of the seed and the maturation and ripening of the fruit would be closely coordinated. Although considerable research has been done on the impact of seed number on final fruit size, surprisingly little attention has been paid to the possible role of the seed in regulation of ripening of the kiwifruit.

One of the features of development of this fruit is the production of the protease actinidin. We do not know how the fruit prevents this active enzyme from digesting fruit tissue. Actinidin can accumulate to form a significant proportion of the soluble protein of each fruit cell. At some stage during the evolution of the kiwifruit it is likely therefore to have served some physiological function. This function is unknown although one hypothesis put forward to explain the presence of proteases in fruit is that by causing diarrhoea in the animals that eat the fruit (including man), these enzymes ensure that seed is spread more evenly.

The kiwifruit is unusual because there is no characteristic change in skin or flesh colour as the fruit ripens. I suspect that Darwin might want to know what was, or is, the animal that eats kiwifruit in the wild, and how did it know that the fruit was ripe and the seed ready to spread? Alternatively he may have concluded that this unusual fruit serves some other function than aiding in the spread of seed.

Acknowledgements

I would like to thank first, the editors for inviting me to write this review, and second, Helen, Katherine and Fiona for their patience and support while it was being written.

Finally I thank, Kevin Davies, Murray Hopping, Garry Burge and David Swain for their constructive comments once it was written.

REFERENCES

Arcus, A.C. (1959) Proteolytic enzyme of *Actinidia chinensis*. *Biochimica et Biophysica Acta*, **33**, 242–244

Arpaia, M.L., Labavitch, J.M., Greve, C. and Kader, A.A. (1987) Changes in the cell wall components of kiwifruit during storage in air or controlled atmosphere. *Journal of the American Society of Horticultural Science*, **112**, 474–481

Baker, E.N. (1973) Preliminary crystallographic data for actinidin, a thiol protease from *Actinidia chinensis*. *Journal of Molecular Biology*, **74**, 411–412

Baker, E.N. (1977) The structure of actinidin, a proteolytic enzyme. *Chemistry in New Zealand*, **41**, 90–94

Baker, E.N. (1980) Structure of actinidin, after refinement at 1.7 Å resolution. *Journal of Molecular Biology*, **141**, 441–484

Baker, E.N., Boland, M.J., Calder, P.C. and Hardman, M.J. (1980) The specificity of actinidin and its relationship to the structure of the enzyme. *Biochimica et Biophysica Acta*, **616**, 30–34

Beever, D.J. and Hopkirk, G. (1990) Fruit development and fruit physiology. In *Kiwifruit science and management*, Ray Richards, in association with the New Zealand Society for Horticultural Science, Auckland

Bhattacharyya, M.K., Smith, A.M., Ellis T.H.N., Hedley, C. and Martin, C. (1990) The wrinkled-seed character of pea described by Mendel is caused by a transposon–like insertion in a gene encoding starch-branching enzyme. *Cell*, **60**, 115–122

Ben-Aire, R., Gross, J. and Sonego, L. (1982) Changes in ripening-parameters and pigments of the Chinese gooseberry (kiwi) during ripening and storage. *Scientia Horticultura*, **18**, 65–70

Bowen, J.H., Lowe, R.G. and MacRae, E.A. (1988) The effect of pre-harvest treatment with ethrel on the starch content of kiwifruit. *Scientia Horticultura*, **35**, 251–258

Carne, A. and Moore, C.H. (1978) The amino acid sequence of the tryptic peptides from Actinidin, a proteolytic enzyme from the fruit of *Actinidia chinensis*. *Biochemical Journal*, **173**, 73–83

Coombe, B.G. (1976) The development of fleshy fruits. *Annual Review of Plant Physiology*, **27**, 507–528

Darwin, C.R. (1859) *The Origin of Species*, Murray, London

Davies, K.M., Hobson, G.E. and Grierson, D. (1988) Silver ions inhibit the ethylene-stimulated production of ripening-related mRNAs in tomato. *Plant, Cell and Environment*, **11**, 729–738

Ferguson, A.R. (1990a) The genus *Actinidia*. In *Kiwifruit science and management*, Ray Richards, in association with the New Zealand Society for Horticultural Science, Auckland

Ferguson, A.R. (1990b) The kiwifruit in China. In *Kiwifruit science and management*, Ray Richards, in association with the New Zealand Society for Horticultural Science, Auckland

Ferguson, A.R. and Bollard, E.G. (1990) Domestication of the kiwifruit. In *Kiwifruit science and management*, Ray Richards, in association with the New Zealand Society for Horticultural Science, Auckland

Ford, I. (1971) Harvesting and maturity of chinese gooseberries. *The Orchardist of New Zealand*, **44**, 129–130

Fuke, Y. and Matsuoka, H. (1984) Studies on the physical and chemical properties of kiwifruit starch. *Journal of Food Science*, **49**, 620–622

Fuke, Y., Sasago, K. and Matsuoka, H. (1985) Determination of chlorophylls in kiwifruit and their changes during ripening. *Journal of Food Science*, **50**, 1220–1223

Grant, J.A. and Ryugo, K. (1984) Influence of within-canopy shading on fruit size, shoot growth, and return bloom in kiwifruit. *Journal of American Society of Horticultural Science*, **109**, 799–802

Harman, J.E. (1981) Kiwifruit maturity. *The Orchardist of New Zealand*, **54**, 126–130

Harris, S. (1981) Ethylene and kiwifruit. *The Orchardist of New Zealand*, **54**, 105

Hopkirk, G.,Beever, D.J. and Triggs, C.M. (1986) Variation in soluble solids concentration in kiwifruit at harvest. *New Zealand Journal of Agriculture Research*, **29**, 475–484

Hopkirk, G., Snelgar, W.P., Horne S.F. and Manson, P.J. (1989) Effect of increased preharvest temperature on fruit quality of kiwifruit (*Actinidia deliciosa*). *Journal of Horticultural Science*, **64**, 227–237

Hopping, M.E. (1976a) Structure and development of fruit and seeds in chinese gooseberry (*Actinidia chinensis* Planch.). *New Zealand Journal of Botany*, **14**, 63–68

Hopping, M.E. (1976b) Effect of exogenous auxins, gibberellins and cytokinins on fruit development in chinese gooseberry (*Actinidia chinensis* Planch.). *New Zealand Journal of Botany*, **14**, 69–75

Hopping, M.E. (1986) Kiwifruit. In *CRC Handbook of Fruit Set and Development*, CRC Press, Boca Raton, Florida

Hopping, M.E. and Hacking, N.J.A. (1983) A comparison of pollen application methods for the artificial pollination of kiwifruit. *Acta Horticulturae*, **139**, 41–50

Hyodo, H., Aizawa, S. and Ao, S.(1987) Ethylene formation during ripening of kiwifruit. *Journal of Japanese Society of Horticultural Science*, **56**, 352–355

Iwahori, S., Tominaga, S. and Yamasaki, T. (1988) Stimulation of fruit growth of kiwifruit, *Actinidia chinensis* Planch., by N-(2-chloro-4-pyridyl) N′-phenylurea, a diphenylurea-derivative cytokinin. *Scientia Horticultura*, **35**, 109–115

Lodge, N. and Robertson, G.L. (1990) Processing of kiwifruit. In *Kiwifruit science and management*, Ray Richards, in association with New Zealand Society for Horticultural Science, Auckland

McDowall, M.A. (1970) Anionic protease from *Actinidia chinensis*; preparation and properties of the crystalline protease. *European Journal of Biochemistry*, **14**, 214–221

MacRae, E.A. and Redgwell, R.J. (1990) Partitioning of ^{14}C-photosynthate in developing kiwifruit. *Scientia Horticultura*, **44**, 83–95

Matsumoto, S., Ohara, T, and Luh, B.S. (1983) Changes in chemical constituents of kiwifruit during post-harvest ripening. *Journal of Food Science*, **48**, 607–611

Mitchell, T., Porter, A.J.R. and John, P. (1988) Authentic activity of the ethylene forming enzyme observed in membranes obtained from kiwifruit (*Actinidia deliciosa*). *New Phytologist*, **109**, 313–319

Okuse, I. and Ryugo, K. (1981) Compositional changes in the developing Hayward kiwifruit in California. *Journal of American Society of Horticultural Science*, **106**, 73–76

Podivinsky, E., Forster, R.L.S. and Gardner, R.C. (1989) Nucleotide sequence of actinidin, a kiwifruit protease *Nucleic Acids Research*, **17**, 8363–8366

Podivinsky, E., Snowden, K.C., Keeling, J., Lin, E. and Gardner, R.C. (1991) Expression of actinidin, a kiwifruit cysteine protease. Abstract for The Second International Symposium on Kiwifruit, Palmerston North, February 18–21

Praekelt, U.M., McKee, R.A. and Smith, H. (1988) Molecular analysis of actinidin, the cysteine proteinase of *Actinidia chinensis*. *Plant Molecular Biology*, **10**(3), 193–202

Pratt, H.K. and Reid, M.S. (1974) Chinese gooseberry: seasonal patterns in fruit growth and maturation, ripening, respiration and the role of ethylene. *Journal of the Science of Food and Agriculture*, **25**, 747–757

Pyke, N.B. and Alspach, P.A. (1986) Inter-relationships of fruit weight, seed number and seed weight in kiwifruit. *New Zealand Journal of Agricultural Science*, **20**, 153–156

Redgwell, R.J., Melton, L.D. and Brasch, D.J. (1988) Cell-wall polysaccharides of kiwifruit (*Actinidia deliciosa*): chemical features in different tissue zones of the fruit at harvest. *Carbohydrate Research*, **182**, 241–258

Reid, M.S., Heatherbell, D.A. and Pratt, H.K. (1982) Seasonal patterns in chemical composition of the fruit of *Actinidia chinensis*. *Journal of American Society of Horticultural Science*, **107**, 316–319

Scienza, A., Visai, C., Conca, E. and Valenti, L. (1983) Relazione tra lo sviluppo, la maturazione del frutto e la presenza di ormoni endogeni in *Actinidia chinensis*. In *Atti del II Incontro Frutticolo SOI sull'Actinidia*, Udine, 1983, Udine, Italy, Centro Regionale per la Sperimentazione Agraria per il Friuli-Venezia Giulia e Sezione Frutticoltura della SOI. pp. 401–421

Scott, K.J., Spraggon, S.A. and McBride, R.L. (1986) Two new maturity tests for kiwifruit. *CSIRO Food Research Quarterly*, **46**, 25–31

Soda, I., Hasegawa T., Suzuki, T. and Ogura, N. (1986) Detection of poly-galacturonase in kiwifruit during ripening *Journal of Biological Chemistry*, **50**, 3191–3192

Soda, I., Hasegawa, T., Suzuki, T. and Ogura, N. (1987a) Changes of poly-uronides during ripening. *Agricultural and Biological Chemistry*, **51**, 581–582

Soda, I., Hasegawa, T. and Suzuki, T. (1987b) Changes in hemicelluloses during after ripening of kiwifruit. *Journal of Agricultural Science, Japan*, **31**, 261–264

Sugimoto, Y., Yamamoto, M., Abe, K. and Fuwa H. (1988) Developmental changes in the properties of kiwifruit starches (*Actinidia chinensis* Planch.) *Journal of Japanese Society of Starch Science*, **35**, 1–10

Walton, E.F. and de Jong T.M. (1990) Estimating the bioenergetic cost of a developing kiwifruit berry and its growth and maintenance respiration components. *Annals of Botany*, **66**, 417–424

Wegrzyn, T. and McHale, R. (1991) Alpha-amylases in kiwifruit. Abstract for The Second International Symposium on Kiwifruit, Palmerston North, February 18–21

Wright, H.B. and Heatherbell, D.A. (1967) A study of respiratory trends and some associated physico-chemical changes of chinese gooseberry fruit, *Actinidia chinensis* (yang-tao) during the later stages of development. *New Zealand Journal of Agricultural Research*, **10**, 405–414

Young, H. (1977) Identification of cytokinins from natural sources by gas-liquid chromatography/mass spectrometry. *Analytical Biochemistry*, **79**, 226–233

Young, H. and Paterson, V.J. (1985) The effects of harvest maturity, ripeness and storage on kiwifruit aroma. *Journal of the Science of Food and Agriculture*, **36**, 352–358

Young, H., Paterson, V.J. and Burns D.J.W. (1983), Volatile aroma constituents of kiwifruit. *Journal of the Science of Food and Agriculture*, **34**, 81–85

Chapter 8

Mango

C. Lizada

8.1 INTRODUCTION

The mango (*Mangifera indica* Linn.), which is a dicotyledonous fruit of the family Anacardiaceae, originated in the Indo-Burmese region (Subramanyam *et al.*, 1975; Tjiptono *et al.*, 1984). It is produced principally in the developing countries of the tropics, with total world production estimated at 15.06 million tonnes (FAO, 1989).

Mango cultivars may be classified into two groups Indian or Indo-Chinese, based mainly on peel pigments and sensory characteristics of the pulp. Most of the Indian varieties, which possess stronger aroma and more intense peel colouration, are monoembryonic and require asexual propagation methods for consistent reproduction of the cultivar. On the other hand, most of the cultivars grown in the South-east Asian region are largely polyembryonic (Kusumo *et al.*, 1984). Due to the contrasting flavours of these groups, populations accustomed to the taste of Indo-Chinese varieties perceive the Indian types as 'medicinal' or possessing a 'turpentine' flavour.

Subramanyam *et al.* (1975) consider two other groups: first, mangoes of India, with character intermediate between the polyembryonic Indo-Chinese varieties and second, hybrids developed in Florida and Hawaii.

Trade in mango has been limited by the highly perishable nature of this fruit. Ripening cannot be delayed sufficiently to allow for long-distance transport. The fruit are also highly susceptible to disease, extremes of temperature and physical injury. Thus, most of the post-harvest technologies for these fruits are designed for disease control and protection against injury during packaging and transport. Technologies for longer-term storage, such as controlled or modified atmospheres, which have been used commercially for temperate fruits have not been

Biochemistry of Fruit Ripening. Edited by G. Seymour, J. Taylor and G. Tucker. Published in 1993 by Chapman & Hall, London. ISBN 0 412 40830 9

applied successfully to the mango. Despite the numerous studies on the physiology and biochemistry of this fruit, little is understood with respect to the fundamental processes relevant to the design of appropriate technologies to enhance marketable life.

India, which produces the largest volume, exports mango largely in the processed form as juice, puree, slices or pickles; in contrast, the Philippines and Mexico mainly export the fresh fruit.

8.2   FRUIT DEVELOPMENT AND HARVEST MATURITY

The final quality of the mango depends not only on the physiological processes occurring during ripening, but also on processes during fruit development and maturation.

Considerable effort has been put into identifying reliable indices of harvest maturity, as this affects subsequent ripening rate, ripe fruit quality (Peacock et al., 1986; Medlicott et al., 1988), response to various postharvest treatments (Esguerra and Lizada, 1990) and processing quality (Kapur et al., 1985).

Mango fruit growth follows a simple sigmoid pattern (Akamine and Goo, 1973; Lakshminarayana, 1973; Mendoza, 1981; Kasantikul, 1983; Tandon and Kalra, 1983; Lam et al., 1982). The period of rapid growth is characterized by an increase in alcohol-insoluble solids, principally starch (Mendoza et al., 1972; Tandon and Kalra, 1984), the accumulation of which is accompanied by increases in amylase activity in 'Dashehari' (Tandon and Kalra, 1983) and 'Haden' varieties (Fuchs et al., 1980). Mattoo and Modi (1969) reported the presence of an amylase inhibitor in the unripe fruit which might account for starch accumulation despite the increase in amylase activity during maturation. The increase in dry matter has been recommended for use as an index of maturity in 'Kensington Pride' (Baker, 1984) and might very well be the basis for the use of specific gravity as a maturity index in 'Alphonso' (Subramanyan et al., 1976), 'Dashehari' (Kapur et al., 1985) and 'Carabao' varieties (Cua and Lizada, 1990).

Titratable acidity correlated very well with days after flower induction in the 'Carabao' mango (Del Mundo et al., 1984a), decreasing in the fully mature fruit to <44.8 meq.100 g^{-1}. In 'Alphonso', titratable acidity increased from the sixth to the tenth week after fruit set and steadily declined thereafter as the fruit matured (Lakshminarayana, 1973).

Although endogenous ethylene and its induction of some ripening processes appear to be involved in the later stages of maturation in the mango (see below), attempts to accelerate maturation and enhance uniformity of the process with ethephon sprays proved unsuccessful (Lertpruk, 1983; Del Mundo et al., 1984b). The failure to manipulate maturation with exogenous ethylene is consistent with the postulate that the outer tissues are more resistant to ethylene action (Cua and Lizada, 1990).

As the mango fruit matures, bloom develops as wax is deposited on the peel (Kosiyachinda *et al.*, 1984). Although bloom may diminish, depending on the handling the fruit undergoes, some of it persists through ripening. The cuticular layer in the fully mature 'Carabao' mango is well defined and its ultrastructure cannot be altered even by overnight soaking or rubbing with methanol or chloroform. Electron-microscopy of peel sections revealed an apparent flattening of the outer layers of wax, which consequently assumed an amorphous appearance (Lizada and Kawada, unpublished). Within the smooth zones are recessed pockets revealing crystalline wax consisting of fringed platelets. As the samples were taken from fruits subjected to the usual packing operations for export, it might be assumed that the amorphous regions originated from previously crystalline platelets. The chemical nature of mango wax deposits, as well as changes during maturation, await characterization.

8.3 RIPENING PROCESSES

8.3.1 Ethylene production

A mango fruit approaching full maturity exhibits a significant decrease in starch and a distinct yellow colour in the pulp (Mendoza *et al.*, 1972; Medlicott *et al.*, 1988; Wang and Shiesh, 1990). In several varieties these changes are accompanied by a decline in pulp rupture force during the later stages of maturation (Cua and Lizada, 1990; Seymour *et al.*, 1990). These observations indicate that some changes associated with ripening appear to be induced prior to harvest maturity, and point to the possibility of ethylene production in the maturing fruit prior to detachment.

'Carabao' mangoes were found to produce ethylene prior to full maturity (Cua and Lizada, 1990). The internal ethylene level reached a peak of $0.35\ \mu l.l^{-1}$ at 110 days after flower induction, beyond which it declined as the fruit approached full maturity ten days later. Measurement of ethylene production in the different portions of the fruit showed that the highest rate ($1.25\ nl.g^{-1}.h^{-1}$) occurred in the outer mesocarp, despite the observation that the highest ethylene-forming enzyme (EFE) activity could be measured in the peel ($>20\ nl\ C_2H_4g^{-1}.h^{-1}$). The levels of the EFE substrate, 1-aminocyclopropane-1-carboxylic acid (ACC), were comparable in all portions of the fruit.

Despite the observation that the pre-harvest levels of ethylene in the 'Carabao' mango were higher than those in harvested fruits at colour break, no respiratory increase was induced by the former. Burg (1962) reported that mature but unripe mangoes had high ethylene levels ($1.87\mu l.l^{-1}$), even while attached to the tree. He suggested that this ethylene is rendered ineffective by a ripening inhibitor from the parent plant. This idea of an inhibitor is consistent with the stimulation of respiration observed after detachment in fruit harvested at different stages during maturation (Lakshminarayana, 1973; Mendoza, 1981). Lakshminarayana further observed an earlier onset of the climacteric as the fruit matured beyond the ninth week after fruit set.

In the 'Carabao' mango, at least, there is apparently a differential effect of this postulated inhibitor on the various processes associated with maturation and/or ripening (e.g. climacteric respiration and carotenogenesis). There appear to be differences also in the degree of inhibition between the inner and outer portions of the mesocarp, such that the former appears to ripen further despite comparable levels of ethylene in the entire mesocarp (Cua and Lizada, 1990).

Hardly any ethylene can be detected in the fully mature 'Carabao' mango, but ethylene production resumes as the fruit approaches colour break (Lizada and Cua, 1990). As with the ethylene production associated with maturation, postharvest ethylene production is accompanied by an increase in both ACC synthase (EC 4.4.1.14) and the ethylene-forming enzyme (EFE).

In most tissues examined, the K_m for ACC synthase is in the micromolar range (Yu et al., 1979; Hyodo et al., 1985), while that for EFE exhibits a range of values.

8.3.2 Pigments

Peel colour is an important criterion of acceptability of the mango (Satyan et al., 1986). During ripening the peel colour gradually changes from green to orange/yellow. Some cultivars develop a reddish blush which has been attributed to anthocyanins, while others, e.g. 'Harumanis' and 'Katchamita', retain most of the green colour, even at the full ripe stage.

In 'Tommy Atkins' Medlicott et al. (1986) observed a rapid destruction of chlorophyll, with chlorophyll a preferentially degraded relative to chlorophyll b. A more rapid loss in chlorophyll a is typically observed in senescence (Simpson et al., 1976) or chemical degradation of extracted chlorophylls (Peiser and Yang, 1977). Medlicott et al. (1986) also reported differing patterns of change in carotenoids and anthocyanins, with the former increasing during ripening. In contrast, the anthocyanin levels gradually declined, indicating that mere unmasking accounts for the increased prominence of blush in some cultivars.

Peel colour development is accompanied by ultrastructural changes associated with chloroplast-to-chromoplast transition. The thylakoid membrane systems in the peel of 'Alphonso' and 'Tommy Atkins' gradually break down, while osmiophilic globules enlarge and increase in number (Medlicott *et al.*, 1986; Parikh *et al.*, 1990). The loss of granal membrane integrity is associated with chlorophyll degradation, while the appearance of osmiophilic globules accompanies increases in carotenoid levels.

The principal carotenoids reported in ripe 'Alphonso' mango peel were β-carotene, xanthophyll esters and xanthophylls. β-carotene and auroxanthin were found to constitute 55% and 12.5%, respectively, of the peel carotenoids in 'Tommy Atkins' (Medlicott, 1985). Thus far, the only anthocyanin pigment identified in the peel is peonidin-3-galactoside, which was extracted by Proctor and Creasy (1969) from 'Haden'.

Pulp carotenoids continue to increase in the detached fruit as ripening proceeds (Chaudhary, 1950; John *et al.*, 1970), with the carotenoid level in the ripe fruit varying among cultivars (Table 8.1). In the fully ripe 'Badami' and 'Alphonso' mangoes, β-carotene constituted more than 50% of the total carotenoids, with phytofluene the next most abundant (Jungalwala and Cama, 1963; John *et al.*, 1970). In 'Tommy Atkins', β-carotene also constituted about two-thirds of the pulp carotenoids (Medlicott, 1985). The predominant xanthophyll in this variety was found to be violoxanthin.

The synthesis of carotenoids in the mango appears to proceed via the same biosynthetic pathway established in other species. In a series of studies Modi *et al.* (1965) and Mattoo *et al.* (1968) presented evidence for the role of mevalonic acid and geraniol in mango carotenogenesis. As in the peel, the synthesis of carotenoids in the pulp is accompanied by changes in the ultrastructure of plastids. As ripening proceeds, tubular

Table 8.1 Carotenoid levels in the ripe pulp of some mango cultivars.

Cultivar	Carotenoids (mg.100 g^{-1} fresh weight)	Reference
Carabao	2.75	Morga *et al.*, 1979
Nam Dorkmai	4.78	Kasantikul, 1983
Badami	8.92	John *et al.*, 1970
Dashehari	5.44	Mann and Singh, 1976
Alphonso	4.76	Ramana *et al.*, 1984
	8.06	Subramanyam *et al.*, 1976
Haden	6.82	Vazquez-Salinas and Lakshminarayana, 1986
Irwin	3.23	"
Kent	5.46	"
Keitt	3.87	Medlicott, 1985
Tommy Atkins	>5	Medlicott *et al.*, 1986
Kensington	5.06	Mitchell *et al.*, 1990

structures visible in the plastids of unripe fruit are lost, while osmiophilic globules increase in size and number in the 'Alphonso' mango (Parikh *et al.*, 1990).

Succinic acid 2,2-dimethylhydrazide, which has been used to enhance peel colour in other fruits, had no distinct effect on the pulp carotenoids of 'Alphonso' mango (Subramanyam and Sebastian, 1970). However, a 5-minute dip in water at 53°C resulted in a slight (7%) increase in total carotenoids. Increases in pulp carotenoids as a result of elevated temperatures have also been reported in 'Tommy Atkins' by Medlicott *et al.* (1985). Subjecting 'Carabao' mangoes to vapour heat treatment or a 10-minute hot water dip at 52–55°C enhances peel colour intensity and results in a more complete disappearance of the green colour of the peel (Lizada *et al.*, 1986; Esguerra and Lizada, 1990).

8.3.3 Carbohydrate metabolism

The starch that has accumulated in the maturing fruit is rapidly lost during ripening (Selvaraj *et al.*, 1989; Morga *et al.*, 1979; Subramanyam *et al.*, 1976), and this loss is evident in the chloroplast where the starch granules become progressively smaller as ripening proceeds. Starch granules completely disappear in the ripe fruit (Parikh *et al.*, 1990; Medlicott *et al.*, 1986), which usually contains negligible levels of starch (Morga *et al.*, 1979; Fuchs *et al.*, 1980).

Starch hydrolysis in the ripening mango has been associated with amylase activity (Fuchs *et al.*, 1980), which exhibits the properties of both α-(EC 3.2.1.1) and β-(EC 3.2.1.2) amylases. The complete disappearance of starch may be attributed to an upsurge of amylase as ripening is completed. Fuchs *et al.* (1980) also reported the presence of a proteinaceous inhibitor detected during the electrophoresis of amylase, which they suggested might be the inhibitor earlier observed by Mattoo and Modi (1969, 1970, as cited by Subramanyam *et al.*, 1975).

As a consequence of starch hydrolysis, total sugars increase during ripening, with glucose, fructose and sucrose constituting most of the monosaccharides (Selvaraj *et al.*, 1989). The total sugar content of the ripe 'Carabao' mango is one of the highest reported, with values exceeding 20% (Peacock and Brown, 1984). However, the lower sugar contents reported for other varieties such as 'Golek' (Lam *et al.*, 1982) might simply reflect differences in the degree of ripeness when optimum eating quality is attained.

Sugars constitute 91% of the soluble solids from the mesocarp of the ripe 'Ngowe' mango (Brinson *et al.*, 1988). Non-reducing sugars, principally sucrose, increase in the later stages of ripening (Selvaraj *et al.*, 1989; Subramanyam *et al.*, 1976; Shashirekha and Patwardhan, 1976; Fuchs *et al.*, 1980). This is consistent with the high activity of the gluconeogenic enzyme fructose-1,6-diphosphatase (EC 3.1.3.11) in the ripe fruit as reported by Rao and Modi (1976) and Kumar and Selvaraj (1990) in several mango cultivars.

In most of the varieties examined fructose was the predominant reducing sugar. Along with an observation that pentoses exhibited a five-fold increase during ripening, this was considered as suggestive of an increase in the oxidative pentose phosphate pathway, which generates the necessary reducing equivalents (NADPH) for biosynthetic processes. Increases were also reported in glucose-6-phosphate dehydrogenase (EC 1.1.1.49) and 6-phosphogluconate dehydrogenase (EC 1.1.1.44).

Mango ripening is accompanied by increases in gluconeogenic enzymes. In several Indian varieties, glucose-6-phosphatase (EC 3.1.3.9) was observed to increase up to the three-quarter-ripe stage, while fructose-1,6-diphosphatase showed increased activity as the fruits ripened from the three-quarter to the full ripe stage (Kumar and Selvaraj, 1990). In abscisic acid-treated mango, enhanced ripening was similarly accompanied by increases in these enzymes (Parikh *et al.*, 1990). These increases could be inhibited by treatment with cycloheximide, indicating *de novo* synthesis of these enzymes during ripening.

8.3.4 Structural polysaccharides and textural changes

Pronounced softening during ripening limits the marketable life of the mango. Softening is accompanied by cell wall disruption (Parikh *et al.* 1990), with the middle lamella appearing as an electron-translucent area in electron micrographs of the ripe fruit (Medlicott, 1985).

In most of the varieties examined an increase in water-soluble polysaccharides has been observed during ripening (Tandon and Kalra, 1984; Lazan *et al.*, 1986; Brinson *et al.*, 1988). In contrast, Roe and Bruemmer (1981) reported a decline in water-soluble polysaccharides in ripening 'Keitt' mangoes, which they attributed to the possibility of extensive polymer degradation such that the products become soluble in ethanol.

Brinson *et al.* (1988) examined cell wall constituents of 'Ngowe' mango and reported a decline in uronic acid content as a percentage weight of the whole wall, i.e. from 25% in the unripe to 19.1% in the ripe fruit. In contrast, the uronic acid content of the water-soluble polysaccharides from these cell wall preparations increased from only 7% in the unripe to 90% in the ripe fruit. Moreover, galactose and arabinose each constituted about 30% of the water-soluble polysaccharides of the cell walls of the unripe fruit. A higher uronic acid content of the water-soluble polysaccharides was observed in the ripe compared to the unripe mesocarp. These observations were interpreted to indicate that during ripening:

1. Mango cell walls are degraded, releasing the combined mono-saccharides of the pectin complex;
2. The resulting water-soluble pectic materials in the cell walls lose arabinose and galactose accounting for the galacturonan-rich poly-saccharide in the mesocarp.

In a more recent study (Tucker and Seymour, 1991), using precautions to prevent cell wall degrading activity during wall preparation and fractionation, less extensive breakdown of mango pectic polymers was reported than observed by Brinson *et al.* (1988). However, there was still a substantial increase in soluble pectin during ripening and a marked loss of galactose residues.

Most determinations of polygalacturonase (EC 3.2.1.15;PG) in the ripening mango have shown low activities as measured by the reducto-metric method. Lazan *et al.* (1986) reported that viscometric assays gave very low activity, although the pattern of change correlated well with that observed using the reductometric method. Brinson *et al.* (1988) failed to detect endoPG activity, while Medlicott (1985) reported very little loss in viscosity in his assays for PG. Thus, although it appears that an exoPG (EC 3.2.1.67) might be involved in cell wall degradation, the low activities measured make it difficult to account for the rather pronounced softening that the mango undergoes during ripening. It must be pointed out that all these assays use polygalacturonic acid as substrate; use of pectic materials isolated from the mango cell wall might give different results.

In 'Carabao' mango, considerable pectinmethylesterase (EC 3.1.1.11; PME) activity could be measured during ripening, increasing as the fruit approaches the half-ripe (50% yellow peel colour) stage and declining thereafter. A similar pattern was observed in some of the varieties examined by Selvaraj and Kumar (1989).

The inner mesocarp of the 'Carabao' mango exhibits softening ahead of the outer mesocarp (Cua and Lizada, 1989). The difference in the degree of softening between portions of the mesocarp is evident even in the ripe fruit, and this difference appears to be variety-dependent (Chaplin *et al.*, 1990). The jelly seed disorder (Van Lelyveld and Smith, 1979) and premature ripening around the seed (Winston, 1984) might be extreme examples of such differences.

8.3.5 Organic acids

Titratable acidity declines as the mango ripens, dropping from 48 meq.100 g^{-1} in the preclimacteric to 5.6 meq.100 g^{-1} in the post-climacteric 'Badami' mango (Shashirekha and Patwardhan, 1976). Similar patterns have been reported for other varieties (Morga *et al.*, 1979; Medlicott and Thompson, 1985; Selvaraj *et al.*, 1989). The predominant acid is citrate, with malate and succinate also found in significant quantities (Shashirekha and Patwardhan, 1976). Although both citrate and succinate consistently decline in all varieties examined, malate exhibits different patterns of changes in different cultivars. In a study of mitochondrial enzymes of the tricarboxylic acid cycle, Baqui *et al.* (1974) reported that citrate synthase (EC 4.1.3.7) decreases during ripening, while isocitrate dehydrogenase (EC 1.1.1.42) and succinate dehydrogenase (EC 1.3.99.1) each increase. These changes are consistent with the observed decline in citrate and succinate.

Isocitrate lyase (EC 4.1.3.1) has been reported to decline rapidly at the early stages of ripening, with a concomitant decrease in glyoxylate, which was found to inhibit this activity (Baqui *et al.*, 1977).

Malic enzyme has been purified from mango pulp (Krishnamurthy and Patwardhan, 1971). The changes in its activity during ripening paralleled those of respiration during the climacteric. This has been reported in 'Haden' (Dubery *et al.*, 1984), 'Dadomia' (Baqui *et al.*, 1977) and 'Pairi' (Krishnamurthy *et al.*, 1971). In 'Alphonso', cytosolic malate dehydrogenase (EC 1.1.1.37) increased during ripening (Parikh *et al.*, 1990).

An increase in citrate cleavage enzyme (EC 4.1.3.8) during ripening has also been reported (Mattoo and Modi, 1970). This activity could be stimulated by a crude fatty acid extract from mango. Mattoo and Modi further suggested that, along with the increases in the oxidative pentose phosphate pathway enzymes, the malic enzyme and malate dehydrogenase, the citrate cleavage enzyme provides the reducing equivalents for synthesis. Its products, acetyl CoA and oxaloacetate, might also be utilized for synthetic processes during ripening.

In 'Pairi' mango, oxaloacetate and α-ketoglutarate reached a maximum prior to the climacteric rise, progressively declining to very low levels (Krishnamurthy *et al.*, 1971). In contrast, aspartate and glutamate exhibited a minimum immediately prior to the climacteric peak, and, although the levels increased slightly at the postclimacteric stage, they were only about 50% and 39%, respectively, of the maximum preclimacteric aspartate and glutamate levels. In 'Badami', aspartate and glutamate, in addition to arginine and lysine levels, are higher at the preclimacteric stage than at either the climacteric or postclimacteric stage (Shashirekha and Patwardhan, 1976). These patterns might be related to the requirement for protein synthesis during ripening (Parikh *et al.*, 1990).

8.3.6 Lipid metabolism

The total lipid increased in the ripening 'Alphonso' mango, as evidenced by the increase in ether-extractable components of the pulp (Bandyopadhyay and Gholap, 1973a). Similar patterns were also obtained from the other Indian cultivars that Selvaraj *et al.* (1989) examined. The glyceride content in 'Alphonso' also increased, while the fatty acid profile changed through ripening. The more unsaturated fatty acids, i.e. palmitoleate and linolenate were found in higher levels in the ripe relative to unripe fruits. These changes were found to correlate with aroma and flavour. In a later study, Bandyopadhyay and Gholap (1973b) reported that the cultivars with stronger aroma and flavour had lower palmitate/palmitoleate ratios.

Applying [2-^{14}C]acetate to 'Alphonso' mangoes, Gholap and Bandyopadhyay (1980) found that acetate was maximally incorporated into saturated fatty acids, while the radioactivity in [1-^{14}C]palmitic acid was incorporated largely in hydroxy fatty acids. Hydroxy fatty acids are precursors of lactones, which are major aroma constituents in 'Alphonso' mangoes (Engel and Tressl, 1983; Idstein and Schrier, 1985).

Mitochondria from unripe 'Dadomia' fruits oxidized fatty acids and this was stimulated by glyoxylate (Baqui *et al.*, 1977). The capacity of the mitochondria to oxidize fatty acids increased in the preclimacteric and climacteric fruits. When used as substrates stearic and oleic acids elicited the highest increases. The products of the β-oxidation of fatty acids are utilized in the synthesis of both carotenoids and terpenoid volatiles.

Selvaraj (1989) examined some enzymes involved in the production of carbonyl volatiles and reported the presence of lipase (EC 3.1.1.3), lipoxygenase (EC 1.13.11.12) and alcohol dehydrogenase (EC 1.1.1.71; aldehyde forming) activities in several Indian cultivars. All of these activities were high at the unripe stage, and declined when the fruits were beyond the half-ripe stage.

8.3.7 Volatile constituents

One can distinguish between cultivars of mango on the basis of flavour and aroma. The volatile constituents of the 'Alphonso', which is considered to have a very strong aroma among the Indian cultivars (Selvaraj, 1989), have been examined by two laboratories (Engel and Tressl, 1983; Idstein and Schrier, 1985) and both identified (Z)-ocimene as a major component. However, the two analyses differed in the approximate concentrations of β-myrcene and (E)-ocimene.

According to Engel and Tressl (1983) characteristic mango flavour could not be attributed to any single component; however, the typical green aroma of the unripe mango might be attributed to cis-ocimene and β-myrcene (Gholap and Bandyopadhyay, 1975), while dimethylstyrene has been described as having mango character (MacLeod and Pieris, 1984).

Distinct varietal differences can be attributed to volatile components unique to each variety. In a comparison of 'Alphonso' with 'Baladi' MacLeod and Pieris identified (Z)-3-hexenyl esters as being responsible for the fresh, green fruity note of the former but these cannot be detected in the latter. The C_9-lipid oxidation product, (E)-2-nonenal, could only be detected in 'Baladi' and might be responsible for the melon-like flavour of this cultivar. Other notable differences include:

1. The level of limonene, which was 40 μl.l^{-1} in Baladi, in contrast to 300 nl.l^{-1} in 'Alphonso';
2. The presence of the hydrocarbons α-gurjunene, germacrene D, bicyclogermacrene, τ- and δ-cadinene and α-selinene and the corresponding oxygenated sesquiterpenoids in 'Baladi', but not in 'Alphonso';
3. The relatively high levels of the ethyl esters of even-numbered fatty acids from C_2 to C_{16} in 'Baladi';
4. The relatively high levels of C_6 aldehydes and alcohols in 'Alphonso'.

8.3.8   Phenolics

Mango latex exudes from the freshly harvested fruit, but this exudate decreases during ripening. The major component of latex in freshly harvested 'Alphonso' mango was found to be 5-[2(Z)-heptadecenyl] resorcinol, which is structurally similar to known phenolic allergens found in species of the *Anacardiaceae* family (Bandyopadhyay *et al.*, 1985). Alkenylresorcinols have also been detected in the peel and flesh of freshly harvested mangoes and appear to be the basis for resistance to fungal pathogens (Prusky, 1990).

Astringency remains perceptible in the Carabao mango until the table-ripe stage is attained, and the progressive loss of astringency is associated with a loss in total phenolic content (Tirtosoekotjo, 1984). On the other hand, the 'Pico', which, like the 'Carabao' mango, contains >0.1% total phenolics and is non-astringent at the ripe stage, shows no significant decrease in phenolic content during ripening. Selvaraj and Kumar (1989) reported differences in tannin content among Indian cultivars, although all showed declining levels in the course of ripening. Gallotannins, which have been detected in the peel and pulp of the mango fruit, were also observed to decrease to negligible levels in the ripe fruit.

Polyphenoloxidase (EC 1.10.3.1) (PPO) showed an increase in the course of ripening in 'Malgoa' and 'Harumanis' (Lazan *et al.*, 1986), and the increase occurred concomitantly with the decline in ascorbic acid levels. In the freshly harvested fruit, the activity measured was <0.1 A_{420nm} min^{-1}.g^{-1} fresh weight, using catechol as substrate. In contrast, the PPO activity, measured by using pyrogallol as substrate, generally declined in the course of ripening in several Indian varieties including the 'Alphonso' (Selvaraj and Kumar, 1989).

The properties of PPO isolated from ripe 'Haden' mangoes have been characterized (Park *et al.*, 1981). The enzyme was found to possess no monophenol oxidase activity and was specific for *o*-diphenolic substrates, showing the highest activity in the presence of catechol as substrate. It exhibited a pH optimum range of 5.6–6.0, and was most effectively inhibited by sulphite. As with PPO isolated from other fruits, some inhibition could be effected by ascorbic acid. Two isoenzymes acting on catechol were detected.

Earlier, Joel *et al.* (1978) reported the presence of two PPOs in the green mango fruit, one possessing typical *o*-diphenol oxidase activity, and the other, *m*-phenol oxidase (laccase; EC 1.10.3.2) activity. The latter was found only in the secretory duct cavities, and could, therefore, be detected also in the exudates from the cut pedicel of the freshly harvested fruit. The authors pointed out that the presence of laccase appears to be characteristic of species belonging to the Anacardiaceae family.

REFERENCES

Akamine, E.K. and Goo, T. (1973) Respiration and ethylene production during ontogeny of fruit. *Journal of the American Society for Horticultural Science*, **98**, 381–383

Baker, I.W. (1984) Mango maturity investigations. *Proceedings of the First Australian Mango Workshop*, pp. 271–273

Bandyopadhyay, C. and Gholap, A.S. (1973a) changes in fatty acids in ripening mango pulp (variety Alphonse). *Journal of Agricultural and Food Chemistry*, **21**, 496–497

Bandyopadhyay, C. and Gholap, A.S. (1973b) Relationship of aroma and flavour characteristics of mango (*Mangifera indica* L.) to fatty acid composition. *Journal of the Science of Food and Agriculture*, **24**, 1497–1503

Bandyopadhyay, C., Gholap, A.S. and Mamdapur, V.R. (1985) Characterisation of alkenyl resorcinol in mango (*Mangifera indica*) latex. *Journal of Agricultural and Food Chemistry*, **33**, 377–379

Baqui, S.M., Mattoo, A.K. and Modi, V.V. (1974) Mitochondrial enzymes in mango fruit during ripening. *Phytochemistry*, **13**, 2049–2055

Baqui, S.M., Mattoo, A.K. and Modi, V.V. (1977) Glyoxylate metabolism and fatty acid oxidation in mango fruit during development and ripening. *Phytochemistry*, **13**, 2049–2055

Brinson, K., Dey, P.M., John, M.A. and Pridham, J.B. (1988) Post harvest changes in *Mangifera indica* mesocarp cell walls and cytoplasmic polysaccharides. *Phytochemistry*, **27**, 719–723

Burg, S.P. and Burg, E.A. (1962) Role of ethylene in fruit ripening. *Plant Physiology*, **37**, 179–189

Chaplin, G.R., Lai, S.C. and Buckley, M.J. (1990) Differential softening and physico-chemical changes in the mesocarp of ripening mango fruit. *Acta Horticulturae*, **269**, 169–179

Chaudhary, M.T. (1950) Carotenoid pigments of different varieties of mangoes: changes during ripening. *Journal of the Science of Food and Agriculture*, **1**, 173–177

Cua, A.U. and Lizada, M.C.C. (1989) Ethylene production on the "Caraboa" mango (*Mangifera indica* L.) fruit during maturation and ripening. Symposium on Tropical Fruit in International Trade, Honolulu, Hawaii

Cua, A.U. and Lizada, M.C.C. (1990) Ethylene production in the "Carabao" mango (*Mangifera indica* L.) fruit during maturation and ripening. *Acta Horticulturae*, **269**, 169–179

Del Mundo, C.R., Lizada, M.C.C., Mendoza, D.B.Jr. and Garcia, N.L. (1984a) Indices for harvest maturity in "Caraboa" mangoes. *Postharvest Research Notes*, **1**, 13–14

Del Mundo, C.R., Mendoza, D.B.Jr. and Lizada, M.C.C. (1984b) Preharvest treatment of "Caraboa" mangoes with ethephon. *Postharvest Research Notes*, **1**, 13–14

Dubery, I.A., Van Rensburg, L.V. and Schabart, J.C. (1984) Malic enzyme activity and related biochemical aspects during ripening of irradiated mango fruit. *Phytochemistry*, **23**, 1383–1386

Engel, K.H. and Tressl, R. (1983) Studies on the volatile components of two mango varieties. *Journal of Agricultural and Food Chemistry*, **31**, 796–801

Esguerra, E.B. and Lizada, M.C.C. (1990) The postharvest behaviour and quality of "Caraboa" mango subjected to vapour heat treatment. *ASEAN Food Journal*

FAO Production Yearbook (1989) Food and Agricultural Organisation. Rome, Italy

Fuchs, Y. Pesis, E. and Zauberman, G. (1980) Changes in amylase activity, starch and sugar contents in mango fruit pulp. *Scientia Horticulturae*, **13**, 155–160

Gholap, A.S. and Bandyopadhyay, C. (1975) Comparative assessment of aromatic principles of ripe Alphonso and Langros mangoes. *Journal of Agricultural and Food Science*, **12**, 262–263

Gholap, A.S. and Bandyopadhyay, C. (1980) Fatty acid biogenesis in ripening mango (*Mangifera indica* cv. Alphonso). *Journal of Agricultural and Food Chemistry*, **28**, 839–841

Hyodo, H., Tanaka, K. and Yoshisaka, J. (1985) Induction of 1-aminocyclopropane-1-carboxylic acid synthase in wounded mesocarp tissue of winter squash (*Cucurbita maxima*) fruit and the effects of ethylene. *Plant Cell Physiology*, **26**, 161–168

Idstein, H. and Schrier, P. (1985) Volatile constituents of "Alphonso" mango (*Mangifera indica*). *Phytochemistry*, **24**, 2313–2316

Joel, D.M., Marbach, I. and Mayer, A.M. (1978) Laccase in *Anacardiacea*. *Phytochemistry*, **17**, 796–797

John, J., Subbarayan, C. and Cama, H.R. (1970) Carotenoids in three stages of ripening in mango. *Journal of Food Science*, **35**, 262–265

Jungalwala, F.B. and Cama, H.R. (1963) Carotenoids in mango (*Mangifera indica*) fruit. *Indian Journal of Chemistry*, **1**, 36–40

Kapur, K.L., Verma, R.A. and Tripathi, M.P. (1985) The effect of maturity and processing on quality of pulp slices. *Indian Food Packer*, **39**, 60–67

Kasantikul, D. (1983) Studies on growth and development, biochemical changes and harvesting indices for mango (*Mangifera indica* L.) cultivar "Nam Dorkmai". MS Thesis, Kasetsast University, Bangkok, Thailand

Kosiyachinda, S., Lee, S.K. and Poernomo (1984) Maturity indices for harvesting of mango. In *Fruit development, postharvest, physiology and marketing in ASEAN.* (eds D.B. Mendoza, Jr. and R.H.B. Wills), ASEAN Food Handling Bureau, Kuala Lumpur, pp. 33–34

Krishnamurthy, S. and Patwardhan, M.V. (1971) Properties of malic enzyme (decarboxylating) from the pulp of mango fruit (*Mangifera indica*). *Phytochemistry*, **10**, 1811–1815

Krishnamurthy, S., Patwardhan, M.V. and Subramanyam, H. (1971) Biochemical changes during ripening of the mango fruit. *Phytochemistry*, **10**, 2577–2581

Kumar, R. and Selvaraj, Y. (1990) Fructose-1-6-bisphosphatase in ripening mango (*Mangifera indica* L.) fruit. *Indian Journal of Experimental Biology*, **28**, 284–286

Kusumo, S.L., Vangnai, V., Yong, S.K. and Namuco, L.O. (1984) *Commercial mango cultivars in ASEAN Mango.* ASEAN Food Handling Bureau, (eds, J.B. Mendoza, Jr. and R.B.H. Wills), pp. 12–20

Lakshminarayana, S. (1973) Respiration and ripening patterns in the life cycle of the mango fruit. *Journal of Horticultural Science*, **48**, 227–233

Lam, P.F., Ng, K.H., Omar, D. and Talib, Y. (1982) Physical, physiological and chemical changes of Golek after harvest. Proceedings of Workshop on Mango and Rambutan. ASEAN Postharvest Training College, Laguna, Phillipines. pp.96–112

Lazan, H., Ali, Z.M.A., Lee, K.W., Voon, J. and Chaplin, G.R. (1986) The potential role of polygalacturonase in pectin degradation and softening in mango fruit. *ASEAN Food Journal*, **2**, 93–95

Lertpruk, S. (1983) Postharvest behaviour of "Carabao" mango (*Mangifera indica* L.) fruit sprayed with ethephon as a preharvest inducer of maturation. M.S. Thesis, University of the Philippines at Los Banos, College, Laguna, Phillipines

Lizada, M.C.C., Agravante, J.U. and Brown, E.O. (1986) Factors affecting postharvest disease control in "Carabao" mango subjected to hot water treatment. *Philippine Journal of Crop Science*, **11**, 153–161

Lizada, M.C.C. and Cua, A.U. (1990) The postharvest behaviour and quality of "Carabao" mangoes subjected to a vapour heat treatment. *ASEAN Food Journal*, **5**, 6–11

MacLeod, A.J. and Pieris, N.M. (1984) Comparison of the volatile components of some mango cultivars. *Phytochemistry*, **23**, 361–366

Mann, S.S. and Singh, R.N. (1976) The cold storage life of Dashehari mangoes. *Scientia Horticulturae*, **5**, 249–254

Mattoo, A.K. and Modi, V.V. (1969) Ethylene and ripening of mangoes. *Plant Physiology*, **44**, 308–31

Mattoo, A.K. and Modi, V.V. (1970) Partial purification and properties of enzymic inhibitors from unripe mangoes. *Enzymologia*, **39**, 237–247

Mattoo, A.K., Modi, V.V. and Reddy, V.V.R. (1968) Oxidation and carotenogenesis regulating factors in mangoes. *Indian Journal of Biochemistry*, **5**, 111–114

Medlicott, A.P. (1985) Mango fruit ripening and the effects of maturity, temperature and gases. Ph.D. Thesis. The Polytechnic, Wolverhampton (CNNA)

Medlicott, A.P. and Thompson, A.K. (1985) Analysis of sugars and organic acids in ripening mango fruits (*Mangifera indica* L. var. Keitt) by high performance liquid chromatography. *Journal of the Science of Food and Agriculture*, **36**, 561–566

Medlicott, A.P., Bhogol, M. and Reynolds, S.B. (1986) Changes in peel pigmentation during ripening of mango fruit (*Mangifera indica* var. Tommy Atkin). *Annals of Applied Biology*, **109**, 651–656

Medlicott, A.P., Reynolds, S.B., New, S.W. and Thompson, A.K. (1988) Harvest maturity effects on mango fruit ripening. *Tropical Agriculture*, **65**, 153–157

Mendoza, D.B. (1981) Developmental physiology of "Carabao" mango (*Mangifera indica* L.) fruits. Ph.D. Thesis, University of the Philippines at Los Banos

Mendoza, D.B., Javier, F.B. and Pantastico, E.B. (1972) Physico-chemical studies during growth and maturation of "Carabao" mango. *Animal Husbandry and Agricultural Journal*, **7**, 33–36

Mitchell, G.E., McLauchlan, R.L.,Beattie, T.R., Banos, C. and Gillen, A.A. (1990) Effect of gamma irradiation on the carotene content of mangoes and red capsicums. *Journal of Food Science*, **55**, 1185–1186

Modi, V.V., Reddy, V.V.R. and Shah, D.V. (1965) Carotene precursors in mangoes. *Indian Journal of Experimental Biology*, **3**, 145–146

Morga, N.S., Lustre, A.O., Tunac, M.M., Balogot, A.H. and Soriano, M.R. (1979) Physicochemical changes in Philippine Caraboa mangoes during ripening. *Food Chemistry*, **4**, 225–234

Parikh, H.R., Nair, G.M. and Modi, V.V. (1990) Some structural changes during ripening of mangoes (*Mangifera indica* var. Alphonso) by abscisic acid treatment. *Annals of Botany*, **65**, 121–127

Park, Y.K., Sato, H.H., Almeida, T.D. and Moretti, R.H. (1981) Polyphenoloxidase of mango (*Mangifera indica* var. Haden). *Journal of Food Science*, **45**, 1619–1621

Peacock, B.C. and Brown, B.I. (1984) Quality comparison of several mango varieties. *Proceedings First Australian Mango Research Workshop*, pp. 334–339

Peacock, B.C., Murray, C., Kosiyachinda, S., Kasittrakakul, M. and Tansiriyakul, S. (1986) Influence of harvest maturity of mangoes on storage potential and ripe fruit quality. *ASEAN Food Journal*, **2**, 99–103

Peiser, G.D. and Yang, S.F. (1977) Chlorophyll destruction by bisulphite oxygen system. *Plant Physiology (Bethesda)*, **60**, 277–281

Proctor, J.T.A. and Creasy, L.L. (1969) The anthocyanin of mango fruit. *Phytochemistry*, **8**, 210

Prusky, D. (1990) Mango diseases: an overview. *Acta Horticulturae*, **291**, 279–287.

Ramana, N.P., Prasad, B.A., Malikarjuaradhya, S., Patwardhan, M., Ananthakrishma, S.M., Rajpoot, N.C. and Subramanyan, L. (1984) Effect of calcium carbide on ripening and quality of Alphonso mangoes. *Journal of Food Science and Technology*, **21**, 278–282

Rao, N.N. and Modi, V.V. (1976) Fructose 1-6, diphosphatase (E.C.3.1.3.11) from *Mangifera indica*. *Phytochemistry*, **15**, 1437–1440

Roe, B. and Bruemmer, J.M. (1981) Changes in pectic substances and enzymes during ripening and storage of "Keitt" mangoes. *Journal of Food Science*, **46**, 186–189

Satyan, S.H., Chaplin, G.R., Willcox, M.E. (1986) An assessment of fruit quality of various mango cultivars. *Proceeding of the First Australian Mango Research Workshop*, Cairn, Queensland, pp. 324–333

Selvaraj, Y. (1989) Studies on enzymes involved in the biogenesis of lipid derived volatiles in ripening mango (*Mangifera indica* L.) fruit. *Journal of Food Biochemistry*, **12**, 289–300

Selvaraj, Y. and Kumar, R. (1989) Studies on fruit softening enzymes and polyphenol oxidase activity in ripening mango (*Mangifera indica* L.) fruit. *Journal of Food Science and Technology*, **26**, 218–222

Selvaraj, Y., Kumar, R. and Pal, D.K. (1989) Changes in sugars, organic acids, amino acids, lipid constituents and aroma characteristics of ripening mango (*Mangifera indica* L.) fruit. *Journal of Food Science and Technology*, **26**, 308–313

Seymour, G.B., N'Diaye, M., Wainwright, H. and Tucker, G.A. (1990) Effect of cultivar and harvest maturity on ripening of mangoes during storage. *Journal of Horticultural Science*, **65**, 479–483

Shashirekha, M.S. and Patwardhan, M.V. (1976) Changes in amino acids, sugars and non-volatile organic acids in a ripening mango fruit (*Mangifera indica*, Badami variety). *Lebensmittel Wissenschaft Technologie*, **9**, 369–370

Simpson, K.L., Lee, T., Rodrigues, D.B. and Chichester, C.O. (1976) In *Chemistry and Biochemistry of Plant Pigments*. (ed T.W. Goodwin), Vol. 1, Academic Press, pp. 779–843

Subramanyam, H. and Sebastian, K. (1970) Effect of succinic acid and 2, 2-dimethyl hydroxide on carotene development in "Alphonso" mango. *Horticultural Science*, **5**, 160–161

Subramanyam, H., Krishnamurthy, S. and Parpia, H.A.B. (1975) Physiology and biochemistry of mango fruit. *Advances in Food Research*, **21**, 223–305

Subramanyan, H., Gowri, S. and Krishnamurthy, S. (1976) Ripening behaviour of mango fruits graded on specific gravity basis. *Journal of Food Science and Technology*, **13**, 84–86

Tandon, D.K. and Kalra, S.K. (1983) Changes in sugars, starch and amylase activity during development of mango fruit cv. Dashehari. *Journal of Horticultural Science*, **58**, 449–453

Tandon, D.K. and Kalra, S.K. (1984) Pectin changes during the development of mango fruit cv. Dashehari. *Journal of Horticultural Science*, **59**, 283–286

Tirtosoekotjo.R.A. (1984) Ripening behaviour and physiochemical characteristics of "Carabao" mango (*Mangifera indica* L.) treated with acetylene from calcium carbide. Ph.D. Thesis. University of the Philippines, Los Banos

Tjiptono, P., Lam, P.E. and Mendoza, D.B.Jr. (1984) Status of the mango industry in ASEAN in mango. ASEAN Food Handling Bureau, pp. 1–11

Tucker, G.A. and Seymour, G.B. (1991) Cell wall degradation during the ripening of mango fruit. *Acta Horticulturae*, **291**, 454–460

Van Lelyveld, L.J. and Smith, J.H.E. (1979) Physiological factors in the maturation and ripening of mango (*Mangifera indica* L.) fruit in relation to the jelly seed physiological disorder. *Journal of Horticultural Science*, **54**, 283–287

Vazquez-Salinas, C. and Lakshminarayana, S. (1985) Compositional changes in mango fruit during ripening at different storage temperatures. *Journal of Food Science*, **50**, 1646–1648

Wang, T.T. and Shiesh, C.C. (1990) Fruit growth, development and maturity indices of "Irwin" mango in Taiwan. *Acta Horticulturae*, **269**, 189–196

Winston, E.C. (1984) Mango varietal selection trials in wet tropics. *Proceedings of the First Australian Mango Research Workshop*. Cairns, Queensland

Yu, Y-B., Adams, D.O. and Yang, S.F. (1979) ACC synthase a key enzyme in ethylene biosynthesis. *Archives of Biochemistry and Biophysics*, **198**, 280–286

Chapter 9

Melons

G.B. Seymour and W.B. McGlasson

9.1 INTRODUCTION

This chapter has been confined to the dessert melons of the genera *Cucumis* (muskmelon) and *Citrullus* (watermelon). These fruits are members of the Cucurbitaceae which is a large family that includes cucumbers, squashes and pumpkins (Whitaker and Davis, 1962). The physiology and biochemistry of dessert melons was last reviewed by Pratt (1971).

The melons are an old crop thought to have originated in the topical Africa/Middle-East region. Melons spread from this region to the Mediterranean, the Indian sub-continent and throughout Asia. The fruit are characterized by their diversity of size, shape, skin appearance and flesh colour. They are grown mainly outdoors during the warmer months of the year, but in some countries they are grown in glass or plastic houses. World production of dessert melons in 1990 was estimated to be 38 million tonnes (FAO, 1990).

9.2 BOTANY

Evidence relating to the evolution and taxonomy of the genera *Cucumis* and *Citrullus* has been surveyed by Mallick and Masui (1986). It is considered that the African species *Cucumis metuliferus* is ancestral to the rest of the genus. *Cucumis sativus* (cucumbers and gherkins) evolved from *Cucumis hardwickii* which diverged from the main line of *Cucumis* evolution at an early stage. The fruit of *Cucumis sativus* are non-climacteric and are typically used as vegetables; they will not be considered further in this chapter.

Biochemistry of Fruit Ripening. Edited by G. Seymour, J. Taylor and G. Tucker. Published in 1993 by Chapman & Hall, London. ISBN 0 412 40830 9

Cucumis melo L. includes all of the dessert melons referred to as musk-melons. *Cucumis melo* is a polymorphic species and has been further classified into groups. The *reticulatus* group includes the netted or rough-skinned melons and the *inodorus* group includes the smooth-skinned types such as Honeydew. Smith and Welch (1964) considered that these groups do not exist as botanical varieties in nature and are only for horticultural convenience, since cultivars of the two hybridize readily. The muskmelons show an enormous diversity of fruit types varying in flesh colour from green, pink to orange; skin colour from green, white, yellow, orange to grey; in skin texture from smooth to netted; and in size from about 500 g to several kg. In the USA the common name for netted melons with orange flesh is Cantaloupe. Some common names for muskmelons include Honeydew, Persian, Crenshaw, Casaba, Spanish, Santa Claus, Hami and Charentais.

Watermelons, *Citrullus lanatus* (Thunb.) Mansf., show less diversity than the muskmelons. Skin colour is generally a shade of mottled green and flesh colour is typically red. Fruit size varies widely as in musk-melons. Seedless triploid strains have been bred.

Cucumis and *Citrullus* are generally monoecious, but may be andro-monoecious. The flowering and fruiting habits have been described by Jones and Rosa (1928) and McGlasson and Pratt (1963). The melon is classed as an inferior berry or 'pepo'. It is indehiscent with the fleshy floral tube adnate to the pericarp. Melon fruits may have solid flesh derived from the placentae, as in watermelon, or have a central cavity as in the muskmelons with the flesh derived from the pericarp. The carpellary structure of the melon is especially apparent in cross-sections of young fruit.

9.3 PHYSIOLOGICAL CHANGES DURING DEVELOPMENT AND RIPENING

The general pattern of fruit growth and differences between cultivars in muskmelons is illustrated in Fig. 9.1. The pattern of growth in water-melons is similar to that in muskmelons (Pratt, 1971).

9.3.1 Maturation

Netted muskmelons (*reticulatus* group), e.g. PMR 45, abscise from the vine upon reaching maturity. This type of melon is still growing rapidly up to the time of abscission. In contrast the Honeydew melon (*inodorus* group) develops more rapidly than the netted 'Cantaloupe' type at first and then almost ceases to grow until it ripens and senesces. For the normal growth pattern to be achieved in melons adequate pollination

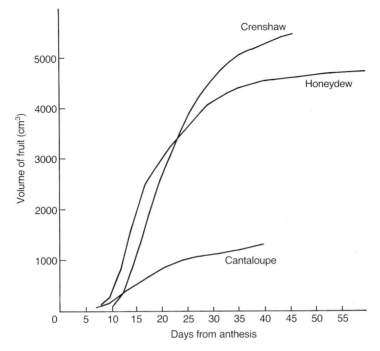

Fig. 9.1 Growth curves for three cultivars of muskmelon. Each point is the average of five fruit. (Redrawn from Pratt, 1971.)

must be followed by fertilization and normal development of the ovules (Pratt, 1971).

The timing of maturation and the onset of ripening in melons depends upon the fruit type. In netted melons the onset of ripening is accompanied by the development of an abscission layer (Pratt, 1971; Kendall and Ng, 1988). At this stage, referred to as 'full-slip', the melon can easily be pulled or slipped from the vine (Sykes, 1990). It is important not to harvest these Cantaloupe melons prior to development of the abscission layer because they will not develop the full flavour. However, quality may be reduced if harvest is greatly delayed. Honeydew fruits are harvested by cutting from the vine and are difficult to choose by appearance alone, for they show few changes as they mature in the field. Horticultural maturity in Honeydew melons is attained 35–37 days after anthesis, but self-ripening requires about 47 days and commercial harvests may need ethylene treatment (Pratt et al., 1977). There are many melon cultivars which have netted or smooth skins and patterns of growth and maturation intermediate between Cantaloupe and Honeydew types. Many of these melons do not 'slip' and other characteristics are used to determine harvest maturity.

Maturity in watermelons is difficult to assess, colour of the ground spot being probably one of the best objective measures of maturity (Pratt, 1971).

Fruit quality in muskmelons is generally related to high internal sugar levels and good flavour. Soluble solids content of netted and Honeydew melons should be at least 9% or 10% for good dessert quality (Pratt, 1971; Hardenburg *et al.*, 1986; Sykes, 1990). In watermelon, principal requirements for the market are flesh crispness, good flesh colour and sweetness (Pratt, 1971).

Netted muskmelons have a shorter storage life than Honeydew types. The netted melons have a storage life of about 14 days even under cool, high humidity conditions. Storage conditions of 2–5°C and 95% relative humidity are recommended for netted melons, with lower temperatures causing chilling injury (Hardenburg *et al.*, 1986). Honeydew melons can be stored at 7° to 10°C and 90% relative humidity for three weeks, but are chilled at temperatures below 5°C (Hardenburg *et al.*, 1986). The intact (closed) epidermis of Honeydew as compared with the fissured epidermal tissue of netted melons may be associated with the differences in storage life, particularly with respect to significant losses of fruit moisture (Lester, 1988).

Watermelons are not adapted to long storage. Recommended storage conditions are between 10° and 15°C and 90% relative humidity. The fruit should keep under these conditions for 2–3 weeks. Chilling injury may occur at temperatures below 10°C (Hardenburg *et al.*, 1986), although small watermelon cultivars, e.g. Minilee, may be more resistant to chilling (Risse *et al.*, 1990).

9.4   BIOCHEMICAL CHANGES DURING DEVELOPMENT AND RIPENING

9.4.1   Respiration and ethylene production

There is wide variation in ripening behaviour among melons. Netted melons tend to have a rapid climacteric at, or near, the time of fruit maturity and abscission, with the interval between the preclimacteric minimum and the climacteric peak being 24 to 48 hours. In Honeydew and Casaba melons the climacteric may extend over several days or may be absent (Lyons *et al.*, 1962; Kitamura *et al.*, 1975; Pratt *et al.*, 1977; Nukaya *et al.*, 1986; Kendall and Ng, 1988).

Watermelons have been reported to show both climacteric (Mizuno and Pratt 1973) and non-climacteric behaviour (Elkashif *et al.*, 1989), though the biochemical changes relating directly to the respiratory climacteric in melons have not been studied in detail.

Ethylene production

A large rise in ethylene production accompanies ripening in netted Cantaloupe type and Honeydew muskmelons (Pratt, 1971; Pratt *et al.*, 1977; Liu *et al.*, 1985). Netted muskmelon fruit produce appreciable amounts of ethylene at or near harvest, while non-netted types may not produce ethylene until as late as 20 days postharvest. Hybrids between these types of melon were intermediate in their rate and time of ethylene production, demonstrating that ethylene production is under genetic control (Kendall and Ng, 1988). Ethylene biosynthesis in muskmelons follows the pathway from methionine to ethylene vis *S*-adenosylmethionine and 1-aminocyclopropane-1-carboxylic acid (ACC) (Liu *et al.*, 1985). Exogenous ethylene will induce ripening in musk-melons, but the response is a function of physiological age of the fruit, tissue temperature, duration of treatment and concentration applied (Pratt, 1971). Liu *et al.* (1985) demonstrated that ethylene stimulated the development of ethylene-forming enzyme (EFE) activity in precli-macteric 'Cantaloupe' muskmelon, hence increasing the capacity of the tissue to convert ACC to ethylene. The conversion of ACC to ethylene in netted muskmelon can be restricted by holding the fruit for 3 hours at 45°C or by storage at 4°C, but these temperature restrictions are reversible (Dunlap *et al.*, 1990). Recently (Ververidis and John, 1991) EFE activity has been recovered for the first time *in vitro* using muskmelon fruit. It is a soluble enzyme whose activity *in vitro* depends on exclusion of O_2 from the preparation medium, and on the addition of Fe^{2+} and ascorbate to the aerobic reaction medium.

Pratt (1971) notes that stimulation of ripening in most muskmelon cultivars is not required since the fruits are adequately self ripening. However, in Honeydew ethylene may be applied to induce uniform ripening after harvest, but fruits must have an acceptable soluble solids content at this time since they have no starch reserves (Pratt, 1971; Bianco and Pratt, 1977; Hardenburg *et al.*, 1986).

Watermelon fruits are very sensitive to ethylene and it has been reported that exposure to concentrations as low as 1 ml.l^{-1} caused placen-tal tissue deterioration and rendered the fruit unfit for consumption (Risse and Hatton, 1982). The effects of exogenous ethylene on water-melon have been used to indicate the possible non-climacteric behaviour of this fruit. Elkashif *et al.* (1989) treated immature, preripe (pink) and ripe (red) fruit of three cultivars of watermelon with ethylene or pro-pylene, or stored the fruit in air alone. Melons of all maturation stages held in air showed little textural change throughout storage and produced little ethylene. Exposure to ethylene resulted in extreme placental tissue and rind softening. Respiration rate was enhanced by ethylene treat-ment, although the respiratory activity returned to normal on removal of

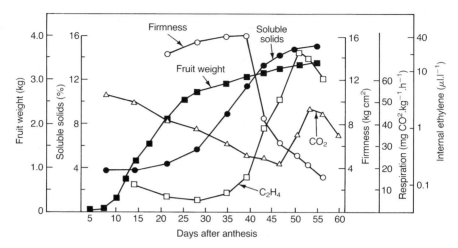

Fig. 9.2 Changes in growth, soluble solids, ethylene production, respiration and flesh firmness of Honeydew melons harvested at different ages. Those values believed to be affected by endogenous ethylene are plotted with open symbols; those not affected are plotted with solid symbols. (Redrawn from Pratt *et al.*, 1977.)

ethylene and a propylene treatment did not result in the stimulation of autocatalytic ethylene production. The increase in respiration only in the presence of ethylene and the lack of autocatalytic ethylene production are typical of the behaviour of non-climacteric fruit. The authors (Elkashif *et al.*, 1989) proposed that increased respiration and ethylene observed in ripening watermelons by earlier workers were the result of senescence-related pathogen proliferation, rather than normal ripening metabolism.

The interrelationship between ethylene production, the climacteric rise in respiration and various other aspects of ripening in Honeydew muskmelons is shown in Fig. 9.2.

9.4.2 Carbohydrate metabolism

High sugar content is an important quality attribute of ripe melons. Sugar accumulation begins during fruit development. Melons of both the netted Cantaloupe type and Honeydew fruit showed a similar pattern of sugar accumulation, with a rapid rise in the accumulation of sugars as the fruit reached full size. In fully ripened Honeydew melons sugars may comprise as much as 16% of the juice (Pratt, 1971; Bianco and Pratt, 1977). The principal sugar accumulated within these muskmelon fruits and in most watermelons is sucrose (Pratt, 1971; Bianco and Pratt, 1977; Mutton *et al.*, 1981; Hubbard *et al.*, 1989), although high levels of fructose may be present in some watermelon cultivars (Elmstrom and Davis, 1981). Muskmelon fruits have little or no starch so they require a con-current supply of photoassimilates from leaves for sugar accumulation during development and ripening (Pratt, 1971; Hubbard *et al.*, 1989, 1990). A drastic reduction of translocate from the source to the melon fruit, by detaching the fruit or by complete defoliation, severely reduces sucrose accumulation (Hubbard *et al.*, 1990). Also high melon soluble solids are associated with slow ripening and high leaf area (Welles and Buitelaar, 1988). Therefore sugar content of ripe muskmelons is very much dependent on allowing sufficient time for sugar accumulation to occur prior to harvest (Bianco and Pratt, 1977).

Sugar concentrations can vary in different parts of muskmelon and watermelon fruits (Pratt, 1971). In a survey of the distribution of sugars in ripe watermelon fruit two cultivars were examined, Charleston Gray and Jubilee, and the fruit were sampled for sugars in five regions. Fructose was found to be the primary sugar in all regions in Jubilee, while sucrose was the primary sugar in Charleston Gray (Chisholm and Picha, 1986).

Studies on the biochemical basis of sucrose accumulation have been undertaken mainly on muskmelons. Sucrose accumulation in musk-melon fruits almost certainly arises mainly from carbohydrates trans-located into the fruit during ripening. Muskmelons synthesize stachyose and raffinose in their leaves and these sugars are translocated to sink tissues such as the fruits. These carbohydrates may be the source utilized for sucrose synthesis (Hubbard *et al.*, 1989, 1990). There is now good evidence that the accumulation of sucrose is regulated by the activities of both acid invertase (EC 3.2.1.26) and sucrose phosphate synthase (SPS) (EC 2.4.1.14). Acid invertase activity and sucrose concentration are normally inversely related in sink tissues. In the phase of rapid sucrose accumulation in muskmelon acid invertase activity declined with a concomitant increase in SPS activity (Schaffer *et al.*, 1987; Hubbard *et al.*, 1989, 1990; Ranwala *et al.*, 1991). The latter enzyme had higher activity in sweet than in non-sweet genotypes, providing further evidence for its role in sucrose accumulation. A role for the enzyme sucrose synthase (EC 2.4.1.13) in sucrose synthesis in melon has not been established (McCollum *et al.*, 1988; Hubbard *et al.*, 1989).

9.4.3   Organic acids

Early work reviewed by Pratt (1971) reported the presence of both citric and malic acid in some muskmelon cultivars, but indicated that citric acid was not detected in watermelons. Leach *et al.* (1989) found that in all of the cultivars of *Cucumis melo* they analysed, citric acid was the major component of the organic acid fraction. In work on watermelons

(Chisholm and Picha, 1986), cvs Charleston Gray and Jubilee, malic and citric were the major organic acids found in both cultivars. Malic acid was found at higher concentrations throughout the fruits than citric acid and both acids were more highly concentrated in the heart or blossom end of the fruit.

9.4.4 Volatiles

Good aroma is one of the important quality attributes in melons (Yabumoto and Jennings, 1977). The volatile compounds likely to be responsible for aroma development, particularly in muskmelons, have

Table 9.1 Compounds identified in volatiles of muskmelons. (After Horvat and Senter, 1987; Wyllie and Leach, 1990.)

acetaldehyde	propane-1, 2-diol diacetate
ethanol	butane-2, 3-diol diacetate
methyl acetate	butane-2, 3-diol acetate
ethyl acetate	propanoate
isopropyl acetate	ethyl (methylthio) propanoate
ethyl propanoate	3-(methylthio) propyl acetate
ethyl isobutanoate	ethyl octanoate
propyl acetate	benzaldehyde
temethyl butanoate	nonyl acetate
butyl acetate	2-nonenal
methyl 2-methylbutanoate	nonanol
ethyl butanoate	(Z)-6-nonenyl acetate
ethyl 2-methylbutanoate	ethyl decanoate
2-methyl-1-propanol	(Z,Z)-3,6-nonadienyl acetate
2-methylbutyl acetate	benzyl acetate
n-butanol	phenylethyl acetate
pentyl acetate	methyl dodecanoate
2-methylbutanol	ethyl dodecanoate
butyl butanoate	megastigma-4,6,8-triene
ethyl hexanoate	phenylpropyl acetate
2-methylbutenyl acetate	sesquiterpene hydrocarbon
hexyl acetate	benzyl propionate
cis-3-hexanyl acetate	3-phenylpropyl acetate
n-hexanol	β-ionone
heptyl acetate	cinnamyl acetate
methyl (methylthio) acetate	isoeugenol
ethyl (methylthio) acetate	butane-2,3-diol acetate butanoate
octyl acetate	isomer of 3,4-dimethoxyacetophenone
butane-2,3-diol diacetate	
2-(methylthio) ethyl acetate	

been the subject of a number of recent investigations. Volatiles identified from ripening Cantaloupe muskmelons are shown in Table 9.1. The volatile ester composition of ripe Cantaloupe and Honeydew melons are found to be very similar except for ethyl butyrate, which was more abundant in Cantaloupe samples (Yabumoto *et al.*, 1978).

Compounds which are thought to be particularly representative of muskmelon aroma include ethyl-2-methyl butyrate, ethyl butyrate, ethyl hexanoate, hexyl acetate, 3-methyl butyl acetate, benzyl acetate, (Z)-6-nonenyl acetate, (E)-6-nonenol, (Z,Z)-3, 6-nonadienol and (Z)-6-nonenal (Kemp *et al.*, 1972a, b; Buttery *et al.*, 1982; Horvat and Senter, 1987). (2,2)-3,6-nonadienol has also been identified from watermelon (Kemp *et al.*, 1974). Wyllie and Leach (1990) suggested that the thioesters found in the Golden Crispy cultivar of muskmelon contribute to the unique aroma of this melon (also refer to Homatidu *et al.*, 1989).

The exact pathway of biogenesis for flavour volatiles in melon is still not known. However, in studies on Cantaloupe melon these volatiles appear to fall into groups according to their patterns of emission during ripening, and there is good agreement within groups as regards structure (Fig. 9.3). These data may indicate a common pathway of bio-

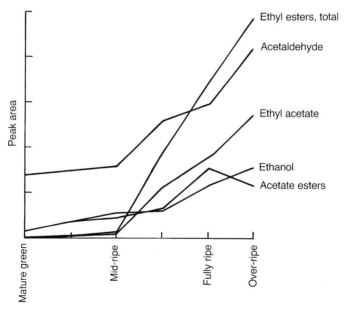

Fig. 9.3 Volatile production as related to fruit maturation in cantaloupe melons. (After Horvat and Senter, 1987.)

synthesis for the various compounds in each group. Yabumoto *et al* (1978) suggest the following scheme for the biogenesis of ethyl and acetate esters in melon:

1.   Pyruvate → acetaldehyde → ethanol → ethyl esters
2.   Pyruvate → acetyl CoA → acetate esters

The thioesters are probably derived from methionine (Wyllie and Leach, 1990).

9.4.5   Cell wall degradation

Softening in muskmelon has been reported to be accompanied by degradation of the fruit's cell walls, although Lester (1988) indicated that loss of mesocarp membrane integrity may also be important. Cell wall changes during ripening in muskmelons include an increase in soluble pectin, a decrease in pectin molecular size, loss of wall galactosyl residues and changes in the molecular size of hemicellulose polymers (Gross and Sams, 1984; McCollum *et al.*, 1989). The mechanisms by which these events are brought about and their relationship to texture changes in ripening muskmelons is not well understood. However, the decrease in polyuronide molecular size in muskmelon was apparently not the result of polygalacturonase (EC 3.2.1.15) activity, which has not been detected in muskmelons (Hobson, 1962; Lester and Dunlap, 1985; McCollum *et al.*, 1989). The loss of galactosyl residues from the walls of ripening muskmelon (Gross and Sams, 1984) may result from the action of a galactosidase/galactanase. Fils-Lycaon and Buret (1992) have recently examined changes in the activity of several glycosidases from the mesocarp tissue of muskmelon. They reported a significant increase in the activity of glycosidases during ripening, including the activity of β-D-galactosidase. However, some caution is required in assigning all β-D-galactosidase activity as a cell wall galactan degrading enzyme if artificial substrates are used to assess this enzyme activity. Other cell wall-related enzyme activities detected in muskmelon include pectinesterase (EC 3.1.1.11) and cellulase (EC 3.2.1.4). The activity of both these enzymes remained constant or decreased during the developmental period when significant reductions in texture were taking place (Lester and Dunlap, 1985).

Cell wall changes in watermelon are apparently less pronounced than in muskmelon unless the fruit are treated with ethylene. Elkashif and Huber (1988a,b) reported that watermelons harvested over-ripe or harvested ripe and stored for as long as 180 days showed only limited pectin degradation. On exposure to ethylene, however, harvested watermelon fruits exhibited marked softening accompanied by significant pectin breakdown and a rise in polygalacturonase activity. Watermelon fruit are reported to be rich in α-mannosidase (EC 3.2.1.24) and β-N-acetyl-hexosaminidase (Nakagawa *et al.*, 1988).

9.4.6   Pigments

The flesh colour of melons is important to the consumer. Early work on pigments of muskmelon (reviewed by Pratt, 1971) demonstrated that in orange-fleshed muskmelons the predominant pigment was β-carotene (84.7% of total carotenoids present). Also present were δ-carotene (6.8%), α-carotene (1.2%), phytofluene (2.4%), phytoene (1.5%), lutein (1.0%), violaxanthin (0.9%) and traces of other carotenoids. Pigment changes during growth and ripening in orange-fleshed muskmelon were followed by Reid *et al.* (1970). Chlorophyll content fell during fruit development (probably due to dilution throughout growth), with a marked decline during the respiratory climacteric. In contrast to the pattern of chlorophyll change, the carotenoid content of the flesh increased steadily from low levels in immature fruits. The carotenoids begin to increase at least ten days prior to the climacteric rise in respiration, with flesh colour development beginning at the centre of the fruits and progressing outward through the pericarp until the flesh is uniformly orange at full maturity. Therefore, the gradual rise in caroten-oid content observed in developing orange-fleshed muskmelon fruits does not reflect a uniform increase in carotenoids throughout all cells in the fruit, but primarily an increase in the amount of tissue containing high levels of carotenoids (Reid *et al.*, 1970).

   In a more recent study Flügel and Gross (1982) reported on pigment and plastid changes during ripening of the green-fleshed muskmelon 'Galia'. They observed relatively low levels of chorophyll and caroten-oids in the flesh with a gradual decrease in both types of pigment during ripening. Yellowing in the peel was not due to an increasing carotenoid content and peel chlorophylls were completely degraded during the ripening process. A partial decrease of total carotenoids in the peel also took place. The carotenoid pattern of the peel and flesh of 'Galia' corresponded to that of chloroplast carotenoids and did not change during development (Table 9.2). Changes in the plastids were confined to plastids in the outer mesocarp and exocarp layers.

   Pigment changes in watermelon fruits have received less attention than those of muskmelon. Pratt (1971) in reviewing earlier work notes that the pigment responsible for the red colour is lycopene. An orange-fleshed type contained prolycopene as the major carotenoid pigment with smaller amounts of phytoene, phytofluene, β-carotene, δ-carotene, proneurosporene, lycopene and traces of other poly-*cis*-lycopenes. In a recent study Watanabe *et al.* (1987) reported the main pigment com-ponents in different flesh coloured watermelon cultivars to be (a) red-

Table 9.2 Quantitative changes in the carotenoid pattern in the exocarp and mesocarp of C. melo. cv Galia during ripening. (From Flügel and Gross, 1982.)

Carotenoids	Exocarp		Mesocarp	
μg.g⁻¹ fresh weight	unripe	ripe	unripe	ripe
β-carotene	136.0	24.0	71.0	37.8
Cryptoxanthin	–	7.0	–	4.0
Lutein	399.0	140.0	142.6	54.7
Chrysanthemaxanthin (trans)	–	13.0	–	3.4
Chrysanthemaxanthin (cis)	18.0	11.0	–	4.0
Zeaxanthin	17.0	15.0	4.2	5.3
Isolutein (trans)	5.0	18.0	10.2	10.1
Isolutein (cis)	–	12.0	–	4.2
Antheraxanthin	15.0	11.0	9.8	4.0
Luteoxanthin (trans)	24.0	22.0	17.9	14.8
Luteoxanthin (cis)	5.0	7.0	–	–
Violaxanthin (trans)	12.0	6.0	36.1	26.8
Violaxanthin (cis)	14.0	17.0	19.3	15.0
Neoxanthin (trans)	105.0	11.0	16.1	1.9
Neoxanthin (cis)	–	–	17.2	2.9
Neochrom	15.0	4.0	3.5	1.1
Unknown mixture	–	2.0	2.1	–
Total carotenoids	770.0	320.0	350.0	190.0

fleshed – lycopene, (b) orange-fleshed – β-carotene, (c) yellow-fleshed – xanthophylls and β-carotene.

9.4.7 Lipids

Forney (1990) has examined changes in the polar lipid, fatty acid composition of ripening Honeydew muskmelons. In the peel he found slight changes during development, but when the fruit began to ripen the composition of fatty acids changed rapidly, with the ratio of unsaturated-to-saturated fatty acids doubling during ripening. The changes in the flesh polar lipid fatty acids were less pronounced than those in the peel. These changes in peel lipid composition may relate to increases in chilling tolerance reported to occur with ripening or solar exposure in melons (Forney, 1990).

Loss of tissue membrane integrity may be one of several biochemical events leading to texture changes in ripening muskmelons (Lester, 1988). The enzyme lipoxygenase (EC 1.13.11.12) may have a key role in this process by catalysing the degradation of unsaturated fatty acids, thereby releasing free radicals that can damage biological membranes (Eskin *et al.*, 1977). Recently Lester (1990) reported the presence of lipoxygenase activity in hypodermis tissue of 30-day postanthesis or older 'Perlita' muskmelon fruit. However, the lack of lipoxygenase activity in the mesocarp tissue indicates that this enzyme is not responsible for the breakdown of membrane integrity in this tissue during ripening.

9.5 POSTHARVEST DISEASES AND DISORDERS

Netted melons are especially susceptible to fungal diseases during market distribution. Since melons are commonly grown as a ground crop they are readily infected by a range of soil-borne fungi. These organisms may lodge in the skin net and enter the fruit through breaks in the skin, especially wounds incurred during harvesting. Melons that 'slip' also have a large exposed wound, the stem scar, which is an important point of entry for decay organisms.

Since muskmelons are normally harvested when the climacteric ripening pattern has begun, they have a limited capability for wound repair. To maintain high quality, growers must handle melons carefully, apply approved fungicides that are effective against the array of disease-causing fungi and cool the fruit quickly (Wade and Morris, 1982, 1983). Although Honeydew melons are less susceptible to pathogenic diseases than netted melons, similar postharvest handling procedures and treatments are recommended.

Lester and Bruton (1986) examined the influence of water loss on the perishability of netted melons. Shrink wrapping in plastic film greatly reduced the rate of water loss and prolonged the storage life at 4°C of melons which had been treated with fungicide. The reduced rates of water loss appeared to delay senescence of both pericarp and mesocarp tissue.

Browning of the vein tracts in melon has often been regarded as a symptom of chilling injury. Honeydew have been reported to suffer chilling injury when stored for two weeks or more below 4°C (Lipton, 1978). Symptoms appear as a reddish-tan surface discolouration. Treatment of Honeydew with 1 ml.l^{-1} ethylene at 20°C before storage at 2.5°C for 2.5 weeks reduced both the incidence and severity of chilling injury (Lipton and Aharoni, 1979). Solar yellowing of the skin of Honeydew melons reduces the susceptibility of the fruit to chilling injury at 2.5°C (Lipton *et al.*, 1987). ACC accumulates in both the skin and underlying tissue of melons stored for 2.5 weeks at 2.5°C. The presence of solar yellowing reduced the accumulation of ACC both in the exposed skin on the top of the melon and in the skin on the bottom of the fruit. Other disorders apart from chilling injury include blossom end rot which has been reported in watermelon. This physiological disorder is considered to be related to a localized deficiency of calcium (Shear, 1975).

9.6 GENETIC IMPROVEMENT

The melon family contains enormous genetic diversity for size, shape, skin and flesh colour, skin texture, low and high sugar levels and aroma, as well as climacteric and non-climacteric patterns of ethylene production. Kendall and Ng (1988) reported that fruit of hybrids bred from these climacteric and non-climacteric types experienced a delayed climacteric in ethylene production compared with the climacteric parent, and concluded that non-climacteric genotypes may be useful for genetically increasing storage life. Considerable potential exists for improving the postharvest attributes of melons by utilizing natural variation in traditional breeding programmes.

The use of molecular techniques is likely to lead to a greater understanding of the regulation of ripening in melons and other fruits. Work is in progress to identify genes encoding enzymes involved in ethylene biosynthesis (C. F. Watson, personal communication). A putative EFE cDNA has been isolated and work is underway to express this gene in yeast. The identity of the gene will be confirmed if the transformed yeast cells convert ACC to ethylene (Hamilton *et al.*, 1991). Down-regulation of the EFE gene in transgenic tomatoes, using antisense RNA, has led to these antisense fruit having greatly reduced ethylene production and delayed ripening (Hamilton *et al.*, 1990). The successful transformation of *Cucumis melo* and the regeneration of transformed plants using *Agrobacterium* gene transfer (Dong *et al.*, 1991) means that antisense technology could be used to reduce ethylene production in melons. Such transgenic fruit may have improved storage and handling characteristics.

REFERENCES

Bianco, V.V. and Pratt, H.K. (1977) Compositional changes in muskmelon during development and in response to ethylene treatment. *Journal of the American Society for Horticultural Science* **102**, 127–133

Buttery, R.G., Seifert, R.M., Ling, L.C., Soderstrom, E.L., Ogawa, J.M. and Turnbough, J.G. (1982) Additional aroma components of Honeydew melon. *Journal of Agricultural and Food, Chemistry* **30**, 1208–1211

Chisholm, D.N. and Picha, D.H. (1986) Distribution of sugars and organic acids within ripe watermelon fruit. *Hortscience*, **21**, 501–503

Dong, Z.Z., Yang, M.Z., Jia, S.R. and Chua, N.H. (1991) Transformation of melon (*Cucumis melo* L.) and expression from the cauliflower mosaic virus promoter in transgenic melon plants. *Bio/technology*, **9**, 858–863

Dunlap, J.R., Lingle, S.E. and Lester, G.E. (1990) Ethylene production in netted muskmelon subjected to postharvest heating and refrigerated storage. *Hortscience*, **25**, 207–209

Elkashif, M.E. and Huber, D.J. (1988a) Electrolyte leakage, firmness and scanning electron microscope studies of water melon fruit treated with ethylene. *Journal of the American Society for Horticultural Science*, **113**, 378–381

Elkashif, M.E. and Huber, D.J. (1988b) Enzymic hydrolysis of placental cell wall pectins and cell separation in watermelon (*Citrullus lanatus*) fruits exposed to ethylene. *Physiologia Plantarum*, **73**, 432–439

Elkashif, M.E., Huber, D.J. and Brecht, J.K. (1989) Respiration and ethylene production in harvested watermelon fruit: Evidence for nonclimacteric respiratory behaviour. *Journal of the American Society for Horticultural Science*, **144**, 81–85

Elmstrom, G.W. and Davis, P.L. (1981) Sugars in developing and mature fruits of several watermelon cultivars. *Journal of the American Society for Horticultural Science*, **106**, 330–333

Eskin, N.A.M., Grossman, S. and Pinksy, A. (1977) Biochemistry of lipoxygenase in relation to food quality. *CRC Critical Reviews in Food Science and Nutrition*, **9**, 1–40

FAO (1990) Food and Agriculture Organization of the United Nations, Production Year Book, Vol. 44, 1990

Fils-Lycaon, B. and Buret, M. (1992) Changes in glycosidase activities during development and ripening of melon. *Postharvest Biology and Technology*, **1**, 143–151

Flügel, M. and Gross, J. (1982) Pigment and plastid changes in mesocarp and exocarp of ripening muskmelon, *Cucumis melo* cv. Galia. *Angew. Botanik*, **56**, 393–406

Forney, C. (1990) Ripening and solar exposure alter polar lipid fatty acid composition of 'Honey Dew' muskmelons. *Hortscience*, **25**, 1262–1264

Gross, K.C. and Sams, C.E. (1984) Changes in cell wall neutral sugar composition during fruit ripening: A species survey. *Phytochemistry*, **23**, 2457–2461

Hamilton, A.J., Lycett, G.W. and Grierson, D. (1990) Antisense gene that inhibits synthesis of the hormone ethylene in transgenic plants. *Nature*, **346**, 284–287

Hamilton, A.J., Bonzayen, M. and Grierson, D. (1991) Identification of a tomato gene for ethylene-forming enzyme by expression in yeast. *Proceedings of the National Academy of Sciences*, **88**, 7434–7437

Hardenburg, R.E., Watada, A.E. and Wang, C.Y. (1986) Commercial storage of fruits, vegetables and florist and nursery stocks. USDA-ARS Handbook 66. 136 pp

Hobson, G.E. (1962) Determination of polygalacturonase in fruits. *Nature*, **195**, 804–805

Homatidu, V., Karvouni, S. and Dourtoglou, V. (1989) Determination of characteristic aroma components of Cantaloupe, *Cucumis melo*, using multidimensional gas chromatography (MGGC). In *Flavours and Off-Flavours*, (ed G. Charalambous), Proceedings of the 6th International Flavour Conference, Rethymnon, Crete, Greece, 5–7 July 1989. Elsevier Science Publishers, Amsterdam

Horvat, R.J. and Senter, S.D. (1987) Identification of additional volatile compounds from Cantaloupe. *Journal of Food Science*, **52**, 1097–1098

Hubbard, N.L., Huber, S.C. and Pharr, D.M. (1989) Sucrose phosphate synthase and acid invertase as determinants of sucrose concentration in developing muskmelon (*Cucumis melo* L.) fruits. *Plant Physiology*, **91**, 1527–1534

Hubbard, N.L., Pharr, D.M. and Huber, S.C. (1990) Sucrose metabolism in ripening muskmelon fruit as affected by leaf area. *Journal of the American Society for Horticultural Science*, **115**, 798–802

Jones, H.A. and Rosa, J.T. (1928) *Truck Crop Plants*. McGraw-Hill, New York

Kemp, T.R., Stolz, L.P. and Knavel, D.E. (1972a) Volatile components of muskmelon fruit. *Journal of Agricultural and Food Chemistry*, **20**, 196–198

Kemp, T.R., Knavel, D.E. and Stolz, L.P. (1972b) Cis-6-nonenal: A flavour component of muskmelon fruit. *Phytochemistry*, **11**, 3321–3322

Kemp, T.R., Knavel, D.E., Stolz, L.P. and Lundin, R.E. (1974) 3-6-nonadien-1-ol from *Citrullus vulgaris* and *Cucumis melo*. *Phytochemistry*, **13**, 1167–1170

Kendall, S.A. and Ng, T.J. (1988) Genetic variation of ethylene production in harvested muskmelon fruits. *Hortscience*, **23**, 759–761

Kitamura, T., Umemoto, T. and Iwata, T. (1975) Studies on the storage of melon fruits. II. Changes in respiration and ethylene production during ripening with reference to cultivars. *Journal of Japanese Society of Horticultural Science*, **44**, 197–203

Leach, D.N., Sarafis, V., Spooner-Hart, R. and Wyllie, S.G. (1989) Chemical and biological parameters of some cultivars of *Cucumis melo*. *Acta Horticulturae*, **247**, 353–357

Lester, G. (1988) Comparisons of 'Honey Dew' and netted muskmelon fruit tissues in relation to storage life. *Hortscience*, **23**, 180–182

Lester, G. (1990) Lipoxygenase activity of hypodermal- and middle-mesocarp tissues from netted muskmelon fruit during maturation and storage. *Journal of the American Society for Horticultural Science*, **115**, 612–615

Lester, G.E. and Bruton, B.D. (1986) Relationship of netted muskmelon fruit water loss to postharvest storage life. *Journal of the American Society for Horticultural Science*, **111**, 727–731

Lester, G.E. and Dunlap, J.R. (1985) Physiological changes during the development and ripening of Perlita muskmelon fruits. *Scientia Horticulturae*, **26**, 323–331

Lipton, W.J. (1978) Chilling injury of 'Honey Dew' muskmelons: Symptoms and relation to degree of ripeness at harvest. *Hortscience*, **13**, 45–46

Lipton, W.J. and Aharoni, Y. (1979) Chilling injury and ripening of 'Honey Dew' muskmelons stored at 2.5°C or 5°C after ethylene treatment at 20°C. *Journal of the American Society for Horticultural Science*, **104**, 327–330

Lipton, W.J., Peterson, S.J. and Wang, C.Y. (1987) Solar radiation influences solar yellowing, chilling injury and ACC accumulation in 'Honey Dew' melons. *Journal of the American Society for Horticultural Science*, **112**, 503–505

Liu, Y., Hoffman, N.E. and Yang, S.F. (1985) Promotion by ethylene of the capability to convert 1-aminocyclopropane-1-carboxylic acid to ethylene in preclimacteric tomato and Cantaloupe fruits. *Plant Physiology*, **77**, 407–411

Lyons, J.M., McGlasson, W.B. and Pratt, H.K. (1962) Ethylene production, respiration and internal gas concentration in Cantaloupe fruits at various stages of maturity. *Plant Physiology*, **37**, 31–36

Mallick, M.F.R. and Masui, M. (1986) Origin, distribution and taxonomy of melons. *Scientia Horticulturae*, **28**, 251–261

McCollum, T.G., Huber, D.J. and Cantliffe, D.J. (1988) Soluble sugar accumulation and activity of related enzymes during muskmelon fruit development. *Journal of the American Society for Horticultural Science*, **113**, 399–403

McCollum, T.G., Huber, D.J. and Cantliffe, D.J. (1989) Modification of polyuronides and hemicellulose during muskmelon fruit softening. *Physiologia Plantarum*, **76**, 303–308

McGlasson, W.B. and Pratt, H.K. (1963) Fruit set patterns and fruit growth in Cantaloupe (*Cucumis melo* L. var *reticulatus* Naud). *Proceedings of American Society for Horticultural Sciences*, **83**, 495–505

Mizuno, S. and Pratt, H.K. (1973) Relations of respiration and ethylene production to maturity in watermelon. *Journal of American Society of Horticultural Science*, **98**, 614–617

Mutton, L.L., Cullis, B.R. and Blakeney, A.B. (1981) The objective definition of eating quality in rock melons (*Cucumis melo*). *Journal of the Science of Food and Agriculture*, **32**, 385–391

Nakagawa, H., Enomoto, N., Asakawa, M. and Uda, Y. (1988) Purification and characterization of α-mannosidase and β-N-acetylhexosaminidase from watermelon fruit. *Agricultural and Biological Chemistry*, **52**, 2223–2230

Nukaya, A., Ishida, H., Shigeoka, H. and Ichikawa, K. (1986) Varietal difference in respiration and ethylene production in muskmelon fruits. *Hortscience*, **21**, 853

Pratt, H.K. (1971) Melons. In *Biochemistry of Fruits and their Products*, (ed A.C. Hulme), Vol 2, Academic Press, London.

Pratt, H.K., Goeschl, J.D. and Martin, F.W. (1977) Fruit growth and development, ripening and role of ethylene in the 'Honey Dew' muskmelon. *Journal of the American Society for Horticultural Science*, **102**, 203–210

Ranwala, A.P., Iwanami, S. and Masuda, H.(1991) Acid and neutral invertases in the mesocarp of developing muskmelon. *Plant Physiology*, **96**, 881–886

Reid, M.S., Lee, T.H., Pratt, H.K. and Chichester, C.O. (1970) Chlorophyll and carotenoid changes in developing muskmelons. *Journal of the American Society for Horticultural Science*, **95**, 814–815

Risse, L.A. and Hatton T.T. (1982) Sensitivity of watermelons to ethylene during storage. *Hortscience*, **17**, 946–948

Risse, L.A., Brecht, J.R., Sargent, S.A., Locascio, S.J., Crall, J.M. and Elmstrom, G.W. (1990) Storage characteristics of small watermelon cultivars. *Journal of the American Society for Horticultural Science*, **115**, 440–443

Schaffer, A.A., Aloni, B. and Fogelman, E. (1987) Sucrose metabolism and accumulation in developing fruit of *Cucumis*. *Phytochemistry*, **26**, 1883–1887

Shear, C.B. (1975) Calcium-related disorders of fruits and vegetables. *Hortscience*, **10**, 361–365

Smith, P.G. and Welch, J.E. (1964) Nomenclature of vegetables and condiment herbs grown in the United States. *Proceedings of the American Society for Horticultural Sciences*, **84**, 534–548

Sykes, S. (1990) Melons: New varieties for new and existing markets. *Agricultural Science*, **3**, 32–35

Ververidis, P. and John, P. (1991) Complete recovery *in vitro* of ethylene-forming enzyme activity. *Phytochemistry*, **30**, 725–727

Wade, N.L. and Morris, S. C. (1982) Causes and control of Cantaloupe postharvest wastage in Australia. *Plant Diseases*, **66**, 549–552

Wade, N.L. and Morris, S.C. (1983) Efficacy of fungicides for postharvest treatment of muskmelon fruits. *Hortscience*, **18**, 344–345

Watanabe, K., Saito, T., Hirota, S. and Takahashi, B. (1987) Carotenoid pigments in red, orange and yellow-fleshed fruits of watermelon (*Citrullus vulgaris*) *Journal of Japanese Society of Horticultural Sciences*, **56**, 45–50

Welles, G.W.H. and Buitelaar, K.(1988) Factors affecting soluble solids content of muskmelon, *Cucumis melo* L. *Netherlands Journal of Agricultural Science*, **3**, 239–246

Whitaker, T.W. and Davis, G.N. (1962) *Cucurbits: Botany, Cultivation and Utilisation*. Interscience Publishers Inc., New York

Wyllie, S. G. and Leach, D. N. (1990) Aroma volatiles of *Cucumis melo* cv Golden Crispy. *Journal of Agricultural and Food Chemistry*, **38**, 2042–2044

Yabumoto, K. and Jennings, W. G. (1977) Volatile constituents of Cantaloupe, *Cucumis melo*, and their biogenesis. *Journal of Food Science*, **42**, 32–37

Yabumoto, K. Yamaguchi, M. and Jennings, W.G. (1978) Production of volatile compounds by muskmelon, *Cucumis melo*. *Food Chemistry*, **3**, 7–15

Chapter 10

Pineapple and papaya

R. E. Paull

10.1 THE PINEAPPLE – AN INTRODUCTION

The pineapple (*Ananas comosus* (L.) Merrill) is a terrestrial member of the diverse family Bromeliaceae. The pineapple fruit shares the distinction of being selected, developed and domesticated by peoples in tropical America in prehistoric times (Collins, 1968). The selection of seedless cultivars led to asexual propagation. The place of origin of these cultivars includes the countries of south-eastern Brazil, Paraguay and northern Argentina (Baker and Collins, 1939). *Ananas* (2n = 50) is distinguished from other Bromeliaceae by having syncarpous fruit of numerous sessile flowers connate with their subtending bracts and with one another, and in bearing a terminal crown of reduced leaves (Collins, 1968). The other edible relative is *A. bracteatus*. Of the many pineapple cultivars, Smooth Cayenne is the major commercial cultivar, though other cultivars are grown on a small scale (Collins, 1968; Samuels, 1970). There has, however, been selection with only limited breeding work (Collins, 1968; Chan, 1986; Cabot, 1989).

10.1.1 Fruit development

The flowering, fruit growth and development of pineapple has been recently reviewed (Bartholomew and Paull, 1986). Inflorescence development is initiated naturally by shortened daylength and cool nights with considerable variation in the exact conditions required depending on the cultivar. Under commercial cropping conditions, growth regulators such as ethylene are used to force flower development (Bartholomew and Criley, 1983). Pollination is unnecessary for fruit development in the self-incompatible commercial cultivars, though *A. comosus* is the only self-

Biochemistry of Fruit Ripening. Edited by G. Seymour, J. Taylor and G. Tucker. Published in 1993 by Chapman & Hall, London. ISBN 0 412 40830 9

incompatible species known in the genus. In a commercial situation, Ethrel (2-chloroethyl phosphonic acid) is used to stimulate shell de-greening and foster ripening (Poignant, 1971; Crochon *et al.*, 1981).

The pineapple fruit is composed of core, fruitlets, the collective flesh and fruit shell. Bract, calyx, and ovary tissue of the sessile flowers have become fused within and between fruitlets during development, to form the collective fruit (Okimoto, 1948). No floral abscission occurs, so the withered style, stamens, and petals can be found on a mature fruitlet. The large bract subtending each fruitlet is fleshy and widened at its base and bends over the flattened calyx surface, covering half of the fruitlet. The fruit mass increases in a continuous sigmoid fashion (Fig. 10.1) once the inflorescence has been initiated. Cell division is completed prior to anthesis; all further development is the result of cell enlargement. Fruit

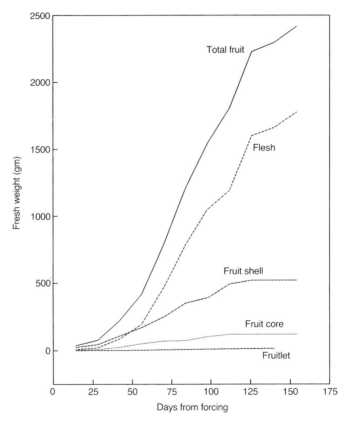

Fig. 10.1 Growth of major parts of the pineapple fruit. (Redrawn from Sideris and Krauss, 1938.)

mass increases about 20-fold from the time of flowering until maturation (Singleton, 1965; Teisson and Pineau, 1982). The flesh mass increase continues up to maturity, with the rate of fruit shell development being slow during the last 40 days of fruit growth (Fig. 10.1). The number of fruitlets comprising a fruit varies widely with plant condition and environmental conditions; a typical Smooth Cayenne fruit for canning has about 150 fruitlets, which produces a mature fruit weighing about 2.2 kg (Tay, 1977). The fused nature of the fruitlets means that the flesh of the fruit is not sterile, but contains yeasts and bacteria (Rohrbach and Apt, 1986). The crown probably has no direct effect on the growth of the fruit (Senanayake and Gunasena, 1975).

10.1.2   Harvesting, storage and handling

Fruit maturity is judged on the extent of fruit 'eye' flatness and skin yellowing, and in Hawaii a minimum reading of 12% total soluble solids is required for fresh fruit (Anon, 1968). Fruit specific gravity at the edible ripe stages varies with season and may have a place as a maturity index (Singleton, 1957; Smith, 1984). Full ripe fruit are unsuitable for transporting to distant markets and a less mature grade is selected (Dull, 1971). Immature fruit are not shipped since they do not develop good flavour, have low Brix and are more prone to chilling injury (Rohrbach and Paull, 1982). Temperatures in the range 7.5° to 12°C are recommended for storage, with relative humidity of 70–90%. Half-ripe Smooth Cayenne fruit can be held for about two weeks at 7.5–12.5°C and still have about one week of shelf life (Dull, 1971). The maximum storage life at 7°C is about four weeks (Paull and Rohrbach, 1985).

Fruit bruising is a major problem during harvesting and packing. This injury is normally confined to the impact side of the fruit. The damaged flesh appears slightly straw-coloured, and becomes lead-grey with time (Keetch, 1978). Chilling injury symptoms occur and can be caused by preharvest or postharvest exposure to temperatures less than 10–12°C (Akamine et al., 1975; Keetch and Balldorf, 1979). The symptoms develop when the fruit is returned to physiological temperature (18–30°C) (Paull and Rohrbach, 1985). Susceptible fruit are generally lower in ascorbic acid and sugars and are opaque (Teisson et al., 1979b; Abdullah and Rohaya, 1983; Paull and Rohrbach, 1985). Partial control of chilling injury symptom development has been achieved by waxing, polyethylene bagging (Paull and Rohrbach, 1982, 1985; Rohrbach and Paull, 1982; Abudullah et al., 1985), heat treatments (Akamine et al., 1975; Akamine, 1976), controlled atmospheres (Abdullah et al., 1985; Paull and Rohrbach, 1985), and ascorbic acid application (Sun, 1971).

10.2   PINEAPPLE – PHYSIOLOGY AND BIOCHEMISTRY

When the fruit is in the half-yellow stage it is regarded as ripe. At this stage, fruit weight is near maximum (Wardlaw, 1937) and Brix and titratable acidity have reached their maximum. Fruit development and composition changes during growth have been reviewed (Dull, 1971; Teisson and Pineau, 1982; Bartholomew and Paull, 1986). The present review concentrates on the marked changes in flesh composition that occur in the three to seven weeks prior to and at the half-yellow shell colour stage (Dull, 1971; Tay, 1977; Teisson and Pineau, 1982).

10.2.1   Pigment changes

Shell chlorophyll levels showed little change until the final 10 to 15 days before full ripeness, but then declined (Gortner, 1965). Shell carotenoid pigments remained reasonably constant during this phase only, declining slightly before rising again as the fruit senesced. Flesh carotenoids increased during these final 10 days before the full ripe stage (Gortner, 1965; Teisson and Pineau, 1982). Cryptoxanthin is the major carotenoid, being three times more prevalent than the β-carotenoids, while no α-carotenoids have been detected (Wills et al., 1986). These changes occur in both harvested fruit and those left on the plant.

10.2.2   Texture changes

There are no marked changes in fruit texture during ripening, though water loss can lead some to reduction in fruit firmness. Senescence-related loss of membrane integrity leads to water-soaked, translucent flesh that tends to be softer.

10.2.3   Respiration and ethylene

Pineapple fruit have a moderate respiration rate, producing around 22 ml $CO_2.kg^{-1}.h^{-1}$ at 23°C. No dramatic respiratory or biochemical change occurs during ripening (Dull et al., 1967), with ethylene production increasing as each fruitlet ages but with no pronounced peak. The absence of a peak in ethylene production and lack of relationship of respiration with pronounced biochemical ripening changes supports the conclusion of a non-climacteric pattern of development (Dull et al., 1967).

10.2.4   Developmental changes

The pH of pineapple juice declines as the fruit approach the fully ripe stage (Teisson and Pineau, 1982), ranging from 3.9 to 3.7, and increasing only as the fruit senesce, while titratable acidity shows opposite trends (Singleton and Gortner, 1965; Teisson and Pineau, 1982). The acidity of flesh (pH ca. 3.8) increases distally from the central core where it is 4 meq.100.ml^{-1} to 10 meq.100 ml^{-1} at the periphery of the flesh (Huet, 1958). Titratable acidity declines during storage at 10°C (Fig. 10.2a), though it shows little change if the fruit are not stored.

A major portion (65–70%) of the total non-volatile acids occurs as free acids (Chan *et al.*, 1973a; Teisson and Combres, 1979). The pineapple fruit has two major non-volatile organic acids; citric and malic (Chan *et al.*, 1973a). Citric acid levels increase during fruit development, reaching a maximum before malate and before full ripeness. Fruit malic acid levels do not change after harvest or during and after storage. During storage

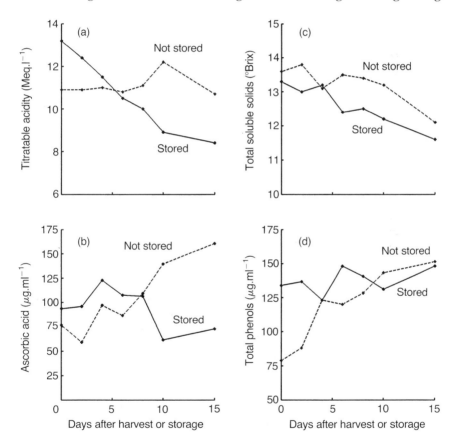

Fig. 10.2 Postharvest changes in flesh composition of pineapple fruit allowed to ripen at 22°C or stored for two weeks at 10°C then allowed to ripen at 22°C. (a) titratable acidity; (b) ascorbic acid; (c) total soluble solids; (d) total phenols. (From Paull, 1988, unpublished results.)

at 7°C, citric acid increased about 25%, then declined slightly when fruit were held at 22.5°C. While in unstored fruit, citric acid did not change.

Ascorbic acid content of ripe pineapple fruit varies from 40 to 1100 μM total juice and does not contribute substantially to titratable acidity. There is, however, a positive correlation (r = 0.956) between ascorbic acid and total acidity (Hamner and Nightingale, 1946). Its level in the fruit varies positively with the amount of solar radiation received by the fruit while still attached to the plant (Singleton and Gortner, 1965). After harvest, ascorbic acid levels remain reasonably constant or increase (Fig. 10.2b), but decline if fruit are stored at 8°C for two to four weeks and then removed to 22°C (Teisson et al., 1979a; Teisson and Combres, 1979; Paull and Rohrbach, 1982). Internal browning is a minor problem if fruit ascorbic acid content is greater than 500 μM (Teisson and Combres, 1979).

Total soluble solids gradually increase showing a change from 2.7 to 9.0°Brix in the last six weeks prior to the full ripe stage (Lodh et al., 1972). Fruit sugars increase through to senescence, unless the fruit is harvested (Kelly, 1911). The more mature, basal tissue of the fruit, tends to be sweeter and can show up to 4 °Brix greater sugar than the flesh near the crown end (Miller and Hall, 1953). Starch is not accumulated in the fruit (Dull, 1971) and this could explain the absence of dramatic changes in total soluble sugars postharvest.

The major juice sugars are sucrose, glucose, and fructose (Gawler, 1962). Peak sucrose concentration is attained at full ripeness, then declines; fructose and glucose continue to increase (Singleton and Gortner, 1965; Tay, 1977). There is a decline in total soluble solids if fruit are stored for two weeks at 10°C then removed to 22°C or held at 22°C without storage (Fig. 10.2c). There is little change in the proportion of the individual sugars.

The phenolic acids; p-coumaric, ferulic, and sinapic, have been tentatively identified in pineapple fruit. These acids, except sinapic, were in fruit showing internal browning (van Lelyveld and de Bruyn, 1977). Changes in phenols, especially postharvest and during storage, have been associated with chilling injury (Teisson, 1977; Paull and Rohrbach, 1985), the total phenol levels declining as chilling-induced browning developed (Teisson et al., 1979a). Total phenol and diphenol levels increase in unstored fruit postharvest (Fig. 10.2d), but levels of the individual phenols apparently vary (Teisson et al., 1979a).

The production of volatile compounds by the pineapple fruit increases during fruit development on the plant and after harvest (Flath, 1986). A wide range of volatiles have been identified (Table 10.1). These include esters, lactones, aldehydes, ketones, alcohols and a group of miscellaneous compounds (Dupaigne, 1970; Flath and Forrey, 1970; Silverstein, 1971; Flath, 1986). Hydrolysis with β-glucosidase (EC 3.2.1.21) showed

Table 10.1 Identified volatile constituents of headspace pineapple crown and whole intact fruit and blended pulp. (After Takeoka *et al.*, 1989., Wu *et al.*, 1991.)

Compound	Crown*	Whole Intact Fruit* (% area)	Blended pineapple pulp % area*	Free+ µg/kg	Bound+ µg/kg
Alcohols					
ethanol	+	1.56			
2-butoxyethanol				74	24
2-phenylethanol		0.30			
2-methylpropanol		0.09			
3-methylbutanol		0.55		23	–
2-pentanol				7	–
3-methyl pentan-2-ol				9	–
1-hexanol		0.05		–	12
(Z)-3-hexanol	+		0.01		
(3 hydroxyphenyl) ethyl alcohol				–	11
linalool		0.04			
eugenol				–	18
4-terpineol		0.07			
α-terpineol		0.07			
Aldehydes					
pentanal	+				
hexanal	+	0.02		10	–
(E)-2-hexenal	+				
(Z)-3-hexenal	+				
benzaldehyde				11	9
p-hydroxybenzaldehyde				45	6
vanillin				23	–
syringaldehyde				80	27
Esters					
ethyl acetate	+	5.91		470	–
diethyl carbonate		0.04	0.02		
methyl propanoate		0.14			
proply acetate		0.08		6	–
methyl 2-methylpropanoate		0.21			
ethyl propanoate		0.12			
ethyl 2-methylpropanoate		0.06			
methyl 3-methylthiopropanoate		1.47	17.41		
ethyl 3-methylthiopropanoate		0.08	3.19		
ethyl propenoate				10	–
2-methylpropyl acetate		0.42	0.04		

Table 10.1   (*contd*)

Compound	Crown*	Whole Intact Fruit* (% area)	Blended pineapple pulp		
			% area*	Free+ μg/kg	Bound+ μg/kg
3-methylthiopropyl acetate			0.10		
butyl acetate		0.03			
methyl butanoate		5.29	0.55	26	–
ethyl butanoate		0.70	0.43		
methyl 2-methylbutanoate		11.42	1.03	70	–
methyl 3-methylbutanoate		0.09			
ethyl 2-methylbutanoate		0.63	0.29		
ethyl 3-methylbutanoate		tr	0.01		
methyl 3-hydroxybutanoate		0.05			
methyl 3-acetoxybutanoate		0.27	0.35	210	–
3-methylbutyl acetate		0.19	0.17		
2-methylbutyl acetate			0.08		
3-methylbut-2-enyl acetate			tr		
methyl pentanoate		0.82	0.24		
ethyl pentanoate		0.04	0.10		
methyl 4-methyl pentanoate				141	–
dimethyl malonate		0.18	0.06	105	–
methyl hexanoate		36.57	14.72		
ethyl hexanoate		1.14	6.80		
methyl (E)-2-hexenoate			0.10		
methyl (E)-3-hexenoate		0.31	0.51		
methyl (Z)-3-hexenoate		0.12	0.05	5	–
methyl 3-hydroxyhexanoate		0.16	0.03	12	–
ethyl 3-hydroxyhexanoate			0.05	52	168
methyl 3-acetoxyhexanoate		0.66	15.17	1071	–
methyl 4-acetoxyhexanoate			0.56	193	–
methyl 5-acetoxyhexanoate		0.45	1.33	676	–
ethyl 3-acetoxyhexanoate		0.05	2.24		
ethyl 4-acetoxyhexanoate			0.19	76	–
ethyl 5-acetoxyhexanoate			0.16	52	–
geranyl hexanoate			0.04		
δ-heptanoate				–	34
methyl heptanoate		0.76	0.20		
ethyl heptanoate			0.10		
methyl octanoate		10.27	6.82	34	–
ethyl octanoate		0.15			
methyl (E)-2-octenoate		0.05	0.03		
methyl (E)-3-octenoate		0.16			
methyl (Z)-4-octenoate		0.89	0.77		
ethyl (Z)-4-octenoate			0.38		
methyl 3-acetoxyoctanoate			0.28	116	–

Table 10.1   (*contd*)

Compound	Crown[*]	Whole Intact Fruit[*] (% area)	Blended pineapple pulp % area[*]	Free[+] μg/kg	Bound[+] μg/kg
methyl 4-acetoxyoctanoate			0.80	6	–
methyl 5-acetoxyoctanoate		0.27	2.36	129	–
ethyl 3-acetoxyoctanoate			0.09	13	–
ethyl 4-acetoxyoctanoate			0.09	42	–
ethyl 5-acetoxyoctanoate			0.70		
methyl nonanoate		0.08	0.13		
ethyl nonanoate			0.04		
methyl decanoate		0.11	0.38		
methyl (Z)-4-decenoate		0.42	3.09		
ethyl tetradecanoate			0.02		
ethyl hexadecanoate			0.02		
ethyl dodecanoate			0.06		
methyl 2, 4-hexadienoate			0.06		
2-phenylethyl acetate		0.18			
methyl phenylacetate			0.10		
Ketones					
2-pentanone				12	–
2-heptanone		tr			
2,5-dimethyl-4-methoxy-3(2H)-furanone			0.05		
2,5-dimethyl-4-hydroxy-3(2H)-furanone				700	491
Hydrocarbons					
chloroform		0.32			
α-pinene	+				
β-pinene	+				
myrcene		0.30			
limonene		0.22			
(E)-β-ocimene		0.20	0.13		
(Z)-β-ocimene			0.02		
styrene	+	0.34			
acetoin		0.12			
1-(E,Z)-3,5-undecatriene		0.39	0.70		
1-(E,E)-3,5-undecatriene		0.03			
1,3,5,8-undecatetraene		1.42			
1-(E,E,Z)-3,5, 8-undecatetraene			0.06		
1-(E,Z,Z)-3,5, 8-undecatetraene					
α-copaene		1.60	1.72		

Table 10.1   *(contd)*

Compound	Crown[*]	Whole Intact Fruit[*] (% area)	Blended pineapple pulp % area[*]	Free[+] μg/kg	Bound[+] μg/kg
β-copaene		0.17	0.16		
α-gurjunene			0.19		
α-muurolene		0.46	1.51		
γ-muurolene		0.27	0.75		
δ-cadinene		0.15	0.51		
β-selinene		0.16			
α-selinene		0.16	0.38		
cyclocopacamphene			0.13		
(sesquiterpene)		1.53	0.14		
(sesquiterpene)		0.03	0.11		
Lactones					
γ-hexalactone			0.06	–	45
δ-hexalactone				26	–
γ-octalactone				–	8
δ-octalactone				99	226
γ-nonalactone				–	6
γ-decalactone				–	6
γ-dodecalactone			0.04		
Miscellaneous					
acetic acid				109	–
cinnamic acid				–	65
hexanoic acid				23	11
phenol				54	–
4-allyl-2, 6-dimethoxyphenol				–	31
methyl cyclohexane			0.14		
dimethyl hexane			0.21		

[*] After Takeoka *et al.*, 1989.
[+] After Wu *et al.*, 1991.
+ = detected, – = not detected

that a number of phenols, aldehydes, alcohols, acids and many lactones are glycosidically bound (Wu *et al.*, 1991) and upon hydrolysis give a pineapple-like aroma. The volatile oil content of pineapple flesh is higher in summer fruit, and is associated with greater amounts of ethyl alcohol and ethyl acetate (Haagen-Smit *et al.*, 1945). Evidence for the importance of individual components and the role of glycosidically-bound volatiles to the fruit aroma and flavour is still lacking.

Pineapple juice has been shown to contain neutral polysaccharides which are predominantly galactomannans (Chenchin and Yamamoto, 1978). This gum is readily hydrolysed by commercial cellulase, hemicellulase and pectinase preparations (Chenchin *et al.*, 1984). Total lipids declined at maturity, as did the squalene content. Phospholipids, total sterol and acid values increased near maturity (Selvaraj *et al.*, 1975). Free amino acids in the juice were at a minimum during the middle of fruit development (Gortner and Singleton, 1965), with the exception of free methionine, found at low levels until the onset of ripening where it increased to 0.7 mM at senescence.

10.2.5 Enzyme changes

Extraction and characterization of enzymes from pineapple fruit has concentrated on those enzymes likely to be involved in enzymatic browning reactions, namely polyphenoloxidase (EC 1.10.3.1) and protease; bromelain (EC 3.4.22.4), and the family of polypeptide inhibitors of this protease. Polyphenoloxidase activity is low at harvest but in chilling-sensitive fruit increases dramatically following storage at 8°C (van Lelyveld and de Bruyn, 1977; Teisson *et al.*, 1979a). The polyphenoloxidase has a pH optimum near 5.0 and a temperature optimum near 45°C (Teisson, 1977); the enzyme is stable to heat when extracted, but loses over 50% of its activity following 20 minutes' exposure to 60°C *in vivo* (Teisson, 1977), which would explain the reduction in chilling-induced internal browning in heated fruit (Akamine *et al.*, 1975). Thiol and bisulphite also inhibit polyphenoloxidase activity. Peroxidase (EC 1.11.1.7) activity has been reported to fall steadily during fruit development (Gortner and Singleton, 1965) reaching a minimum of one-third of its initial value during ripening. Acid peroxidase (pH optimum 5.0) did not appear to be related to the development of chilling injury symptoms (Teisson, 1977), though it did decline during storage (Teisson *et al.*, 1979a). Some purification has been accomplished (Teisson, 1977) of both peroxidase and the polyphenoloxidase.

Ascorbic acid oxidase (EC 1.10.3.3) may play a role in internal browning of pineapple fruit following chilling, but no activity of this enzyme was detected in a study by van Lelyveld and de Bruyn, (1977). Ammonium sulphate-precipitated protein from pineapples has been shown to have ascorbic acid oxidase, activity. This activity has a high pH optimum (>8) and no activity is detected below pH 6.0 (Teisson, 1977). These assay conditions may explain published discrepancies. Catalase (EC 1.11.1.6) and indole acetic acid oxidase have also been detected, but changes in activity during ripening or storage have not been reported. Indole acetic acid oxidase has a pH optimum of 3.5, which is at a variance with other fruit (Teisson, 1977).

Pineapple acetone powders contain proteases and a family of poly-peptide protease inhibitors (Reddy *et al.*, 1975). Protease activity appears abruptly after flowering and remains high during fruit development (Gortner and Singleton, 1965) and subsequently declines during ripening (Gortner and Singleton, 1965; Lodh *et al.*, 1973). Two minor proteases have been isolated from pineapple fruit (Ota *et al.*, 1972). However, the major cysteine protease is bromelain, a glycoprotein found in both fruit and stem. Fruit bromelain has a higher molecular weight (31kDa) than stem bromelain (28kDa), while isoelectric points and pH optima also differ being pI 4.6, versus 9.6 and pH 8.0 versus 5 to 6 for fruit and stem bromelain, respectively. The amino terminal end of the fruit bromelain has an additional alanine (Yamada *et al.*, 1976) and the amino acid sequence around the reactive cysteine residue is the same in both enzymes. However, fruit bromelain shows only 20% cross reactivity with stem bromelain antibody (Sasaki *et al.*, 1973).

10.3 PAPAYA – AN INTRODUCTION

Carica papaya L., *C. candamarencis* Hook f., *C. pentagonia* and *C. monoica* are members of the small family (*Caricaceae*) having four genera and 31 species. These four species are grown for their fruit. They are soft woody, usually dioecious trees with latex vessels throughout, have large, deeply-lobed leaves and long petioles. Papaya, papaw and paw paw (*C. papaya*) (2n = 18) is cultivated throughout the tropics for its fruit (Purseglove, 1968). The fruit are eaten green or ripe, fresh or in salads. They are also used for making juice, jam, crystallized fruit and canned. The ripe fruit are also cooked before eating. Latex is collected following scratching of the skin of green papaya on the tree. The dried latex is further purified for use as a meat tenderizer, the active principle is the protease; papain. Andean Mountain papaya (*C. candamarencis*) is cultivated at high elevations for its small fruit, eaten stewed or as preserves (Everett, 1952). The Ecuadorian babaco (*C. pentagonia*) grows between 2000 to 3000 metres above sea level, and is also eaten fresh.

10.3.1 Fruit development

The fleshy papaya is a berry, 5 to 40 cm long and can weigh to over 5 kg. The fruit shape may be spherical or oblong depending upon whether from pistillate or hermaphrodite flowers. The common, commercial 'Solo' type fruit from hermaphrodite flowers is pear-shaped, cylindrical or grooved (Fig. 10.3). The fruit is normally composed of five longit-udinal carpels united laterally with the flesh surrounding a five-angled large central cavity where numerous seeds are attached to placentas in

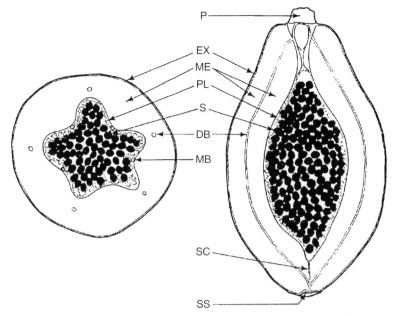

Fig. 10.3 Papaya fruit anatomy showing the location of the longitudinal placental, dorsal and marginal vascular bundles. P – peduncle; EX – exocarp; ME – mesocarp; PL – placenta; DB – dorsal bundle; MB – marginal bundle; S – seeds; SC – stigma canal; SS – stigma scar.

parietal positions (Fig. 10.3). Seeds are attached to the placenta by 0.5- to 1.0-cm stalks. Mature seeds are dark grey or black and each is enclosed in a sarcotesta. The edible flesh (1.5 to 4 cm thick) changes from white in immature fruit to red or yellow during ripening (Nakasone, 1986). The skin is thin and smooth, turning from green or greenish yellow to yellow or orange when ripe.

Transverse sections of the gynecium show dorsal vascular bundles alternating with marginal bundles (Roth and Clausnitzer, 1972). The dorsal vascular bundles develop into the main vascular supply (Storey, 1967). There is a network of latex tubes throughout the pericarp. The laticifers develop close to the vascular bundles (Roth and Clausnitzer, 1972) (Fig. 10.3).

Fruit development from pollination to ripeness takes 168 to 182 days in Hawaii, with an extra two weeks delay during winter (Nakasone, 1986). The variation ranges from 173 days from fruit set to ripeness when grown under 30° day/20°C night to 282 days at 24° day/12°C night (Kuhne and Allan, 1970; Allan *et al.*, 1987). Flower abortion occurs soon after anthesis with most fruit drop (<6 cm) occurring two weeks later (Ong, 1983). Fruit length shows a double sigmoid growth (Fig. 10.4a)

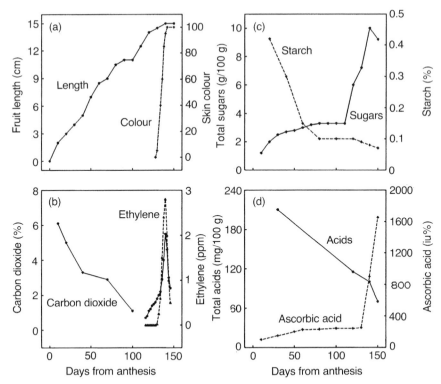

Fig. 10.4 Growth of papaya fruit length (cv. Sunrise) after anthesis (a; redrawn from Ong, 1983) and changes in fruit cavity content (b) of carbon dioxide and ethylene of cv. Solo versus change in skin colour (redrawn from Wardlaw and Leonard, 1936; Akamine and Goo, 1979). Changes in flesh total sugars (c; redrawn from Chan *et al.*, 1979, cv. Waimanalo; Selvaraj *et al.*, 1982b, cv. Sunrise; Chittiraichelvan and Shanmugavelu, 1979, cv. Co. 2) and (d) total acidity and ascorbic acid (redrawn from Selvaraj *et al.*, 1982b cv. Sunrise; Chittiraichelvan and Shanmugavelu, 1979, cv. Co. 2).

with an initial lag period (Ong, 1983). A sigmoid pattern for increase in fruit volume is found (Kuhne and Allan, 1970). The shape of the growth curve varies considerably depending upon month of fruit set and cultivar. Fruit set during warmer weather did show a similar pattern for volume as found for the increase in fruit length (Ong, 1983). Low temperatures (<15°C) during the early phase of growth significantly delayed growth and reduced fruit size, possibly due to reduced cell division. In gynecia of less than 1 mm in diameter, all tissues are composed of meristematic cells. Nearly mature smaller fruit can develop without pollination (Schaffner, 1935), there is, however, a positive correlation between number of seeds and fruit weight (Allan, 1969; Chittiraichelvan and Shanmugavelu, 1979).

The green fruit is well known for the abundance of milky latex containing papain. The instant exudation of latex upon slight injury to the skin is indicative of a network of latex tubes throughout the pericarp. There appears to be considerable pressure in the tubes as latex can immediately spurt out to as far as 20 cm when a large green fruit is pricked with a pointed object (Nakasone, 1986). The latex content diminishes as fruit begin to ripen and there is no visible latex at the fully ripe stage (Chen, 1964). The latex vessels near the epidermis apparently break during ripening and release the latex to the surface of the fruit. This exuded latex disrupts the integrity of the fruit cuticle (Paull and Chen, 1989).

10.3.2   Harvesting, storage and handling

Papaya fruit are not harvested until the skin colour shows some yellowing (Akamine and Goo, 1971). Less mature fruit ripen poorly with lower total soluble solids and therefore need to be culled before packing. A minimum grade value of 11.5% total soluble solids is used in Hawaii (Anon, 1990), with a total soluble solids change of 1.5% to 2% if harvested as skin colour begins to change (Akamine and Goo, 1971). Efforts have been made to objectively measure fruit maturity with a non-destructive, physical measure. Delayed light emission intensity (Forbus et al., 1987; Forbus and Chan, 1989), body transmission spectroscopy at three wavelengths (Birth et al., 1984) and reflectance measurement (Hunter 'b' value) (Couey and Hayes, 1986) have been evaluated. The change of exuding latex colour from white to colourless has also been suggested as a measure of harvest maturity (Traub et al., 1935). The large variation in papaya fruit acoustic properties, at the same stage of ripeness precludes the use of ultrasound for sorting fruit (Hayes and Chignon, 1982).

Papaya shipments arriving at terminal markets have a range of disorders associated with mechanical injury, over-ripeness and parasitic diseases (Cappellini et al., 1988). The inter-relationship between mechanical injury and storage disorders (i.e. chilling injury) and the incidence of postharvest diseases has not been fully determined (Somner and Mitchell, 1978; Alvarez and Nishijima, 1987; Nishijima et al., 1990).

The papaya fruit at colour-turning stage can be stored at temperatures as low as 7°C for less than 14 days and it will ripen normally (Thompson and Lee, 1971; Chen and Paull, 1986). At 7°C temperature, decay during storage is less than at 12° or 13°C (Arisumi, 1956; El-Tomi et al., 1974). Chilling injury symptoms include skin scald, hard lumps in the pulp around the vascular bundles and water soaking of flesh (Thompson and Lee, 1971; El-Tomi et al., 1974; Chen and Paull, 1986). Fruits become progressively less susceptible to chilling stress as they ripen (Chen and

Paull, 1986). Symptoms of chilling injury occur after 14 days at 5°C for mature green fruit and 21 days for 60% yellow fruit. The decrease in susceptibility has been related to the stage of the fruit climacteric (Chan, 1988). Prior to the appearance of visible chilling injury symptoms, an increase in electrolyte leakage and ethylene production occurs (Chan *et al.*, 1985).

The optimum temperature for fruit ripening is between 22.5° and 27.5°C, with fruit taking 10 to 18 days to reach full skin yellowing from the colour-break stage (An and Paull, 1990). Ripening rate varies with cultivar (Zhang and Paull, 1990) with a range of 7 to 16 days from the colour break stage (Fig. 10.4a). The rate of softening differs between cultivars with respect to the rate of respiration, and ethylene production, skin degreening and flesh colour development (Zhang and Paull, 1990). Exogenous ethylene application dramatically increases the rate of fruit softening and flesh colour development (Nazeeb and Broughton, 1978; An and Paull, 1990). Ethylene-treated papayas ripened faster and more uniformly individually and as a cohort in terms of skin degreening, softening and flesh colour development (An and Paull, 1990). Since fruit ripen from the inside outwards, the effect of ethylene treatment is to accelerate the rate of ripening of the mesocarp tissue nearer the skin.

10.4   PAPAYA – PHYSIOLOGY

Papaya is a climacteric fruit (Wardlaw, 1936; Jones and Kubota, 1940; Akamine, 1966; Selvaraj *et al.*, 1982a). Mature green fruit exhibit a typical respiratory curve with an apparently extended preclimacteric minimum period (Fig. 10.4b). Fruit harvested at the colour-turning stage having already passed the preclimacteric minimum. The increase in ethylene production parallels the respiration rise (Fig. 10.4b) and reaches a maximum at the same time as the respiratory climacteric (Paull and Chen, 1983). During ripening the seed cavity carbon dioxide level can reach nearly 16% and oxygen levels *ca.* 0.5% (Wardlaw and Leonard, 1938). Lower amounts of carbon dioxide (3–6%) (Fig. 10.4b) are found in the smaller Solo type fruit (Akamine and Goo, 1979; Paull and Chen, 1989). The fruit central cavity can develop a negative pressure (Wardlaw and Leonard, 1938) which is probably associated with change in flesh gas transfer as it becomes water-soaked due presumably to loss of cellular compartmentation. Cellular electrolyte leakage does increase during ripening (Chan *et al.*, 1985). The concentration of DNA and RNA show little change during ripening. Protein content increases up to the preclimacteric then declines throughout ripening (Pal and Selvaraj, 1987).

10.4.1 Carbohydrates

In the early stages of fruit development, glucose is the major fruit sugar. At 110 days after anthesis, when the fruit seeds and pulp begins to change colour, there is a dramatic change in sugar content (Fig. 10.4c) (Chen, 1964; Chan et al., 1979; Chittiraichelvan and Shanmugavelu, 1979; Selvaraj et al., 1982a). Sucrose increased from 18% of the total sugar to 80% of the sugars about 135 days after anthesis (Chan et al., 1979) as a result of movement from leaves to fruit. These results are at odds with many earlier reports of ripe papaya sugar composition (Jones and Kubota, 1940; Selvaraj et al., 1982a), which indicated much lower amounts of sucrose and higher percentages of glucose and fructose. The difference is due to a failure to allow for invertase (EC 3.2.1.26) activity during tissue homogenization (Chan and Kwok, 1976), and which has been shown to increase 15-fold during fruit ripening (Pal and Selvaraj, 1987). Trace quantities of sedoheptulose have also been reported (Ogata et al., 1972).

Starch declines during fruit development (Fig. 10.4c) to 0.1% dry weight in ripe papaya mesocarp tissue (Chan et al., 1979; Selvaraj et al., 1982a). The small increase (1–2%) in total soluble sugars during the ripening of detached fruit support this conclusion (Chen, 1964; de Arriola et al., 1975). Amylase activity has been reported in papaya fruit (Tan and Weinheimer, 1976; Pal and Selvaraj, 1987). The activity is high at 1–2 weeks after anthesis declining with a peak of activity occurring 80 to 110 days after anthesis (Tan and Weinheimer, 1976); a second activity peak occurs as the fruit approaches maturity, with the activity decreasing during fruit ripening (Tan and Weinheimer, 1976; Pal and Selvaraj, 1987). Some starch grains have been detected by electron microscopy near the epidermis of the mature green fruit (Sanxter, 1990).

10.4.2 Organic acids

The titratable acidity declines (Fig. 10.4d) during fruit growth to reach 1.54 meq.100g^{-1} with a pH in the range 5.0–5.5 in ripe papaya puree. This is made up of 0.28 meq.100g^{-1} ascorbic acid which together with malic, citric acid and α-ketoglutaric acid is 85% of the total titratable acidity. Total volatile acid contributes 8% to the total titratable acidity (Chan et al., 1971; Selvaraj et al., 1982b). Malic and citric acid are formed in about equal amounts; being ten times more abundant than α-ketoglutaric acid, malonic acid, fumaric and succinic acid (Chittiraichelvan and Shanmugavelu, 1979). Non-volatile acidity, especially citric acid, decreases by two-thirds during fruit maturation, with little change during the ripening phase (Chen, 1964; Chittiraichelvan and Shanmugavelu, 1979; Selvaraj et al., 1982a). The titratable acidity has been reported to increase slightly during ripening (de Arriola et al., 1975, 1980). This increase could be partly associated with free galacturonic acid which increases during ripening; there is little increase in isolated organic acids (Thomas and Beyers, 1979; Chittiraichelvan and Shanmugavelu, 1979; Selvaraj *et al.*, 1982a).

Ascorbic acid quadruples during fruit ripening (Fig. 10.4d) (Selvaraj *et al.*, 1982b) to 55 mg.100g^{-1} (Orr *et al.*, 1953; de Arriola *et al.*, 1975). The low contribution of acidity to fruit flavour has meant there have been few studies of the enzymes involved in organic acid metabolism. Malic dehydrogenase however, declines during fruit maturation and ripening, as does NADH reductase (Pal and Selvaraj, 1987).

10.4.3   Flesh pigments

Carotenoids (Pro-Vitamin A) differ between the yellow- and red-fleshed papaya (Yamamoto, 1964). The red-fleshed fruit has 63.5% of the carotenoids content as lycopene but this compound was found to be absent in yellow-fleshed fruit. Other carotenoids present were β-carotene, δ-carotene, cryptoxanthin monoepoxide and cryptoxanthin. The total cartenoids content increases up to 14-fold from the mature green stage (Fig. 10.4a) to nearly 4 mg.100g^{-1} at full ripe stage (Selvaraj *et al.*, 1982a; de Arriola *et al.*, 1975) while skin chlorophyll declined to one-sixteenth of its content in ripe compared to immature fruit (Birth *et al.*, 1984).

10.4.4   Flavour

The examination of volatile flavour concentrates from papaya fruit by GC-MS has led to many compounds being identified (Table 10.2) from up to 200 components eluted (Flath and Forrey, 1977; MacLeod and Pieris, 1983; Heidlas *et al.*, 1984; Schreier *et al.*, 1985; Flath *et al.*, 1990). The amount of each component varied both with cultivar and locality (MacLeod and Pieris, 1983), with linalool as the major component. There were smaller amounts of benzyl isothiocyanate and phenylacetonitrile with butyric, hexanoic and octanoic acid and their methyl esters being minor components. The amounts and relative content of volatiles has been shown to vary with stage of ripeness (Katague and Kirch, 1965; Flath *et al.*, 1990); for example, linalool production increased nearly 400-fold with only a 7-fold increase in benzyl isothiocyanate (Table 10.2) during ripening (Flath *et al.*, 1990). The content of phenolic compounds in the flesh fell to one-quarter the harvest value during ripening (Tan and Lam, 1985). Very little free linalool or isothiocyanate were present in intact tissue, but were produced following injury (Heidlas *et al.*, 1984; Schreier *et al.*, 1985). The acid and methyl esters of butyric, hexanoic and octanoic acid were also produced following maceration due to enzymatic activity (Chan *et al.*, 1973b). The formation of the free acid is probably the major cause of the off-odour and off-flavour in puree. Numerous esters

Table 10.2 Papaya fruit headspace volatile components from mature green (MG), colour-break (CB), quarter-ripe (QR) and full-ripe (FR) fruit. (After Flath *et al.*, 1990.)

	Amount volatiles trapped* (µg)			
Component	MG	CB	QR	FR
Alcohols				
linalool oxide A			0.06	4.51
linalool oxide B			0.03	4.07
linalool	0.58	1.84	2.09	191.78
4-terpineol				0.12
Aldehydes				
hexanal	0.14	0.43	0.15	0.21
heptanal	0.04	0.14		0.03
benzaldehyde	0.02	0.04	0.03	0.24
octanal	0.13	0.59	0.25	0.12
nonanal	0.63	2.27	0.71	0.51
decanal	0.35	1.17	0.42	0.32
Esters				
ethyl acetate	4.35	5.33	1.37	3.32
ethyl butyrate				0.01
prop-2-yl butyrate				0.05
methyl hexanoate				0.05
γ-hexalactone				0.04
methyl octanoate				0.05
ethyl benzoate				0.05
methyl salicylate	0.02	0.03	0.01	0.01
butyl hexanoate				0.06
ethyl octanoate				0.09
γ-octalactone				0.08
methyl geranate				0.06
triacetin	0.02	0.01	0.01	0.02
butyl benzoate				0.07
3-methylbutyl benzoate				0.03
Heteroatoms				
methyl thiocyanate	0.03	0.05	0.05	0.03
phenylacetonitrile	0.88	2.34	1.86	2.34
benzyl isothiocyanate	0.29	0.57	0.39	2.19
Hydrocarbons				
myrcene				0.60
α-phellandrene				0.03
terpene				0.09
α-terpinene				0.02
β-phellandrene				0.02
limonene				0.08

Table 10.2   *(contd)*

	Amount volatiles trapped* (μg)			
Component	MG	CB	QR	FR
(Z)-β-ocimene				0.37
(E)-β-ocimene				0.35
γ-terpinene				0.01
terpinolene				0.12
sesquiterpene		0.06	0.05	0.02
sesquiterpene		0.09	0.07	0.01
sesquiterpene		0.07	0.04	0.01
sesquiterpene		0.03		
caryophyllene	0.24	0.56	0.43	0.15
germacrene D		0.17	0.08	0.02
pentadecane	0.07		0.09	0.04
sesquiterpene	0.04	0.08	0.03	
Ketones				
pentane-2, 4-dione	0.01			
4-hydroxy-4-methyl-pentan-2-one	0.15	0.20	0.04	0.05
heptan-2-one				0.03
6-methylhept-5-en-2-one		0.01		0.02
geranylacetone		0.10	0.10	0.07

* From air, at 1 l min^{-1} for 120 min.

and monoterpenes appeared only in ripe fruit at low levels (Flath *et al.*, 1990). Only one papaya volatile, methyl benzoate, is described as having papaya qualities on odour assessment (MacLeod and Pieris, 1983). The Hawaii variety had very little methyl butanoate (0.06%) (Flath and Forrey, 1977) while the Sri Lankan variety had 48.3% (MacLeod and Pieris, 1983). The sweaty odour quality of some papaya cultivars is due probably to the production of methyl butanoate (MacLeod and Pieris, 1983).

The mountain papaya showed at least 55 components of which 45 have been identified in the headspace volatiles (Morales and Duque, 1987). The major aroma components are ethyl butyrate, butanol, ethyl acetate, methyl butyrate and butyl acetate. This analysis differs from an earlier report in the relative amounts of the components and absence of others (Idstein *et al.*, 1985), due possibly to cultivar or growing region. Thirty-seven volatile compounds were identified in babaco headspace, with aliphatic esters accounting for 84% of total area on the GC trace. Ethyl butanoate and ethyl hexanoate are the major components (Bartley, 1988); an earlier report had found alcohols to be the major constituent (Shaw *et al.*, 1985).

10.5   PAPAYA – BIOCHEMISTRY

10.5.1   Cell wall degrading enzymes

Papaya fruit soften to an edible stage in 6 to 12 days when harvested at the colour break stage. Water-soluble pectin declined as a percentage of total fresh weight rapidly after the 70 to 80% skin yellow stage (Chen, 1964), with no decline in acid-soluble pectin content. This result agrees in part with other published results (Shetty and Dubash, 1974; de Arriola *et al.*, 1975; Lazan *et al.*, 1991) who found an initial increase in water-soluble pectin during ripening. The decline in water-soluble pectin could be due to its degradation to ethanol-soluble material. There are problems in reconciling these results with the changes in cell wall degrading enzymes during ripening (Paull and Chen, 1983). The peak in xylanase (EC 3.2.1.32) and polygalacturonase (EC 3.2.1.15) activity occur when the fruit has 40–60% skin yellowing. Unlike many other fruit, the activity increases rapidly at this stage, then apparently declines. Pectin methyl esterase (EC 3.1.1.15) and CMC-cellulase both continue to increase as the fruit ripen, only declining as the fruit becomes over-ripe and begins to break down. β-galactosidase (EC 3.2.1.23) activity doubles during ripening (Lazan *et al.*, 1991); its role in softening however, is unclear.

Although early in fruit development, four esterase isoenzymes are detected, only two appear to be present when the fruit starts to ripen (Tan and Weinheimer, 1976). One of these bands may be pectin methyl esterase that is present at the start of ripening and continues to increase in activity during ripening (Paull and Chen, 1983; Pal and Selvaraj, 1987). The average activity in ripe fruit is 0.013 meq.min^{-1} g^{-1} (Aung and Ross, 1965). Pectin methyl esterase has an optimum pH between 7.5 and 8.0 is stimulated by NaCl (Chang *et al.*, 1965; Lourenco and Catutani, 1984) and inhibited by sucrose, 10% sucrose giving *ca.* 10% inhibition of esterase activity (Chang *et al.*, 1965). The enzyme is heat stable, retaining *ca.* 80% of its original activity after 60 minutes at 50°C. Kinetic data suggest that polygalacturonic acid is a competitive inhibitor (Lourenco and Catutani, 1984). The enzyme is reported to have a molecular weight of 53 kDa.

The polygalacturonase activity in papaya increases in the endocarp during fruit ripening (Chan *et al.*, 1981; Paull and Chen, 1983), peaking when the fruit has 40–60% skin yellowing. The highest activity is found in the placenta, with activity decreasing outwardly to the exocarp (Chan *et al.*, 1981; Lazan *et al.*, 1991). Heat treatment used for insect disinfestation may injure the fruit, causing areas of the flesh to fail to soften (Chan *et al.*, 1981). This injury may be associated with a disruption of wall-degrading enzymes, two polygalacturonases have been isolated; an endo-acting form, with a molecular weight *ca.* 164 kDa, and an exo-acting form (EC 3.2.1.67) with a molecular weight of *ca.* 34 kDa (Chan and Tam, 1982). The pH optimum for both partially purified enzymes is

4.6, with an optimum temperature of 45°C. Up to 80% of total activity is lost when the enzyme is exposed to 85°C for 10 minutes (Chan *et al.*, 1981; Chan and Tam, 1982).

10.5.2   Ethylene synthesis

Studies on ethylene production in papaya have involved measurement of ethylene-forming enzyme (EFE) activity in different tissues of the fruit during ripening. The highest level of EFE activity was found in the exocarp of 75% ripe fruit (Chan *et al.*, 1990). In mature green fruit and non-senescent tissue of ripening fruit, the majority was in the placental and dorsal bundle. As the fruit ripen from the inside outwards, there is a progressive decline in EFE activity in the placental tissue and a rise and then decline of EFE in dorsal bundles. The level of EFE remains constant until the late stages of ripening (Paull and Chen, 1990). Chilling (Chan *et al.*, 1985) and heat treatments used for insect disinfestation (Chan, 1986; Paull and Chen, 1990) both lead to temporary decline in EFE activity (Paull and Chen, 1990). The level of 1-amino-1-cyclopropane carboxylic acid (ACC), the substrate for EFE, is initially low in fruit mesocarp tissue during ripening, increasing three-fold when the peak of ethylene synthesis occurs (Paull and Chen, 1990). The results suggest that at the peak of ethylene synthesis, EFE activity may be limiting.

10.5.3   Isothiocyanate metabolism

The fruit latex contains *ca.* 10% of the dry weight as a thioglucoside, e.g. benzylglucosinate (Tang, 1971). This compound is hydrolysed by the enzyme thioglucosidase (EC 3.2.3.1), which is not found in the latex (Tang, 1973), to benzyl isothiocyanate (BITC). The concentration of benzyl isothiocyanate decreased in the fruit pulp, but increased in the seeds as the fruit matured and ripened (Tang, 1971). Free BITC is detected in fruit head space analysis, implying it is a normal flavour metabolite (Tang, 1971). Injury to the fruit induces the enzymatic hydrolysis of benzylglucosinate. It has been suggested that BITC (46 μM) inhibits ethylene synthesis (by 60%) in papaya tissue discs (Patil and Tang, 1974); however, there have been no subsequent reports of BITC inhibiting ethylene synthesis in other fruit tissue. Benzyl isothiocyanate's possible role in fruit could be as inhibitor of papain (Tang, 1974), resistance to disease (Patil *et al.*, 1973), and as an inhibitor of fruit fly egg lay and development (Seo and Tang, 1982).

10.5.4   Proteases

The latex of papaya from leaves, petioles, stems and fruit, contains a number of basic thiol proteases including papain (EC 3.4.22.2), chymopapains A and B (EC 3.4.22.6), and papaya peptidase A, all with pI > 8.75. Much of the earlier confusion about the number of proteases present in papaya latex concerns the use of dried, rather than fresh, latex. The latex also contains cellulase (EC 3.2.1.4), lysozyme, (EC 3.2.1.17), and glutamine cyclotransferase.

Papain has been the most intensively studied of these enzymes because of its importance in the brewing, food and pharmaceutical industries. The enzyme is very stable and easily extracted. The polypeptide chain of papain contains 212 amino acid residues, and has a molecular weight of 23.35 kDa. There are X-ray diffraction data (Kamphuis et al., 1984), as well as complete sequence data for papain (Mitchel et al., 1970) and chymopapain A (Robinson, 1975). The essential residue is cysteine 25 with the six other cysteines forming disulphide bridges (Baker and Drenth, 1987). For papain, serine and trytophan, and for chymopapain, glutamine and tryptophan occur on either side of the cysteine active site. Asparagine and alanine occur around the histidine residue of both papain and chymopapain (Brocklehurst et al., 1987). The chymopapains are glycoproteins.

The precise function of these latex proteases is unknown. Protection against insects and fungi are felt to be possible functions, since the digestive juice of a predatory insect may catalyse the conversion of propapain to the active enzyme (Lowe, 1976). During fruit ripening the protease activity in mesocarp tissue declines (Paull and Chen, 1983) as the latifiers begin to break down. Intriguingly, the latex contains a low molecular weight polypeptide papain inhibitor which binds more effectively at a pH above 5.5 (Baines et al., 1982).

10.5.5 Other enzymes

The breakdown of sucrose to fructose and glucose due to invertase (EC 3.2.1.26) activity increases in ripening papaya, paralleling the increase in respiration (Pal and Selvaraj, 1987). Invertase has been partially purified and characterized from papaya fruits (Chan and Kwok, 1976; Lopez et al., 1988), and has an optimum pH of ca. 4.5, and an optimum temperature of 40°C (Chan and Kwok, 1976; Lopez et al., 1988). The K_m is ca. 4.2 mM (Lopez et al., 1988). The apparent molecular weight as determined by gel filtration was 275 kDa (Chan and Kwok, 1976) but only 52 kDa by gel electrophoresis (Lopez et al., 1988). The much higher value obtained by gel filtration was possibly due to aggregation at low salt concentration. Invertase is not inhibited by glucose, but is inhibited by fructose (Lopez et al., 1988) and iodoacetamide (Chan and Kwok, 1976); 90% of the invertase activity is lost after exposure of the enzyme preparation to 60°C for two minutes (Chan and Kwok, 1976).

Papaya contains at least four isoenzymes of an acid phosphatase (Tan and Weinheimer, 1976), the activity of the different isozymes declining to a minimum as the fruit approaches maturity. The partially purified enzyme (EC 3.1.3.2) has a pH optimum of 6.0, temperature optimum of 37°C, K_m of 1.0 mM and apparent molecular weight of 120 kDa. The enzyme loses 70% of its activity when heated at 60°C for eight minutes. The enzyme activity increases during fruit ripening (Pal and Selvaraj, 1987) and may play a role in the generation of off-flavour and odour during puree processing (Carreno and Chan, 1982). Adenosine triphosphate levels and ribonuclease activity decline during ripening (Pal and Selvaraj, 1987). Leucine aminopeptidase declines after anthesis, increasing again as the fruit approaches maturity, then declines as the fruit ripens (Tan and Weinheimer, 1976).

Catalase (EC 1.11.1.6) increases during fruit maturation and ripening (Pal and Selvaraj, 1987). The enzyme has been purified 32-fold from ripe fruit (Chan *et al.*, 1978). The enzyme has a pH optimum of 6.1 and is stable when frozen, though inactivated by acidification to pH 3.5. It has an apparent molecular weight of 160 kDa. The enzyme is rapidly inactivated by heat, losing 90% of activity after three minutes at 60°C.

Peroxidase (EC 1.11.1.7) activity is present in fruit throughout development (Tan and Weinheimer, 1976) and reaches a minimum at the start of the climacteric, then increases during ripening (Pal and Selvaraj, 1987; Da Silva *et al.*, 1990). The amounts of the different isoenzymes (n = 6) vary during fruit development (Tan and Weinheimer, 1976), with at least four peroxidase isoenzymes being present in quarter-ripe fruit (Sawato, 1969). These can be separated into soluble and ionically-bound forms (Da Silva *et al.*, 1990) having molecular weighs of 41 kDa and 54 kDa, respectively.

Light and light plus exogenous nitrate induced nitrate reductase (EC 1.7.99.4) activity in the exocarp of green and mature green fruit (Menary and Jones, 1974). As the fruit ripened, the activities greatly declined in the endocarp, mesocarp and exocarp, with a nine-fold decline in the endocarp from green stage. No activity was detected in the seed. This decline is probably associated with a lack of endogenous nitrate for reduction. Preliminary evidence suggests that xylem transport to the fruit ceases as the fruit reaches the mature green stage (Menary and Jones, 1974).

The activity of phenylalanine ammonia lyase (EC 4.3.1.5) in papaya fruit is two-fold higher 24 hours after irradiation (100 krad), then declines (Tan and Lam, 1985). The activity in untreated controls initially increases following harvest, then remains constant during ripening, while the phenol content declines. Polyphenoloxidase (EC 1.10.3.1) has been reported to rise during the late ripening stage (Pal and Selvaraj, 1987) and could account for the decline in phenol content (Tan and Lam, 1985) that is observed during ripening. These phenolic compounds, especially the chlorogenic acid content, show a relationship to fungal resistance in the fruit (Tan *et al.*, 1982).

REFERENCES

Abdullah, H., and Rohaya, M.A. (1983) The development of black heart disease in Mauritius pineapple (*Ananas comosus* cv. Mauritius) during storage at lower temperatures. *MARDI Research Bulletin*, **11**, 309–319

Abdullah, H., Rohaya, M.A. and Zaipun, M.Z. (1985) Effect of modified atmoshere on black heart development and ascorbic acid contents in 'Mauritius' pineapple (*Ananas comosus* c. 'Mauritius') during storage at low temperature. *ASEAN Food Journal*, **1**, 15–18

Akamine, E.K. (1966) Respiration of fruits of papaya (*Carica papaya* L.var. Solo) with reference to effect of quarantine disinfestation treatments. *Proceedings of American Society of Horticultural Science*, **89**, 231–236

Akamine, E.K. (1976) Postharvest control of endogenous brown spot in fresh Australian pineapples with heat. *HortScience*, **11**, 586–588

Akamine, E.K. and Goo, T. (1971) Relationship between surface color development and total soluble solids in papaya. *HortScience*, **6**, 567–568

Akamine, E.K. and Goo, T. (1979) Concentrations of carbon dioxide and ethylene in the cavity of attached papaya fruit. *HortScience*, **14**, 138–139

Akamine, E.K., Goo, T., Steepy, T., Greidanus, T. and Iwaoka, N. (1975) Control of endogenous brown spot of pineapple in postharvest handling. *Journal American Society of Horticultural Science*, **100**, 60–65

Allan, P. (1969) Effect of seeds on fruit in *Carica papaya*. *Agroplantae*, **1**, 163–170

Allan, P., McChlery, J. and Biggs, D. (1987) Environmental effects on clonal female and male *Carica papaya* L. plants. *Scienta Horticulturae*, **32**, 221–232

Alvarez, A.M. and Nishijima, W.T. (1987) Postharvest diseases of papaya. *Plant Disease*, **71**, 681–686

An, J.F. and Paull, R.E. (1990) Storage temperature and ethylene influence on ripening of papaya fruit. *Journal American Society of Horticultural Science*, **115**, 949–953

Anon. (1968) Wholesale Standards for Hawaiian-grown pineapple. Hawaii Department of Agriculture, Marketing Division

Anon. (1990) Standards for Hawaii-grown papaya. Hawaii Department of Agriculture, Marketing Division

Arisumi, T. (1956) Test shipments of papaya with special reference to storage decay control. Hawaii Agriculture Experimental Station Technical Bulletin, No. 29

Aung, T. and Ross, E. (1965) Heat sensitivity of pectinesterase activity in papaya puree and of catalase-like activity in passion fruit juice. *Journal of Food Science*, **30**, 144–147

Baines, B.S., Kierstan, M.P.J. and Brocklehurst, K. (1982) Polypeptide inhibitors of papain and bromelain. *Biological Society Transactions*, **10**, 171–172

Baker, E.N. and Drenth, J. (1987) The thiol proteases: structure and mechanism. In *Biological macromolecules and assemblies* (eds F.A. Jurnak and A. McPherson), John Wiley & Sons, New York, pp. 314–368

Baker, K.F. and Collins, J.L. (1939) Notes on the distribution and ecology of *Ananas* and *Pseudoananas* in South America. *American Journal of Botany*, **26**, 697–702

Bartholomew, D.P and Criley, R.A. (1983) Tropical fruit and beverage crops. In *Plant Growth Regulating Chemicals*, Vol. 2., (ed L.G. Nickell), CRC Press, Boca Raton, Fla, pp. 1–11

Bartholomew, D.P. and Paull, R.E. (1986) Pineapple. In *CRC Handbook of Fruit Set and Development*, (ed. S.P. Monselise) CRC Press, Boca Raton, Florida, pp. 371–388

Bartley, J.P. (1988) Volatile flavor components in the headspace of babaco fruit (*Carica pentagonia*). *Journal of Food Science*, **53**, 138–140

Birth, G.S., Dull, G.G., Magee, J.B., Chan, H.T. and Cavaletto, C.G. (1984). An optical method for estimating papaya maturity. *Journal American Society of Horticultural Science*, **109**, 62–66

Brocklehurst, K., Willenbrock, F. and Salih, E. (1987) Cysteine proteinases. In *Hydrolytic enzymes*, (eds A. Neuberger and K. Brocklehurst), Elsevier Science Public, BV, pp. 39–152

Cabot, C. (1989) Amélioration génétique de l'ananas. II. Objectifs du programme de création varieale entrepris en Cote d'Ivoire et techniques utilisées pour sa réalisation. *Fruits*, **44**, 183–191

Cappellini, R.A., Ceponis, M.J., and Lightner, G.W. (1988) Disorders in apricot and papaya shipments to the New York Market, 1972–1985. *Plant Disease*, **72**, 366–368

Carreno, R. and Chan, H.T. (1982) Partial purification and characterization of an acid phosphatase from papaya. *Journal of Food Science*, **47**, 1498–1500

Chan, H.T. (1986) Effects of heat treatments on the ethylene forming enzyme system in papayas. *Journal of Food Science*, **51**, 581–583

Chan, H.T. (1988) Alleviation of chilling injury in papaya. *HortScience*, **23**, 868–870

Chan, H.T., Chang, T.S.K., Stafford, A.E. and Brekke, J.E. (1971) Nonvolatile acids of papaya. *Journal of Agricultural and Food Chemistry*, **19**, 263–265

Chan, H.T., Chenchin, E. and Vonnahme, P. (1973a) Nonvolatile acids in pineapple juice. *Journal of Agricultural and Food Chemistry*, **21**, 208–210

Chan, H.T., Flath, R.A., Forrey, R.R., Cavaletto, C.G., Nakayama, T.O.M. and Brekke, J.E. (1973b) Development of off-odors and off-flavors in papaya puree. *Journal of Agricultural and Food Chemistry*, **21**, 566–570

Chan, H.T., Coney, H.M. and Sakai, W.S. (1990) Distribution of the ethylene forming enzyme in ripening *Carica papaya*. Proceedings of 7th World Congress of Food Science, Singapore. In *Trends in Food Processing* (eds A.H. Ghee, N. Lodge and O.K. Lian), Singapore Institute of Food Science and Technology

Chan, H.T., Hibbard, K.L., Goo, T. and Akamine, E.K. (1979) Sugar composition of papayas during fruit development. *HortScience*, **14**, 140–141

Chan, H.T. and Kwok, S.C.M. (1976) Isolation and characterization of β-fructofuranosidase from papaya. *Journal of Food Science*, **41**, 320–323

Chan, H.T., Sanxter, S. and Couey, H.M. (1985) Electrolyte leakage and ethylene production induced by chilling injury of papayas. *HortScience*, **20**, 1070–1072

Chan, H.T. and Tam, S.Y.T. (1982) Partial separation and characterization of papaya endo- and exo-polygalacturonase. *Journal of Food Science*, **47**, 1478–1483

Chan, H.T., Tam, S.Y.T. and Koide, R.T. (1978) Isolation and characterisation of catalase from papaya. *Journal of Food Science*, **43**, 989–991

Chan, H.T., Tam, S.Y.T. and Seo, S.T. (1981). Papaya polygalacturonase and its role in thermally injured ripening fruit. *Journal of Food Science*, **46**, 190–191,197

Chan, Y.K. (1986) Differential compatibility in a diallel cross involving three groups of pineapple [*Ananas comosus* L. (Merr.)]. *MARDI Research Bulletin*, **14 (1)**, 23–27

Chang, L.W.S., Morita, L.L. and Yamamoto, H.Y. (1965) Papaya pectinesterase inhibition by sucrose. *Journal of Food Science*, **30**, 218–222

Chen, N.K.L. (1964) Some chemical changes during the post-harvest ripening of papaya fruit. *Bot. Bull. Acad. Sinica*, **V**, 89–99

Chen, N.M. and Paull, R.E. (1986) Development and prevention of chilling injury in papaya fruit. *Journal American Society of Horticultural Science*, **111**, 639–643

Chenchin, K.L. and Yamamoto, H.Y. (1978) Isolation characterization and enzymic hydrolysis of pineapple gum. *Journal of Food Science*, **43**, 1261–1263

Chenchin, K., Yugawa, A. and Yamamoto, H.Y. (1984) Enzymic degumming of pineapple and pineapple mill juices. *Journal of Food Science*, **49**, 1327–1329

Chittiraichelvan, R. and Shanmugavelu, K.G. (1979) A study on the correlation of fruit weight and volume with seed weight and number in Co. 2 papaya (*Carica papaya* L.) *Indian Journal of Horticulture*, **35**, 222–224

Collins, J.L. (1968). *The pineapple*. Leonard Hill, London.

Couey, H.M. and Hayes, C.F. (1986) Quarantine procedure for Hawaiian papaya using fruit selection and a two stage hot-water immersion. *Journal of Economics and Entomology*, **79**, 1307–1314

Crochon, M., Tisseau, R., Teisson, C. and Huet, R. (1981). Effet d'une application d'Ethrel avant la récolte sur la qualité gustative de ananas de Côte d'Ivoire. *Fruits*, **36**, 409–415

Da Silva, E., Lourenco, E.J. and Neves, V.A. (1990) Soluble and bound peroxidases from papaya fruit. *Phytochemistry*, **29**, 1051–1056

de Arriola, M.C., de Madrid, M.C. and Rolz, C. (1975) Algunos cambios fisicos y quimicos de la papaya durante su almacenamiento. *Proceedings of Tropical Region of the American Society of Horticultural Science*, **19**, 97–109

de Arriola, M.C., Calzada, J.F., Menchu, J.F., Rolz, C. and Garcia, R. (1980) Papaya. In *Tropical and subtropical fruits*, (eds S. Nagy and P.E. Shaw), AVI Publishing Inc., Westport, Conn., pp. 316–340

Dull, G.G. (1971) The pineapple: general. In *The biochemistry of fruits and their products*, Vol. 2, (ed A.C. Hulme) Academic Press, London. pp. 303–331

Dull, G.C., Young, R.E. and Biale, J.B. (1967) Respiratory patterns in fruit of pineapple, *Ananas comosus* detached at different stages of development. *Physiologia Plantarum*, **20**, 1059–1065

Dupaigne, P. (1970) L'arome de l'ananas. *Fruits*, **25**, 793–805

El-Tomi, A.L., Abou Aziz, A.B., Abdel-Kader, A.S. and Abdel-Wahab, F.K. (1974) The effect of chilling and non-chilling temperatures on the quality of papaya fruit. *Egyptian Journal of Horticulture*, **1**, 179–185

Everett, P. (1952) The Mountain Paw Paw: a giant herbaceous plant. *New Zealand Journal of Agriculture*, **84**, 12

Flath, R.A. (1986). Pineapple. In *Tropical and subtropical fruits: composition, properties and uses* (eds. S. Nagy and P.E. Shaw) AVI Publishing, Inc., Westport, Conn., pp. 157–183

Flath, R.A. and Forrey, R.R. (1970) Volatile components of Smooth Cayenne pineapple. *Journal of Agricultural and Food Chemistry*, **18**, 306–309

Flath, R.A. and Forrey, R.R. (1977) Volatile components of papaya (*Carica papaya* L. Solo variety). *Journal of Agricultural and Food Chemistry*, **25**, 103–109

Flath, R.A., Light, D.M., Jang, E.B., Mon T.R. and John, J.O. (1990) Headspace examination of volatile emissions form ripening papaya (*Carica papaya* L., Solo variety). *Journal of Agricultural and Food Chemistry*, **38**, 1060–1063

Forbus, W.R. and Chan, H.T. (1989) Delayed light emission as a means of predicting papaya susceptibility to fruit fly infestation. *Journal American Society Horticulture Science*, **114**, 521–525

Forbus, W.R., Seuter, S.D. and Chan, H.T. (1987) Measurement of papaya maturity by delayed light emission. *Journal of Food Science*, **52**, 356–360

Gawler, J.H. (1962) Constituents of canned Malayan pineapple juices. I. Amino acids, non-volatile acids, sugars, volatile carbonyl compounds and volatile acids. *Journal of the Science of Food and Agriculture*, **13**, 57–61

Gortner, W.A. (1965) Chemical and physical development of the pineapple fruit. IV. Plant pigment constituents. *Journal of Food Science*, **30**, 30–32

Gortner, W.A. and Singleton, V.L. (1965) Chemical and physical development of the pineapple fruit. III. Nitrogenous and enzyme constituents. *Journal of Food Science*, **30**, 24–29

Haagen-Smit, A.J., Kirchner, J.G., Prater, A.N. and Deasy, C.L. (1945) Chemical studies of pineapple (*Ananas sativa* Lindl.) I. The volatile flavor and odor constituents of pineapple. *Journal of the American Chemical Society*, **67**, 1646–1652

Hamner, K.C. and Nightingale, G.T. (1946) Ascorbic acid content of pineapples as correlated with environmental factors and plant composition. *Food Research*, **11**, 535–541

Hayes, C.F. and Chignon, H.T.G. (1982) Acoustic properties of papaya. *Journal of Texture Studies*, **13**, 397–402

Heidlas, J., Lehr, M., Idstein, H. and Schreier, P. (1984) Free and bound terpene compounds in papaya (*Carica papaya*, L.) fruit pulp. *Journal of Agricultural and Food Chemistry*, **32**, 1020–1021

Huet, R. (1958) La composition chimique de l'ananas. *Fruits*, **13**, 183–197

Idstein, H., Keller, T. and Schreier, P. (1985) Volatile constituents of Mountain Papaya (*Carica candamarencis*, syn c. pubescens Lenne et Koch) Fruit. *Journal of Agricultural and Food Chemistry*, **33**, 663–666

Jones, W.S. and Kubota, H. (1940) Some respirational changes in the papaya fruit during ripening and the effects of cold storage on these changes. *Plant Physiology*, **15**, 711–717

Kamphuis, I.G., Kalk, K.H., Swarte, M.B.A. and Drenth, J. (1984) Structure of papain refined at 1.65 A resolution. *Journal of Molecular Biology*, **179**, 233–256

Katague, D.B. and Kirch, E.R.(1965) Chromatographic analysis of volatile components of papaya fruits. *Journal of Pharmaceutical Science*, **54**, 891–894

Keetch, D.P. (1978) Bruising of pineapples. Farming in South Africa, Bulletin H11/1978

Keetch D.P. and Balldorf, D.B. (1979) The incidence of certain pineapple fruit blemishes in the eastern cape and border. *Citrus & Subtropical Fruit Journal*, **551**, 12–15

Kelly, W.P. (1911) A study of the composition of Hawaiian pineapples. *Journal of Industrial Engineering and Chemistry*, **3**, 403–405

Kuhne, F.A. and Allan, P. (1970) Seasonal variation in fruit growth of *Carica papaya* L. *Agroplantae*, **2**, 99–104

Lazan, H., Mohd. Ali, Z., Liang, K.S. and Yee, K.L. (1991) Polygalacturonase activity and variation in ripening of papaya fruit with tissue depth and heat treatment. *Physiologia Plantarum*, (in press)

Lodh, S.B., Selvaraj, Y., Chadha, K.L., and Melanta, K.R. (1972) Biochemical changes associated with growth and development of pineapple fruit variety Kew II. Changes in carbohydrate and mineral constituents. *Indian Journal of Horticulture*, **29**, 287–291

Lodh, S.B., Divakar, N.G., Chadha, K.L., Melanta, K.R. and Selvaraj, Y. (1973) Biochemical changes associated with growth and development of pineapple fruit variety kew. III. Changes in plant pigments and enzyme activity. *Indian Journal of Horticulture*, **30**, 381–383

Lopez, M.E., Vattuone, M.A. and Sampietro, A.R. (1988) Partial purification and properties of invertase from *Carica papaya* fruits. *Phytochemistry*, **27**, 3077–3081

Lourenco, E.J. and Catutani, A.T.(1984) Purification and properties of pectin-esterase from papaya. *Journal of the Science of Food and Agriculture*, **35**, 1120–1127

Lowe, G. (1976) The cysteine proteinases. *Tetrahedron*, **32**, 291–302

MacLeod, A.J. and Pieris, N.M. (1983) Volatile components of papaya (*Carica papaya* L.) with particular reference to glucosinolate products. *Journal of Agricultural and Food Chemistry*, **31**, 1005–1008

Menary, R.C. and Jones, R.H. (1974). Nitrate accumulation and reduction in papaw fruits. *Australian Journal of Biological Sciences*, **25**, 531–542

Miller, E.V. and Hall, G.D. (1953) Distribution of total soluble solids, ascorbic acid, total acid, and bromelain activity in the fruit of the natal pineapple (*Ananas comosus* L. Merr.). *Plant Physiology*, **28**, 532–534

Mitchel, R.E., Chaiken, I.M. and Smith, E.L. (1970) The complete amino acid sequence of papain. *Journal of Biological Chemistry*, **245**, 3485–3492

Morales, A.L. and Duque, C. (1987) Aroma constituents of the fruit of the Mountain papaya (*Carica pubescens*) from Colombia. *Journal of Agricultural and Food Chemistry*, **35**, 538–340

Nakasone, H.Y. (1986) Papaya. In *CRC Handbook of fruit set and development*, (ed. S.P. Monoelise), CRC Press, Boca Raton, Florida, pp. 277–301

Nazeeb, M. and Broughton, W.J. (1978) Storage conditions and ripening of papaya 'Bentong' and 'Taiping'. *Scientia Horticulturae*, **9**, 265–277

Nishijima, W.T., Eberosle, S. and Fernandez, J.A. (1990) Factors influencing development of postharvest incidence of *Rhizopus* soft rot of papaya. *Acta Horticulturae*, **269**, 495–502

Ogata, J.N., Kawano, Y., Bevenue, A. and Casarett, L.J. (1972) The ketoheptose content of some tropical fruits. *Journal of Agricultural and Food Chemistry*, **20**, 113–115

Okimoto, M.C. (1948) Anatomy and histology of the pinapple inflorescence and fruit. *Botanical Gazette*, **110**, 217–231

Ong, H.T. (1983) Abortion during the floral-fruit development in *Carica papaya* in Serdang, Malaysia. *Pertanika*, **6**, 105–107

Orr, K.J., Dennings, H. and Miller, C.D. (1953) The sugar and ascorbic acid content of papaya in relation to fruit quality. *Food Research*, **18**, 532–537

Ota, S., Horie, K., Hagino, F., Hashimoto, C. and Date, H. (1972) Fractionation and some properties of the proteolytically active components of bromelains in the stem and the fruit of the pineapple plant. *Journal of Biochemistry*, **71**, 817–830

Pal, D.K. and Selvaraj, Y. (1987) Biochemistry of papaya (*Carica papaya* L.) fruit ripening: changes in RNA, DNA, protein and enzymes of mitochondrial, carbohydrate, respiratory and phosphate metabolism. *Journal of Horticultural Science*, **62**, 117–124

Patil, S.S. and Tang, C.S. (1974) Inhibition of ethylene evolution in papaya pulp tissue by benzyl isothiocyanate. *Plant Physiology*, **53**, 585–588

Patil, S.S., Tang, C.S. and Hunter, J.E. (1973) Effect of benzyl isothiocyanate treatment on the development of postharvest rots in papaya. *Plant Disease and Reproduction*, **57**, 86–89

Paull, R.E. and Chen, N.J. (1983) Postharvest variation in cell wall-degrading enzymes of papaya (*Carica papaya* L.) during fruit ripening. *Plant Physiology*, **72**, 382–385

Paull, R.E. and Chen, N.J. (1989) Waxing and plastic wraps influence water loss from papaya fruit during storage and ripening. *Journal of the American Society of Horticultural Science*, **114**, 937–942

Paull, R.E. and Chen, N.J. (1990) Heatshock response in field-grown ripening papaya fruit. *Journal of the American Society Horticultural Science*, **115**, 623–631

Paull, R.E. and Rohrbach, K.G. (1982) Juice characteristics and internal atmosphere of waxed 'Smooth Cayenne,' pineapple fruit. *Journal of the American Society of Horticultural Science*, **107**, 448–452

Paull, R.E. and Rohrbach, K.G. (1985) Symptom development of chilling injury in pineapple fruit. *Journal of the American Horticultural Science*, **110**, 100–105

Poignant, A. (1971) La maturation controlée de l'ananas II – L'ethrel et son action au cours des phases ascendante et descendante de la maturité. *Fruits*, **26**, 23–35

Purseglove, J.W. (1968) *Tropical crops: Dicotyledons.*, Vol. 1, J. Wiley & Sons, Inc. New York, pp. 45–51

Reddy, M.N., Keim, P.S., Heinrickson, R.L. and Kezdy, F.J. (1975) Primary structural analaysis of sufhydryl protease inhibitors from pineapple stem. *Journal of Biological Chemistry*, **250**, 1741–1750

Robinson, G.W. (1975) Isolation and characterization of papaya peptidase A from commercial chymopapain. *Biochemistry*, **14**, 3695–3700

Rohrbach, K.G. and Apt, W.J. (1986) Nematode and disease problems of pineapples. *Plant Disease*, **70**, 81–87

Rohrbach, K.G. and Paull, R.E. (1982) Incidence and severity of chilling induced browning of waxed 'Smooth Cayenne' pineapple. *Journal of the American Society Horticultural Science*, **107**, 453–457

Roth, I. and Clausnitzer, I. (1972) Desanollo y anatomica del fruto y de la semella de *Carica papaya* L. (Lechosc). *Acta Botanica Venezuela*, **7**, 187–206

Samuels, G. (1970) Pineapple cultivars 1970. *Proceedings of Tropical Regions of American Society of Horticultural Science*, **14**, 13–24

Sanxter, S.R. (1990) Ontogeny and senescence of photosynthetic activity in the exocarp of *Carica papaya* L. M.Sc. Thesis, University of Hawaii, 110 pp

Sasaki, M., Kato, T. and Iida, S. (1973) Antigenic determinant common to four kinds of thiol proteases of plant origin. *Journal of Biochemistry*, **74**, 635–637

Sawato, M. (1969) Changes in isozyme pattern and kinetics of heat inactivation of peroxidasic enzyme of papaya following gamma irradiation. University of Hawaii Master's Thesis

Schaffner, J.H. (1935) Artificial parthenocarpy. *Journal of Heredity* **26**, 261–262

Schreier, P., Lehr, M., Heidlas, J. and Idstein, H. (1985) Volatiles from papaya (*Carica papaya* L.) fruit: Indication of precusors of terpene compounds. *A. Lebensm.-Unters. Forsch.*, **180**, 297–303

Selvaraj, Y., Divakar, N.G., Subhas Chandler, M., Chadha, K.L. and Melanta, K.L. (1975) Biochemical changes associated with growth and development of pineapple variety Kew. IV. Changes in major lipid constituents. *Indian Journal of Horticulture*, **32**, 64–67

Selvaraj, Y., Pal, D.K., Subramanyam, M.D. and Iyer, C.P.A. (1982a) Fruit set and the developmental pattern of fruits of five papaya varieties. *Indian Journal of Horticulture*, **39**, 50–56

Selvaraj, Y., Pal, D.K., Subramanyam, M.D. and Iyer, C.P.A. (1982b) Changes in the chemical composition of four cultivars of papaya (*Carica papaya* L.) during growth and development. *Journal of Horticultural Science*, **57**, 135–143

Senanayake, Y.D.A. and Gunasena, H.P.M. (1975) A study on the influence of crown leaves on fruit growth of pineapple *Ananas comosus* (L.) Merr. cv. Kew. *Journal of National Agricultural Society of Ceylon*, **12**, 106–114

Seo, S.T. and Tang, C.S. (1982) Hawaiian fruit flies (*Diptera: Taphritidae*): toxicity of benzyl isothiocyanate against eggs of 1st instars of three species. *Journal of Economics and Entomology*, **75**, 1132–1135

Shaw, G.J., Allen, J.M. and Visser, F.R. (1985) Volatile flavor components of babaco fruit (*Carica pentagonia*, Heiborn.). *Journal of Agricultural and Food Chemistry*, **33**, 795–797

Shetty, S.R. and Dubash, P.J. (1974) Relationship of pectin content of papaya fruit to its firmness and maturity. *Indian Food Packer*, **28 (2)**, 14–16

Sideris C.P. and Krauss, B.H. (1938) Growth phenomena of pineapple fruits. *Growth*, **2**, 181–196

Silverstein, R.M. (1971) The pineapple: Flavour. In *The biochemistry of fruits and their products*, vol. 2, (ed. A.C. Hulme), Academic Press, London, pp. 325–331

Singleton, V.L. (1957) The nature of the separation produced by floatation of Cayenne pineapple fruits. *Pineapple Research Institute News,,* **5**, 1–10 (Private document)

Singleton, V.L. (1965) Chemical and physical development of the pineapple fruit. I. Weight per fruitlet and other physical attributes. *Journal of Food Science*, **30**, 98–104

Singleton, V.L. and Gortner, W.A. (1965) Chemical and physical development of pineapple fruit III. Nitrogenous and enzyme constituents. *Journal of Food Science*, **30**, 24–29

Smith, L.G. (1984) Pineapple specific gravity as an index of eating quality. *Tropical Agriculture*, **61**, 196–199

Somner, N.F. and Mitchell, F.G. (1978) Relation of chilling temperatures to postharvest *Alternia* rot of papaya fruit. *Proceedings of Tropical Regions of American Society of Horticultural Science*, **22**, 40–47

Storey, W.B. (1967) Theory of the derivation of the unisexual flower of *Cariaceae*. *Agron. Trop. (Maracay, Venez.)*, 273–321

Sun, S-K. (1971) A study of black heart disease of the pineapple fruits. *Plant Protection Bulletin, Taiwan*, **13**, 39–48

Takeoka, G., Battery, R.G., Flath, R.A. Teranishi, R. Wheeler, E.L. Wieczorek, R.G. and Guentert M. (1989). Volatile constituents of pineapple (*Ananas comosus* [L] Merr). In *Flavor Chemistry: Trends and Development*. (eds R. Teranishi, R.G. Battery and F. Shahidi), American Chemical Society, Washington D.C., pp. 223–237

Tan, S.C. and Lam, P.F. (1985) Effect of gamma irradiation of PAL activity and phenolic compounds in papaya (*Carica papaya* L.) and mango (*Mangifera indica* L.) fruits. *ASEAN Food Journal*, **1**, 134–136

Tan, S.C., Teo, S.W. and Abd Ghani, A. (1982) Factors affecting fungal resistance in papaya fruit. *Sains Malaysiana*, **11**, 21–31

Tan, S.C. and Weinheimer, E.A. (1976) The isozyme patterns of developing fruit and mature leaf of papaya (*Carica papaya*, L.). *Sains Malaysiana*, **5**, 7–14

Tang, C.S. (1971) Benzyl isothiocyanate of papaya fruit. *Phytochemistry*, **10**, 117–121

Tang, C.S. (1973) Localization of benzyl glucosinolate and thioglucosidase in *Carica papaya* fruit. *Phytochemistry*, **12**, 769–773

Tang, C.S. (1974) Benzyl isothiocyanate as a naturally occurring papain inhibitor. *Journal of Food Science*, **39**, 94–96

Tay, T.H. (1977) Fruit ripening studies on pineapple. *MARDI Research Bulletin*, **4**, 29–34

Teisson, C. (1977) Le brunissement interne de l'ananas. Docteur dies sciences naturelles these, Presentée à la Faculté des Sciences de l'Université d'Abidjan, 184 pp

Teisson, C. and Combres, J.C. (1979) Le brunissement interne de l'ananas. III. Symptomatologie. *Fruits*, **34**, 315–339

Teisson, C., Martin-Prevel, P. and Marchal, J. (1979a) Le brunissement interne de l'ananas. IV. Approache biochimique du phenomene. *Fruits*, **34**, 315–339

Teisson, C., Lacoeuilhe, J.J. and Combres, J.C. (1979b) Les brunissement interne de l'ananas. V. Recherches des moyen de lutte. *Fruits*, **34**, 399–415

Teisson, C. and Pineau, P. (1982) Quelques donneés sur les dernières phases du developpement de l'ananas. *Fruits*, **37**, 741–748

Thomas, C. and Beyers, M. (1979) Gamma irradiation of subtropical fruits. III. A comparison of the chemical changes occurring during normal ripening of mangoes and papaya with changes produced by gamma irradiation. *Journal of Agricultural and Food Chemistry*, **27**, 157–163

Thompson, A.K. and Lee, G.R. (1971) Factors affecting the storage behaviour of papaya fruit. *Journal of Horticultural Science*, **46**, 511–516

Traub, H.P., Robinson, T.R. and Stevens, H.E. (1935) Latex test for maturity of papaya fruit. *Science*, **82**, 569–570

van Lelyveld, L.J. and de Bruyn, J.A. (1977) Polyphenols, ascorbic acid and related enzyme activities associated with black heart in Cayenne pineapple fruit. *Agrochemophysica*, **9**, 1–6

Wardlaw, C.W. (1936) Studies in tropical fruits. II. Observations on internal gas concentrations in fruits. *Annals of Botany*, **50**, 655–676

Wardlaw, C.W. (1937) Tropical fruits and vegetables. An account of their storage and transport. *Tropical Agriculture*, **24**, 288–298

Wardlaw, C.W. and Leonard, E.R. (1938) Studies in tropical fruits. III. Preliminary observations on pneumatic pressures in fruit. *Annals of Botany NS*, **2**, 301–315

Wills, R.B.H., Lim, J.S.K. and Greenfield, H. (1986) Composition of Australian foods. 31. Tropical and Sub-Tropical fruit. *Food Technology of Australia*, **38**, 118–123

Wu, P., Kuo, M.C., Hartman, T.G., Rosen, R.T. and Ho, C.T. (1991) Free and glycosidically bound aroma compounds in pineapple (*Ananas comosus* L. Merr.) *Journal of Agricultural and Food Chemistry*, **39**, 170–172

Yamada, F., Takahashi, N., and Murachi, T. (1976) Purification and characterization of a proteinase from pineapple fruit, fruit bromelain FA2. *Journal of Biochemistry*, **79**, 1223–1234

Yamamoto, H.Y. (1964) Comparison of the carotenoids in yellow and red-fleshed *Carica papaya*. *Nature*, **201**, 1049–1050

Zhang, L.X. and Paull, R.E. (1990) Ripening behavior of papaya genotype. *HortScience*, **25**, 454–455

Chapter 11

Pome fruits

M. Knee

11.1 INTRODUCTION

The pome fruits, apples (*Malus pumila*) and pears (*Pyrus communis*) constitute one of the four or five major classes of fruit in terms of world production and is one of the two major groups of temperate fruits (the other being grapes). The pome group also includes quince (*Cydonia oblonga*), oriental pear (*Pyrus serotina*), medlar (*Mespilus germanica*), and many other wild species in the Rosaceae.

The edible portion of a pome fruit develops through the enlargement of the receptacle which encloses and fuses with the ovary. Among a huge number of cultivars of apple and pear, commercial production is restricted to about ten important apple and about five pear varieties, worldwide; these fruit bearing 'scions' are always grafted onto a genetically distinct rootstock. Although there is a large pool of variation, breeding has made slow progress and crop performance has been advanced through intensive management of all aspects of production. This involves physical operations, such as pruning, hand thinning and harvest, and chemical treatments such as nutrients, growth regulators, pesticides and fungicides; thus these fruits may receive more chemical inputs than many other crops. As with other crops, fertilizers increase yields, but high use of nitrogen and potassium leads to fruit with poor keeping quality and fertilizer inputs have been drastically reduced (Atkinson *et al.*, 1980). The increased use of herbicides for control of grass and weeds has also altered the nutritional balance of orchard systems (Hipps and Perring, 1989).

Apples and pears are harvested commercially before they become ripe for eating. Fear of loss through abscission, senescence and pathogen attack encourages growers to harvest immature fruit. Pears deteriorate

Biochemistry of Fruit Ripening. Edited by G. Seymour, J. Taylor and G. Tucker. Published in 1993 by Chapman & Hall, London. ISBN 0 412 40830 9

rapidly after the onset of ripening, but apples are commonly harvested just as they begin to ripen. After many years of research on maturity indicators there is no generally agreed method to define the best time of harvest. Measurements of colour, sugar content, flesh firmness and staining of starch by iodine are all in use (Knee *et al.*, 1989). Machine harvesting has been another intensively researched topic, but hand harvesting still accounts for most of the labour input in apple and pear production; indeed, the availability of appropriately priced labour can be a decisive factor in the profitability of fruit growing in a particular area.

In all major apple and pear growing areas the bulk of the crop is stored for several months to maintain availability for as long as economically desirable. For many pear varieties refrigerated storage at $-1°C$ is so effective in delaying ripening that no other treatment is necessary. Apples are more likely to be injured at this temperature and are usually stored at $0°$ to $4°C$. Long term storage of apples nearly always involves 'controlled atmospheres' (CA), high carbon dioxide or low oxygen, or a combination of the two. CA conditions are achieved in a sealed room by fruit respiration or by flushing with gases. The atmospheres retard fruit ripening and delay death through senescence or pathogen attack. However, fruit may be damaged when there is insufficient oxygen to sustain aerobic respiration, or by the toxicity of carbon dioxide. The technology of fruit storage has become complex and expensive, and pome fruits account for most of the applications of this technology. If apples could be stored at lower temperatures, or if the ripening of the fruit could be manipulated at the gene level the costs of CA could be avoided. Pome fruit storage usually also involves some chemical treatments. Fungicides are applied to control decay initiated at sites of mechanical damage or by latent infections from the field (Swinburne, 1983). Many apple varieties develop non-pathogenic lesions on the skin during storage. These 'scald' and spot disorders are controlled by anti-oxidants, such as diphenylamine and ethoxyquin (Meigh, 1970). Since the biochemistry of scald is at least partly understood it may be a more realistic objective for control at the gene level than more fundamental changes in storage technology.

Some apple cultivars are grown specifically for processing, but low grade apples rejected for fresh consumption are also used. Fresh fruit may be used directly in pie and sauce manufacture, or apples may be partially processed and kept for future manufacture of foods. Sliced apple may be frozen, thermally processed in cans, or treated with sulphite and held in refrigeration. Other major processing operations involve extraction of juice, for direct consumption, or for alcoholic fermentation. Attaining maximum yield and stable clarity of juice are major concerns of the industry. A limited trade in apple pectin for various food uses is a by-product of juice extraction.

11.2 PHYSIOLOGY

11.2.1 Physical properties

Cell separation occurs during the growth of apples so that about 25% of the volume of a mature fruit is air space between cells (Khan and Vincent, 1990). Low cell-to-cell contact area (Vincent 1989) may contribute to the osmotic fragility of the cells (Simon, 1977) and may be related to the low calcium content of apples. Junction zones in the middle lamella should be a major site of calcium deposition; pears have higher calcium contents than apples and contain less than 5% air space. In apples the air spaces tend to form radial 'canals' through the cortex (Khan and Vincent, 1990) and they continue to increase in volume during fruit storage and ripening, so that the whole fruit increases in volume (Hatfield and Knee, 1988). This increase in air space in apples may account for the decline in flesh firmness during growth and this can be difficult to separate from ripening-related changes in fruit texture. However, if apples are harvested in the preclimacteric state, further softening is associated with the onset of the climacteric rise. Cell separation proceeds in many apples until they attain a dry or 'mealy' texture, at which stage the consumer's teeth pass between cells, without breaking them to release juice.

The major pathway for gas diffusion into pome fruits is through the lenticels, which are scattered over the fruit surface. The skin seems to be the main barrier to diffusion of gases to or from cortical cells (Burg and Burg, 1965; Cameron and Yang, 1982; Knee, 1991a). Diffusive resistance declines during the later stages of apple fruit development (Knee *et al.*, 1990) but increases when apples are kept in a dry atmosphere (Hatfield and Knee, 1988). The gradient between internal and external concentrations of gases must be remembered whenever the relation between (external) ethylene, oxygen or carbon dioxide and fruit physiology is being studied.

11.2.2 Respiration climacteric

Apples and pears are both climacteric fruit and indeed the climacteric pattern of respiration was first described for apples (Kidd and West, 1924). Carbon dioxide production and oxygen uptake increase by 50 to 100% during ripening, apparently without any change in respiratory quotient (Fidler and North, 1967). Ethylene production increases about 1000-fold and, at the same time as the respiratory rise (Reid *et al.*, 1973). Softening, colour change from green to yellow, formation of cuticular waxes and synthesis of aroma compounds all seem to be associated with the climacteric.

Removing fruit from the tree is often said to advance the onset of the climacteric. Defoliation and girdling of spurs promote a rise in ethylene in attached apple fruits (Sfakiotakis and Dilley, 1973) and it seems that leaves produce 'tree factors' which inhibit ripening. However, if individual apples are separated to avoid cross diffusion of ethylene, detachment may have little effect (Knee *et al.*, 1983a); furthermore, apples on the tree seem to be as sensitive to ethylene as detached fruit (Knee, 1989). Ethylene synthesis rises after harvest of apple fruits at almost any stage of development. In the first month after flowering, treatment with ethylene has no effect on the time of rise, but the rise can be advanced by ethylene treatment of later-harvested fruits. As the natural onset of the climacteric is approached, lower concentrations of ethylene will advance the rise in ethylene synthesis (Knee *et al.*, 1987).

Refrigeration below 10°C suppresses the respiration climacteric in both apples and pears (Fidler and North, 1967), but advances the onset of ethylene production in most pear cultivars and some apples (Knee *et al.*, 1983b). Other ripening processes may be slow under these conditions (or non-existent in pears at –1°C), but a chilling treatment may help to promote rapid and uniform ripening on return to a higher temperature. Some pear cultivars (e.g. 'Passe Crassane') require chilling for ripening to occur (Ulrich, 1961); ethylene treatment can substitute for chilling, but may not yield such good quality fruit. Sometimes fruits can experience enough chilling during growth on the tree to initiate ripening and this may lead to problems, especially in pear storage (Wang *et al.*, 1972). It is difficult to detect effects of ethylene on apples or pears in refrigerated storage and this led to the view that ethylene has no role in ripening under these conditions. Ethylene may often be non-limiting under storage conditions, but this does not exclude its role as the immediate trigger of fruit ripening (Knee, 1985a).

In many commercial cultivars of apple, ethylene production will accelerate under almost any conditions of storage within 10 days after harvest. This may occur because a population including some ethylene-producing fruits is enclosed with limited ventilation. However, even if apples are isolated with their own air supply or if ethylene is removed from the atmosphere in bulk storage, the time of rise is delayed only by a few days for most cultivars. Thus commercial storage controls the progress of ripening rather than its initiation. Fruit cultivars with a chilling requirement for induction of ethylene synthesis can remain preclimacteric for several weeks at 20°C (Knee *et al.*, 1983b); this has not been exploited commercially because it is very difficult to avoid chance exposure to ethylene which would induce ripening at 20°C. In a few other cultivars the onset of ethylene production can be long delayed (Knee, 1985a). The best example is 'Gloster 69', which requires prolonged treatment with high concentrations of ethylene to induce the onset of rapid ethylene production. It can be stored in the preclimacteric state for several months (Knee and Tsantili, 1988).

11.3   BIOCHEMISTRY

11.3.1   Compositional changes

Carbohydrates and organic acids

Carbohydrate is transported to developing pome fruits as sorbitol and converted mainly to fructose and starch with some glucose and sucrose (Berüter, 1985). Some sorbitol persists throughout the life of the fruit and it can even increase when apples are held at 0°C (Fidler and North, 1968). Starch hydrolysis usually begins in the later stages of fruit growth but before the onset of the climacteric; this contributes to a further increase in free sugars (Berüter, 1985; Knee *et al.*, 1989). Later, sucrose is slowly hydrolysed to form more glucose and fructose (Whiting, 1970).

In apples and pears the major organic acid is malic acid; some apples contain appreciable amounts of citric acid and quinic acid is a major component of pears (Ulrich, 1970). Malic acid is metabolized to a greater extent than the others and may fall by 50% during the life of a fruit. It is a major substrate of respiration and this accounts for the respiratory quotient of 1.1 or higher which is typical of these fruits (Fidler and North, 1967).

Cell walls

Apple fruit cell walls consist mainly of cellulose and pectin, with some hemicellulose and a very small amount of extensin (Knee and Bartley, 1981). Hemicelluloses purified from apples share compositional and structural features with xyloglucans isolated from other primary cell walls (Knee, 1973a; Voragen *et al.*, 1986). Compositional changes during ripening are restricted to the pectic polymers and there is no evidence of changes in the cellulose or hemicellulose (Bartley, 1976).

According to one account the pectin consists of two separate polymers, a rhamnogalacturonan and a homogalacturonan, and these are both at least 70% methyl-esterified. The rhamnogalacturonan carries side chains of arabinose and galactose residues and may constitute much of the primary wall matrix; the homogalacturonan may form the middle lamella (Knee and Bartley, 1981). Alternatively de Vries *et al.* (1982) concluded that rhamnogalacturonan and homogalacturonan represented regions of a single polymer. The galactose side chains of the rhamno-galacturonan are lost before the onset of ethylene production, when apples are left on the tree (Knee, 1973b) or in low oxygen storage (Bartley, 1977), but this loss occurs with other ripening changes in detached apples under normal storage conditions (Knee, 1973b). Homogalacturonan is solubilized during ripening; viscosity and sedimentation analysis of soluble and wall-bound pectin did not reveal major changes in their degree of polymerization during fruit ripening (Knee, 1978a). If the homogalacturonan and rhamnogalacturonan are linked in the cell wall, ripening should involve cleavage to liberate high molecular weight fragments. There is physico-chemical evidence for increased mobility of pectic carboxyl groups in fruit cell walls on ripening (Irwin *et al.*, 1984). In principle, this means only that some restraint on movement was lost, although the authors eliminated a number of possible explanations and concluded that the change resulted from cleavage of polygalacturonate.

Removal of calcium ions promotes cell separation in alcohol-extracted apple tissue, and solubilization of homogalacturonan from cell walls (Knee, 1978a). Infiltration with calcium salts in neutral buffers reverses the early stages of softening in pears, and much of the softening in apples after prolonged storage (Knee, 1982c; Stow, 1989). Removal of calcium from junction zones between pectin molecules in the middle lamella is a possible mechanism of cell separation and fruit softening, but there is no direct evidence that it occurs.

The cell walls of pears have a similar monomeric composition to those of apples (Ahmed and Labavitch, 1980a), except that the 'stone cells' of pears are lignified. Pears are a classic source of xylan (Chanda *et al.*, 1951); however, the available information does not exclude the presence of this hemicellulose in apples or the presence of xyloglucan in pears. Like apples, pears contain homogalacturonan and rhamnogalacturonan, and these do not seem to be linked covalently (Dick and Labavitch, 1989). Compositional changes are restricted to the pectic polymers. In contrast to the situation in apple, the rhamnogalacturonan carries less galactosyl residues and these do not change during ripening. Fragments representing both the homogalacturonan and the rhamnogalacturonan increase in the soluble fraction of ripening pears (Ahmed and Labavitch, 1980a).

Incorporation of ^{14}C from methionine into pectin increases on ripening of both apples and pears (Knee, 1978b, 1982a). However, the galacturonosyl residue content of the cell walls remains constant or declines. The role of pectin synthesis in ripening is still an open question.

Pigments

The pigments of pome fruits consist of anthocyanins, chlorophylls and carotenoids. The main anthocyanin in apples is idaein (cyanidin-3-

galactoside), but the pigments in pears do not seem to have been identified (Timberlake, 1981). Anthocyanin synthesis occurs during fruit growth, whereas colour changes during ripening depend mainly on the simultaneous disappearance of chlorophylls a and b (Knee, 1972, 1980a). Throughout apple fruit development the carotenoids are represented by those typical of photosynthetic higher plant tissue, β-carotene, lutein, violaxanthin, neoxanthin and cryptoxanthin. Carotene declines during ripening but the xanthophylls, particularly lutein and violaxanthin increase substantially (Knee, 1972; Gross et $al.$, 1978); the newly-synthesized pigments accumulate as mono- and diesters, mainly with palmitate and oleate. The appearance of these esters is one of the earliest events in fruit ripening, and may precede the rise in ethylene synthesis associated with the climacteric (Knee, 1972, 1988a).

Lipids

Pigment changes are one indication of the development from a photo-synthetic chloroplast to a non-photosynthetic chromoplast. Another is the loss of galactolipid and its associated linolenyl moieties from the lipid fraction, since these are typical chloroplast membrane constituents (Galliard, 1968). Phospholipids and fatty acyl groups typical of other cell membranes remain constant or increase slightly during apple ripening. However, [^{14}C]acetate incorporation by apple cortex into various phospholipids increases approximately ten-fold from pre- to post-climacteric; this indicates that turnover of lipid increases, even though composition does not change (Bartley, 1985). The sterol:phospholipid ratio influences the fluidity of cell membranes, but in contrast with an earlier observation (Galliard, 1968), the sterol content of apple tissue was found not to change with ripening (Bartley, 1986).

In animal tissues lipid peroxidation during ageing is accompanied by formation of a fluorescent pigment, 'lipofuscin'. This pigment was reported to accumulate in ripening pears (Maguire and Haard, 1975); however, the extraction protocol was later shown to generate material with similar fluorescence properties from xanthophylls and there was no evidence of the presence of lipofuscin in post-climacteric apples (Knee, 1982b).

Hulme and Rhodes (1971) review the older work on the cuticular components of apples. Recent research has focused on the sesquiterpene hydrocarbon, α-farnesene, because of its suspected role in the development of scald. Traces of farnesene are present on the surface of pre-climacteric fruits and it increases rapidly on ripening. Damage to cell membranes probably occurs through free radical chain reactions during the spontaneous oxidation of farnesene to 'conjugated trienes' (Huelin and Coggiola, 1970). The initiation of these reactions may be influenced by the levels of 'natural antioxidants' which are also present in the surface wax (Sal'kova and Zvyagintseva, 1981; Meir and Bramlage, 1988).

Flavour compounds

Apple and pear flavours depend upon complex mixtures of organic compounds, many of which are synthesized during the climacteric phase. Simple aliphatic esters are quantitatively the most important compounds and are formed mainly from 2- to 6-carbon alcohols and acids. They are usually saturated and may include branched 4- and 5-carbon units. Butyl ethanoate, 2-methyl butyl ethanoate and hexyl ethanoate are typical constituents (Dimick and Hoskin, 1983). Traces of free alcohols corresponding to those in the esters are usually detectable in the vapour phase; they may be more prominent in the cell contents since their air–water partition coefficients are an order of magnitude lower than for the corresponding esters (Knee and Hatfield, 1976). The saturated aliphatic esters often contribute a generic fruity aroma. Other kinds of volatile compound occur in trace quantities or in a restricted range of cultivars; some of these compounds may be 'character impact compounds' for a particular cultivar. For example, alkyl 2,4-decadienoates occur in some pear cultivars (Jennings and Sevenants, 1964) and 4-methoxy(propenyl benzene) gives a spicy flavour to some apples (Williams *et al.*, 1977). Terpenoid compounds are represented among apple volatiles by linalool and its epoxide, as well as farnesene (Dimick and Hoskin, 1983). Acetaldehyde is produced by senescent apples but higher aldehydes may not be produced by intact fruit. Hexanal and trans-2-hexenal are formed on tissue disruption, these can be dominant compounds giving a 'green' flavour to immature fruits (Flath *et al.*, 1967).

11.3.2 Pathways and enzymes

Reserve carbohydrates

The acid invertase (EC 3.2.1.26) found in immature apple fruits is mainly wall-bound (Yamaki and Ishikawa, 1986), and would allow the tissue to utilize sucrose. In more mature fruits sorbitol dehydrogenase (EC 1.1.1.14) is predominant, allowing the utilization of the major translocated carbohydrate, sorbitol, for synthesis of the major sugar accumulated, fructose (Berüter, 1985; Yamaki and Ishikawa, 1986). Sorbitol oxidase has also been detected as a minor cell-wall bound activity in apples (Yamaki and Ishikawa, 1986).

A number of enzymes could be involved in the conversion of starch to sugars that occurs towards the end of the growth phase in apples and pears. β-Amylase (EC 3.2.1.2) activity is present throughout development of Passe Crassane pears, whereas α-amylase (EC 3.2.1.1) activity increases on ripening after most of the starch has disappeared (Latché et al., 1975). It is possible that starch phosphorylase (EC 2.4.1.1) which has been detected in apples (Clements, 1970), is more important in starch mobilization than the amylases (Preiss, 1982).

Organic acids

Organic acids could accumulate through metabolism of imported carbohydrate and amino acids and by CO_2 fixation, since apple fruits contain phosphoenol pyruvate (PEP) carboxylase (EC 4.1.1.31) activity (Blanke et al., 1987). Organic acids can also be metabolized. Slices of apple tissue decarboxylate malate to pyruvate with an NADP malic enzyme (ME, EC 1.1.1.40) (Dilley, 1966), pyruvate to acetaldehyde with pyruvic carboxylase (EC 4.1.1.1), and reduce the acetaldehyde to ethanol with alcohol dehydrogenase (ADH, EC 1.1.1.1) (Hulme and Rhodes, 1971). Since apple ADH accepts NADPH as a cofactor (Bartley and Hindley, 1980), NADP can be regenerated so that ME action can continue. The soluble ME is likely to be involved in malate metabolism in the intact fruit, but the pyruvate produced would be metabolized by pyruvic dehydrogenase (EC 1.2.4.1) rather than by pyruvic carboxylase. The NADPH produced by ME action could be used for synthetic purposes; alternatively it could be oxidized by the external NAD(P)H oxidase of the mitochondria (Lance, 1981). Overall, ME may provide a means whereby cells can maintain TCA cycle operation with an input of malate (Chapter 1). The metabolism of malate by tissue slices increases as apples pass through the climacteric (Hulme and Rhodes, 1971), so that there is a correlation between stimulation of CO_2 production by malate applied to slices and whole fruit respiration (Knee, 1971). Earlier reports of an increase in ME activity during ripening (Hulme et al., 1965) were not confirmed in recent work (Knee et al., 1989). The increase in malate metabolism also occurs in peel tissue excised from preclimacteric apples, but has not been correlated with changes in ME activity (Rhodes et al., 1968).

Various authors have tried to relate the effects of high CO_2 concentration used in storage of apples, to effects on mitochondrial enzymes. Apples in high CO_2 atmospheres accumulate succinate (Hulme, 1956), and CO_2 inhibits succinate dehydrogenase (EC 1.3.99.1) from other plant sources (Bendall et al., 1960). Shipway and Bramlage (1973) argue for a more general effect of CO_2 on mitochondrial enzymes. In further research on pears, the activities of cytoplasmic ATP-dependent phosphofructokinase (EC 2.7.1.11) and the pyrophosphate-fructose-6-phosphate phosphotransferase (EC 2.7.1.90) were both shown to decline under 10% CO_2 (Kerbel et al., 1988)

Cell wall metabolism

The only cell wall degrading enzymes which have been convincingly demonstrated in apple fruits are pectinesterase (EC 3.1.1.11), exo-polygalacturonase (exo-PG, EC 3.2.1.67), β-galactosidase (EC 3.2.1.23) (Knee and Bartley, 1981) and β-1,4 glucanase (EC 3.2.1.4) (Abeles and Takeda, 1990). The absence of endo-polygalacturonase (endo-PG, EC 3.2.1.15) activity from apples is consistent with the high molecular weight of the soluble pectin which accumulates on ripening (Knee, 1978a). The exo-PG releases uronic acid from apple cell-walls and, with pectinesterase, it could account for losses of uronic acid during ripening (Bartley, 1978). Loss of galactose residues from the cell wall of apple could be catalysed by the β-galactosidase since it releases galactose from polymers and from plant cell walls (Bartley, 1974). Most of the activity remains in the wall fraction after extraction in 5 mM phosphate buffer and the wall-bound activity can be detected by incubation of apple discs with 4-nitrophenyl-β-D-galactoside (Bartley, 1977). Activity is present in unripe (preclimacteric) fruit and increases by about 50% on ripening (Bartley, 1977).

European pears contain an exo-PG which may be similar to the apple enzyme, and an endo-PG (Pressey and Avants, 1976). The endo-PG can release uronic acid from fruit cell walls (Pressey and Avants, 1976) and the rise in its activity on ripening accounts for the simultaneous decline in degree of polymerization of pectin in the fruit (Bartley *et al.*, 1982). Ahmed and Labavitch (1980b) found several other glycosidases in pears, but did not detect endo-β-1,4 glucanase. Yamaki and Matsuda (1977) detected endoglucanases with pH optima at pH 5.5 and pH 7.0 in Oriental pears (*Pyrus serotina*). The acidic form predominated in immature fruit, whereas the neutral form increased on ripening as did a β-glucosidase (EC 3.2.1.21). Endo-PG and pectinesterase activities were also present in immature fruit and their activities declined to low levels before rising again on ripening. β-Galactosidase activity also increased with ripening, but this enzyme was present throughout fruit development. The finding of endoglycosidase activities in immature fruits is exceptional, although most other authors have not examined fruit so early in development.

Phenol metabolism

Phenylalanine ammonia-lyase (PAL, EC 4.3.1.5) has been studied in apples mainly in relation to anthocyanin synthesis in the peel (Faragher and Chalmers, 1977). However, a green apple variety can contain as much PAL as a red variety, and in a red variety the rise in PAL occurs after development of colour (Blankenship and Unrath, 1988). Phytochrome regulation of PAL activity in apples may not be an adequate explanation of the light requirement for anthocyanin synthesis (Siegelman and Hendricks, 1958). Other enzymes specific to the anthocyanin pathway are probably coordinately induced. Ethylene induces a further increase in enzyme activity in tissue held in the light and various forms of stress probably enhance PAL activity via an effect on ethylene production (Faragher and Brohier, 1984). When PAL activity changes there is often an opposite change in the activity of a PAL inactivating system (Tan, 1980).

In addition to a complex mixture of phenolic compounds, apples contain polyphenoloxidase (PPO, EC 1.10.3.1) activity, so that when tissue is cut or disrupted, browning occurs. The complexity of substrates and possible modifications of the enzyme after tissue disruption (Smith and Montgomery, 1985), may account for the lack of correlation between browning and total phenol or enzyme content (Coseting and Lee, 1987). Moulding et al. (1988) described properties of peroxidase (EC 1.14.18.1) in relation to discoloration of apple products. As is common with plant tissues, they found multiple forms of the enzyme with optima around pH 5.2 and pH 5.9. They drew attention to the heat stability of peroxidase and were able to detect activity in a commercial apple juice. According to Goodenough et al. (1983), the pH optimum of PPO is close to the natural pH of apple juice and this enzyme also showed high heat stability.

Aliphatic metabolism

Apple alcohol dehydrogenase activity decreases with carbon number through the aldehyde series, but the activity is adequate for rates of alcohol production required for synthesis of volatile esters in ripening fruit (Bartley and Hindley, 1980). Another enzyme involved in ester metabolism in apple, carboxylic ester hydrolase (EC 3.1.1.1), exists in multiple forms with a range of molecular weights and specificities for substrates from acetate to hexanoate (Bartley and Stevens, 1981). A distinct lipolytic acylhydrolase (EC 3.1.1.3) was detected after a different preparation procedure. Goodenough and Entwhistle (1982) reported a single esterase of high molecular weight whose activity increased during fruit development, but subsequent investigation revealed nearly constant activity (Knee et al., 1989). Esterase activity is particularly high in the peel, so that the volatiles diffusing out of an apple are enriched in alcohols (Knee and Hatfield, 1976). Apple tissue and whole apples can inter-convert exogenously-supplied aldehydes, alcohols and esters at any stage of ripening (Knee and Hatfield, 1981). The absence of ester synthesis in immature apples and in low-oxygen stored fruit implies that more remote precursors are not being metabolized. Apple tissue metabolizes fatty acyl precursors by β-oxidation to shorter chain compounds and by reduction to alcohols; however it is not clear whether endogenous volatile esters are formed from lipid precursors or *de novo* from acetate (Bartley et al., 1985).

Lipid metabolism

Lipoxygenase (EC 1.13.11.12) occurs in apples and pears in the microsomal fraction (Kim and Grosch, 1978, 1979) and at high specific activity, although apple is a poor source in fresh weight terms. Activity of this enzyme is higher in the core and peel of the fruit than in the cortex and it has been associated with membrane disruption leading to core browning disorders (Feys *et al.*, 1980). The 13-hydroperoxides formed by lipoxygenase action on linoleic or linolenic acids (Kim and Grosch, 1979) could be cleaved to hexanal and 2-*trans*-hexanal by hydroperoxide lyase. This may account for the presence of these aldehydes after tissue disruption, but the occurrence of *trans*-2-hexenyl acetate in apple volatiles (Dimick and Hoskin, 1983) suggests that the enzyme may be active *in vivo*. A hydroperoxide lyase isolated from pears was specific for linoleate 9-hydroperoxides and formed nonenal and oxononanoic acid (Kim and Grosch, 1981).

Ethylene synthesis

A crucial link in the elucidation of the pathway of ethylene synthesis (Chapter 1) was the discovery of the intermediate 1-amino-cyclopropane-1-carboxylic acid (ACC) in apple tissue (Adams and Yang, 1979). The ACC synthase (EC 4.4.1.14) in apples has recently been purified and the amino acid sequence around the active site determined; this 48 kDa protein is immunologically distinct from the tomato enzyme (Yip *et al.*, 1990, 1991). Apples also contain ACC malonyl transferase activity (Mansour *et al.*, 1986) and malonylation may regulate the low rate of ethylene synthesis during growth of the fruit (Knee, 1985b). The increase in ethylene production when pears are held at low temperature is associated with an increase in wound-inducible ethylene production and ACC synthase (EC 4.4.1.14) activity. At the same time, these wound effects become increasingly insensitive to transcriptional inhibitors. ACC synthase activity in the whole fruit remains low until fruit has ripened for at least two days at a higher temperature (Knee, 1987).

11.4   REGULATION

11.4.1   Role of ethylene

The rise in ethylene production that accompanies the climacteric appears to be an all important regulatory event in the development of pome fruits. However, it is evident that other regulatory events precede and follow this rise and there may be other aspects of regulation unrelated to ethylene. Before the rise, apple fruits become more sensitive to ethylene and it is envisaged that the fruit arrives at a stage when it responds to the low endogenous level of ethylene (Knee *et al.*, 1987). A long-standing hypothesis proposes that the initiation of ripening is regulated by the balance between ethylene and indolylacetic acid (IAA). Infiltration of pear fruits with IAA or synthetic auxin analogues arrested ripening, whereas auxin antagonists and metabolites of IAA promoted ripening (Frenkel and Dyck, 1973; Frenkel and Haard, 1973; Frenkel *et al.*, 1975). The pears used in many of these experiments had been kept for some time in refrigerated storage so that ethylene production was probably initiated; auxin treatment further stimulated ethylene production (Frenkel and Dyck, 1973). A fluorometric assay indicated that endogenous IAA levels increased in apples before the rise in ethylene production and declined as the rise occurred (Mousdale and Knee, 1981). Thus it is plausible that high auxin levels could inhibit the progress of ripening, but its initiation requires another explanation.

The induction of ethylene synthesis in pears and some apple cultivars by refrigeration implies that there could be an environmental signal for ripening in these fruits. Little is known about the transduction of this signal. Fruits are not more sensitive to ethylene at inductive temperatures and the effect of chilling may be independent of ethylene action (Knee, 1988b). Recent results indicate that transcriptional changes occur in pears under refrigeration and these are followed by further changes when fruits are moved to a higher temperature that permits the progress of ripening (Wilson *et al.*, 1990).

Treatment of immature pears with low ethylene concentrations can cause softening without induction of rapid ethylene synthesis (Wang *et al.*, 1972). Similarly, chlorophyll degradation, volatile ester synthesis, pectin solubilization and softening can occur in 'Gloster 69' apples after prolonged storage without the induction of rapid ethylene synthesis (Knee and Tsantili, 1988). On the other hand, when endogenous ethylene synthesis was inhibited in apples by aminoethoxyvinylglycine (AVG), ACC synthase activity, a respiratory rise and softening could be induced by ethylene treatment (Bufler, 1984). Thus it appears that high ethylene concentrations can promote, but may not be required for, many ripening processes. High ethylene concentrations could ensure that all ripening processes occur synchronously under normal conditions. The only processes that are dependent on rapid ethylene synthesis may prove to be respiration and ethylene synthesis itself. Interestingly, ethylene treatment increased respiration in immature pears (Wang *et al.*, 1972) and Gloster apples (Knee and Tsantili, 1988) before it stimulated ethylene synthesis. Climacteric respiration seems to be a consequence of high ethylene levels, rather than a central event in fruit ripening.

11.4.2   Metabolic control

There may be further levels of control after ethylene regulation. When pears are stored at $-1°C$, ethylene synthesis occurs but does not lead to ripening. In apples ethylene accumulates to $100 \, \mu l.l^{-1}$ or more, and ripening occurs at storage temperatures, but CA conditions control the consequences of ethylene action. CA storage inhibits fruit respiration (Fidler and North, 1967) but it is not clear how this occurs or whether it is a cause or consequence of the inhibition of ripening processes. Possible sites of action of CO_2 directly on respiratory metabolism have been mentioned above. The oxygen concentrations employed in fruit storage should not inhibit cytochrome oxidase directly and the delay in response of respiration to changes in oxygen concentration suggests that this is an indirect response (Knee, 1980b). One possibility is that the tissue responds to incipient anoxia by closing down energy consuming metabolism and the reduced energy demand leads to lower respiratory flux (Knee, 1991b). Even processes that seem to be purely degradative may have an energy demand. For example, softening of pears seems to be a consequence of polygalacturonase (PG) action, but it can be arrested by transfer of the tissue to a nitrogen atmosphere, which does not affect PG activity in the short term (Knee, 1982c). CA effects on pear ripening can be reproduced by infiltration with mannose; this causes conversion of phosphate to mannose-6-phosphate, which is not further metabolized, and the depletion of phosphate pools leads to a general restriction of energy metabolism (Watkins and Frenkel, 1987). Thus it is clear that inhibition of respiration could result in inhibition of ripening.

11.4.3   Maintenance of cellular integrity

Sooner or later in the life of the fruit the point is reached where orderly cell function can not be maintained. This may occur sooner when the tissue is under some kind of stress, low temperature, adverse atmosphere or nutritional imbalance. Studies on irradiation of pears show a diminishing capacity for repair of mitochondrial function as ripening proceeds (Romani *et al.*, 1968). Oxidative damage to cellular constituents has already been mentioned as a cause of cell death. This kind of process is often considered as an alternative to genetically controlled ripening and senescence. However, cell membrane integrity and function must be maintained actively and are themselves subject to genetic control. It is possible that enzymes involved in cellular maintenance are down-regulated as ripening proceeds.

11.4.4   Future prospects for control

Future strategies of control of pome fruit ripening could attempt to intervene at various levels from the involvement of ethylene, through ripening metabolism to the development of physiological disorders. Manipulation could involve control of the expression of regulatory proteins through biotechnology. Such manipulation will require development of methods for transformation, selection and regeneration of pome fruit plants. In addition it is necessary to identify targets for manipulation; this review has not found many examples of enzymes showing large differences in activity at different stages of fruit development. ACC synthase is the only obvious example, but the identity of many of the enzymes responsible for cell wall, pigment and flavour metabolism remains to be discovered. A recent study has revealed changes in expression of a small number of mRNAs and proteins in ripening apples (Lay-Yee et al., 1990).

These new approaches to control of quality and postharvest deterioration will supplant only some aspects of existing technology and will need to be integrated with other practices. Thus it is important to integrate an understanding of the physiological basis of existing technology such as refrigeration and CA storage with the development of new technology.

REFERENCES

Abeles, F.B. and Takeda, F. (1990) Cellulase activity in ripening strawberry and apple fruits. *Scientia Horticulturae* **42**, 269–275

Adams, D.O. and Yang, S.F. (1979) Ethylene biosynthesis: identification of 1-aminocyclopropane-1-carboxylic acid as an intermediate in the conversion of methionine to ethylene. *Proceedings of National Academy of Sciences (USA)*, **76**, 170–174

Ahmed, A.E.R. and Labavitch, J.M. (1980a) Cell wall metabolism in ripening fruit I. Cell wall changes in ripening 'Bartlett' pears. *Plant Physiology*, **65**, 1009–1013

Ahmed, A.E.R. and Labavitch, J.M. (1980b) Cell wall metabolism in ripening fruit II. Changes in carbohydrate-degrading enzymes in ripening 'Bartlett' pears. *Plant Physiology*, **65**, 1014–1016

Atkinson, D., Jackson, J.E., Sharples, R.O. and Waller, W.M. (1980) *Mineral Nutrition of Fruit Trees*, Butterworths, London

Bartley I.M. (1974) β-Galactosidase activity in ripening apples. *Phytochemistry*, **13**, 2107–2111

Bartley, I.M. (1976) Changes in the glucans of ripening apples. *Phytochemistry*, **15**, 625–626

Bartley, I.M. (1977) A further study of β-galactosidase activity in apples ripening in store. *Journal of Experimental Botany*, **28**, 943–948

Bartley, I.M. (1978) Exo-polygalacturonase of apple. *Phytochemistry*, **17**, 213–216

Bartley, I.M. (1985) Lipid metabolism of ripening apples. *Phytochemistry*, **24**, 2857–2859

Bartley, I.M. (1986) Changes in sterol and phospholipid composition of apples during storage at low temperature and low oxygen concentration. *Journal of the Science of Food and Agriculture*, **37**, 31–36

Bartley, I.M. and Hindley, S.J. (1980) Alcohol dehydrogenase of apple. *Journal of Experimental Botany*, **31**, 449–459

Bartley, I.M. and Stevens, W.M. (1981) Carboxylic ester hydrolases of apple. *Journal of Experimental Botany*, **32**, 741–751

Bartley, I.M., Knee, M. and Casimir, M.A. (1982) Fruit softening 1. Changes in cell wall composition and endo-polygalacturonase in ripening pears. *Journal of Experimental Botany*, **33**, 1248–1255

Bartley, I.M., Stoker, P.G., Martin, A.D.E., Hatfield, S.G.S. and Knee, M. (1985) Synthesis of aroma compounds by apples supplied with alcohols and methyl esters of fatty acids. *Journal of the Science of Food and Agriculture*, **36**, 567–574

Bendall, D.S., Ranson, S.L. and Walker, D.A. (1960) Effects of carbon dioxide on the oxidation of succinate and reduced diphosphopyridine nucleotide by *Ricinus* mitochondria. *Biochemical Journal*, **76**, 221–225

Berüter, J. (1985) Sugar accumulation and changes in the activities of related enzymes during development of the apple fruit. *Journal of Plant Physiology*, **121**, 331–341

Blanke, M.M., Hucklesby, D.P., Notton, B.A. and Lenz, F. (1987) Utilization of bicarbonate by apple fruit phosphoenolpyruvate carboxylase. *Phytochemistry*, **26**, 2475–2476

Blankenship, S.M. and Unrath, C.R. (1988) PAL and ethylene content during maturation of Red and Golden Delicious apples. *Phytochemistry*, **27**, 1001–1003

Bufler, G. (1984) Ethylene-enhanced l-aminocyclopropane-l-carboxylic acid synthase activity in ripening apples. *Plant Physiology*, **75**, 192–195

Burg, S.P. and Burg, E.A. (1965) Gas exchange in fruits. *Physiologia Plantarum*, **18**, 870–874

Cameron, A.C. and Yang S.F. (1982) A simple method for the determination of resistance to gaseous diffusion in plant organs. *Plant Physiology*, **70**, 21–23

Chanda, S.K., Hirst, E.L. and Percival, E.G.V. (1951) The constitution of a pear cell-wall xylan. *Journal of the Chemical Society*, **1951**, 1240–1246

Clements, R.L. (1970) Protein patterns in fruits. In *The Biochemistry of Fruits and Their Products*, Vol. 1, (ed A.C. Hulme), Academic Press, London, pp. 159–178

Coseting, M.Y. and Lee, C.Y. (1987) Changes in apple polyphenoloxidase and polyphenol concentrations in relation to degree of browning. *Journal of Food Science*, **52**, 985–989

Dick, A.J. and Labavitch, J.M. (1989) Cell wall metabolism in ripening fruit IV. Characterization of the pectic polysaccharides solubilized during softening of 'Bartlett' pear fruit. *Plant Physiology*, **89**, 1394–1400

Dilley, D.R. (1966) Purification and properties of apple fruit malic enzyme. *Plant Physiology*, **41**, 214–220

Dimick, P.S. and Hoskin, J.C. (1983) Review of apple flavor – State of the art. *CRC Critical Reviews in Food Science and Nutrition*, **18**, 387–409

Faragher, J.D. and Brohier, R.L. (1984) Anthocyanin accumulation in apple skin during ripening: regulation by ethylene and phenylalanine ammonia-lyase. *Scientia Horticulturae*, **22**, 89–96

Faragher, J.D. and Chalmers, D.J. (1977) Regulation of anthocyanin synthesis in apple skin III. Involvement of phenylalanine ammonia-lyase. *Australian Journal of Plant Physiology*, **4**, 133–141

Feys, M., Naesens, W., Tobback, P. and Maes, E. (1980) Lipoxygenase activity in apples in relation to storage and physiological disorders. *Phytochemistry*, **19**, 1009–1011

Fidler, J.C., and North, C.J. (1967) The effect of storage on the respiration of apples I. The effect of temperature and concentration of carbon dioxide and oxygen on the production of carbon dioxide and uptake of oxygen. *Journal of Horticultural Science*, **42**, 189–206

Fidler, J.C. and North, C.J. (1968) The effect of conditions of storage on the respiration of apples IV. Changes in concentration of possible substrates of respiration, as related to production of carbon dioxide and uptake of oxygen by apples at low temperatures. *Journal of Horticultural Science*, **43**, 429–439

Flath, R.A., Black D.R., Guadagni, DG., McFadden, W.H. and Schultz, T.H. (1967) Identification and organoleptic evaluation of compounds in Delicious apple essence. *Journal of Agricultural and Food Chemistry*, **15**, 29–35

Frenkel, C. and Dyck, R. (1973) Auxin inhibition of ripening in Bartlett pears. *Plant Physiology*, **51**, 6–9

Frenkel, C. and Haard, N.F. (1973) Initiation of ripening in Bartlett pear with an antiauxin, alpha (*p*-chlorophenoxy) isobutyric acid. *Plant Physiology*, **52**, 380–384

Frenkel, C., Haddon, V.R. and Smallheer, J.M. (1975) Promotion of softening and ethylene synthesis in Bartlett pears by 3-methylene oxindole. *Plant Physiology*, **56**, 647–649

Galliard, T. (1968) Aspects of lipid metabolism in higher plants II. The identification and quantitative analysis of lipids from the pulp of pre- and post-climacteric apples. *Phytochemistry*, **7**, 1915–1922

Goodenough, P.W. and Entwhistle, T.G. (1982) The hydrodynamic properties and kinetic constants with natural substrates of the esterase from *Malus pumila* fruit. *European Journal of Biochemistry*, **127**, 145–149

Goodenough, P.W., Kessel, S., Lea, A.G.H. and Loeffler, T. (1983) Mono- and diphenolase activity from fruit of *Malus pumila*. *Phytochemistry*, **22**, 359–363

Gross, J., Zachariae, A., Lenz, F. and Eckhardt, G. (1978) Carotenoid changes in the peel of the "Golden Delicious" apple during ripening and storage. *Zeitschrift fur Pflanzenphysiologie.*, **89**, 321–332

Hatfield, S.G.S. and Knee, M. (1988) Effects of water loss on apples in storage. *International Journal of Food Science and Technology*, **23**, 575–583

Hipps, N.A. and Perring, M.A. (1989) Effects of soil management systems and nitrogen fertiliser on the firmness and mean fruit weight of Cox's Orange Pippin apples at harvest. *Journal of the Science of Food and Agriculture*, **48**, 507–510

Huelin, F.E. and Coggiola, I.M. (1970) Superficial scald, a functional disorder of apples V. Oxidation of farnesene and its inhibition by diphenylamine. *Journal of the Science of Food and Agriculture*, **21**, 44–48

Hulme, A.C. (1956) Carbon dioxide injury and the presence of succinic acid in apples. *Nature*, **178**, 218–219

Hulme, A.C. and Rhodes, M.J.C. (1971) Pome fruits. In *The Biochemistry of Fruits and Their Products*, Vol. 2 (ed A.C. Hulme), Academic Press, London, pp. 333–373

Hulme, A.C., Jones, J.D. and Wooltorton, L.S. C. (1965) The respiration climacteric in apple fruits. Biochemical changes occurring during the development of the climacteric in fruit on the tree. *New Phytology*, **64**, 152–157

Irwin, P.L., Pfeffer, P.E., Gerasimowicz, W.V., Pressey, R. and Sams, C.E. (1984) Ripening-related perturbations in apple cell wall nuclear spin dynamics. *Phytochemistry*, **23**, 2239–2242

Jennings, W.G. and Sevenants, M.R. (1964) Volatile esters of Bartlett pear. III. *Journal of Food Science*, **29**, 158–163

Kerbel, E.L., Kader, A.A. and Romani, R.J. (1988) Effects of elevated CO_2 concentrations on glycolysis in intact 'Bartlett' pear fruit. *Plant Physiology*, **86**, 1205–1209

Khan, A.A. and Vincent, J.F.V. (1990) Anisotropy of apple parenchyma. *Journal of the Science of Food and Agriculture*, **52**, 455–466

Kidd, F. and West, C. (1924) The course of respiratory activity throughout the life of an apple. *Annual Report of the Food Investigations Board*, London, 27–32

Kim, I.S. and Grosch, W. (1978) Lipoxygenasen aus Birnen, Erd- und Stachelbeeren: partielle Reinigiung und Eigenschaften. *Zeitschrift fur Lebensmittel. Untersuchung und Forschung.*, **167**, 324–326

Kim, I.S. and Grosch, W. (1979) Partial purification of a lipoxygenase from apples. *Journal of Agricultural and Food Chemistry*, **27**, 243–246

Kim, I.S. and Grosch, W. (1981) Partial purification and properties of a hydroperoxide lyase from fruit of pear. *Journal of Agricultural and Food Chemistry*, **29**, 1220–1225

Knee, M. (1971) Ripening of apples during storage II. Respiratory metabolism and ethylene synthesis in Golden Delicious apples during the climacteric and under conditions stimulating commercial storage practice. *Journal of the Sciences of Food and Agriculture*, **22**, 368–71

Knee, M. (1972) Anthocyanin, carotenoid and chlorophyll changes in the peel of Cox's Orange Pippin apples during ripening on and off the tree. *Journal of Experimental Botany*, **23**, 184–96

Knee, M. (1973a) Polysaccharides and glycoproteins of apple fruit cell walls. *Phytochemistry*, **12**, 637–653

Knee, M. (1973b) Polysaccharide changes in cell walls of ripening apples. *Phytochemistry*, **12**, 1543–1549

Knee, M. (1978a) Properties of polygalacturonate and cell cohesion in apple fruit cortical tissue. *Phytochemistry*, **17**, 1257–1260

Knee, M. (1978b) Metabolism of polymethylgalacturonate in apple fruit cortical tissue during ripening. *Phytochemistry*, **17**, 1261–1264

Knee, M. (1980a) Methods of measuring green colour and chlorophyll content of apple fruit. *Journal of Food Technology*, **15**, 493–500

Knee, M. (1980b) Physiological responses of apple fruits to oxygen concentrations. *Annals of Applied Biology*, **96**, 243–253

Knee, M. (1982a) Fruit softening II. Precursor incorporation into pectin by pear tissue slices. *Journal of Experimental Botany*, **33**, 1256–1262

Knee, M. (1982b) Attempted isolation of fluorescent ageing pigments from apple fruits. *Journal of the Sciences of Food and Agriculture*, **33**, 209–212

Knee, M. (1982c) Fruit softening III. Requirement for oxygen and pH effects. *Journal of Experimental Botany*, **33**, 1263–1269

Knee, M. (1985a) Evaluating the practical significance of ethylene in fruit storage. In *Ethylene and Plant Development* (eds J.A. Roberts and G.A. Tucker), Butterworth Press, London, pp. 297–315

Knee, M. (1985b) Metabolism of 1-aminocyclopropane-1-carboxylic acid during apple fruit development. *Journal of Experimental Botany*, **36**, 670–678

Knee, M (1987) Development of ethylene biosynthesis in pear fruits at −1°C. *Journal of Experimental Botany*, **38**, 1724–1733

Knee, M. (1988a) Carotenol esters in developing apple fruits. *Phytochemistry*, **27**, 1005–1009

Knee, M. (1988b) Effects of temperature and daminozide on the induction of ethylene synthesis in two varieties of apple. *Journal of Plant Growth Regulation*, **7**, 111–119

Knee, M. (1989) Control of post-harvest action of ethylene on tree fruits. *Acta Horticulturae*, **239**, 417–426

Knee, M. (1991a) Rapid measurement of diffusion of gas through the skin of apple fruits. *HortScience*, **26**, 885–887

Knee, M. (1991b) Fruit metabolism and practical problems of fruit storage under hypoxia and anoxia. In *Plant Life Under Oxygen Deprivation*, (eds M.B. Jackson, D.D. Davies and H. Lambers), SPB Academic Publishing, The Hague, Netherlands, pp. 229–243

Knee, M. and Bartley, I.M. (1981) Composition and metabolism of cell wall polysaccharides in ripening fruits. In *Recent advances in the biochemistry of fruits and vegetables* (eds J. Friend and M.J.C. Rhodes), Academic Press, London, pp. 133–148

Knee, M. and Hatfield, S.G.S. (1976) A comparison of methods for measuring the volatile components of apple fruits. *Journal of Food Technology*, **11**, 485–493

Knee, M. and Hatfield, S.G.S. (1981) The metabolism of alcohols by apple fruit tissue. *Journal of the Science of Food and Agriculture*, **32**, 593–600

Knee, M. and Tsantili, E. (1988) Storage of 'Gloster 69' apples in the preclimacteric state for up to 200 days after harvest. *Physiologia Plantarum*, **74**, 499–503

Knee, M., Smith, S.M. and Johnson, D.S. (1983a) Comparison of methods for estimating the onset of the respiration climacteric in unpicked apples. *Journal of Horticultural Science*, **58**, 521–526

Knee, M., Looney, N.E., Hatfield, S.G.S. and Smith, S.M. (1983b) Initiation of rapid ethylene synthesis by apple and pear fruits in relation to storage temperature. *Journal of Experimental Botany*, **34**, 1207–1212

Knee, M., Hatfield, S.G.S. and Bramlage, W.J. (1987) Response of developing apple fruits to ethylene treatment. *Journal of Experimental Botany*, **38**, 972–979

Knee, M., Hatfield, S.G.S. and Smith, S.M. (1989) Evaluation of various indicators of maturity for harvest of apple fruit intended for long-term storage. *Journal of Horticultural Science*, **64**, 413–419

Knee, M., Hatfield, S.G.S. and Farman, D. (1990) Sources of variation in firmness and ester content of 'Cox' apples stored in 2% oxygen. *Annals of Applied Biology*, **116**, 617–623

Lance, C. (1981) Cyanide-insensitive respiration in fruits and vegetables. In *Recent Advances in the Biochemistry of Fruits and Vegetables*, (eds J. Friend and M.J.C. Rhodes), Academic Press, London, pp. 63–87

Latché, A., Pech, J.C., Diarra, A. and Fallot, J. (1975) Facteurs susceptibles de contrôler l'amylolyse chez la pomme et la poire. In *Facteurs et Regulation de la Maturation des Fruits* (ed. R. Ulrich), CNRS, Paris, pp. 299–306

Lay-Yee, M., DellaPenna, D. and Ross, G.S. (1990) Changes in mRNA and protein during ripening in apple fruit (*Malus domestica* Borkh. cv Golden Delicious). *Plant Physiology*, **94**, 850–853

Maguire, Y.P. and Haard, N.F. (1975) Fluorescent product accumulation in ripening fruit. *Nature*, **258**, 599–600

Mansour, R., Latché, A., Vaillant, V., Pech, J.C. and Reid, M.S. (1986) Metabolism of 1-aminocyclopropane-1-carboxylic acid in ripening apple fruits. *Physiologia Plantarum*, **66**, 495–502

Meigh, D.F. (1970) Apple scald. In *The Biochemistry of Fruits and Their Products*, vol. 1, (ed. A.C. Hulme), Academic Press, London, pp. 556–570

Meir, S. and Bramlage, W.J. (1988) Antioxidant activity of 'Cortland' apple peel and susceptibility to superficial scald after storage. *Journal of the American Society of Horticultural Science*, **113**, 412–418

Moulding, P.H., Singleton, D.E., McLellan, K.M. and Robinson, D.S. (1988) Purification and heat stability of Cox's apple pulp peroxidase isoenzymes. *International Journal of Food Science and Technology*, **23**, 343–351

Mousdale, D.M.A. and Knee, M. (1981) Indolyl-3-acetic acid and ethylene levels in ripening apple fruits. *Journal of Experimental Botany*, **32**, 753–758

Preiss, J. (1982) Regulation of the biosynthesis and degradation of starch. *Annual Review of Plant Physiology*, **33**, 431–454

Pressey, R. and Avants, J.K. (1976) Pear polygalacturonases. *Phytochemistry*, **15**, 1349–1351.

Reid, M.S., Rhodes, M.J.C. and Hulme, A.C. (1973) Changes in ethylene and CO_2 during the ripening of apples. *Journal of the Science of Food and Agriculture*, **24**, 971–979

Rhodes, M.J.C., Wooltorton, L.S.C., Galliard, T. and Hulme, A.C. (1968) Metabolic changes in excised fruit tissue I. Factors affecting the development of a malate decarboxylation system during the aging of discs of preclimacteric apples. *Phytochemistry*, **7**, 1439–1451

Romani, R.J., Yu, I.K., Ku, L.L., Fisher, K. and Dehgan, N. (1968) Cellular senescence, radiation damage to mitochondria and the compensatory response in ripening pear fruits. *Plant Physiology*, **43**, 1089–1096

Sal'kova, E.G. and Zvyagintseva, Y.V. (1981) Natural antioxidants in the coating of the apple and their isolation and chromatographic separation. *Prikladnaya Biokhimiya iMikrobiologiya.*, **17**, 293–299

Sfakiotakis, E.M. and Dilley, D.R. (1973) Internal ethylene concentrations in apple fruits attached to or detached from the tree. *Journal of the American Society of Horticultural Science*, **98**, 501–503

Shipway, M.R. and Bramlage, W.J. (1973) Effects of carbon dioxide on activity of apple mitochondria. *Plant Physiology*, **51**, 1095–1098

Siegelman, H.W. and Hendricks, S.B. (1958) Photocontrol of anthocyanin synthesis in apple skin. *Plant Physiology*, **33**, 185–190

Simon, E.W. (1977) Leakage from fruit cells in water. *Journal of Experimental Botany*, **28**, 1147–1152

Smith, D.M. and Montgomery, M.W. (1985) Improved methods for the extraction of polyphenol oxidase from d'Anjou pears. *Phytochemistry*, **24**, 901–904

Stow, J. (1989) The involvement of calcium ions in maintenance of apple fruit tissue structure. *Journal of Experimental Botany*, **40**, 1053–1057

Swinburne, T.R. (1983) Quiescent infections in post-harvest diseases. In *Post-Harvest Pathology of Fruits and Vegetables*, (ed. C. Dennis), Academic Press, London, pp. 1–21

Tan, S.C. (1980) Phenylalanine ammonia-lyase and the phenylalanine ammonia-lyase inactivating system: effects of light, temperature and mineral deficiencies. *Australian Journal of Plant Physiology*, **7**, 159–167

Timberlake, C.F. (1981) Anthocyanins in fruits and vegetables. In *Recent Advances in the Biochemistry of Fruits and Vegetables*, (eds J. Friend and M.J.C. Rhodes), Academic Press, London, pp. 221–247

Ulrich, R. (1961) Temperature and maturation: pears require preliminary cold treatment. *Recent Advances in Botany*, **1961**, 1172–1176

Ulrich, R. (1970) Organic acids. In *The Biochemistry of Fruits and Their Products*, vol. 1, (ed. A.C. Hulme), Academic Press, London, pp. 89–118

Vincent, J.F.V. (1989) Relationship between density and stiffness of apple flesh. *Journal of the Sciences of Food and Agriculture*, **47**, 443–462

Voragen, A.G.J., Schols, H.A. and Pilnik, W. (1986) Structural features of the hemicellulose polymers of apples. *Zeitschrift für Lebensmittel Untersuchung und Forschung*, **183**, 105–110

de Vries, J.A., Rombouts, F.M., Voragen, A.G.J. and Pilnik, W. (1982) Enzymic degradation of apple pectins. *Carbohydrate Polymers*, **2**, 25–33

Wang, C.Y., Mellenthin, W.M. and Hansen, E. (1972) Maturity of Anjou pears in relation to chemical composition and reaction to ethylene. *Journal of the American Society of Horticultural Science*, **97**, 9–12

Watkins, C.B. and Frenkel, C. (1987) Inhibition of pear fruit ripening by mannose. *Plant Physiology*, **85**, 56–61

Whiting, G.C. (1970) Sugars. In *The Biochemistry of Fruits and Their Products*, Vol. 1, (ed. A.C. Hulme), Academic Press, London, pp. 1–32

Williams, A.A., Tucknott, O.G. and Lewis, M.J. (1977) 4-methoxyallylbenzene, an important aroma component of apples. *Journal of the Science of Food and Agriculture*, **28**, 185–190

Wilson, I.D., Tucker, G.A., Knee, M. and Grierson, D. (1990) Changes in mRNA during low temperature storage and ripening of pears. *Phytochemistry*, **29**, 2407–2409

Yamaki, S. and Ishikawa, K. (1986) Roles of four sorbitol related enzymes and invertase in the seasonal alteration of sugar metabolism in apple tissue. *Journal of the American Society of Horticultural Science*, **111**, 134–137

Yamaki, S. and Matsuda, K. (1977) Changes in the activities of some cell wall-degrading enzymes during development and ripening of Japanese pear fruit (*Pyrus serotina* Rehder *var.* culta Rehder). *Plant & Cell Physiology*, **18**, 81–93

Yip, W.K., Dong, J.G., Kenny, J.W., Thompson, G.A. and Yang, S.F. (1990) Characterization and sequencing of the active site of l-aminocyclopropane-l-carboxylate synthase. *Proceedings of the National Academy for Sciences, USA*, **87**, 7930–7934

Yip, W.K., Dong, J.G. and Yang, S.F. (1991) Purification and characterization of 1-aminocyclopropane-1-carboxylate synthase from apple fruits. *Plant Physiology*, **95**, 251–257

Chapter 12

Soft fruit

K. Manning

12.1 INTRODUCTION

The classification and morphological characteristics of soft fruits have previously been described in Hulme (1971). As a group the soft fruits include various berries, currants and also the strawberry, a botanical false fruit. Economically, strawberries, raspberries and blackcurrants are the most important of the soft fruits (Table 12.1). They are valued as a fresh product, but in general have a short postharvest shelf life. As their name implies, most soft fruits lack a firm texture, even when freshly picked; therefore the handling of soft fruits commercially is kept to a minimum to reduce damage, with refrigeration being widely used during transportation and storage to slow down fruit softening. As a result, the majority of soft fruits are processed into products ranging from frozen and canned whole fruit, through jams and conserves to juices and essences used as flavourings and colourings for other food products.

Thus, the usefulness of soft fruits as a fresh food depends primarily upon the textural changes within the fruit tissue. An understanding of the biochemistry of softening for this group of fruits in particular could have significant commercial importance in the future.

12.2 PHYSIOLOGY

12.2.1 General features

Soft fruits share some common features during their development with a wide diversity of fleshy fruits, both edible and inedible. An initial phase

Biochemistry of Fruit Ripening. Edited by G. Seymour, J. Taylor and G. Tucker. Published in 1993 by Chapman & Hall, London. ISBN 0 412 40830 9

Table 12.1 World production of soft fruits in 1988. (From FAO Production Yearbook, 1988, No. 42.)

	Strawberries		Raspberries[a]		Currants[b]	
	Tonnes	%	Tonnes	%	Tonnes	%
World	2 267 000	100	365 900	100	585 200	100
Africa	25 950	1.14	–	–	–	–
Egypt	20 000c	0.88	–	–	–	–
S. Africa	5707	0.25	–	–	–	–
N & C America	678 800	29.94	32 000	8.75	45	0.01
Canada	29 500c	1.30	19 000c	5.19	–	–
Mexico	120 000d	5.29	–	–	–	–
USA	529 300	23.35	13 000c	3.55	45c	0.01
S America	13 650	0.60	–	–	–	–
Argentina	5500c	0.24	–	–	–	–
Brazil	2200c	0.10	–	–	–	–
Peru	3000c	0.13	–	–	–	–
Venezuela	2000c	0.09	–	–	–	–
Asia	312 400	13.78	–	–	58	0.01
China	8000	0.35	–	–	–	–
Israel	13 000c	0.57	–	–	–	–
Japan	208 300c	9.19	–	–	–	–
Korean Rep	37 000c	1.63	–	–	–	–
Turkey	45 000	1.99	–	–	58c	0.01

Europe	1 100 000	48.52	189 100	51.68	483 600	82.64
Austria	16 300^c	0.72	–	–	26 300^c	4.49
Belgium-Lux	23 000^d	1.01	436^c	0.11	4000^d	0.68
Bulgaria	15 180	0.67	4082	1.12	142	0.02
Czechoslovakia	24 830	1.10	1000^c	0.27	34 900	5.96
Denmark	9000^d	0.40	180^c	0.05	4000^d	0.68
Finland	9000^c	0.40	300^c	0.08	2000^c	0.34
France	95 070	4.19	6700	1.83	10 000^c	1.71
German DR	28 430	1.25	–	–	31 950	5.46
Germany FR	54 830	2.42	28 040	7.66	141 700	24.21
Greece	6000	0.26	–	–	–	–
Hungary	13 100^c	0.58	19 300^c	5.27	19 600^c	3.35
Ireland	4000^d	0.18	100^c	0.03	225^c	0.04
Italy	190 700	8.41	1900	0.52	480	0.08
Netherlands	27 000^d	1.19	400^c	0.11	2000^d	0.34
Norway	18 000^c	0.79	3200^c	0.87	17 500	2.99
Poland	249 300	11.00	43 420	11.87	165 300	28.25
Portugal	2000^c	0.09	–	–	–	–
Romania	39 000^c	1.72	–	–	–	–
Spain	171 000^d	7.54	–	–	–	–
Switzerland	5400	0.24	–	–	–	–
United Kingdom	55 000^d	2.43	27 000^d	7.40	23 000^d	3.93
Yugoslavia	43 420	1.92	53 070	14.50	472^c	0.08

Table 12.1 (*contd*)

	Strawberries		Raspberries[a]		Currants[b]	
	Tonnes	%	Tonnes	%	Tonnes	%
Oceania	9400	0.41	2781	0.76	1535	0.26
Australia	5000[c]	0.22	481[c]	0.13	635[c]	0.11
New Zealand	4400[c]	0.19	2300[c]	0.63	900[c]	0.15
USSR	127 000[c]	5.60	142 000[c]	38.81	100 000[c]	17.09

[a] Some data on raspberries (*Rubus idaeus*) may include other berries of the genus *Rubus* such as blackberries, loganberries and dewberries.
[b] Data on blackcurrants includes white currants and red currants.
[c] FAO estimate.
[d] Unofficial figure.

of growth and enlargement is followed by a maturation phase during which the fruit acquires the capacity to ripen. Ripening itself is defined by a set of physico-chemical changes characteristic of each fruit. These usually involve transitions in colour arising from the degradation of existing pigments and the synthesis of new and often intensely coloured pigments, texture changes resulting in tissue softening and sometimes liquefaction, and the production of highly distinctive flavours and aromas which affect the palatability of the fruit. Changes involving senescence occur throughout development and are most obvious in the later stages as the fleshy tissues disintegrate, leaving the seeds to survive. Ripening is thus part of a continuous developmental process in which several physiological phases may overlap.

12.2.2 Macroscopic changes

Of the soft fruits the strawberry (genus *Fragaria*) is perhaps the best studied in terms of its physiology and biochemistry. This undoubtedly is due to its unusual structure as a false fruit, having the true fruits or achenes on the outside of a fleshy receptacle and attached to it through vascular connections. Much of the information presented in this chapter inevitably relates to the strawberry but where appropriate, other soft fruits will be referred to.

Growth

Fruit of the strawberry grow rapidly, with full size being attained approximately 30 days after anthesis, depending upon conditions. The kinetics of fruit growth in the strawberry appear to vary with cultivar, some cultivars having a single phase of sigmoidal growth (Woodward, 1972; Forney and Breen, 1985a; Stutte and Darnell, 1987) and others having biphasic growth patterns (Archbold and Dennis, 1984; Perkins-Veazie and Huber, 1987). In one cultivar cited, Ozark Beauty, both types of growth pattern have been reported (Mudge *et al.*, 1981; Veluthambi *et al.*, 1985). However, the plants used in these studies were grown under different lighting conditions and were sampled at different frequencies, factors which could have affected their observed growth characteristics. Perkins-Veazie and Huber (1987) suggested that bimodal receptacle growth was related to the development of the endosperm and embryo within the achenes, the second period of accelerated growth coinciding with embryo maturation and encompassing receptacle ripening. A strawberry variant genotype in which pollination does not lead to receptacle growth (Veluthambi *et al.*, 1985), and in which achene development may be abnormal, could be of value in elucidating the role of achenes in receptacle development.

Fruit size at maturity is influenced by the position of the fruit on the inflorescence (cyme) and decreases in the order: primary > secondary > tertiary (Moore *et al.*, 1970). Although relative growth rates in primary and secondary fruit were similar (Stutte and Darnell, 1987), differences in fruit size were correlated with a longer lag period following pollination in secondary fruit growth. The delay in secondary fruit growth appears to be physiologically determined, and to be independent of environmental conditions. Removal of primary fruit results in increased secondary fruit weight (Stutte and Darnell, 1987), indicating that inter-fruit competition occurs on the cyme, akin to apical dominance in shoots.

Differences in final fruit size are also known to be genetically determined and correlate with the number and size of developed achenes (Moore *et al.*, 1970). Variations in fruit size among cultivars appears to be additionally determined by the growth-promoting activity of individual achenes.

Changes in fruit shape in certain soft fruits may be physiologically significant. In the raspberry, for example, which has many drupelets attached to a central receptacle, an abscission zone forms between each drupelet and its point of attachment to the receptacle and causes fruit detachment in the later stages of ripening. Burdon and Sexton (1990) consider that receptacle expansion contributes to cell separation at the abscission zone creating stresses which lead to drupelet abscission.

Development and ripening

In strawberry the time taken for fruit to become fully ripe is closely related to temperature (Perkins-Veazie and Huber, 1987) and can vary between 20 and 60 days. This probably reflects an overall effect of temperature on metabolic rate as determined by enzyme activity.

For most soft fruits ripening occurs rapidly and fruit retain optimum condition for a relatively short time. Ripening in the strawberry is invariably marked by simultaneous changes in colour, flavour and texture in normal fruit (Fig. 12.1). However, a mutant cultivar of *Fragaria ananassa*, 'White Carter', lacks the red pigment anthocyanin and is thus white when fully ripe. This cultivar shows otherwise normal ripening behaviour in respect of texture and flavour. The metabolic block in colour formation has not been identified and this mutant could assist investigations of the biochemistry of colour formation in the fruit.

12.2.3   Microscopic changes

In a careful and detailed description of strawberry fruits during development Knee *et al.* (1977) showed that the growth of the receptacle after petal fall was initially due to a combination of cell division and cell

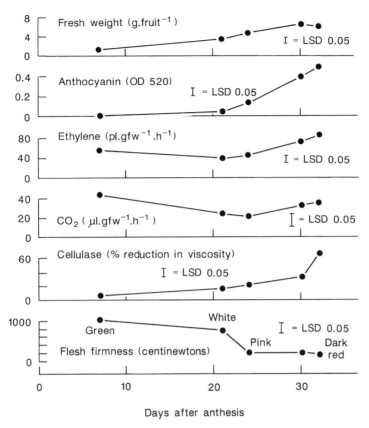

Fig. 12.1 Changes in fruit weight, anthocyanin content, ethylene and CO_2 production, cellulase activity and flesh firmness during strawberry fruit development. (After Abeles and Takeda, 1990.)

expansion. Cell proliferation was complete at seven days and thereafter growth was by means of an increase in cell volume, amounting to about 1000-fold during development. Cell enlargement was accompanied by major changes in the cell wall and subcellular structure. At petal fall, cells in the receptacle had dense cell walls and small vacuoles; starch grains were present in the plastids and Golgi apparatus and ribosomes were abundant. During development the cell walls became swollen and increasingly diffuse and the plastids lost their starch grains, eventually degenerating in ripe fruit. Although changes in the plastids might be considered a senescence characteristic, the mitochondria in ripe fruit appeared normal. The increasing hydration and disorganization of the cell wall and the greater solubility of the middle lamella and wall matrix during ripening accompanied fruit softening.

12.3 BIOCHEMISTRY

12.3.1 Composition

Most of the earlier data on composition is still accurate and relevant today and the reader is referred to Hulme (1971) for much of this information. However, advances in analytical techniques in recent years have improved the quantitative aspects of analysis and this section will concern itself with significant new information.

Sugars

Important in flavour, sugars are one of the main soluble components in soft fruit and provide energy for metabolic changes. For strawberry, raspberry and blackcurrant, sucrose, glucose and fructose account for more than 99% of the total sugars in ripe fruit, with sorbitol, xylitol and xylose occurring in trace amounts (Makinen and Söderling, 1980). Glucose and fructose were present in equivalent concentrations in ripe strawberry fruit (2.3 and 2.2 g.100 g^{-1} fresh weight, respectively), and accounted for about 83% of the total sugars (Wrolstad and Shallenberger, 1981). In the same study glucose, fructose and sucrose were present in raspberry fruit at virtually equal concentrations, approximately 2.0 g.100 g^{-1} fresh weight. For both raspberry and strawberry the content of individual sugars in ripe fruit and the total sugar content varied with the season, but the glucose: fructose ratio and the content of individual sugars as a percentage of the total sugar varied to only a small extent with geographical location, variety and season.

Few data are available on changes in sugar content during the ripening of soft fruit. Forney and Breen (1986) showed that in strawberry, sucrose is present at a very low level in fruit for the first 10 days post-anthesis, but thereafter increased rapidly to reach a maximum at the turning stage before declining sharply in over-ripe fruit. For the first 10 days, glucose declined and fructose remained constant on a fresh weight basis, then both increased. On a per fruit basis total sugar increases throughout strawberry fruit development (Woodward, 1972).

Acids

Like sugars, organic acids are important flavour components and sugar/acid ratios are often used as an index of consumer acceptability and quality in fruits. Acids can affect flavour directly and are also important in processing, since they affect the formation of off-flavours and the gelling properties of pectin. In addition, acids regulate cellular pH and may influence the appearance of fruit pigments within the tissue.

Of the soluble solids, acids are second to sugars. In the strawberry, the non-volatile organic acids are quantitatively the most important in determining fruit acidity. Numerous volatile organic acids have been qualitatively identified in strawberry (Mussinan and Walradt, 1975), and some may be important aroma constituents of the fruit.

Few studies have been made on the changes in acid content in soft fruits during ripening. In the strawberry, total acidity expressed on a fresh weight basis increases modestly to a maximum in mature green fruit before declining more rapidly in the later stages of ripening (Spayd and Morris, 1981). However, on a per fruit basis acid content increases considerably during development and ripening (Woodward, 1972). Reyes *et al.* (1982) showed that the loss of total acidity in overripe fruit was mainly due to a reduction in the content of malic acid. In raspberry too acid content fell during ripening (Sjulin and Robbins, 1987).

Cell wall components

Cell structure, and particularly the nature of cell wall polymers, is believed to determine the physical properties of plant tissues. The problems of how changes in the texture of fruit tissues during ripening are related to cell wall composition has exercised researchers for many years. The rapidity and extent with which soft fruits, in particular, lose their firmness during ripening is a major factor determining fruit quality and postharvest shelf life. However, this has received relatively little attention in soft fruits in general, the strawberry being the most extensively studied. In this fruit, the complex relationship between carbohydrate composition, cell structure and the physical property of the whole tissue is further complicated by increases in cell volume which continue throughout ripening. Furthermore, the net synthesis of polyuronides in the ripening fruit (Woodward, 1972; Huber, 1984) could mask any changes occurring in cell wall polymers laid down earlier in cell development.

Most of the reports on cell wall composition in strawberry have concentrated on the pectic substances, the principal carbohydrate polymer found in the middle lamella. A high proportion of freely-soluble polyuronides exists in the receptacle at petal fall and in ripe fruit (Knee *et al.*, 1977). The pectic substances are relatively loosely bound at all stages of development judging by the ease with which they can be solubilized by EDTA (Woodward, 1972; Knee *et al.*, 1977; Huber, 1984). Huber (1984) found that the proportion of total polyuronide which was soluble increased from 30% in green fruit to 65% in ripe fruit. There was, however, little alteration in the average molecular size of these soluble polymers during development. The increased hydration of the cell wall (Knee *et al.*, 1977) during ripening is consistent with increased polyuronide solubility. Huber (1984) suggested that newly-synthesized poly-uronides added to the wall might affect its structural integrity. This could arise if the newly-synthesized polyuronides were less firmly bound to the cell wall.

Cell wall structure could be determined by the way in which poly-uronides interact with other carbohydrate polymers. Ripening in the strawberry is accompanied by increases in the proportions of the neutral sugars associated with the soluble polyuronides, notably rhamnose, and to a lesser degree arabinose and galactose (Huber, 1984). The increases in these sugars, which can be linked to the polygalacturonan backbone via the rhamnosyl moiety (McNeil *et al.*, 1980), indicate that alterations may occur during ripening in the way in which carbohydrates are cross-linked in the cell wall. While covalent linkages are important in cell wall structure, ionic stabilization between carboxyl groups in pectin molecules by Ca^{2+} (Neal, 1965) are also important and this depends upon the degree of esterification of the pectin.

Xylose is the predominant cell wall neutral sugar in blueberry, blackberry, raspberry and strawberry, the latter having the highest loss of total neutral wall residues (mainly arabinose and galactose) during ripening (Gross and Sams, 1984). Knee *et al.* (1977) found an increase in xylose, mannose and glucose residues in soluble cell wall fractions during ripening of strawberry fruit and suggested that hemicellulosic polysaccharides may be either degraded or released from inter-polymer bonds. Huber (1984) found little alteration in the sugar composition of the hemicellulose fraction from strawberry, but a significant reduction occurred in its average molecular size with ripeness.

Little is known about alterations in cellulose – the major cell wall carbohydrate – during ripening of soft fruit. One report apparently shows a 60% reduction in cellulose content in strawberry fruit during ripening (Spayd and Morris, 1981). However, when recalculated on a per fruit basis the data show that the content of cellulose increases during fruit development by about 2.5-fold.

Pigments

The presence of pigments in fruits is significant for several reasons. Fruits, and particularly soft fruits, many of which are highly coloured, may have evolved pigments to assist in seed dispersal by attracting predators. Some present day cultivars of many fruits, however, may have complements of pigments which are the result of breeding and colour selection. Pigments are important aesthetic components, they are natural indicators of fruit ripeness and some pigments including carotenoids and flavonoids have vitamin activity.

Much of what is known about the pigment composition of soft fruits can be found in Hulme (1971). Wrolstad (1976) has catalogued some of the physico-chemical properties of the anthocyanins found in various fruits.

The perceived colour of fruits can be influenced by co-pigmentation, for example anthocyanins may form complexes with flavones, causing bathochromic colour shifts from red to blue (Asen *et al.*, 1972) and an increase in colour intensity. Fruits such as raspberries may change hue between the ripe and over-ripe stages simply as a result of increased pigment concentration (Sagi *et al.*, 1974). In many fruits minor pigments such as carotenoids will be masked by the more intense pigments such as anthocyanins. During strawberry ripening there is destruction of chlorophyll, consistent with the disappearance of the chloroplasts and a decrease in the content of carotenoids (Gross, 1982).

Phenolics

Phenolics comprise a diverse group of substances including the secondary plant metabolites polyphenols (tannins), proanthocyanidins (condensed tannins) and esters of hydroxybenzoic acids and hydroxy-cinnamic acids. Phenolics are normal constituents of fruits (Hulme, 1971; Foo and Porter, 1980, 1981; Foo, 1981; Schuster and Herrmann, 1985) affecting their taste, palatability and nutritional value. They are responsible for astringency owing to their interaction with the proteins and mucopolysaccharides in saliva (Ozawa *et al.*, 1987). Typical of many edible fruits, soft fruits lose astringency during ripening, although this is not necessarily correlated with a decrease in soluble polyphenol content. Ozawa *et al.* (1987) demonstrated that polysaccharides can disrupt the binding of polyphenols to proteins and suggested that in some fruits the loss of astringency during ripening might be due to interactions between soluble pectins and polyphenols.

Volatile constituents

The flavour quality of food is a combination of the sensory impressions of taste detected by the tongue and aroma sensed by·the nose. Information on the flavour chemistry of natural and processed fruit products is of great importance to the food industry for determining optimal harvesting dates and storage conditions, and for the production of essences. The aroma of soft fruits such as the strawberry is an important component of their flavour. The relative abundance of individual volatiles from a fruit is a 'finger print' of a particular cultivar and species. Extensive studies have been made on the qualitative composition of volatiles from strawberry; esters, alcohols and carbonyls are important for its fruity flavour. Sulphur-containing compounds, while present in lower concentrations than some other volatiles, were also found to be important in strawberry flavour (Dirinck *et al.*, 1981). These authors also detected important flavour differences between cultivars.

The mixture of volatiles from strawberry is complex, with over two hundred compounds having been identified (Tressl *et al.*, 1969). Since volatiles differ in their organoleptic properties only a relatively few of these are likely to contribute significantly to flavour. Of these, furaneol (2,5-dimethyl-4-hydroxy-3(2H)-furanone) is one of the most important, although it is also found in many other fruits (Pickenhagen *et al.*, 1981). In one of the few studies of flavour changes during strawberry maturation Yamashita *et al.* (1977) demonstrated an increase in volatile fatty acids. Since unripe fruit lack any characteristic strawberry flavour it must be assumed that the volatile constituents of ripe fruit which have organoleptic properties are not present during early fruit development.

12.4 METABOLIC PATHWAYS

There is generally a dearth of information on the metabolic pathways and enzymes in soft fruits as they relate to ripening, even for strawberry. This situation not only reflects the relatively low economic value of soft fruit compared with apples, for example, but also the technical difficulties encountered in isolating enzymes and unravelling the biochemistry of recalcitrant fruit such as the strawberry on account of its high levels of polyphenols and soluble pectins which interfere with protein purification.

12.4.1 Sugar metabolism

The processes involved in the accumulation of sugars in fleshy fruits remain poorly understood. The strawberry has received some attention because the fruit are relatively strong sinks for dry matter accumulation, typically accounting for about 40% of the total dry weight of greenhouse (Forney and Breen, 1985b) or field grown plants (Olsen *et al.*, 1985). During fruit development an increasing proportion of ^{14}C-photoassimilate in leaves was found to be exported to the receptacle with a decreasing proportion of this ^{14}C going to the achenes (Nishizawa and Hori, 1988). Sucrose is the main carbohydrate translocated to the fruit (Forney and Breen, 1985c) and fruit growth rates correlate with rates of sucrose uptake (Forney and Breen, 1985a). Sucrose appears to be unloaded apoplastically and hydrolysed in the free space before uptake (Forney and Breen, 1986) and this may account for the greater accumulation of glucose and fructose over sucrose in ripe fruit. Arnold (1968) postulated that the presence of the enzyme invertase (EC 3.2.1.26) in the cell wall could assist in the utilization of sucrose in growing tissues. In a study of soluble and ionically-bound (cell wall) invertase in strawberry fruit Poovaiah and Veluthambi (1985) found high soluble invertase activity at anthesis which decreased rapidly during development, except for a small increase at the time of ripening. They concluded this form was not positively correlated with receptacle growth. However, they observed that a salt-extracted invertase, believed to be cell wall bound, increased during the initial auxin-induced growth of the receptacle, indicating that this form of the enzyme may be important in the sink activity of the fruit.

12.4.2 Cell wall degradation

As already mentioned, the chemical basis of cell wall structure is uncertain. The role of enzymes in cell wall changes leading to softening is equally poorly characterized. An early report of the presence of endo-polygalacturonase (EC 3.2.1.15; PG) in strawberry, an enzyme implicated in the softening of several other types of fruit, by Gizis (1964) has not been consistently substantiated by later workers (Neal, 1965; Barnes and Patchett, 1976; Archer, 1979; Huber, 1984; Abeles and Takeda, 1990). Using a sensitive and specific bioassay for PG, Al-Jamali (1972) detected the enzyme in extracts of strawberry fruit at the onset of softening and a transient rise in PG activity following wounding of the fruit has been observed (Al-Jamali and Dostal, 1976). The disputed presence of PG may partly be due to differences in assay method. Furthermore, Al-Jamali (1972) reported that homogenates of unripe green fruit apparently contained a specific inhibitor of PG activity. If an inhibitor is present in strawberry it does not appear to influence the activity of PG from other fruits (Huber, 1984). Polymethylgalacturonase, the enzyme which is active towards pectin rather than polygalacturonic acid, does not appear to be present in strawberry either (Barnes and Patchett, 1976; Abeles and Takeda, 1990). Thus the view prevails that loss of firmness in strawberry does not occur by the enzymic hydrolysis of the non-esterified galacturonosyl residues of the rhamnogalacturonan backbone.

Pectinmethylesterase (EC 3.1.1.11; PME) activity from strawberry has been reported (Gizis, 1964; Neal, 1965; Archer, 1979; Barnes and Patchett, 1976), the latter finding that PME activity approximately doubled from the small green stage to the ripe stage and then fell to its initial level in over-ripe fruit. Increased pectin methylation in the later stages of fruit ripening would reduce the number of Ca^{2+} cross-links resulting in lowered stabilization of the pectin.

Endo 1,4-β-D-glucanase (cellulase) (EC 3.2.1.4) appears to be the most likely candidate found so far in the enzymic softening of strawberry fruit. Barnes and Patchett (1976) detected cellulolytic activity using a sensitive viscometric assay in fruit at the ripe and over-ripe stages. Abeles and Takeda (1990) observed a six-fold increase in cellulase activity between the green and dark red stages of ripening (Fig. 12.1). In spite of the close temporal correlation between cellulase activity and softening there is little microscopic evidence for major changes in the cellulose matrix during fruit development (Al-Jamali, 1972) and it remains to be shown that the enzyme can significantly degrade native strawberry fruit cellulose.

Few, if any, enzymic studies on softening have been done in other soft fruit. Cellulase activity may be implicated in fruit which abscise, such as raspberry and blackcurrant. In the raspberry there is a well-defined abscission zone between the base of each drupelet and its point of attachment to the receptacle (MacKenzie, 1979). Abscission, measured by fruit retention strength, is generally found to correlate with ethylene production rates and arises from differential growth of cells in regions around the abscission zone (Burdon and Sexton, 1990). Abscission in tissues from other species is known to be accelerated by ethylene and cellulase is induced in cells forming the abscission zones in these tissues (Sexton *et al.*, 1985). It is believed that cellulase activity is responsible for cell separation in these regions, but it has yet to be established if cellulase has a role in abscission for fruit such as the raspberry.

12.4.3 Respiration

Most, if not all, soft fruit are classically defined as non-climacteric on account of a lack of the rise in respiratory activity during ripening. Ethylene production, which is also used as a basis for this classification, is generally lower in non-climacteric than climacteric fruits. Iwata *et al.* (1969a) have challenged this simple distinction between climacteric and non-climacteric fruits based upon respiration rates. From respiration studies of tissue homogenates they grouped fruits into three types:

1. Those in which respiration gradually decreases throughout ripening as exemplified by the orange, a typical non-climacteric fruit (Biale and Young, 1981);
2. Those in which respiration rates increase temporarily with full ripeness occurring after the peak of respiration typical of climacteric fruit such as the tomato;
3. Those in which respiration is at a maximum at the ripe to overripe stages as found in strawberries.

While this system of classification has not been widely adopted it is generally recognized that some fruits are intermediate between climacteric and non-climacteric fruits.

The postulated role of climacteric respiration in supplying additional metabolic energy for synthetic purposes and for utilizing catabolite products may be inappropriate for the strawberry, which shows an intense ripening phase without an apparent requirement for additional metabolic energy. However, Ben-Arie and Faust (1980) reported a three-fold increase in ATPase (EC 3.6.1.3) activity on a fresh weight basis in strawberry fruit between the green and dark red stages, indicating that extra energy for transport processes may be required during this phase of development.

12.4.4   Pigments

The anthocyanin pigments found in soft fruit are flavonoids derived from phenylpropanoid secondary metabolites. The biosynthesis of flavonoids is one of the most studied biochemical pathways in plants but there have until recently been few studies on the enzymes of these pathways in soft fruits.

Fruit anthocyanin synthesis is believed to share a common pathway from the primary metabolic precursor phenylalanine (Fig. 12.2). Phenylalanine ammonia-lyase (PAL) (EC 4.3.1.5), the first enzyme in the pathway, has a pivotal role in directing synthesis towards secondary metabolites and has been most studied. Aoki *et al.* (1970) reported that PAL activity extracted from ripe strawberry fruit was much higher than that from green fruit. A close correlation was subsequently found between PAL activity and anthocyanin levels, which increase during ripening (Hyodo, 1971). However, anthocyanin synthesis continued after PAL activity had reached a maximum. In contrast to apple fruit (Faragher and Chalmers, 1977), light was not required for the increase in PAL activity in the strawberry. Given *et al.* (1988a) confirmed the temporal relationship between PAL activity and anthocyanin accumulation in strawberry (Fig. 12.3). They also showed that the activity of the terminal enzyme in the synthesis of pelargonidin-3-glucoside, the principal anthocyanin in strawberry, uridine diphosphate glucose: flavonoid O^3-transferase (UDPGFT) (EC 2.4.1.91) paralleled this increase in anthocyanin. The enzyme UDPGFT would be expected to be relatively specific to anthocyanin synthesis but since PAL activity also increased in parallel it indicates that *trans* cinnamic acid, the product of PAL, is largely committed to pigment production. However, *in vitro* activities of these enzymes were lower (UDPGFT) and higher (PAL) than apparent rates of anthocyanin accumulation in the fruit.

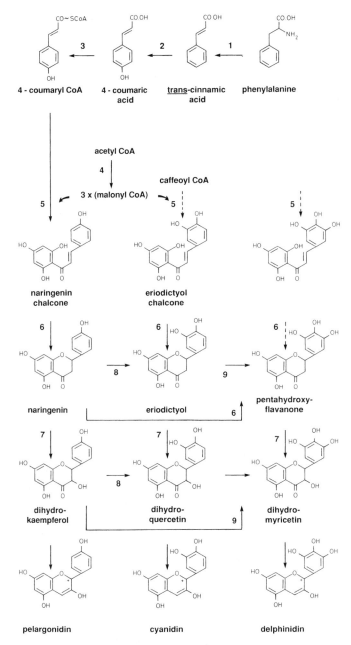

Fig. 12.2   Proposed biosynthetic pathway of anthocyanidins. (1) Phenylalanine ammonia-lyase; (2) cinnamate 4-hydroxylase; (3) 4-coumarate CoA ligase; (4) acetyl CoA carboxylase; (5) chalcone synthase; (6) chalcone isomerase; (7) flavonone-3-hydroxylase; (8) flavonoid 3'-hydroxylase; (9) flavonoid 3',5'-hydroxylase.

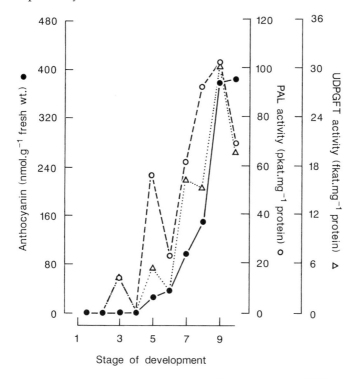

Fig. 12.3   Anthocyanin content and activities of PAL and UDPGFT in ripening strawberry fruit. (From Given *et al.*, 1988a.)

The role of PAL activity in anthocyanin synthesis in the strawberry was examined further by Given *et al.* (1988a) using the competitive inhibitor L-α-aminoxy-β-phenylpropionic acid (L-AOPP). Interestingly, L-AOPP lowered anthocyanin content without affecting fruit softening.

Although natural inactivators of PAL are known (Creasy, 1987), the increase in PAL activity in strawberry was shown not to be due to decreased inactivation of pre-existing enzyme but was the result of *de novo* synthesis (Given *et al.*, 1988b). In contrast to several other species, the enzyme from strawberry is a single isoenzyme having a tetrameric sub-unit structure (Given *et al.*, 1988b). Cross-reactivity with antibodies to PAL from *Phaseolus* indicates considerable homology between the enzymes from these species.

By analogy with other systems the metabolic regulation of PAL activity in soft fruit is likely to be complex (Creasy, 1987). For example, the enzyme is known to be induced and is also probably inactivated by *trans* cinnamic acid, the product of its reaction, and inactivated by endogenous proteases.

12.4.5   Flavour

The biosynthesis of aroma compounds in soft fruits is inevitably complex owing to the large number of volatiles which have been identified. Ripe strawberry fruit are able to metabolize a range of exogenously added aldehydes to their corresponding alcohols and thence to a variety of esters by reaction with endogenous acids (Yamashita *et al.*, 1976). The fruit develop the capacity to esterify 1-pentanol during maturation from immature green to ripe (Yamashita *et al.*, 1977). These reactions are presumably enzymatic, although the enzymes responsible have not been characterized. The metabolism of aroma precursors in strawberry fruit can be modified by cofactors and temperature. Oxygen may be essential for the metabolism of some precursors (Drawert and Berger, 1983).

The question of why so many aroma compounds are present in fruits and of how their synthesis is regulated is an interesting one. It seems unreasonable to expect a different enzyme to catalyse the formation of each compound, as this would require a considerable expenditure of energy for protein synthesis for the production of substances destined to be lost from the fruit. The diversity of aroma substances may simply reflect a lack of absolute specificity of enzymes such as reductases, esterases, oxidases, etc., involved in their formation. Hence several formic acid esters, for example, are found in ripe strawberry fruit, although formic acid is esterified with 1-pentanol at a much slower rate in this tissue than isocaproic acid (Yamashita *et al.*, 1977). Several relatively unspecific enzyme reactions occurring consecutively in a pathway leading to the formation of an alcohol, for instance, might therefore be expected to produce a plethora of different compounds.

12.5   HORMONAL REGULATION OF DEVELOPMENT AND RIPENING

Of the soft fruits the strawberry has been most studied in terms of the regulation of development. Its unique structure with the achenes on the exterior of the fruit has made it a generally useful system to study the role of the seeds in development since the achenes can be removed relatively easily with little physical damage to the remaining parts of the fruit.

12.5.1   Auxin

The classic studies of Nitsch (1950, 1955) using bioassays to estimate levels of endogenous hormones demonstrated that the growth of the strawberry receptacle is regulated by the achenes, that synthetic auxins can restore the growth of receptacles from which the achenes have been removed and that the achenes are a source of indole-3-acetic acid (IAA). Thus the idea arose that the achenes (or seeds) are synthesizers and exporters of auxin to the fruit.

The subsequent development of rigorous analytical techniques enabled the levels of free and conjugated IAA to be determined (Dreher and Poovaiah, 1982; Archbold and Dennis, 1984), although the data are inconsistent. Dreher and Poovaiah (1982) found that free IAA (believed to be the biologically active form of the hormone) was the predominant form in the achenes attaining a maximum level of 3 μg.g^{-1} dry weight at 10 days post-anthesis, with less than 1% of this amount in the receptacle at the same stage of development. Amide-conjugated IAA was reported by these workers to be the principal form of IAA in the receptacle, equivalent to free IAA in the achenes on a per fruit basis. In contrast, Archbold and Dennis (1984) reported substantially higher levels of free (Fig. 12.4) and ester-conjugated IAA in receptacle tissue than Dreher and Poovaiah (1982), but found the level of amide-linked IAA in the receptacle to be much lower at all stages of development. The amide derivative of IAA appears to be the main auxin from day 11 after anthesis, having a bimodal distribution with a peak in concentration at 11 days followed by a decrease, then a further accumulation, in the later stages of fruit maturation.

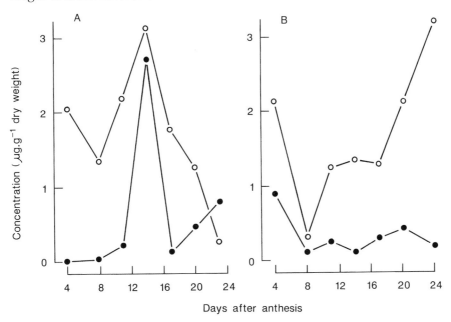

Fig. 12.4 Free indole-3-acetic acid, IAA (A) and free abscisic acid, ABA (B) in strawberry achene (o) and receptacle (•) tissue during fruit development. (From Archbold and Dennis, 1984.)

Although the role of the achenes in fruit growth is undisputed, receptacle growth is not well correlated with endogenous auxin levels and the inter-relationships between the different forms of IAA are incompletely understood. High levels of ester IAA in young achenes may serve as a source of IAA as the achenes develop, supplementing IAA synthesized *de novo*. The accumulation of amide-conjugated IAA in the achenes may either be an end-product of IAA metabolism or a storage form of IAA to be used as a source of free IAA during seed germination.

In a study of the fate of exogenously supplied labelled IAA in receptacles, Lis (1974) reported that labelled IAA-aspartate and IAA-glucose were formed, indicating that the tissue was able to conjugate free IAA. Another study showed that in young developing fruit of the day-neutral cultivar 'Fern', the ethyl ester of IAA, used to induce fruit set, was rapidly metabolized after application, presumably by the action of esterases (Darnell and Martin, 1987). However, the nature of the products was not determined. It is interesting to note that fruit set and development respond to a range of compounds with auxin-like activity in June-bearing and everbearing varieties (Mudge *et al.*, 1981), whereas day-neutral cultivars appear to be more selective (Darnell *et al.*, 1987). These patterns of response may reflect real differences in specificity between cultivars or they may be due to differences in transport or metabolism of these compounds.

Until recently, contradictory results have been obtained concerning the role of the achenes in the ripening of strawberry fruit. Nitsch (1950) and later Guttridge and Nunns (1974) reported that removal of the achenes from receptacles *in situ* did not affect subsequent ripening. On the other hand, Mudge *et al.* (1981) observed that ripening was delayed in parthenocarpic fruit treated with 1-naphthylacetic acid (1-NAA) and suggested that ripening could be manipulated for horticultural purposes by the use of growth regulators. Kano and Asahira (1978) reported that de-achened fruit cultured *in vitro* ripened before intact fruit. In another report, Veluthambi *et al.* (1985) found that ripening behaviour of 1-NAA-treated fruit was comparable to pollinated fruits. Such discrepancies may be due to differences in the concentrations of 1-NAA used, or the methods of application.

Clear evidence for a role of the achenes in ripening has been obtained by Given *et al.* (1988c). This study, using the day-neutral cultivar 'Brighton', showed that removal of the achenes from one half of a large green fruit accelerated ripening in the de-achened half, as measured by anthocyanin accumulation and loss of firmness and chlorophyll (Figs 12.5 and 12.6). The enzyme PAL was induced by this treatment. Furthermore, increases in anthocyanin and PAL activity were prevented by application of the synthetic auxins 1-NAA and 2,4-dichlorophen-oxyacetic acid (2,4-D), but not by the inactive auxin analogue phenoxyacetic acid. The hypothesis was made by Given *et al.* (1988c) that in normal fruit, auxin produced in the achenes inhibits ripening in green fruit and as the fruit develops the level of auxin in the achenes declines, modulating ripening. This implies that the concentration of free IAA in the receptacle falls to a critical level to permit ripening to proceed. It should be noted that high concentrations of various auxins stimulated ripening when applied to de-achened fruit (Guttridge *et al.*, 1977). However, it is not uncommon for supra-optimal levels of auxin to reverse the physiological effects observed at much lower concentrations.

The suggestion that auxin concentration may be the main hormonal factor influencing ripening in the strawberry is supported by the lack of

Fig. 12.5   Effect of removal of the achenes upon ripening of a strawberry fruit. Achenes were removed from one longitudinal half (right) of a mature green fruit *in situ* and the fruit photographed seven days later. Ripening was seen in the right-hand half only.

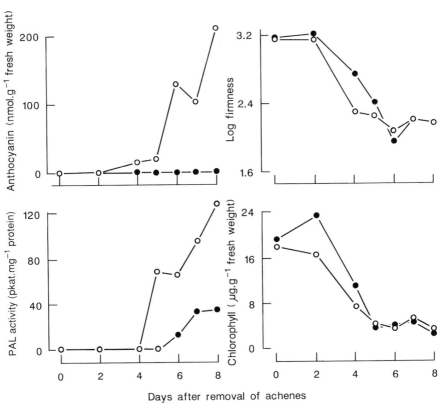

Fig. 12.6 Effect of removal of the achenes on anthocyanin and chlorophyll content, PAL activity and firmness of ripening strawberry fruit. Achenes were removed from one longitudinal half of each of twelve strawberries at the mature green stage. Fruit were sampled at intervals after achene removal and anthocyanin and chlorophyll content, PAL activity and firmness measured in the de-achened (o) and the non-de-achened (●) half of each fruit. (From Given *et al.*, 1988c.)

any apparent involvement of ethylene in this process, as evidenced by the poor correlation with endogenous ethylene production (Knee *et al.*, 1977; Abeles and Takeda, 1990) shown in Fig. 12.1, the unresponsiveness to exogenous ethylene (Iwata *et al.*, 1969b; Hoad *et al.*, 1971; Nestler, 1978) and the insensitivity to inhibitors of ethylene synthesis such as aminoethoxyvinylglycine (AVG) and of ethylene perception including silver and norbornadiene (Given *et al.*, 1988c). This does not exclude the possibility that other factors may have an influence upon ripening. For example, Guttridge *et al.* (1977) demonstrated a reddening effect of fumaric acid, fruit acids and monophenols. The physiological significance of such observations is, however, doubtful.

12.5.2   Gibberellins, cytokinins and abscisic acid

The focus of interest on the role of auxin in strawberry fruit development is due to the ability of this class of growth substance to substitute for normal pollination in the growth of the fruit, but this does not preclude the involvement of other growth substances. Indeed, gibberellin-, cytokinin- and abscisic acid-like activity have been reported in the strawberry (Lis *et al.*, 1978).

The gibberellin GA_3 was reported to act synergistically with 1-naphthaleneacetamide (1-NAAm) in intact fruit cultured *in vitro* (Kano and Asahira, 1978) in promoting growth and ripening. These authors also showed the cytokinin N^6-benzyladenine (BA) suppressed growth and ripening in conjunction with 1-NAAm. They concluded that auxin was the dominant hormone regulating growth and ripening, but that its effects could be modified by other growth substances. The effect of temperature upon achene development in relation to ripening was considered to be related to cytokinin activity in the achene where it is mainly located (Kano and Asahira, 1979). They suggested that ripening in fruit grown at lower temperatures is delayed because high cytokinin-like activity is maintained in the achenes due to their later development. Abscisic acid (ABA) accumulates in the achenes during ripening without increasing substantially in the receptacle (Fig. 12.4). Kano and Asahira (1981) showed that exogenous ABA hastened ripening in receptacles cultured *in vitro*. The results of growth regulator studies performed *in vitro* should, however, be interpreted with caution, since detached fruit fail to reach full size and grow much more slowly than fruit on the plant indicating that essential growth substances might be lacking.

12.5.3   Ethylene

In common with the physiology and biochemistry, the hormonal regulation of development and ripening of soft fruits other than strawberry is little investigated. Most of the studies relating to ethylene have been concerned with the potential use of ethylene-generating substances in improving fruit removal from the parent plant by mechanical harvesting. Ethylene accelerates abscission in a range of plant tissues (Sexton *et al.*, 1985) and the effects of ethylene-releasing compounds have been studied for a number of soft fruits. Van Oosten (1973) found that 2-chloroethyl phosphonic acid (CEPA) promoted ripening in blueberries and gooseberries and enhanced fruit drop in gooseberries, blackcurrants and redcurrants. In the case of raspberries, CEPA did not increase the quantity of fruit recovered by mechanical harvesting but accelerated ripening in this fruit (Fig. 12.7). Ethylene production by raspberry fruit during normal development shows a climacteric rise (Burdon, 1987)

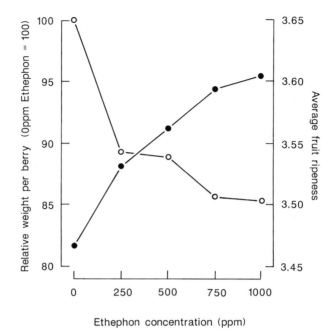

Fig. 12.7 Effects of CEPA concentration on average berry weight (o) and average fruit ripeness (●) of harvested raspberries. CEPA was applied at 4-day intervals during the harvesting period. (After Joliffe, 1975.)

which is believed to stimulate the later abscission of the drupelets, although this may not be the only factor contributing to abscission (Burdon and Sexton, 1990).

12.6   GENE EXPRESSION DURING DEVELOPMENT AND RIPENING

Among the soft fruits the regulation of gene expression during development and ripening has so far been studied exclusively in the strawberry and has been concerned with the mechanism by which auxin may regulate the biochemical events in the fruit. Anthocyanin synthesis, for example, appears to be regulated in ripening strawberry fruit by *de novo* synthesis of PAL (Given *et al.*, 1988b) and inhibited by application of auxin. Soluble and bound forms of invertase may be differentially regulated by auxin during early fruit development (Poovaiah and Veluthambi, 1985). Several examples of auxin-regulated changes in gene expression have been described in developmentally regulated processes in other plant systems (Theologis, 1986), and the induction of enzymes of anthocyanin synthesis and the regulation of their corresponding mRNAs by auxins studied (Ozeki *et al.*, 1990).

Changes in the abundance of specific polypeptides during strawberry fruit development have been reported (Veluthambi and Poovaiah, 1984). Removing achenes from the fruit retarded growth and prevented poly- peptides of 81, 76 and 37 kDa from being formed and supplementing the de-achened receptacles with auxin restored their formation and recept- acle growth. In contrast, 52 and 57 kDa polypeptides were present in de- achened receptacles, but absent in normal fruit and auxin-treated de-achened receptacles. A strawberry variant genotype in which receptacle growth at anthesis depends upon the application of 1-NAA showed a correlation between reduced receptacle growth and the accu- mulation of a 52 kDa polypeptide (Veluthambi *et al.*, 1985). A 52 kDa glycine-rich polypeptide has subsequently been identified in auxin- deprived and non-growing receptacles (Reddy and Poovaiah, 1987). However, a recently isolated and characterized cDNA clone to an auxin- repressed mRNA did not show identity with this glycine-rich poly- peptide (Reddy and Poovaiah, 1990). Two cDNA clones corresponding to two auxin-induced mRNAs have been isolated (Reddy *et al.*, 1990). The levels of these mRNAs correlated with fruit growth and were apparently regulated by endogenous auxin, but were not affected by exogenous ethylene or the ethylene synthesis inhibitor AVG. Thus auxin not only induces but also represses the expression of certain genes in the strawberry during development.

A method developed to isolate nucleic acids from ripe strawberry fruit (Manning, 1991), a particularly recalcitrant tissue, has enabled changes in gene expression to be monitored during ripening. A comparison of *in vitro* translated products from total RNA of receptacles at the white and red stages of ripeness separated by two-dimensional gel electrophoresis shows numerous differences (Fig. 12.8). A similar number of poly- peptides were found to decrease as well as increase during ripening. It follows that in strawberry, at least, fruit ripening is accompanied by changes in gene expression representing both up and down regulation.

For the strawberry we are at an exciting stage in elucidating the molecular events which bring about ripening in this fruit, and how they might be regulated at the hormonal level. Identification of ripening- related genes and their protein products will in due course yield infor- mation on the basic mechanisms underlying the changes in texture, flavour and colour occurring during strawberry fruit ripening. It is expected that this information will lead to a rational approach to mani- pulating these ripening parameters in soft fruit in general for the benefit of consumers and producers alike.

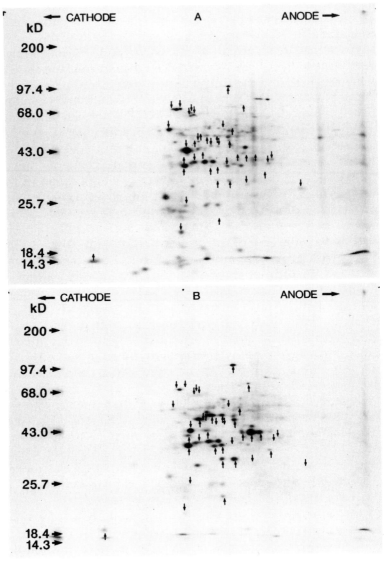

Fig. 12.8 Comparison of *in vitro* translation products from strawberry fruit at the white and red stages of ripeness. Total RNA was isolated from receptacles of white (A) and red (B) fruit and translated *in vitro* in a rabbit reticulocyte system. 35S-labelled polypeptides were separated by two-dimensional polyacrylamide gel electrophoresis using isoelectric focusing in the first dimension (left to right) and size separation in the presence of SDS in the second dimension (top to bottom). The positions of the 35S-labelled products which increase or decrease during ripening are indicated by arrows pointing up or down, respectively. (Unpublished results.)

REFERENCES

Abeles, F.B.and Takeda, F. (1990) Cellulase activity and ethylene in ripening strawberry and apple fruits. *Scientia Horticulturae*, **42**, 269–275

Al-Jamali, A.F. (1972) Cell wall enzymes and senescence in strawberry fruits. PhD dissertation, Purdue University

Al-Jamali, A.F. and Dostal, H.C. (1976) Polygalacturonase activity on consecutively excised strawberry fruit quadrants. *Bulletin of the National History Research Center*, **7**, 127–133

Aoki, S. , Araki, C. , Kaneko, K. and Katayama, O. (1970) L-Phenylalanine ammonia-lyase activities in Japanese chestnut, strawberry, apple fruit and bracken. *Journal of Food Science Technology*, **17**, 507–511

Archbold, D.D. and Dennis, F.G. (1984) Quantification of free ABA and free and conjugated IAA in strawberry achene and receptacle tissue during fruit development. *Journal of the American Society of Horticultural Science*, **109**, 330–335

Archer, S.A. (1979) Pectolytic enzymes and degradation of pectin associated with breakdown of sulphited strawberries. *Journal of the Sciences of Food and Agriculture*, **30**, 692–703

Arnold, W.N. (1968) Selection of sucrose as the translocate of higher plants. *Journal of Theoretical Biology*, **21**, 13–20

Asen, S. , Stewart, R.N. and Norris, K.H. (1972) Co-pigmentation of anthocyanins in plant tissues and its effect on colour. *Phytochemistry*, **11**, 1139–1144

Barnes, M.F. and Patchett, B.J. (1976) Cell wall degrading enzymes and the softening of senescent strawberry fruit. *Journal of Food Science*, **41**, 1392–1395

Ben-Arie, R. and Faust, M. (1980) ATPase in ripening strawberries. *Phytochemistry*, **19**, 1631–1636

Biale, J.B. and Young, R.A. (1981) Respiration and ripening in fruit. Retrospect and prospect. In *Recent Advances in the Biochemistry of Fruits and Vegetables*, (eds J. Friend and M.J.C. Rhodes), Academic Press, London, pp. 1–39

Burdon, J.N. (1987) The role of ethylene in fruit and petal abscission in red raspberry (*Rubus idaeus* L. cv Glen Clova). PhD Thesis, University of Stirling, U.K

Burdon, J.N. and Sexton, R. (1990) Fruit abscission and ethylene production of red raspberry cultivars. *Scientia Horticulturae*, **43**, 95–102

Creasy, L.L. (1987) The role of enzyme inactivation in the regulation of synthetic pathways: A case history. *Physiologia Plantarum*, **71**, 389–392

Darnell, R.L., Greve, L.C. and Martin, G.C. (1987) Synthesis and biological activity of [^{14}C]IAA ethyl ester. *HortScience*, **22**, 97–99

Darnell, R.L. and Martin, G.C. (1987) Absorption translocation and metabolism of Et-IAA in relation to fruit set and growth of day-neutral strawberry. *Journal of the American Society of Horticultural Science*, **112**, 804–807

Dirinck, P.J. , De Pooter, H.L. , Willaert, G.A. and Schamp, N.M. (1981) Flavor quality of cultivated strawberries: the role of the sulphur compounds. *Journal of Agricultural and Food Chemistry*, **29**, 316–321

Drawert, F. and Berger, R.G. (1983) On the biogenesis of aroma compounds in plants and fruits. XXth Communication: Influence of exogenous parameters on aroma biosynthesis in strawberry fruit. *Lebensm.-Wiss. u.-Technol.* , **16**, 209–214

Dreher, T.W. and Poovaiah, B.W. (1982) Changes in auxin content during development in strawberry fruits. *Journal of Growth Regulation*, **1**, 267–276

Faragher, J.D. and Chalmers, D.J. (1977) Regulation of anthocyanin synthesis in apple skin. III Involvement of phenylalanine ammonia-lyase. *Australian Journal of Plant Physiology*, **4**, 133–141

Foo, L.Y. (1981) Proanthocyanidins: gross chemical structures by infra-red spectra. *Phytochemistry*, **20**, 1397–1402

Foo, L.Y. and Porter, L.J. (1980), The phytochemistry of proanthocyanidin polymers. *Phytochemistry*, **19**, 1747–1754

Foo, L.Y. and Porter, L.J. (1981) The structure of tannins of some edible fruits. *Journal of the Sciences of Food and Agriculture*, **31**, 711–716

Forney, C.F. and Breen, P.J. (1985a) Growth of strawberry fruit and sugar uptake of fruit discs at different inflorescence positions. *Scientia Horticulturae*, **27**, 55–62

Forney, C.F. and Breen, P.J. (1985b) Dry matter partitioning and assimilation in fruiting and de-blossomed strawberry. *Journal of the American Society of Horticultural Science*, **110**, 181–185

Forney, C.F. and Breen, P.J. (1985c) Collection and characterization of phloem exudate from strawberry pedicels. *HortScience*, **20**, 413–414

Forney, C.F. and Breen, P.J. (1986) Sugar content and uptake in the strawberry fruit. *Journal of the American Society of Horticultural Science*, **111**, 241–247

Given, N.K., Venis, M.A. and Grierson, D. (1988a) Phenylalanine ammonia-lyase activity and anthocyanin synthesis in ripening strawberry fruit. *Journal of Plant Physiology*, **133**, 25–30

Given, N.K., Venis, M.A. and Gierson, D. (1988b) Purification and properties of phenylalanine ammonia-lyase from strawberry fruit and its synthesis during ripening. *Journal of Plant Physiology*, **133**, 31–37

Given, N.K. , Venis, M.A. and Grierson, D. (1988c) Hormonal regulation of ripening in the strawberry, a non-climacteric fruit. *Planta*, **174**, 402–406

Gizis, E.J. (1964) The isolation and characterisations of the pectic enzymes and the pectic substances of the north west strawberry. PhD Thesis, Oregon State University

Gross, J. (1982) Changes of chlorophylls and carotenoids in developing strawberry fruits (*Fragaria ananassa*) cv. Tenira. *Gartenbauwissenschaft*, **47**, 142–144

Gross, K.C. and Sams, C.E. (1984) Changes in cell wall neutral sugar composition during fruit ripening: a species survey. *Phytochemistry*, **23**, 2457–2461

Guttridge, C.G., Jarrett, J.M., Stinchcombe, G.R. and Curtis, P.J. (1977) Chemical induction of local reddening in strawberry fruits. *Journal of Sciences of Food and Agriculture*, **28**, 243–246

Guttridge, C.G. and Nunns, A. (1974) Promotion of reddening in unripe strawberry fruits by fungal extracts. *Nature*, **247**, 389

Hoad, G.V. , Anderson, H.M., Guttridge, G.C. and Sparks, T.R. (1971) Ethylene and ripening of strawberry fruits. In *Bristol University, Long Ashton Research Station Annual Report*, 33–34

Huber, D.J. (1984) Strawberry fruit softening: the potential roles of polyuronides and hemicelluloses. *Journal of Food Science*, **49**, 1310–1315

Hulme, A.C. (1971) *The Biochemistry of Fruits and Their Products*, vol II, Academic press, New York

Hyodo, H. (1971) Phenylalanine ammonia-lyase in strawberry fruits. *Plant Cell Physiology*, **12**, 989–991

Iwata, T. , Omata, I. and Ogata, K. (1969a) Relationship between the ripening of harvested fruits and the respiratory pattern. II. Respiratory pattern of fruits and its classification. *Journal of the Japanese Society of Horticultural Science*, **2**, 73–80

Iwata, T. , Omata, I. and Ogata, K. (1969b) Relationship between the ripening of harvested fruits and the respiratory pattern. III. Changes of ethylene concentration in fruits and responses to applied ethylene with relation to the respiratory pattern. *Journal of the Japanese Society of Horticultural Science*, **7**, 64–72

Joliffe, P.A. (1975) Effects of ethephon on raspberry fruit ripeness, fruit weight and fruit removal. *Canadian Journal of Plant Science*, **55**, 429–437

Kano, Y. and Asahira, T. (1978) Effects of some plant growth regulators on the development of strawberry fruits in vitro culture. *Journal of the Japanese Society of Horticultural Science*, **47**, 195–202

Kano, Y. and Asahira, T. (1979) Effect of endogenous cytokinins in strawberry fruits on their maturing. *Journal of the Japanese Society of Horticultural Science*, **47**, 463–472

Kano, Y. and Asahira, T. (1981) Roles of cytokinin and abscisic acid in the maturing of strawberry fruits. *Journal of the Japanese Society of Horticultural Science*, **50**, 31–36

Knee, M. , Sargent, J.A. and Osborne, D.J. (1977) Cell wall metabolism in developing strawberry fruits. *Journal of Experimental Botany*, **28**, 377–396

Lis, E.K. (1974) Uptake and metabolism of sucrose-14C and IAA-1-14C in strawberry fruit explants cultivated *in vitro*. *XIX International Horticultural Congress*, 1A:61 (Abstract)

Lis, E.K., Borkowska, B. and Antoszewski, R. (1978) Growth regulators in the strawberry fruit. *Fruit Science Report* V, 17–29

MacKenzie, K.A.D. (1979) The structure of the fruit of the red raspberry (*Rubus idaeus* L.) in relation to abscission. *Annals of Botany (London)*, **43**, 355–362

Makinen, K.K. and Söderling, E. (1980) A quantitative study of mannitol, sorbitol, xylitol and xylose in wild berries and commercial fruits. *Journal of Food Science*, **45**, 367–371

Manning, K. (1991) Isolation of nucleic acids from plants by differential solvent precipitation. *Analytical Biochemistry*, **195**, 45–50

McNeil, M. , Darvill, A.G. and Albersheim, P. (1980) The structure of plant cell walls. X. Rhamnogalacturonan-I, a structurally complex pectic polysaccharide in the walls of suspension cultured sycamore cells. *Plant Physiology*, **66**, 1128–1134

Moore, J.N. , Brown, G.R. and Brown, E.D. (1970) Comparison of factors influencing fruit size in large-fruited and small-fruited clones of strawberry. *Journal of the American Society of Horticultural Science*, **95**, 827–831

Mudge, K.W., Narayanan, K.R. and Poovaiah, B.W. (1981) Control of strawberry fruit set and development with auxins. *Journal of American Society of Horticultural Science*, **106**, 80–84

Mussinan, C.J. and Walradt, J.P. (1975) Organic acids from fresh California strawberries. *Journal of the Agriculture and Food Chemistry*, **23**, 482–484

Neal, G.E. (1965) Changes occurring in the cell walls of strawberries during ripening. *Journal of the Sciences of Food and Agriculture*, **16**, 604–611

Nestler, V. (1978) On the effect of ethephon on strawberry ripening. *Arch. Gartenbau*, **26**, 99–104

Nishizawa, T. and Hori, Y. (1988) Translocation and distribution of ^{14}C-photoassimilates in strawberry plants varying in developmental stages of the inflorescence. *Journal of the Japanese Society of Horticultural Science*, **57**, 433–439

Nitsch, J.P. (1950) Growth and morphogenesis of the strawberry as related to auxin. *American Journal of Botany*, **37**, 211–215

Nitsch, J.P. (1955) Free auxins and free tryptophane in the strawberry. *Plant Physiology*, **30**, 33–39

Olsen, J.L., Martin, L.W., Pelofske, P.J. , Breen, P.J and Forney, C.F. (1985) Functional growth analysis of field grown strawberry. *Journal of the American Society of Horticultural Science*, **110**, 89–93

Ozawa, T. , Lilley, T.H. and Haslam, E. (1987) Polyphenol interactions: astringency and the loss of astringency in ripening fruit. *Phytochemistry*, **26**, 2937–2942

Ozeki, Y., Komamine, A. and Tanaka, Y. (1990) Induction and repression of phenylalanine ammonia-lyase and chalcone synthase enzyme proteins and mRNAs in carrot cell suspension cultures regulated by 2,4-D. *Physiologia Plantarum*, **78**, 400–408

Perkins-Veazie, P. and Huber, D.J. (1987) Growth and ripening of strawberry fruit under field conditions. *Proceedings of Florida State Horticultural Society*, **100**, 253–256

Pickenhagen, W. , Velluz, A., Passerat, J.P. and Ohloff, G. (1981) Estimation of 2, 5-dimethyl-4-hydroxy-3(2H)-furanone (furaneol) in cultivated and wild strawberries, pineapples and mangoes. *Journal of the Sciences of Food and Agriculture*, **32**, 1132–1134

Poovaiah, B.W. and Veluthambi, K. (1985) Auxin regulated invertase activity in strawberry fruits. *Journal of the American Society of Horticultural Science*, **110**, 258–261

Reddy, A.S.N. , Jena P.K. , Mukherjee, S.K. and Poovaiah, B.W. (1990) Molecular cloning of cDNAs for auxin-induced mRNAs and developmental expression of the auxin-inducible genes. *Plant Molecular Biology*, **14**, 643–653

Reddy, A.S.N. and Poovaiah, B.W. (1987) Accumulation of a glycine rich protein in auxin-deprived strawberry fruits. *Biochemical and Biophysical Research Communications*, **147**, 885–891

Reddy, A.S.N. and Poovaiah, B.W. (1990) Molecular cloning and sequencing of a cDNA for an auxin-repressed mRNA: correlation between fruit growth and repression of the auxin-regulated gene. *Plant Molecular Biology*, **14**, 127–136

Reyes, F.G.R. , Wrolstad, R.E. and Cornwell, C.J. (1982) Comparison of enzymic, gas-liquid chromatographic, and high performance liquid chromatographic methods for determining sugars and organic acids in strawberries at three stages of maturity. *Journal of the Association of Official Analytical Chemists*, **65**, 126–131

Sagi, F. , Kollanyi, L. and Simon, I. (1974) Changes in the colour and anthocyanin content of raspberry fruit during ripening. *Acta Alimentaria*, **3**, 397–405

Schuster, B. and Herrmann, K. (1985) Hydroxybenzoic and hydroxycinnamic acid derivatives in soft fruits. *Phytochemistry*, **24**, 2761–2764

Sexton, R. , Lewis, L.N. , Trewavas, A.J. and Kelly, P. (1985) Ethylene and abscission. In *Ethylene and Plant Development*, (eds J.A. Roberts and G. Tucker), Butterworths, London

Sjulin, T.M. and Robbins, J. (1987) Effects of maturity, harvest date, and storage time on postharvest quality of red raspberry fruit. *Journal of the American Society of Horticultural Science*, **112**, 481–487

Spayd, S.E. and Morris, J.R. (1981) Physical and chemical characteristics of puree from once-over harvested strawberries. *Journal of the American Society of Horticultural Science*, **106**, 101–105

Stutte, G.W. and Darnell, R.L. (1987) A non-destructive developmental index for strawberry. *HortScience*, **22**, 218–221

Theologis, A. (1986) Rapid gene regulation by auxin. *Annual Reviews of Plant Physiology*, **37**, 407–438

Tressl, R. , Drawert, F. and Heimann, W. (1969) Gaschromatographisch-massenspektrometrische Bestandsaufnahme von Erdbeer-Aromastoffen. *Z. Naturforsch.*, **B24**, 1201–1202

van Oosten, A.A. (1973) The evaluation of CEPA in relation to harvesting of small fruits. *Acta Horticulturae*, **34**, 391–395

Veluthambi, K. and Poovaiah, B.W. (1984) Auxin-regulated polypeptide changes at different stages of strawberry fruit development. *Plant Physiology*, **75**, 349–353

Veluthambi, K. , Rhee, J.K., Mizrahi, Y. and Poovaiah, B.W. (1985) Correlation between lack of receptacle growth in response to auxin accumulation of a specific polypeptide in a strawberry (*Fragaria ananassa* Duch.) variant genotype. *Plant Cell Physiology*, **26**, 317–324

Woodward, J.R. (1972) Physical and chemical changes in developing strawberry fruits. *Journal of the Sciences of Food and Agriculture*, **23**, 465–473

Wrolstad, R.E. (1976) Colour and pigment analyses in fruit products. *Oregon State University, Agricultural Experimental Station Bulletin*, **624**, 1–17

Wrolstad, R.E. and Shallenberger, R.S. (1981) Free sugars and sorbitol in fruits – a compilation from the literature. *Journal of the Association of Official Analytical Chemistry*, **64**, 91–103

Yamashita, I. , Nemoto, Y. and Yoshikawa, S. (1976) Formation of volatile alcohols and esters from aldehydes in strawberries. *Phytochemistry*, **15**, 1633–1637

Yamashita, I., Iino, K. , Nemoto, Y. and Yoshikawa, S. (1977) Studies on flavor development in strawberries. 4. Biosynthesis of volatile alcohol and esters from aldehyde during ripening. *Journal of Agriculture and Food Chemistry*, **25**, 1165–1168

Chapter 13

Stone fruit

C. J. Brady

13.1 INTRODUCTION

Stone fruit are a diverse group, mostly of the genus *Prunus*, with a characteristic lignified endocarp, a fleshy mesocarp and a thin exocarp or skin. Besides the *Prunus* species the group discussed here includes the olive, *Olea europea*.

The stone fruit or 'drupes' develop from flowers with a superior ovary and have no floral residues around the pedicel. In *Prunus*, there is a characteristic suture that runs down one side of the fruit and this is particularly pronounced in some peach and nectarine varieties. The ventral edge of the seed or pit closes late in development; in cultivars that develop quickly and mature early, the endocarp may not have completed development when the fruit ripen, giving the 'split pit syndrome'.

It was noted in the review of stone fruit by Romani and Jennings (1971) that, despite their commercial importance and their culture in many countries, there were relatively few detailed studies of their biochemistry and physiology. This situation has not changed. Stone fruit attract much research attention, but this is mainly devoted to the cultural and commercial aspects of the array of varieties that are grown. Being grown in a range of environments and being relatively early maturing, breeding to fill particular market niches is feasible. The resulting range of genotypes limits the applicability of studies in depth. A further aspect that limits interest in the molecular details of the development and ripening of stone fruit relates to their relatively short postharvest life. Mostly the fruit are consumed without extended postharvest storage, so there has been a restricted interest in defining the biochemical limits to storage life.

Biochemistry of Fruit Ripening. Edited by G. Seymour, J. Taylor and G. Tucker. Published in 1993 by Chapman & Hall, London. ISBN 0 412 40830 9

13.2   THE *PRUNUS* SPECIES AND THEIR ORIGINS

Prunus species are naturalized in various parts of the world, and the detailed pathways of their evolution are uncertain. In America, *P. chicasa* Michx. the Indian cherry, *P. fasciculata* A.Gray wild almond, *P. ilicifolia* Walp evergreen cherry, *P. maritima* Wangenh. beach plum, *P. pumila* Linn. dwarf cherry and *P. virginiana* Linn. choke cherry are known; some are used locally but none are commercially important. In Europe, *P. avium* Linn. the parent of the sweet cherry, *P. institia* Linn. damson, *P. padus* Linn. bird cherry or hagberry and *P. spinosa* Linn. blackthorn or sloe are grown and utilized locally as food, as preserves or for fermented drinks.

The species that are well known commercially have their origins in Asia or southern Europe. *P. mume* Sieb. and Zucc. is a native of Japan; *P. domestica* Linn., the European plum, is thought to be native to the Caucasus and is now naturalized in Greece and other parts of southern Europe. There are suggestions that the species may have had its origins in Syria. The apricot, *P. armeniaca* Linn. is native to Armenia, Arabia and the higher regions of Central Asia. There are references to wild types near Kabul, in Siberia, and in northern China. Various locations have been given for the origin of *P. amygdalus* Stokes, the almond, including western Asia, the Caspian area and the regions around Lebanon, Kurdistan and Turkistan. Peach, *P. persica* Stokes, has its origins in the orient. It was known in China in the 10th century BC, in Persia in the 4th century BC and in Europe at the beginning of the Christian era. Nectarines were described in Europe in the 16th century AD, and in America in 1720.

13.3   PEACHES AND NECTARINES (*P. persica*)

The biochemistry and physiology of peaches and nectarines are comparable, and apart from the skin tissue, differences within the two groups are likely to be as significant as those between them.

13.3.1   Growth and development

As in other stone fruits there is a distinct pattern of development that was well described by Tukey (1936). Three stages of development are recognized, and growth is described in terms of a double sigmoid pattern. In stage I, cell division is rapid and accompanied towards the end of the period by cell expansion so there is a rapid increase in pericarp volume; stage II is a period of relative quiescence in the pericarp and rapid development of the embryo; in stage III, the endocarp completes its development and the pericarp resumes a rapid increase in volume predominantly but not entirely, due to cell expansion. In early maturing varieties, including the varieties with a low requirement for winter chill, stage II is compressed and endocarp closure may not be complete when the pericarp is mature.

Chalmers and van den Ende (1975) distinguished the pattern of changes in fresh weight from those in dry weight with the transitions between stages occurring earlier for fresh than for dry weight. The transition from stage I to II in fresh weight gain related to a change from pericarp to stone growth (Chalmers and van den Ende, 1977) and was not apparent in the dry weight change. It was suggested that the quiescent stage II in dry weight accumulation was mechanistically related to the increment in embryo dry weight that followed. An increase in soluble sugar concentration in the flesh was detected before the period of rapid growth of the embryo, and the former may have signalled the latter. DeJong and Goudriaan (1989) examined the double sigmoid pattern of fruit growth in terms of daily carbohydrate demands to satisfy growth and respiration. Growth was described as two intersecting regressions of the logarithm of the relative growth rate on time (actually degree days), and differences in maturity could be related to the timing of the shift from the first to the second regression. Carbohydrate requirements were calculated from dry matter accumulation and the regression of respiration rate on relative growth rate. For two cultivars, the carbohydrate demand as grams per fruit per day showed a classical double sigmoid pattern that may explain why vegetative growth tends to accelerate when most of the fruit crop are in stage II. Differences in fruit size between cultivars appear to relate to differences in cell number that are apparent in stage I (Scorza *et al.*, 1991).

Respiration rate and ethylene production are high during stage I when growth is predominantly by cell division. Respiration rates are 150 ml CO_2.kg^{-1}.h^{-1} in the early part of stage I, falling to 30–40 ml kg^{-1}.h^{-1} through stage II and rising gradually through stage III to a climacteric peak of 50–80 ml kg^{-1}.h^{-1}. Ethylene production rates are in the range 0.2 to 0.4 µl.kg^{-1}.h^{-1} early in stage I falling to 0.01 to 0.02 µl.kg^{-1}.h^{-1} in stage II rising late in stage III, as ripening progresses to a rate in excess of 20 µl.kg^{-1}.h^{-1} and sometimes reaching 100 µl.kg^{-1}.h^{-1} (Looney *et al.*, 1974).

Levels of 1-aminocyclopropane-1-carboxylic acid (ACC) reflect the pattern of ethylene evolution through stages I and II. Miller *et al.* (1988) found little increment in the ACC content of mesocarp and seed tissues through the climacteric rise. Consistent with this, ethylene production was stimulated by ACC addition to climacteric fruit (El-Agamy *et al.*, 1981). On the other hand, ACC and malonyl-ACC accumulation were found in the pericarp of 'Redhaven' peaches through stage II, and ACC was observed to rise and malonyl-ACC to decline through stage III

(Tonutti *et al.*, 1991). ACC concentrations in the range of 1 to 4 µl.kg^{-1} were found by Amoros *et al.* (1989) and Tonutti *et al.* (1991). Malonyl-ACC levels were 2 to 12 µmol.kg^{-1} .

Genotypes vary in rates of ethylene production. El-Agamy *et al.* (1981) observed a lower rate of ethylene production by 'Flordagold' than the Fla 3-2 selection and suggested that there may be a relationship between this and the firmness of 'Flordagold'. Biggs (1976) also suggested that firm cultivars have relatively low rates of ethylene evolution through ripening.

Tsuchida *et al.* (1990) reported an increase in abscisic acid in harvested peach fruits and an increase in phaseic acid through ripening and senescence. Abscisic acid concentrations reached a maximum of 4.2 mg.kg^{-1}. Looney *et al.* (1974) observed an increase in abscisic acid commencing about the stage II to III transition and continuing in attached fruits through stage III to concentrations about 50 µg.kg^{-1} . Fruits in stage III ripen rapidly after harvest, and Looney *et al.* suggest that this may relate to the abscisic acid content, implying a role for abscisic acid in the regulation of ripening. They imply further, that since detachment of the fruit at this stage results in precocious ripening, the tree supplies an unstable ripening inhibitor to the fruit. An increase in abscisic acid through the climacteric was also measured by Tsay *et al.* (1984).

Miller *et al.* (1987) measured IAA levels and ethylene production through the development of 'Redhaven' peaches. IAA concentrations were initially – 18 days after anthesis – 20 to 30 µg.kg^{-1} , fell with time and into stage II before rising late in stage II and through stage III. The mesocarp concentrations increased through stage III to 30 µg.kg^{-1}. Ethylene production was related to the IAA concentration except that the rise in IAA content in stages II and III was not reflected as an increase in ethylene production until ripening commenced. Miller and Walsh (1990) measured 2.8 to 6.5 µg.kg^{-1} IAA in the pericarp of ripening fruit and 239 to 1042 µg.kg^{-1} in the seed.

13.3.2   Fruit composition

The nutrient composition of commercial stone fruit is well described in food composition tables and will not be repeated in detail here. Major components that may relate to metabolic controls are, however, pertinent. As in other Rosaceae, the stone fruits translocate carbon as sorbitol as well as sucrose (Manolov *et al.*, 1977), and the fruits contain significant levels of sorbitol. Young fruits store some carbon as starch, but this is used before the fruit enter stage III and starch to sugar conversions are not involved in ripening. Ripe fruit contain significant amounts of free galacturonic acid, presumably a product of pectin breakdown (Ash and Reynolds (1954). The fruit are decidedly acid, with acid levels in excess of 1% as malate and a sap pH sometimes below pH3.5. The malate level decreases, and the citrate concentration increases through ripening, but the overall acid level declines and the pH rises slightly (Kakiuchi *et al.*, 1981).

Soluble solids in mature peaches and nectarines should be in excess of 10% for acceptable fruit quality; typically 65 to 80% of the soluble solids are sugars; sucrose is the dominant sugar and glucose, fructose and sorbitol are significant components. Cultivars differ widely in the glucose, fructose, sorbitol ratios, and such variation may contribute to flavour differences.

Alcohol-insoluble solids in the mature fruit constitute 2.2% to 2.5% of the fresh weight, and this is mostly cell wall material. In ripe fruit, 24% to 29% of the alcohol-insoluble substance was found to be pectin in four peach cultivars (Shewfelt *et al.*, 1971). This percentage did not change consistently through ripening, although the chelate-soluble proportion may increase 2.5-fold. On a fresh weight basis, Ben-Arie and Lavee (1971) measured an increase in total pectin in ripening fruit, and confirmed an increase in chelate-soluble pectin through ripening, and in cold-stored fruit that develop the internal breakdown or 'woolly' symptom. Peach pectins are highly esterified, but there is a slight decrease in esterification in the latter stages of ripening (Ben-Arie and Lavee, 1971; Shewfelt *et al.*, 1971). Using X-ray microanalysis, Burns and Pressey (1987) detected calcium in the pericarp tissue in the range 0.3–0.4 mg.g^{-1} dry cell wall. This low concentration, which did not change with ripening, was related to the high degree of esterification of the pectins. Anatomical and ultrastructural aspects of wall development were described by King *et al.* (1987).

The non-pectin cell wall polymers have received little attention. Gross and Sams (1984) measured decreases in cell wall non-cellulosic neutral sugar of 39% in NY-2603 nectarine and 28% in Loring peach from the pit-hardening to the ripe stage. The loss was mainly through stage III before ripening. Losses of arabinose and galactose accounted for most of the change. Losses of cell wall galactose and arabinose were also measured by Dawson *et al.* (1991) in nectarine fruit in the six days following commercial harvest. The loss of galactose was spread through fractions soluble in chelates, sodium carbonate and guanidium thiocyanate, while the arabinose loss was in the guanidium and 4M potassium hydroxide fractions, indicating that different polymers were involved. A loss of cellulose from ripening peaches was reported by Sistrunk (1985).

The content of phenolic compounds is greatest through stage I and then declines (Kumar, 1987). Phenolic compounds were found by Robertson *et al.* (1988) to vary from 17 to 141 mg% in mature peach fruit across four cultivars. Cultivars of low quality had distinctly higher phenolic content. The prominent phenolic compounds are chlorogenic acid, epicatechin, catechin, cyanidin and caffeic acid derivatives. Minor components are p-hydroxyphenylacetate, quercetin, myricetin, delphinidin, and gallic, genistic and protocatechuic acids (Hermann, 1989; Senter et al., 1989). No marked changes in the phenolic compounds were observed during storage or ripening.

There is an excellent review of the role of lactones in flavour development in peaches and nectarines in Romani and Jennings (1971). The role of gamma and delta lactones in nectarine flavour is emphasized also by Engel et al. (1988a,b) who describe an increase in the length of aliphatic side chains through ripening. An extensive analysis of volatiles in tree-ripened nectarines is presented by Takeoka et al. (1988), who also emphasize the prominence of lactones and other products of the peroxidation of unsaturated fatty acids.

Peach lipids have received little attention. The changes in the pigments during the growth and development of the peach cultivar 'Earligro' were studied by Lessertois and Moneger (1978). Contents of chlorophyll a and b, and the carotenoids were measured at intervals as the fruit developed. Carotenoid concentrations were highest at the earliest samplings (about 70 mg.kg^{-1}), declined through development and declined rapidly through stage III; only traces were present in ripe fruit. b-carotene, cryptoxanthin and its epoxides, zeaxanthin, antheraxanthin, auroxanthin and trollichrome were the prominent carotenoids. The chlorophyll levels were highest (510 mg.kg^{-1}) towards the end of stage I and then declined progressively. The loss of mesocarp chlorophyll is commonly used as an index of maturity (Delwiche and Baumgardner, 1985) to judge commercial harvest. Peach fruit pigments were also studied by Katayama et al., 1971).

13.3.3   The enzymes of the peach fruit

A low protein content, the presence of phenolic compounds and a very high content of soluble pectic compounds present difficulties for those who would study peach fruit enzymes, and there have been relatively few detailed studies reported. The glycolytic, gluconeogenic and mitochondrial enzymes have not attracted detailed study. The enzymes involved in carbon storage have been studied, as have the enzymes that contribute to enzymic browning. Enzymes related to cell wall metabolism have attracted some interest.

In developing fruit, sucrose synthase (EC 2.4.1.13) and invertase (EC 3.2.1.26) activities decrease sharply relative to fruit weight through stage I and there is minimal activity in stage II (Fig. 13.1). Sucrose synthase activity increases as sucrose accumulates through stage III, and there is a slight increase in invertase in the ripening fruit. There is little sucrose phosphate synthase (EC 2.4.1.14) activity in fruit of any maturity.

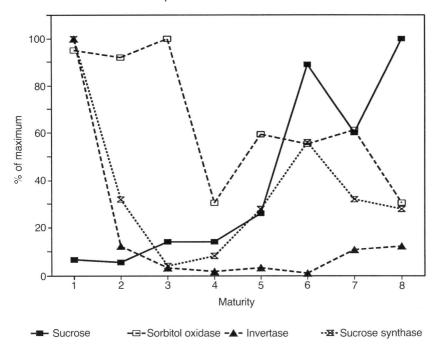

% of maximum

Maturity

—■— Sucrose   —□—Sorbitol oxidase —▲— Invertase   ··⊠·· Sucrose synthase

Fig. 13.1   Changes in soluble carbohydrates, sucrose synthase, acid invertase, and sorbitol oxidase in developing peach fruit. (Modified from Moriguchi *et al.*, 1990.)

Sorbitol oxidase activity is maintained in the range 50–100 μmol.kg^{-1}.h^{-1} throughout fruit development (Moriguchi *et al.*, 1990). Peach fruit sucrose synthase is a tetramer of identical 87 kDa subunits. The purified enzyme has a specific activity of 2 μmol.min^{-1}.mg^{-1}, K_m(fructose) 4.8 mM, K_m(UDPglucose) 0.033 mM, K_m(sucrose) 62.5 mM and K_m(UDP) 0.08 mM (Moriguchi and Yamaki, 1988). Much of the acid invertase in mesocarp tissue is firmly bound to particulates and is not readily solubilized (Ugalde *et al.*, 1988).

ACC synthase (EC 4.4.1.14) and ethylene-forming enzyme (EFE) increase during ripening. In ripening nectarines, the EFE capacity was found to be at least twice the capacity of ACC synthase, although ACC accumulated (Brecht and Kader, 1984c), indicating that compartmentation or the *in vivo* environment modified the activities. In the presence of exogenous ethylene, the synthesis of ACC was inhibited in stage II fruit (Brecht and Kader, 1984a), while the conversion of ACC to ethylene was inhibited in mature fruit.

Four polyphenoloxidase (EC 1.10.3.1) isoforms were recovered from peach fruit extracts by Wong *et al.* (1971a,b) and characterized in terms of substrate preferences. Catechol oxidase (EC 1.10.3.1) and laccase (EC 1.10.3.2) (Harel *et al.*, 1970), and peroxidase (EC 1.11.1.7) and polyphenoloxidase (Flurkey and Jen, 1978) activities were monitored in developing or stored peach fruits. Polyphenoloxidase was partially purified from 'Redhaven' peaches. The purified samples contained two forms of active protein, separable by polyacrylamide gel electrophoresis, and with pIs in the 4.5 to 5.5 range. There was evidence that the enzymes were glycoproteins. A computer video imaging system was used to analyse oxidative browning in developing peach fruit (Stutte, 1989) and related to the local distributions of polyphenol oxidase activities.

Glucosidase and galactosidase activities in developing 'Redhaven' peaches were assessed by Kupferman and Loescher (1980). The activities increased, relative to dry weight or extracted protein, through development particularly galactosidase activity late in stage III.

Pectin-metabolizing enzyme activity has been measured in relation to the internal breakdown symptom that develops in mature peaches stored at low temperature for several weeks. Internal breakdown leads to a 'mealy' or 'woolly' texture and a reduction in juice yield (see Lill *et al.*, 1989 for an extensive discussion of the postharvest physiology of peaches and nectarines). A difference in pectin extraction and a slight difference in pectin esterification led to the hypothesis that the internal breakdown symptoms are an expression of a disruption of the developmental changes late in stage III that normally lead to fruit softening (De Haan, 1959; Watkins, 1964; Ben-Arie and Lavee, 1971). The hypothesis awaits clear definition but it involves these concepts:

1. pectinmethylesterase is present in mature fruits;
2. polygalacturonase is not present in mature fruits;
3. during low temperature storage, pectins are de-esterified;
4. low-ester, high molecular weight pectins accumulate at low temperature forming a hygroscopic matrix.

Pectinmethylesterase

The molecular characteristics of peach fruit pectinmethylesterase (EC 3.1.1.11) are unknown, but there is evidence of three basic proteins of M_r about 32kDa (Glover, unpublished). Levels of activity in the range 10 to 15 milliequivalents.kg^{-1}.h^{-1} are found in mature fruit (Ben-Arie and Lavee, 1971; Ben-Arie and Sonego, 1980). An increase in the level of active enzyme was observed after two weeks of low temperature storage (Ben-Arie and Sonego, 1980; von Mollendorf and de Villiers, 1988a,b), although not by Buescher and Furmanski (1978). There is no understanding of the regulation of pectinmethylesterase activity *in situ*, but clearly the potential measured *in vitro* is not fully expressed in the tissues; thus there is no certainty that an increase in assayed activity reflects a change in substrate esterification in cell walls. The presence of multiple forms of the enzyme raises other possibilities for changes in control through development and/or low temperature storage.

Polygalacturonase

Important to the evolution of the hypothesis of a role for modified pectins contributing to internal breakdown is the repeated observation that the 'woolly' symptoms do not develop if the fruit are allowed to commence ripening before they are stored at low temperature (von Mollendorf and de Villiers, 1988a). As ripening commences, poly-galacturonase (EC 3.2.1.15) activity can be measured (von Mollendorf and de Villiers, 1988b), so the fruit are stored when they have both pectinmethylesterase and polygalacturonase activities, allowing a balanced catabolism of the pectins. Buescher and Furmanski (1978), however, suggested that the unbalanced catabolism of the pectins occurred when fruit were warmed after low-temperature storage, and that there was a delay in polygalacturonase accumulation in stored fruit.

During normal ripening, polygalacturonase activity increases as the fruit soften; as the enzyme accumulates, the solubility of the pectins increases (Pressey *et al.*, 1971) suggesting, but not establishing, a causative connection. Exo- and endo-acting polygalacturonases have been distinguished in ripe peaches (Pressey and Avants, 1973); in a survey of a range of cultivars, the endo-acting enzyme (EC 3.2.1.15) was found in significant amounts only in freestone types; the exo-enzyme (EC 3.2.1.67) had comparable activity in the mesocarp of freestone and clingstone fruit (Pressey and Avants, 1978).

Two forms of exo-polygalacturonase were separated by Downs and Brady (1990). Each form increased through ripening. The enzyme proteins of M_r 66 kDa were separable by ion-exchange chromatography; the proteins were both glycosylated, but antibodies to one form failed to recognize the other. The antibodies provided evidence that the increase during ripening in the activity of this enzyme was due to the synthesis of the enzyme protein (Downs *et al.*, 1992). The exo-acting enzymes are activated by calcium ions and in the presence of 1mM calcium have pH optima in the range 6.0–6.4.

The endo-acting enzyme has not been purified. Lee *et al.* (1990) described a genomic sequence in peaches that has extensive homology with the tomato endo-polygalacturonase sequence. Figure 13.2 compares the structures of the tomato and peach genes. The peach gene is the more compact structure. The introns in the two genes are similarly placed, but the peach introns are shorter and intron 5 of the tomato gene is not found in peach. There is 54% homology between the two genes. There is in ripe,

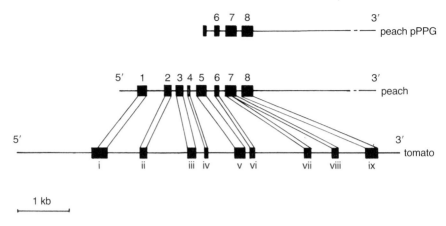

Fig. 13.2 A comparison of the endopolygalacturonase gene in tomato and in peach. Exons are shown as solid bars and introns as lines. The tomato exons (Bird *et al.*, 1988) are labelled i to ix. The peach exons, 1 to 8, are as described in Lee *et al.* (1990) with additional sequence provided by Spiers and Lee (unpublished data).

but not in unripe fruit, an antigen of 47 kDa that is recognized by antibodies raised against tomato endo-polygalacturonase (Lee *et al.*, 1990).

The peach endo-polygalacturonase has maximal activity at pH 4.5, is inhibited by calcium, binds strongly to concanavalin A, indicating glycosylation, and is strongly basic. Both endo and exo-polygalacturonase increase in activity gradually as fruits soften, but the rate of increase in activity accelerates when the fruit are very ripe (Fig. 13.3). In the case of the one exo-polygalacturonase, for which there is immunological evidence, an increment in enzyme protein accompanies the rapid, late increase in activity. The mechanisms of regulation involved in this pattern of enzyme accumulation are otherwise unknown. Regulation is of interest because the pattern of accumulation is quite different from that occurring in tomato and avocado. It is also of interest that although the peach fruit becomes very soft, and a change in pectin solubility accompanies softening and endopolygalacturonase as in tomato, the enzyme level in peach is less than 0.1% of that in tomato.

The molecular details involved in the pre-conditioning treatments that limit internal breakdown in cold-stored peaches will not be apparent until the controls exercised through normal ripening are better understood. It is known that changes occur in the population of messenger RNAs in the mesocarp tissues of peaches during ripening (Callahan *et al.*, 1987, 1989), and that gene expression in fruits is cultivar specific (Morgans *et al.*, 1987). One mRNA that changes is related to pTom13, a

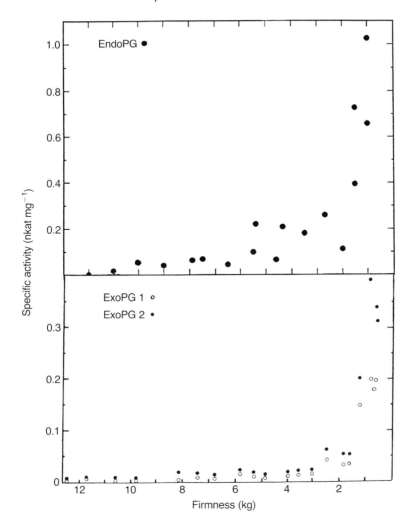

Fig. 13.3   The changing activities of endopolygalacturonase (upper) and exopolygalacturonase (lower) in softening freestone peach fruits. Activities are related to fruit compressibility measured with a penetrometer, with activities recovered from individual fruits presented. (Data in upper panel from G. Orr; data in lower panel modified from Downs *et al.*, 1992.)

clone of the gene for an ethylene-forming enzyme in tomato (Chapter 14; Callahan *et al.*, 1990). Expression of the polygalacturonase genes may also changes through ripening (Lester, unpublished); since allowing some polygalacturonase accumulation before storage appears to reduce symptoms, it might be suggested that it is not the enzyme that is cold sensitive, but the mechanisms that lead to its accumulation. The mechanism involved may extend from the expression of the particular polygalacturonase gene to the complex of controls that result in normal ripening. The latter complex is indicated by evidence that an elevated level of carbon dioxide (a competitive inhibitor of ethylene binding) through the preconditioning period maintains fruit firmness and limits internal breakdown through subsequent cold storage (Nanos and Mitchell, 1991).

An enzyme that reduces the viscosity of carboxymethyl cellulose was measured in peaches (Hinton and Pressey, 1974). The activity developed with maturity and increased during ripening, but the correlation with the rate of softening was poor. The activity was maximal at neutral pH and was stimulated by 0.01 M calcium chloride and by 0.1 M sodium chloride. It is likely that, as in other fruits, more than one enzyme contributes to this activity in peaches and it is unlikely that cellulose is the true substrate. The peach extracts showed no ability to degrade cellulose or cellobiose.

13.3.4   Ripening mutants

Late-maturing and slow-ripening mutants of 'Fantasia' nectarine were described by Brecht *et al.* (1984). The mutant lines softened and developed flesh colour ahead of a delayed and sometimes protracted respiratory and ethylene climacteric. Ripening and autocatalytic ethylene production was advanced by external ethylene. The slow-ripening characteristic appears to be controlled by a single recessive gene (Ramming, 1991). The slow-ripening lines had low levels of ACC and, at the time when the normal type was ripening, was deficient in both ACC synthase and EFE. Wounding did not result in increased ethylene production, but the addition of ACC to wounded tissue did, indicating that EFE but not ACC synthase activities were stimulated by wounding (Brecht and Kader, 1984b).

13.4   PLUM (*P. domestica*)

Like other stone fruit, plums have a triphasic pattern of development. The fruit are rich in sugars which accumulate rapidly as the fruit mature. Sucrose is the main sugar, but glucose, fructose and sorbitol are also important. Sucrose may form 8% of the fresh weight of some fruit. There is considerable variation between cultivars in sugar content and in the proportions of the four major sugars (Vitanov *et al.*, 1988; Roemer, 1990). Total sugar content is often 10–12%, being the major component of the 15–20% of dry matter of the fruit. The acid content of the fruit decreases as the fruit ripen (Pangelova-Shchurkova and Vitanova, 1987).

Caffeoylquinic acids are the major phenolic components with *p*-coumaryl glucose, feruloyl glucose and caffeoyl glucose as significant components (Hermann, 1989). Phenolic compounds increase when harvested fruits are stored (Filgueiras and Chitarra, 1988).

Plums are typical climacteric fruits. Preclimacteric respiration rates of 5 mg.kg^{-1}.h^{-1} and climacteric rates of 40 mg.kg^{-1}.h^{-1} were recorded by Sekse (1988). The preclimacteric rate is unusually low. An ethylene climacteric accompanies the respiratory peak. Applications of ethylene advance anthocyanin and phenolics accumulation and stimulate sugar storage in Santa Rosa plums (Lee and Lee, 1980).

Plums, particularly the high sugar types, store at low temperature more satisfactorily than peaches and nectarines. They are, however, subject to internal breakdown after storage (Kotze *et al.*, 1989). There has been no detailed chemical or biochemical study of internal breakdown in plums. In general, plums have attracted the curiosity of biochemists to a remarkably small degree.

13.5 MUME, OR JAPANESE APRICOT (*P. mume*)

Mume is a climacteric sugar-rich fruit. The climacteric in ethylene production is accompanied by an increase in ACC synthase activity and an accumulation of EFE. Both ACC synthase and EFE activities decrease after cycloheximide treatment demonstrating that they are subject to rapid turnover (Sawamura *et al.*, 1988). The inhibition of ethylene synthesis by exogenous ethylene has been demonstrated in mume fruit discs (Sawamura and Miyazaki, 1989). A role for ethylene in mume fruit senescence was shown by Osajima *et al.* (1987) in a study of the storage of the fruits in modified atmospheres.

Pectins become more soluble as mume fruits ripen and the HCl-soluble pectin fraction decreases. In common with other stone fruit, there is a decrease in malic acid and some increase in citric acid through ripening. Goto *et al.* (1988) measured an increase in the phosphatidyl choline: phosphatidyl ethanolamine ratio as mume fruits matured and related this ratio to an increase in K$^+$ leakage and a decreased sensitivity to chilling injury.

13.6 APRICOT (*P. armeniaca*)

A detailed survey of the composition of the climacteric apricot fruit is given by Joshi *et al.* (1986). The mature apricot fruit contains 10–12% soluble sugar, but only a small portion of this is sucrose. Ketose oligosaccharides, apparently fructosans, have been detected in apricots (Ash and Reynolds, 1954), possibly reflecting transfructosidase activity, but this activity has not been studied in any detail. The accumulation of sugars through development of the apricot fruit was described by Nigam and Sharma (1987), and of amino acids by Nigam and Sharma (1988). Asparagine is the most prominent free amino compound.

Gamma-lactones contribute strongly to the distinctive flavour of the apricot. The lactones were studied in detail by Guichard *et al.* (1990), who related the flavour and aroma of six cultivars to the content of the individual lactones and the balance between them. The biochemistry of the synthesis of the flavour components and of the carotenoid pigments, including vitamin A, awaits detailed elaboration.

Dried apricots suffer textural deterioration in storage, and endogenous polygalacturonase activity had been implicated in this. Polygalacturonase activity has been measured in mature fruits, but not characterized as to its mode of pectin hydrolysis nor its molecular form. The contribution of polygalacturonases or other enzymes to the softening of ripening apricots is not defined.

13.7   ALMOND (*P. amygdalus*)

Pectin substances in the walls of the pericarp dehiscence zone of almond fruits were studied using a fluorescent cytochemical technique (Weis *et al.* 1988). Walls of mesocarp tissue were shown to contain arabinose, galactose, xylose and glucose as well as galacturonic acid, rhamnose and mannose (Morrison *et al.*, 1987). Polygalacturonase, α-1,3-D-glucanase, α-galactosidase (EC 3.2.1.22), β-galactosidase (EC 3.2.1.23), α-arabinosidase (EC 3.2.1.55) and α-mannosidase (EC 3.2.1.24) activities were demonstrated in the mesocarp and related to gum duct formation.

13.8   TART OR SOUR CHERRY (*P. cerasus*)

Greatest interest in the chemistry and biochemistry of *P. cerasus* has been in the flavour volatiles. The fruit of this species has 16% to 20% soluble solids when mature; 60% of the soluble solids are sugars, and fructose is by far the most prominent sugar. The malic acid content is high and decreases somewhat after harvest.

The chlorophyll and carotenoids (Ben and Gaweda, 1987a) and anthocyanins (Ben and Gaweda, 1987b) were studied in developing fruit. Benzaldehyde, benzyl alcohol, benzoic acid, eugenol and monoterpene glycosides were identified in extracts of tart cherries by Schwab and Schreier (1990), and further characterized by Schwab *et al.* (1990).

13.9 SWEET CHERRY (*P. avium*)

While some older studies report that the cherry is a climacteric fruit, more recent studies have failed to detect ripening associated changes in respiration or ethylene production. Sekse (1988) observed a continually declining respiration rate in harvested cherries and no peak of ethylene production. In a study of fruit ripening on the tree, Hartmann (1989) measured a higher rate of ethylene production and a higher content of ACC and its malonate derivative in ripening than in younger fruits, but failed to demonstrate a climacteric pattern. An increase in protein synthesis, of which an increase in polysomes is evidence, is seen in climacteric fruits. Analyses of polysomes in maturing cherry fruits gave no evidence of an increment in the rate of protein synthesis associated with ripening in cherries (Drouet *et al.*, 1983; Fils-Lycaon *et al.*, 1988). Detached fruit in the dark failed to ripen even in the presence of exogenous ethylene. Hartmann *et al.* (1987) classified the cherry as non-climacteric and contrasted its responses to ethylene to those of the climacteric apple and pear fruits.

Cherry fruits accumulate high concentrations of sugars from translocated sucrose and sorbitol. Invertase was partially purified from the mesocarp tissue (Krishman and Pueppke, 1990). Two enzymes were identified by separation through polyacrylamide gels. Both activities contained 63 kDa polypeptides. The partially purified holoenzyme had a molecular weight of 400 kDa and contained 63 kDa subunits. The K_m for sucrose was 4 mM and the pH optimum 5.0.

The variations between cultivars in the contents of amino acids and of anthocyanins were surveyed by Katsarova *et al.* (1987). Pifferi and Cultrera (1974) partially purified two forms of polyphenol oxidase by ammonium sulphate precipitation, ion exchange and gel permeation chromatography and polyacrylamide gel electrophoresis. The forms differed in the molecular size of the active protein and in pH optimum. The enzymes oxidized the anthocyanin, cyanidin-3-glucoside in the presence of pyrocatechol and chlorogenic acid; pyrogallol and *p*-cresol were poor substrates, and hydroquinone and resorcine ineffective substrates, for the enzyme. It was suggested that polyphenol oxidase, in the absence of ascorbic acid was responsible for the destruction of the anthocyanin in cherries. Presumably, tissue disruption is needed to facilitate this reaction as polyphenol oxidase normally has a plastid location.

Lipids in fresh, ripe cherries were studied by Bishop and Wade (1977). Lipids comprised about 0.1% of the fruit weight and were composed of 31% neutral lipid, 24% glycolipid and 45% phospholipid. The neutral lipid fraction contained wax esters, sterol esters, free sterol and triterpene acids. The triterpene acids, which formed about 20% of the neutral lipid were mainly ursolic and oleanolic acid. Acylated sterol glucosides, monoglycosyldiacylglycerol and diglycosyldiacylglycerol formed the glycolipids and phosphatidylcholine and phosphatidyl-ethanolamine accounted for 80% of the phospholipids. Sitosterol formed over 90% of the sterols in different fractions. Fatty acid analyses were made for each fraction and 16:0, 18:1, 18:2 and 18:3 acids were major components in each. The neutral lipids were more and the glycolipids less saturated than the other fractions.

The pectins of cherry fruits were characterized by Barbier and Thibault (1982) and Fils-Lycaon and Buret (1990), and some structural details were determined by Thibault (1983). Fruit softening was associated with an increase in total pectins per fruit. As the fruit softened, and the total pectin content increased, pectin solubility and the ratio of neutral sugar to uronic acid in the soluble pectins also increased. The observations suggest a continuing pectin metabolism that contributes to fruit textural changes and the eventual structural breakdown that occurs in the over-ripe fruit. There are some reports of polygalacturonase activity in ripe cherries (Steele and Yang, 1960) but these have not been confirmed (Fils-Lycaon and Buret, 1990). In assessing cell wall-degrading enzymes in ripe fruit there is the possibility of a microbial activity confusing the result, and it may be that this has contributed to some observations.

13.9.1   Fruit 'cracking'

'Cracking' is an important commercial problem in cherries. It has been associated with the osmotic potential of the tissue and skin properties, but there is no simple relationship between fruit osmotic potential and susceptibility to cracking after rain. Calcium treatments have sometimes been successful in delaying fruit softening and reducing fruit cracking (Lidster *et al.*, 1978, 1979). There is a suggestion that fruit are more susceptible to cracking if growth is particularly rapid in stage III (Pommier, 1987). Gibberellic acid treatment can increase fruit firmness and reduce the tendency for cracking (Kupferman, 1989); GA_3 treatment was shown by Facteau (1982) to increase fruit firmness and decrease water-soluble pectin content. Wade (1988) demonstrated that inhibitors of respiration effectively reduced cracking of cherries in water.

13.10   OLIVE (*OLEA EUROPAEA* L)

The olive is discussed here with the *Prunus* species as it has a similar fruit structure – it is a drupe or stone fruit. The commercial aspects and the composition and processing of this fruit are contained in the excellent review of Fernandez-Diez (1971). The origins of the olive tree pre-date written history, but its natural home now is the Mediterranean area where the bulk of the commercial harvest occurs. It has found a place in other districts with Mediterranean climates, and notably in California.

The growth pattern of the olive fruit is described as sigmoidal. It initially accumulates carbohydrates in the fleshy mesocarp but at a relatively early stage, the soluble carbohydrate content begins to decline and oil is accumulated. In the young fruit, polar lipids form a significant proportion of the total lipids but with increasing maturity the neutral lipid predominates, largely as the triglyceride of oleic acid, but with significant amounts of 16:0, 18:0, and 18:2 fatty acids. The tocopherol content and distribution varies between genotypes (Golubev *et al.*, 1987). Because of its economic importance there is a voluminous literature on the composition and processing of the oil of the olive fruit (for example, Frega and Lercker, 1985). Mannitol is an important component of the soluble carbohydrate pool in the leaves and fruit of the olive, and carbon is translocated as mannitol. Glucose, fructose and mannitol are the main sugars in the flesh of the mature fruit (Wodner *et al.*, 1988).

The fruit maturing after harvest does not show a respiratory or ethylene climacteric, nor is ethylene production induced by ethephon treatment (Goren *et al.*, 1988). Chlorophyll loss is accelerated in the presence of ethylene as in some other non-climacteric fruit. There is a recognition that a climacteric-like response may occur through maturation of the attached fruit (Rugini *et al.*, 1982), and the mature fruit have the capacity to oxidize added ACC to ethylene (Vioque *et al.*, 1980). Ethephon treatment is widely used to induce the abscission of the fruit.

There has been surprisingly little study of lipogenesis in the olive fruit. Marzouk and Cherif (1980) followed the passage of carbon from acetate-1-^{14}C to lipids and noted that more carbon was directed to neutral lipids and to oleic acid as fruits matured. This reflected the changing pool sizes. Donaire *et al.* (1984) related enzymic events to structural changes in the flesh tissue, and noted an increase in glucose-6-phosphate dehydrogenase (EC 1.1.1.49) and lipoxygenase (EC 1.13.11.12) activities when lipids started to accumulate. Carbon from acetate was found to enter the lipid pool more readily than carbon from citrate, but this may depend on the greater capacity for acetate uptake by the tissues. Donaire *et al.* (1984) also observed an increase in the passage of carbon to neutral lipids as the lipid pool increased with fruit maturity. Presumably lipogenesis depends upon the generation of acetyl-CoA and malonyl-CoA in the mitochondria and/or plastids, and the synthesis of acyl carrier protein (ACP) intermediates in the plastids and/or proplastids, but these details do not yet appear to have been evaluated.

Enzymatic browning in harvested fruit has been studied. Sciancalepore (1985), from a study of five varieties, concluded that browning reflected the presence of *o*-diphenols and lipoxygenase and was largely independent of peroxidase activity. The chlorophylls and carotenoids present in olives have been studied in some detail (Minquez-Mosquera et al., 1989; Minguez-Mosquera and Garrida-Fernandez, 1989). In ripening fruit there is a large decrease in chlorophyll and a smaller decrease in carotenoid content, and it is suggested that lipoxygenase is involved in chlorophyll degradation in the fruit and in pressed oil (Minquez-Mosquera et al., 1990) but the evidence for this is not convincing.

Oleuropein is responsible for the bitter taste of olives. Glucosides of oleuropein and its dimethyl derivative accumulate during fruit maturation. Amiot et al. (1989) related esterase activity to the increase in the glucosyl forms.

Catechol oxidase in green fruit was studied in some detail by Sciancalepore and Longone (1983a,b,c). The enzyme purified 260-fold was most active at pH 4.5 with 4-methylcatechol as substrate. Three isoenzymes were separated by column procedures, and eight by electrophoresis.

The role that pectin- and glucan-catabolizing enzymes may play in the softening of olive fruits, including the softening of the pickled fruit has been considered. Pectin esterase activity is present in the mature black fruit, but declines with over-ripeness (Mingula-Mosquera et al., 1978). Polygalacturonase activity was not present in green fruit but was detected in ripe olives (Castillo-Gomez et al., 1978b) and increased when the ripe fruit were stored (Castillo-Gomez et al., 1978a; Minguez-Mosquera, 1982). β-1,4-Glucanase (EC 3.2.1.4) activity was detected in the fruit and activity increased through ripening and in stored fruit (Heredia-Moreno and Fernandez-Bolanos, 1985a,b). Cellulolytic activity was inhibited by cellobiose (K_i 43.8 mM) or by glucose (K_i 471 mM) (Heredia et al., 1989b). Five molecular forms of endoglucanase activity of M_r 18, 23, 32, 44 and 59 kDa were detected, along with exoglucanase and glucosidase activities (Heredia-Moreno and Fernandez-Bolanos, 1989; Heredia et al., 1989a). The press pulp of olives has been studied particularly in relation to its use as stock feed. Gil-Serrano and Tejero-Mateo (1988) described the structure of a typical xyloglucan with terminal galactose residues in olive pulp.

REFERENCES

Amiot, M.J., Fleuriet, A. and Macheix, J.J. (1989) Accumulation of oleuropein derivatives during olive maturation. Phytochemistry, 28, 67–69

Amoros, A., Serrano, M., Riquelme, F. and Romojaro, F. (1989) Levels of ACC and physical and chemical parameters in peach development. Journal of Horticultural Science, 64, 673–677

Ash, A.S.F. and Reynolds, T. (1954) Ketose oligosaccharides in the apricot fruit. Nature (London), 174, 602

Barbier, M. and Thibault, J.F. (1982) Pectic substances of sweet cherry. *Phytochemistry*, **21**, 111–115

Ben, J. and Gaweda, M. (1987a) Changes in the quantity of pigments in the developing tart cherries of North Star and Lutowka cultivars. Part I. Chlorophylls and carotenoids. *Fruit Science Reports*, **14**, 163–170

Ben, J. and Gaweda, M. (1987b) Changes in the quantity of pigments in the developing tart cherries of North Star and Lutowka cultivars. Part II. Growth of fruit sets and amount of anthocyanins. *Fruit Science Reports*, **14**, 171–178

Ben-Arie, R. and Lavee, S. (1971) Pectic changes occurring in Elberta peaches suffering from woolly breakdown. *Phytochemistry*, **10**, 531–538

Ben-Arie, R. and Sonego, L. (1980) Pectolytic enzyme activity involved in woolly breakdown of stored peaches. *Phytochemistry*, **19**, 2553–2555

Biggs, R.H. (1976) Biological basis for firmness in the "Flordagold" peach. *Proceedings of the Florida State Horticultural Society*, **89**, 213–214

Bird, C.R., Smith, C.J.S., Ray, J.A., et al. (1988) The tomato polygalacturonase gene and ripening-specific expression in transgenic plants. *Plant Molecular Biology*, **11**, 651–662

Bishop, D.G. and Wade, N.L. (1977) Lipid composition of sweet cherries. *Phytochemistry*, **16**, 67–68

Brecht, J.K. and Kader, A.A. (1984a) Ethylene production by "Flamekist" nectarine fruit as influenced by exposure to ethylene and propylene. *Journal of the American Society for Horticultural Science*, **109**, 302–305

Brecht, J.K. and Kader, A.A. (1984b) Ethylene production by fruit of some slow-ripening nectarine genotypes. *Journal of American Society for Horticultural Science*, **109**, 763–767

Brecht, J.K. and Kader, A.A. (1984c) Regulation of ethylene production by ripening nectarine fruit as influenced by ethylene and low temperature. *Journal of American Society for Horticultural Science*, **109**, 869–872

Brecht, J.K., Kader, A.A. and Ramming, D.W. (1984) Description and postharvest physiology of some slow-ripening nectarine genotypes. *Journal of American Society for Horticultural Science*, **109**, 596–600

Buescher, R.W. and Furmanski, R.J. (1978) Role of pectinesterase and polygalacturonase in the formation of woolliness in peaches. *Journal of Food Science*, **43**, 264–266

Burns, J.K. and Pressey, R. (1987) Ca^{2+} in cell walls of ripening tomato and peach. *Journal of American Society for Horticultural Science*, **112**, 783–787

Callahan, A., Morgens, P. and Walton, E. (1989) Isolation and *in vitro* translation of RNAs from developing peach fruit. *HortScience*, **24**, 356–358

Callahan, A., Morgens, P.H., Cohen, R.A., Nichols, K.E. and Scorza, R. (1990) Expression of PCH313 during fruit softening and tissue wounding. *HortScience*, **25**, 130–131

Callahan, A., Walton, E., Wydoski, D. and Morgens, P. (1987) mRNA populations change as peach fruit develop. *Plant Physiology*, (supplement), **83**, 124

Castillo-Gomez, J., Minguez-Mosquera, M.I. and Fernandez-Diez, M.J. (1978a) Presence of polygalacturonase and its relation with some products used in the pickling industry. *Grasas Aceitas*, **29**, 97–101

Castillo-Gomez, J., Minguez-Mosquera, M.I. and Fernandez-Diez, M.J. (1978b) Presence of polygalacturonase in the ripe black olive. Factors inducing the enzyme. *Grasas Aceitas*, **29**, 333–338

Chalmers, D.J. and van den Ende, B. (1975) A reappraisal of the growth and development of peach fruit. *Australian Journal of Plant Physiology*, **2**, 623–634

Chalmers, D.J. and van den Ende, B. (1977) The relationship between seed and fruit development in the peach. *Annals of Botany*, **41**, 707–714

Dawson, D.M., Melton, L.D. and Watkins, C.B. (1991) Changes in neutral polysaccharides in the cell walls of nectarines during ripening and in mealy fruit. *Proceedings of Australasian Postharvest Conference*, 19, (abstract)

De Haan, I. (1959) Pectin conversions in peaches during cold storage. *South African Industrial Chemistry*, **2**, 26–34

DeJong, T.M. and Goudriaan, J. (1989) Modeling peach fruit growth and carbohydrate requirements: reevaluation of the double-sigmoid growth pattern. *Journal of the American Society for Horticultural Science*, **114**, 800–804

Delwiche, M.J. and Baumgardner, R.A. (1985) Ground colour as a peach maturity index. *Journal of the American Society for Horticultural Science*, **110**, 53–57

Donaire, J.P., Belver, A., Rodriquez-Garcia, M.I. and Megias, L. (1984) Lipid biosynthesis, oxidative enzyme activities and cellular changes in growing olive fruits. *Revue Experimental Fisiologie*, **40**, 191–203

Downs, C. and Brady, C.J. (1990) Two forms of exopolygalacturonase increase as peach fruits ripen. *Plant Cell and Environment*, **13**, 523–530

Downs, C., Brady, C.J. and Gooley, A. (1992) Exopolygalacturonase protein accumulates late in peach fruit ripening. *Physiologia Plantarum*, **85** 133–140.

Drouet A., Nivet, C. and Hartmann, C. (1983) Polyribosomes from aging apple and cherry fruit. *Plant Physiology*, **73**, 754–757

El-Agamy, S.Z.A., Aly, M.M. and Biggs, R.H. (1981) Ethylene as related to fruit ripening in peaches. *Proceedings of Florida State Horticultural Society*, **94**, 284–289

Engel, K.H., Flath, R.A., Buttery, R.G., Mon, T.R., Ramming, D.W. and Teranishi, R. (1988a) Investigations of volatile constituents in nectarines. 1. Analytical and sensory characterization of aroma components in some nectarine cultivars. *Journal of Agricultural and Food Chemistry*, **36**, 549–553

Engel, K.H., Ramming, D.W., Flath, R.A. and Teranishi, R. (1988b) Investigation of volatile constituents in nectarines. 2. Changes in aroma composition during nectarine maturation. *Journal of Agricultural and Food Chemistry*, **36**, 1003–1006

Facteau, T.J. (1982) Levels of pectic substances and calcium in gibberellic-treated sweet cherry fruit. *Journal of the American Society for Horticultural Science*, **107**, 148–151

Fernandez-Diez, M.J. (1971) The olive. In *Biochemistry of fruit and their products*, (ed A.C. Hulme), Academic Press, London pp. 255–279

Filgueiras, H.A.C. and Chitarra, M.I.F. (1988) Influence of film packaging and storage temperature on the contents of phenolic compounds in the plum Roxa de Delfim Moreira. *Pesquisa Agropec. Brasileira*, **23**, 63–74

Fils-Lycaon, B., Buret, M., Drouet, A. and Hartmann, C. (1988) Ripening and overripening of a non-climacteric fruit: the sweet cherry Bigarreau Napolean. 2. Polyribosomal changes. *Science des Aliments*, **8**, 459–466

Fils-Lycaon, B. and Buret, M. (1990) Loss of firmness and changes in the pectic fractions during ripening and overripening of sweet cherry. *HortScience*, **25**, 777–778

Flurkey, W.H. and Jen, J.J. (1978) Peroxidase and polyphenoloxidase activities in developing peaches. *Journal of Food Science*, **43**, 1826–1828

Frega, N. and Lercker G. (1985) Composition of lipids from the olive drupe during maturation. I. Total lipids and acids. *Agrochimica*, **29**, 300–309

Gil-Serrano, A. and Tejero-Mateo, P. (1988) A xyloglucan from olive pulp. *Carbohydrate Research*, **181**, 278–281

Golubev, V.N., Gusar, Z.D. and Mamedov, E. Sh. (1987) Tocopherols of *Olea europaea. Khimiya. Prirodnyich. Soedinenii.*, pp. 139–140

Goren, R., Nishijima, C. and Martin, G.C. (1988) Effects of external ethylene on the production of endogenous ethylene in olive leaf tissue. *Journal of the American Society for Horticultural Science*, **113**, 778–783

Goto, M., Minamide, T. and Iwata, T. (1988) The change in chilling sensitivity of mume (Japanese apricot, *Prunus mume* Sieb. et Zucc.) depending on maturity at harvest and its relationship to phospholipid composition and membrane permeability. *Journal of the Japanese Society of Horticultural Science*, **56**, 479–485

Gross, K.C. and Sams, C.E. (1984) Changes in cell wall neutral sugar composition during fruit ripening – a species survey. *Phytochemistry*, **23**, 2457–2461

Guichard, E., Kusterman, A. and Mosandl, A. (1990) Chiral compounds from apricots. Distribution of gamma-lactone enantiomers and stereodifferentiation of dihydroactinidiolide using multi-dimensional gas chromatography. *Journal of Chromatography*, **498**, 396–401

Harel, E.A., Mayer, A.M. and Lerner, H.R. (1970) Changes in the levels of catechol oxidase and laccase activity in developing peaches. *Journal of the Science of Food and Agriculture*, **21**, 542–544

Hartmann, C. (1989) Ethylene and fruit ripening of a non-climacteric fruit: the cherry. *Acta Horticulturae*, **258**, 89–96

Hartmann, C., Drouet, A. and Morin, F. (1987) Ethylene and ripening of apple, pear and cherry fruit. *Plant Physiology and Biochemistry*, **25**, 505–512

Heredia-Moreno, A. and Fernandez-Bolanos, J. (1985a) Cellulases in olives and their possible effects in texture changes. II. Cellulolytic activity in Hojiblanca variety. *Grasas Aceitas*, **36**, 98–104

Heredia-Moreno, A. and Fernandez-Bolanos, J. (1985b) Cellulases in olives and their possible effect on texture changes. III. A study of factors which modify their activity. *Grasas Aceites*, **36**, 171–176

Heredia-Moreno, A. and Fernandez-Bolanas, J. (1989) Purification of an endoglucanase from *Olea europaea erolensis*. Agric. Food Chem. Consum. Proc. Eur. Conf. Food Chem. **1**, 355–359

Heredia, A., Fernandez-Bolanos, J. and Guilleni-Bejarano, R. (1989a) Characterization y purification parcial de enzimas celuloloticos en aceitunas. *Grasas Aceitas*, **40**, 190–193

Heredia, A., Fernandez-Bolanos, J. and Guilleni, R. (1989b) Inhibitors of cellulolytic activity in olive fruits (*Olea europaea*, Hojiblanca). *Zeitschrift für Lebensmittel Untersuchung Forschung*, **189**, 216–218

Hermann, K., (1989) Vorkommen und Gehalte der Phenolcarbonsauren in Obst. *Erwerbsobstbau*, **31**, 185–189

Hinton, D.M. and Pressey, R. (1974) Cellulase activity in peaches during ripening. *Journal of Food Science*, **39**, 783–785

Joshi, S., Srivastava, R.K. and Dhar, D.N. (1986) The chemistry of *Prunus armenica*. *British Food Journal*, pp. 74–80

Kakiuchi, N., Tokita, T., Tanaka, K. and Matsuda, K. (1981) Relationship between respiration, ethylene formation, chemical composition and maturation of peaches. *Bulletin of Fruit Tree Research Station*, Yatabe 1, 57–77

Katayama, T., Nakayama, T.O.M., Lee, T.H. and Chichester, C.O. (1971) Carotenoid transformations in ripening apricots and peaches. *Journal of Food Science*, **36**, 804–806

Katsarova, S., Grigorova, S. and Vishanska, Yu. (1987) Amino acids, anthocyanins and aromatic substances in the fruit of some fruit cherry varieties. *Rastenievudni Nauki*, **24**, 91–95

King, G.A., Henderson, K.G. and Lill, R.E. (1987) Growth and anatomical and ultrastructural studies of nectarine fruit wall development. *Botanical Gazette*, **148**, 443–455

Kotze, W.A.G., Nolte, S.H., Dodd, M.C., Gurgon, K.H. and Crouse, K. (1989) Is it possible to restrict the incidence of internal breakdown in plums? *Deciduous Fruit Grower*, **39**, 64–68

Krishman, H.B., and Pueppke, S.G. (1990) Cherry fruit invertase: partial purification, characterization and activity during fruit development. *Journal of Plant Physiology*, **135**, 662–666

Kumar, S. (1987) Changes in phenolic content and polyphenol oxidase activity in developing peach (*Prunus persica* Batsch) fruits. *Plant Physiology and Biochemistry, India*, **14**, 131–135

Kupferman, E.M. (1989) Cherry warehouse survey shows value of GA use. *Good Fruit Grower*, **40**, 10–13

Kupferman, E.M. and Loescher, W.H. (1980) Glycosidase activities and development of fruit mesocarp tissues. *Journal of the American Society for Horticultural Science*, **105**, 452–454

Lee, E., Speirs, J., Gray, J. and Brady, C.J. (1990) Homologies to the tomato endopolygalacturonase gene in the peach genome. *Plant Cell and Environment*, **13**, 513–521

Lee, J.C., and Lee, Y.B. (1980) Physiological study on coloration of plum fruits I. Effect of ethephon on fruit composition and anthocyanin development in Santa Rosa plum (*Prunus salicina*). *Journal of the Korean Society of Horticultural Science*, **21**, 36–41

Lessertois, D. and Moneger, R. (1978) Evolution des pigments pendant la croissance et la maturation du fruit *Prunus persica*. *Phytochemistry* **17**, 411–415

Lidster, P.D., Porritt, S.W. and Tung, M.A. (1978). Texture modification of 'Van' sweet cherries by postharvest calcium dips. *Journal of the American Society for Horticultural Science*, **103**, 527–530

Lidster, P.D., Tung, M.A. and Yada, R.G. (1979) Effects of preharvest and postharvest calcium treatments on fruit calcium content and susceptibility of 'Van' cherries to impact damage. *Journal of the American Society for Horticultural Science*, **104**, 790–793

Lill, R.E., O'Donoghue, E.M. and King, G. (1989) Postharvest physiology of peaches and nectarines. *Horticultural Reviews*, **11**, 413–452

Looney, N.E., McGlasson, W.B. and Coombe, B.G. (1974) Control of fruit ripening in peach *Prunus persica*: action of succinic acid-2, 2-dimethylhydrazide and (2-chloroethyl)phosphonic acid. *Australian Journal of Plant Physiology*, **1**, 77–86

Manolov, P., Borichenko, N. and Rangelov, B. (1977) Sorbitol and free sugars as mobile forms of peach tree assimilate. *Gradinar. Lozar. Nauka*, **114**, 45–51

Marzouk, B. and Cherif, A. (1980) Lipogenesis in the olive. *Revue Francaise Corps Gras*, **27**, 487–491

Miller, A.N., Walsh, C.S. and Cohen, J.D. (1987) Measurement of indole-3-acetic acid in peach fruits (*Prunus persica* L. Batsch cv Redhaven) during development. *Plant Physiology*, **84**, 491–494

Miller, A.N., Krizak, B.A. and Walsh, C.S. (1988) Ethylene evolution and ACC content in the fruit pericarp tissue and seeds during development. *Journal of the American Society for Horticultural Science*, **113**, 119–124

Miller, A. N. and Walsh, C. S. (1990) Indole-3-acetic acid concentration and ethylene evolution during early fruit development in peach. *Plant Growth Regulation* **9**, 37–46

Mingula-Mosquera, M.I., Castilla-Gomez, J. and Fernandez-Diez, M.J. (1978) Presence of pectinesterase and its relation with the softening of some pickling products. *Grasas Aceitas*, **29**, 29–36

Minguez-Mosquera, M.I. (1982) Evolution of pectic constituents and pectolytic enzymes during ripening and storage of the hojiblanca olive. *Grasas Aceitas*, **33**, 327–333

Minguez-Mosquera, M.I., Garrido-Fernandez, J. and Gandul-Rojas, B. (1989) A rapid method for determining pigments in green table olives. *Grasas Aceitas*, **40**, 206–212

Minquez-Mosquera, M.I., Gandul-Rojas, B., Garrido-Fernandez, J. and Gallardo-Guerrero, L. (1990) Pigments in virgin olive oil. *Journal of the American Oil Chemists Society*, **67**, 192–196

Minquez-Mosquera, M. I. and Garrido-Fernandez, J. (1989) Chlorophyll and carotenoid presence in olive fruit (*Olea europaea*). *Journal of Agricultural and Food Chemistry*, **37**, 1–7

Morgens, P., Walton, E., Wydoski, D. and Callahan, A. (1987) Fruit mRNA populations differ among high quality-cold sensitive and low quality-cold hardy peach cultivars. *Plant Physiology (supplement)*, **83**, 124

Moriguchi, T., Sanada, T. and Yamaki, S. (1990) Seasonal variation of some enzymes relating to sucrose and sorbitol metabolism in peach fruit. *Journal of the American Society for Horticultural Science*, **115**, 278–281

Moriguchi, T. and Yamaki, S. (1988) Purification and characterization of sucrose synthase from peach (*Prunus persica*) fruit. *Plant Cell Physiology*, **29**, 1361–136

Morrison, J.C., Greve, L.C. and Labavitch, J.M. (1987) The role of cell wall-degrading enzymes in the formation of gum ducts in almond fruit. *Journal of the American Society for Horticultural Science*, **112**, 367–372

Nanos, G.D. and Mitchell, F.G. (1991) High temperature conditioning to delay internal breakdown development in peaches and nectarines. *HortScience*, **26**, 882–885

Nigam, V.N. and Sharma, S.D. (1987) Changes in the sugar content in developing apricot fruits cv. Newcastle (*Prunus armeniaca* L.). *Fruit Science Reports*, **14**, 115–118

Nigam, V.N. and Sharma, S.D. (1988) Pattern of changes of amino acids in developing fruits of apricot cv. Newcastle (*Prunus armeniaca* L.) *Haryana Journal of Horticultural Science*, **17**, 135–139

Osajima, Y., Wada, K. and Ito, H. (1987) The effects of ethylene and acetaldehyde removing agents and seal packaging with plastic film on the keeping quality of Japanese apricot (*Prunus mume* Sieb. et Zucc.) and Kabosa (*Citrus sphaerocarpa* hort. ex Tanaka) fruits. *Journal of the Japanese Society of Horticultural Science*, **55**, 524–553

Pangelova-Shchurkova, I. and Vitanova, I. (1987) Seasonal changes in the weight and dry matter, organic acid and tannin contents of plum fruits. *Fiziologiia Rastenii*, **13**, 61–67

Pifferi, P.G. and Cultrera, R. (1974) Enzymatic degradation of anthocyanins: the role of sweet cherry polyphenol oxidase. *Journal of Food Science*, **39**, 786–791

Pommier, P. (1987) Léclatement de la cerise. Recherche de nouveaux moyens de lutte. *Arbor. Fruit.*, **34**, 20–23

Pressey, R. and Avants, J.K. (1973) Separation and characterization of endopolygalacturonase and exopolygalacturonase from peaches. *Plant Physiology*, **52**, 252–256

Pressey, R. and Avants, J.K. (1978) Difference in polygalacturonase composition of clingstone and freestone peaches. *Journal of Food Science*, **43**, 1415–1417

Pressey, R., Hinton, D.M. and Avants, K. (1971) Polygalacturonase activity and solubilization of pectin in peaches during ripening. *Journal of Food Science*, **36**, 1070–1073

Ramming, D.W. (1991) Genetic control of a slow-ripening fruit trait in nectarine. *Canadian Journal of Plant Science*, **71**, 601–603

Robertson, J.A., Meredith, F.I. and Scorza, R. (1988) Characteristics of fruit from high- and low-quality peach cultivars. *HortScience*, **23**, 1032–1034

Roemer, K. (1990) Das Zuckermuster verschiedener Obstarten. Teil IV. *Prunus domestica* "Pflaumen". *Erwerbsobstbau*, **32**, 42–46

Romani, R.J. and Jennings, W.G. (1971) Stone fruits. In *Biochemistry of fruits and their products*, (ed A.C. Hulme), Academic Press, London, pp. 411–436

Rugini, E., Bongi, G. and Fontanazza, G. (1982) Effects of ethephon on olive ripening. *Journal of the American Society for Horticultural Science*, **107**, 835–838

Sawamura, M. and Miyazaki, T. (1989) Effects of exogenous ethylene on ethylene production in discs of mume fruits. *Journal of Horticultural Science*, **64**, 633–638

Sawamura, M., Miyazaki, T. and Kusunose, H. (1988) Ethylene biosynthesis in mume (*Prunus mume* Sieb. et Zucc.) fruits *Research Reports Kochi University, Agricultural Science*, **37**, 179–185

Schwab, B. and Schreier, P. (1990) Untersuchungen uber gebundene Aromastoffe in Sauerkirschen (*Prunus cerasus* L.) *Zeitschrift fürLebensmittel-Untersuchung Forschung* **190**, 228–231

Schwab, W., Scheller, G. and Schreier, P. (1990) Glycosidically bound aroma components from sour cherry. *Phytochemistry*, **29**, 607–612

Sciancalepore, V. (1985) Enzymatic browning in five olive varieties. *Journal of Food Science*, **50**, 1194–1195

Sciancalepore, V. and Longone, V. (1983a) Research on catechol oxidase from green olives. 1. Partial purification and some properties of the enzyme. *Industria Alimentica*, **22**, 549–554

Sciancalepore, V. and Longone, V. (1983b) Studies on catechol oxidase from green olives. 2. Isoenzymes. *Industria Alimentica*, **22**, 640–644

Sciancalepore, V. and Longone, V. (1983c) Studies on catechol oxidase from green olives. 3. Electrophoretic analysis. *Industria Alimentica*, **22**, 745–748

Scorza, R., May, L.G., Purnell, B. and Upchurch, B. (1991) Differences in number and area of mesocarp cells between small- and large-fruited peach cultivars. *Journal of the American Society for Horticultural Science*, **116**, 861–864

Sekse, L. (1988) Respiration of plum (*Prunus domestica* L.) and sweet cherry (*P. avium* L.) fruits during growth and ripening. *Acta Agricultura Scandinavica*, **38**, 317–320

Senter, S.D., Robertson, J.A. and Meredith, F.I. (1989) Phenolic compounds of the mesocarp of Cresthaven peaches during storage an ripening. *Journal of Food Science*, **54**, 1259–1260

Sistrunk, W.A. (1985) Peach quality assessment: fresh and processed. In *Evaluation of Quality of Fruits and Vegetables*. (ed H.E. Patee), AVI Publishing, Westport, CT, pp. 1–46

Shewfelt, A.L., Paynter, V.A. and Jen, J.J. (1971) Textural changes and molecular characteristics of pectic constituents in ripening peaches. *Journal of Food Science*, **36**, 573–575

Steele, W.F. and Yang, H.Y. (1960) The softening of brined cherries by polygalacturonase and the inhibition of polygalacturonase in model systems by alkyl aryl sulfonates. *Food Technology*, **14**, 121–126

Stutte, G.W. (1989) Quantitation of net enzymatic activity in developing peach fruit using computer video image analysis. *HortScience*, **24**, 113–115

Takeoka, G.R., Flath, R.A., Guntert, M., and Jennings, W. (1988). Nectarine volatiles: vacuum steam distillation versus head space sampling. *Journal of Agricultural and Food Chemistry*, **36**, 553–560

Thibault, J.F. (1983) Enzymatic degradation and *b*-elimination of the pectin substances in cherry fruits. *Phytochemistry*, **22**, 1567–1572

Tonutti, P., Casson, P. and Ramina, A. (1991) Ethylene biosynthesis during peach fruit development. *Journal of the American Society for Horticultural Science*, **116**, 274–279

Tsay, L-M., Mizuno, S. and Kozukue, N. (1984) Changes in respiration, ethylene evolution and abscisic acid content during ripening and senescence of fruits picked at young and mature stages. *Journal of the Japanese Society of Horticultural Science*, **52**, 458–463

Tsuchida, H., Mizuno, S. and Kozukue, N. (1990) Changes in abscisic and phaseic acids during ripening and senescence of peach fruits. *Journal of the Japanese Society of Horticultural Science*, **58**, 801–805

Tukey, B. (1936) Development of cherry and peach fruits as affected by destruction of the embryo. *Botanical Gazette*, **98**, 1–24

Ugalde, T.D., Chalmers, D.J. and Jerie, P.H. (1988) Intercellular invertase in developing peach mesocarp. *Australian Journal of Plant Physiology*, **15**, 377–383

Vioque, A., Albi, M.A. and Vioque, B. (1980) Production of ethylene from 1-aminocyclopropane-1-carboxylic acid from different olive trees. *Grasas Aceitas*, **31**, 196–199

Vitanov, M., Pangelova-Shchurkova, I. and Vitanova, I. (1988) Comparative studies of the plum varieties Stanley and Gabrovska and their parents. *Genetika i Selektsiia*, **21**, 136–140

Von Mollendorf, L.J. and de Villiers, O.T. (1988a) Physiological changes associated with the development of woolliness in "Peregrine" peaches during low temperature storage. *Journal of Horticultural Science*, **63**, 47–51

Von Mollendorf, L.J. and de Villiers, O.T. (1988b) Role of pectolytic enzymes in the development of woolliness in peaches. *Journal of Horticultural Science,* **63**, 53–58

Wade, N.L. (1988) Effect of metabolic inhibitors on cracking of sweet cherry fruit. *Scientia Horticulturae*, **34**, 239–248

Watkins, J.B. (1964) Changes in the pectic substances of stored Elberta peaches. *Queensland Journal of Agricultural Science*, **21**, 47–58

Weis, K.G., Polito, V.S. and Labavitch, J.M. (1988) Microfluorometry of pectic materials in the dehiscence zone of almonds (*Prunus dulcis* Mill. DA Webb) fruits. *Journal of Histochemistry and Cytochemistry*, **36**, 1037–1041

Wodner, M., Lavee, S. and Epstein, E. (1988) Identification and seasonal changes of glucose, fructose and mannitol in relation to oil accumulation during fruit development in *Olea europaea* L. *Scientia Horticulturae*, **36**, 47–54

Wong, T.C., Luh, B.S. and Whitaker, D.R. (1971a) Isolation and purification of polyphenol oxidase isoenzymes of clingstone peaches. *Plant Physiology*, **48**, 19–24

Wong, T.C., Luh, B.S. and Whitaker, D.R. (1971b) Effect of phloroglucinol and resorcinol on the clingstone peach polyphenoloxidase-catalyzed oxidation of 4-methylcatechol. *Plant Physiology*, **48**, 24–30

Chapter 14

Tomato

G.Hobson and D. Grierson

14.1   INTRODUCTION

On a worldwide scale, the tomato (*Lycopersicon esculentum*) continues to increase in importance for consumption as a fresh crop, for inclusion as a major constituent in many prepared foods, and also for research into the fundamental principles of growth and development in plants. Members of the genus *Lycopersicon* are tolerant of a wide range of both environmental and nutritional conditions. A few of the species have been crossed to provide a large number of varieties directed towards either the production of a single-harvest field crop, or, particularly under protection, a succession of fruit for the fresh market over quite a long timespan. Field tomatoes are amenable to mechanical harvesting through the combined efforts of plant breeders and agricultural engineers, thus allowing vast tonnages to be grown and processed economically. The products are incorporated into a wide range of canned, frozen, preserved or dried foods. In view of their economic importance, and because they are amenable to the techniques of molecular biology and genetic engineering, tomato fruit have been selected for intense study at the molecular level over the last decade.

Books and reviews concerning tomato cultivation (Walls, 1989), development (Varga and Bruinsma, 1986), scientific background (Atherton and Rudich, 1986), history (Davies and Hobson, 1981), distribution of the various species (Taylor, 1986) and genetics (Stevens and Rick, 1986) have been published in the last decade. Earlier reviews concentrated on the composition of tomato fruit (Salunkhe *et al.*, 1974; Herrmann, 1979; Davies and Hobson, 1981); here we shall briefly summarize these aspects and place more emphasis on recent findings that sharpen our perception

Biochemistry of Fruit Ripening. Edited by G. Seymour, J. Taylor and G. Tucker. Published in 1993 by Chapman & Hall, London. ISBN 0 412 40830 9

regarding the major biochemical and genetic control mechanisms governing tomato fruit development and ripening.

14.2   HISTORY AND CLASSIFICATION

The original home of the tomato would seem to be along the western seaboard of South America where the cold ocean currents moderate the mean air temperatures, even close to the equator. In Ecuador and Peru, various species are distributed along the coastal plain and up into the foothills of the Andes. Thus the longitudinal spread of the genus is over 2000 km from central Ecuador to northern Chile, but the average distance from the coast is limited to about 200 km. From this supposed origin of the native tomato in South America, historians consider that tomato seed was carried northwards, possibly in drainage canals, towards Mexico where the fruit became extremely popular and where the native languages contain many synonyms of the word tomato. The conquistadores who arrived from the Old World in the early 16th century returned with seed to Spain. Known since Elizabethan times as the 'love apple', two distinct explanations for the name have been offered. Plants grown in Seville in 1501 had yellow fruit, and seeds were transferred to Morocco and thence to Italy by 1544 where it was naturally referred to as 'pomo dei Mori' or 'Moor's apple'. It is not difficult to imagine the Parisians corrupting the name to 'pomme d'amour' and endowing the fruit with aphrodisiac properties. The other explanation dates from 1554 where the yellow fruit became 'pomi d'oro', 'mala aurea' or 'poma amoris', synonyms that persisted well into the 19th century and translated in this country as 'love apple'. Although the world population has risen roughly in line with tomato consumption, no direct connection between one and the other has ever been established!

The tomato belongs to the family Solanaceae, and previous attempts at classification of the various species have centred around the colour of the fully-ripened fruit (Davies and Hobson, 1981):

1. Eulycopersicon   red, yellow or brown
2. Eriopersicon   green or largely green with purple stripes

Cytogenetically, however, it is probably more meaningful to divide the genus *Lycopersicon*, to which almost all tomato species belong, into '*L. hirsutum*-like' and '*L. peruvianum* and *L. chilense*-like'. The divisions would thus be:

1. *L. hirsutum*-like (all members cross-hybridize relatively easily; (Stevens and Rick, 1986).
   L. hirsutum and *L. hirsutum* var *glabratum*.

Plate 1 *Litchi chinensis* (Lychee)

Plate 2 *Averrhoa carambola* (Starfruit)

Plate 3 *Cyphomandra betacea*
(Tamarillo)

Plate 4 *Passiflora edulis* (Passion fruit)
(below left)

Plate 5 *Nephelium lappaceium*
(Rambutan) (below right)

Plate 6

Plate 7

Ed Kawy

Plate 8 (below)

Size, shape and colour variation in different variaties and mutants of tomato. Unless otherwise stated, mutants are in cv. Ailsa Craig background. (6) *L. cheesmanii*, a wild species that grows in the Galapagos Islands; (7) cv. Ed Kawy, fruit of which can weigh 500 grams; (8) square-shaped tomato in an Ailsa Craig background; (9) cherry tomatoes cv. Gardiner's Delight; (10) mutant *green flesh*; (11) mutant *yellow flesh*; (12) mutant *apricot*; (13) mutant *ripening inhibitor*; (14) mutant *longkeeper*, probably a variant of the slow-ripening mutany *alcobacer*; (15) mutant *green stripe*, in which the stripes turn golden when the fruit ripen; (16) mutant *unpigmented fruit epidermis*, in which the ripening fruit appears pink rather than orange.

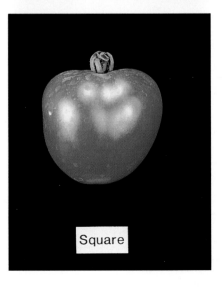

Square

Plate 9

Green flesh

Plate 10

Yellow

Plate 11

Apricot

Plate 12

Ripening inhibiter

Plate 13

Plate 14

Plate 15

Plate 16

Solanum pennellii (there is good evidence for reclassifying it as
L. pennellii (Rick, 1979).
L. esculentum
L. esculentum var cerasiforme
L. cheesmanii
L. pimpinellifolium
L. parviflorum
L. chmielewskii
Solanum lycopersicoides
2. L. peruvianum and L. chilense-like
L. peruvianum
L. peruvianum var humifusum
L. chilense

Within each of these groups, cross-fertilization is relatively easy
although the progenies are sometimes sterile. Cross-pollination between
members of a different group can be achieved only with difficulty
(Stevens and Rick, 1986). While classification based on the final colour of
the fruit is the more convenient system, scientifically the cytological
evidence is the more defensible. The wide diversity of conditions under
which tomatoes will bear a crop and the rise in total production from
year to year is a tribute to the plant breeders who have scoured wild
accessions growing in extraordinary habitats for characters that would
extend the range in the hope that these might then be commercially
exploitable. An indication of the possible sources at the disposal of the
breeder and commonly incorporated in modern cultivars is given in
Fig. 14.1. A study of single gene mutants in a standard genetic background (Darby, 1978; Stevens and Rick, 1986) has been carried out
alongside breeding programmes, and linkage maps have been constructed showing the likely location of many characters on each of the
chromosomes (Stevens and Rick, 1986; Mutschler et al., 1987). More
recently, the classical genetic map has been extended using RFLP
(restriction fragment length polymorphism) markers (Kinzer et al., 1990).

14.3   TOMATO PRODUCTION

Over the last 25 years or so, tomato production has overtaken that of
bananas, pome fruits and, recently, grapes to become at the present time
second only to citrus in terms of weight of crop on an annual basis (Table
14.1). The most important countries involved in tomato growing are the
USA, Russia, Turkey, China, Egypt and Italy. Indications are that 60–65%
of the total crop weight is used for processing, and this is centred in two
main regions, California and Italy. The crop is regarded as easy to grow,

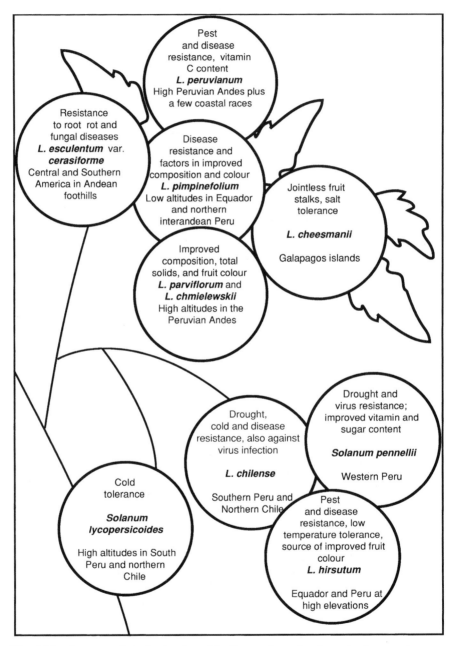

Fig. 14.1 Sources of attributes found in the modern-day tomato, *Lycopersicon esculentum*.

Table 14.1   The average world production of tomatoes in comparison with other popular fruits during the years 1979–1981 compared with 1990.[*]

Commodity	Production in tonnes (000s)		Percentage change
	1979–1981	1990	
Grapes	65 991	59 943	–9
Citrus	57 026	73 195	+28
Tomatoes	51 907	69 304	+34
Apples and pears	43 140	50 103	+16
Bananas	36 849	45 845	+24

[*]Data from Production Yearbook (1990), vol. 44, Food and Agriculture Organisation, Rome, 1991.

perennial, determinate or indeterminate (depending on the variety), very often self-fertile and tolerant of a wide range of environmental and nutritional conditions. The quality and yield of tomatoes, especially those grown under protection, has increased very considerably over the last 20 years as more became known about the interactions between genotype, nutrition and environment. It is now relatively common for long-season, heated crops under glass to yield 150 tonnes.acre^{-1} (360 tonnes.ha^{-1}), while 200 tonnes.acre^{-1} (480 tonnes.ha^{-1}) is not unknown, even in north-west Europe.

14.4   FRUIT QUALITY

14.4.1   Size and shape

After fertilization, the number of cells in a tomato fruit increase dramatically over a period of 2–3 weeks. Thereafter, growth is almost exclusively by cell expansion, as the fruit pericarp and developing seeds accumulate carbohydrates from the parent plant. The average size of different cultivars of tomato fruit can vary by a factor of at least 500, even when conditions do not limit fruit growth (Colour plate 6, 7). The shape is also largely under genetic control, although nutrition and environment can also have an influence. In the more temperate parts of the world, consumers expect tomatoes to be reasonably round and uniform in shape, weighing about 75 grams, slightly wider than deep and having an average of three locules (resulting from crossing a bilocular line (Fig. 14.2A) with a multilocular one (Fig. 14.2B). Cultivars having this type of fruit grow well, even when light levels are limiting. In those areas of the world where radiation is non-limiting, it is possible to grow much larger multilocular fruit of less perfect shape but on short trusses which set

A

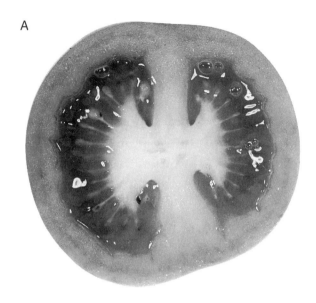

B

Fig. 14.2   Cross-section of (A) a bilocular and (B) a multilocular tomato.

many fewer flowers. Tomatoes grown for processing and canning have a high solids content and are typically pear-shaped; the 'square' tomato having angular sides is still a breeding curiosity (Colour plate 8). Cherry tomatoes (*L. esculentum* var *cerasiforme*) are relatively small, typically weighing between 10 and 35 g (Colour plate 9), and having exquisite flavour potential. They contain a proportion of their sugar content as sucrose, which is unusual in tomato fruit, as well as the normal reducing sugars of glucose and fructose (Picha, 1986; Gough and Hobson, 1990).

14.4.2   Colour

It is entirely possible that the ancestors of the modern tomato remained green when ripe, and would have been unattractive to man. Coloured fruit could have arrived on the scene much later, and it is likely that the first seeds brought back to Europe from Mexico in the 16th century bore yellow fruit, closely followed by red-fruited lines, which subsequently became the dominant form. In modern cultivated varieties chlorophyll, which is located in the chloroplast thylakoids, reaches a peak in concentration relatively early in the growth of the fruit. Ripening occurs towards the end of development when growth has slowed down or ceased; ethylene evolution and respiration begin to rise above a base level, and chloroplasts start to turn into chromoplasts, initially in the locular jelly surrounding the seeds and then in the pericarp, progressively from the blossom-end of the fruit to the stem end. The sequence of pigment changes have been much studied, and reviews have been written by Khudairi (1972), Goodwin (1980), and Grierson and Kader (1986). In normal fruit chlorophyll is replaced by oxygenated carotenes, and xanthophylls. Phytofluene, which is colourless, increases towards ripeness, whereas β-carotene peaks a little before full colour development. Quantitatively by far the most important compounds are phytoene (colourless) and lycopene. Various mutants affecting the colour of fully-developed fruit have been helpful in elucidating the metabolic pathways from acetyl CoA through mevalonic acid to colourless and then coloured pigments (Stevens and Rick, 1986). A failure to complete the breakdown of chlorophyll results in 'green flesh' mutants (Colour plate 10), while yellow fruit contain reduced polyenes but no carotenes (Colour plate 11). *Apricot* mutants have increased carotene and polyenes (Colour plate 12), and this is largely true for the non-ripening mutants such as '*ripening inhibitor*' (Colour plate 13) and '*longkeeper*' (Colour plate 14). A further gene induces vertical green stripes in the epidermis of unripe fruit that turn golden yellow as the tomatoes ripen (Colour plate 15). Most European cultivars have an orange pigment in the skin, thought to be chalconaringenin (Baker *et al.*, 1982), so that the lycopene in the ripened flesh gives an orange-red external colour. A mutant of cv Ailsa Craig having colourless skin (*yy*) thus appears pink when ripe and not orange-red (Colour plate 16); it is common for North American cultivars to have a colourless epidermis. The mutants *'dark green'*, *'crimson'* and *'high pigment'* have higher than normal levels of pigments and have been extensively tested as a means of improving the basic colour of tomato lines (Wann *et al.*, 1985).

It is likely that the degradation products from chlorophyll are re-utilized to form new compounds associated with ripening, some of them colourless, and these first show up as plastoglobuli as the chloroplasts are transformed to chromoplasts (Khudairi, 1972). Eventually, the mature chromoplast contains osmiophilic globules thought to contain carotenoid and xanthophyll pigments, plus boat-shaped lycopene crystals and undulating membrane systems. In addition, novel proteins are synthesized during the transition from chloroplasts to chromoplasts, resulting in distinct protein patterns in each organelle (Bathgate *et al.*, 1985).

14.4.3   Tomato composition

While providing ample vitamin C in fresh fruit, the calorific value of the tomato is particularly modest because of its low dry matter and fat content. The balance between water content and the various constituents of ripe tomato fruit is so dependent on the genotype, the environment in which the plant is grown, the nutritional treatment and, to a minor extent, the nature of the post-harvest treatment, that precise figures can be misleading. Nevertheless, a general indication of the levels of the major and a few minor constituents is given (Table 14.2); much more detail can be obtained from Davies and Hobson (1981). The range of analytical values listed in standard reference works is also shown in the same review. It is worth pointing out that the sugar content, mainly in the locule walls, reaches a peak when the fruit is fully ripe, while the malic acid contribution to the acidity falls quickly as the fruit turn red. The citric acid content is much more stable throughout the ripening period, and much of the acidity is to be found in the locular contents. These differences are particularly important, since the ratio of sugars to acids plays a major role in determining the taste of a tomato, with high sugars and high acids being favoured. Thus, compositional tests must be made on tissue samples having the appropriate proportions of walls to locular juices.

14.4.4   Nutritional regimes

Until a few years ago, experimentation was intense to establish nutritional principles for tomatoes grown in various solid substrates (Adams,

Table 14.2   Typical composition of a ripe cultivated tomato fruit.

Component	% Fresh weight
Dry matter	6.5
Total carbohydrate	4.7
Fat	0.15
Protein-N	0.4
Reducing sugars	3.0
Sucrose	0.1
Total soluble solids	4.5*
Malic acid	0.1
Citric acid	0.2
Fibre	0.5
Vitamin C	0.02
Potassium	0.25

*Determined by refractometry in °Brix (= % sucrose); this represents not only the reducing sugar content but also the organic acids and minor constituents in the sap that contribute to the refractive index (Davies and Hobson, 1981; Hobson and Kilby, 1985).

Table 14.3   Average concentrations of nutrients for optimal growth of tomato plants by hydroponic techniques such as 'Rockwool' or the 'Nutrient Film' technique (NFT).

Element	Concentration in $mg.l^{-1}$	
	Rockwool (Sonneveld and Welles, 1984)	NFT* (Adams and Ho, 1989)
NO_3-N	147	175–200
NH_4-N	7	–
P	47	30–40
K	274	300–400
Ca	150	150–200+
Mg	24	85
S	80	64
Fe	0.6	10–12+
Mn	0.6	1
Zn	0.3	0.3
B	0.2	0.5
Cu	0.03	0.3
Mo	0.05	0.05

* pH of the solution is kept between 5.5 and 6.0; to improve fruit quality, NaCl may be added to increase the conductivity to a minimum of 3 mS.cm^{-1} without loss in yield (Adams, 1987).
+Including the amount introduced from the water supply which could provide up to 50% of this figure.

1986). For maximizing production under protection, solution culture is now widely used, and for this it is possible to recommend more precise levels of nutrients; typical recipes used in conjunction with 'Rockwool' (a spun, largely inert mineral substrate) or recirculating solution culture (usually known as the nutrient-film technique, or NFT) are given in Table 14.3. The effect on fruit quality, composition and yield of the various major nutritional elements has been closely considered by Adams (1986), and visual symptoms of nutritional and physiological disorders of both plants and fruit have been illustrated and described (Hobson *et al.*, 1977; Winsor and Adams, 1987). For a typical mature tomato plant, analysis has shown that about 25% of the total N and K is contained by the leaves, with another 60% being in the fruit. Magnesium is evenly divided between leaf and fruit, but 70% of the calcium is fixed in the leaves compared with 5% in the fruit. In contrast to potassium, very little calcium is translocated from the leaves once it has been assimilated. Calcium is present in normal tomato fruit in limited amounts, and adverse conditions of growth will lead to deficiency symptoms in the form of 'blossom-end' rot (Ho and Adams, 1989). Among the factors limiting the concentration in fruit cells is rapid growth, high salinity in the root zone, low growing temperatures, high humidity, excessive use of ammonium-nitrogen and, of course, inadequate calcium nutrition (Adams, 1990). Some cultivars seem much less efficient at absorbing and translocating calcium in the xylem than others (Martinez *et al.*, 1987).

14.4.5 Taste and flavour

The flavour of a foodstuff is a combination of the volatile components detected by the nose and the taste compounds present in much greater quantities and sensed by the tongue and adjacent tissues. Of the four basic tastes generally used to describe any food, there is little bitterness in a ripening or ripe tomato. The natural saltiness in the fruit is also not an important factor, except where high salinity during growth (used deliberately or as a result of minimal watering of the plants) has increased the dry matter content of the fruit beyond the normal range (Gough and Hobson, 1990). The taste is thus mainly determined by the sweetness induced by the reducing sugars, plus some sucrose in cherry tomatoes, and sourness (equivalent to acidity) caused by the organic acid content. Sugars constitute between 65% and 70% of the total soluble solids in a tomato fruit, and an accurate estimate of the latter property can be made using a refractometer (Hobson and Kilby, 1985). The soluble solids depend to a large extent on the rate of starch accumulation during the rapid growth phase of development (Dinar and Stevens, 1981; Ho and Hewitt, 1986). The harvesting of tomatoes before full ripeness has an effect not only on the peak sugar content but also on the development of a full flavour spectrum, thus affecting consumer acceptability (Picha, 1986; Stevens, 1986). The volatiles, making up the tomato aroma, complement the taste components to give the flavour of the whole fruit. More than 400 substances contribute to the odour, but no single, or simple combination of compounds, has a smell reminiscent of the ripe fruit. Identified constituents fall into the following classes: hydrocarbons, phenols, ethers, aldehydes, alcohols, ketones, esters, lactones, sulphur compounds, amines and a wide variety of heterocyclic molecules (Petro-Turza, 1987). A mixture of *cis*-3-hexenal, *trans*-2-hexenal, hexanal, β-ionone, 1-penten-3-one, 3-methyl butanal, *cis*-3-hexenol and 3-methyl butanol was considered 'very similar' to the flavour of a freshly-sliced tomato (Buttery *et al.*, 1987). The 'GC-sniff method' whereby components of fresh tomato aroma are separated by gas-chromatography and then assessed for odour characteristics has been exploited by both Hayase *et al.* (1984) and McGlasson *et al.* (1987). Such studies have provided direct evidence in support of long-held beliefs that vine-ripening is preferable to artificial methods with under-ripe fruit, flavour is eroded if the fruit are refrigerated and not allowed to recover before consumption, and field-grown tomatoes have more flavour than those grown in glasshouses. In view of the delays implicit in marketing and the fact that tomatoes are kept (and even refrigerated) for up to a week before being consumed, much of the crop is prevented from reaching its full flavour potential or is eaten after it has passed its prime.

14.5 ETHYLENE SYNTHESIS AND ACTION

Ethylene plays an important role in the initiation and continuation of ripening in all climacteric fruits, including the tomato. A comprehensive review of the biosynthesis and regulation of ethylene production is available (Yang and Hoffman, 1984) and is summarized in Chapter 1. During the biosynthetic cycle, fragmentation of *S*-adenosyl methionine (SAM) leads to the formation of 1-aminocyclopropane-1-carboxylic acid (ACC) from C-3 and C-4 of methionine. In turn, ACC is oxidatively degraded to ethylene, CO_2 and HCN. Regulation of the activity of the two enzymes specific for ethylene biosynthesis, ACC synthase (EC 4.4.1.14) and ACC oxidase (also known as the 'ethylene-forming enzyme', EFE), plays a critical role in ethylene production. Fruit ripening and various environmental stresses such as wounding, water deficit and chilling temperatures stimulate ethylene formation through the induction of these enzymes. Although ACC synthase is generally considered as the rate-limiting step in the biosynthetic pathway, regulation of EFE is also important. During tomato fruit growth and development up to the start of the climacteric respiration rise, internal ethylene concentrations are less than $1\mu l.1^{-1}$. This basic level is referred to as 'system 1' production (McMurchie et al., 1972), and is shown by non-climacteric, mutant tomatoes as well as normal unripe fruit. At the onset of ripening, ethylene synthesis begins to increase. This occurs about 1–2 days before any external colour change at the blossom-end of green fruit becomes noticeable and precedes the synthesis of enzymes such as PG (Grierson and Tucker, 1983; Su et al. 1984). The rise in ethylene production then becomes auto-catalytic, eliciting a climacteric respiration rise and triggering many, but not all subsequent changes leading to full ripeness (Lyons and Pratt, 1964; Sawamura et al., 1978; Jeffery et al., 1984). This constitutes 'system 2' evolution, which rises to a peak and then falls away towards and beyond full ripeness. Although research on the hormone has been carried out for more than 50 years, facts concerning the nature of the ethylene receptor, the control of tissue sensitivity and the mode of action at the molecular level are still hard to gather.

The application of exogenous ethylene to mature green tomatoes at $100\mu l.1^{-1}$ for 24–48 hours at 20–25°C is sufficient to induce ripening, but less mature fruit need continuous exposure to the hormone, or one of its analogues such as propylene, for longer periods to ensure ripening (McGlasson et al., 1975). For the induction of natural ripening, there is probably no critical value that the internal ethylene must reach, although an upturn in the production rate of the hormone seems to be necessary for ripening, probably complemented by a change in the sensitivity of the tissue. Inhibition of ethylene synthesis inhibits ripening. This can be accomplished chemically using aminooxyacetic acid (AOA) or aminoethoxyvinyl glycine (AVG), which both interfere with the conversion of SAM to ACC (Yang and Hoffman, 1984).

Interestingly, a small number of chemicals inhibit the perception of ethylene and thereby prevent ripening. For instance, the gaseous cyclic olefin 2,5-norbornadiene displays competitive inhibition of ethylene action (Sisler, 1982). Silver ions cause the release of wound ethylene, but ripening is nonetheless severely inhibited (Atta-Aly et al., 1987; Tucker and Brady, 1987) because the presence of the hormone is not perceived presumably since silver blocks the receptor (Davies et al., 1988). Eventually, silver-treated fruit ripen either by sequestering the inhibitor or by the synthesis of fresh binding-sites (Davies et al., 1990). More recently, transgenic tomatoes have been produced in which ethylene synthesis and ripening are both greatly reduced, either by inhibiting the production of ACC synthase or EFE with antisense genes, or by metabolizing ACC to prevent its conversion to ethylene (refer to section 14.12).

14.6   TEMPERATURE STRESSES AND THEIR EFFECTS ON RIPENING

Tomato fruit are chilling-sensitive, inasmuch as below about 11°C unripe or incipiently-ripening samples not only colour more slowly than in ambient conditions but also display injury symptoms caused by exposure to low temperatures. However, once fruit have ripened, they can be kept for a week or two under cold conditions without damage except that, without a period of recovery under normal conditions, the flavour profile will be severely distorted (Cheng and Shewfelt, 1988). The injury sustained by susceptible fruit will depend on a combination of several factors – the stage of development, whether the tomato has been conditioned towards cold or not, and the integral of the degrees below 11°C and time (degree-days). Extensive chilling damage often results in firm patches of tissue on an otherwise red fruit, while milder conditions lead to premature softening (Hobson, 1987). Lycopene synthesis and ethylene production by unripe fruit following exposure to various amounts of cold-stress have been described, with the former being more severely affected than the latter (Watkins et al., 1990). Changes in eight mRNAs homologous to ripening specific cDNA clones have been investigated and, in general, most of these fell in concentration following chilling, except the mRNA for the 'ethylene-forming enzyme' (EFE) (Watkins et al., 1990; Hamilton et al., 1991). Chilling stimulated expression of this mRNA, but on transfer back to 24°C, levels declined rapidly.

Exposure of mature green tomatoes to high temperatures attenuates some of the normal ripening processes. Postharvest heat treatments are emerging as commercially-acceptable ways for inhibiting fruit ripening and promoting disease-resistance (Klein and Lurie, 1991). It is well-known that above 30°C lycopene formation is inhibited, although this is restored on return to moderate temperatures. Tomato fruit from cultivars not having the even-ripening (uu) gene respond on exposure to hot conditions by forming a hard, suberized ring of tissue ('greenback') around the shoulders of the fruit (Hobson et al., 1977). More recent studies have confirmed that lycopene formation and ethylene production are affected by heat stress and there is also a general inhibition of ripening throughout the fruit at high temperatures. Picton and Grierson (1988) exposed incipiently-ripening tomatoes to 35°C, and followed pigment changes and the expression of twelve ripening-related genes, including that for polygalacturonase (PG), using appropriate cDNA clones as hybridization probes. Although some protein synthesis continued, the expression of most mRNAs was inhibited. Exogenous ethylene neither stimulated PG mRNA nor restored ripening. These investigations showed that high temperatures lead to the formation of 'heat-shock' proteins at the expense of others that would normally promote ripening, synthesize ethylene, and maintain sensitivity towards the hormone. In further studies it was shown that the supply of exogenous ethylene did not restore autocatalytic production of the hormone, ACC synthesis, normal pigment synthesis or even the expected softening rate (Yang *et al.*, 1990). That ACC synthase activity was affected by heat more rapidly than the ethylene-forming enzyme was shown by Biggs *et al.* (1988), with recovery on the return of the fruit to more moderate temperatures.

14.7   THE MECHANISM OF TOMATO RIPENING

Of all climacteric fruits, arguably most is known about the molecular mechanisms controlling the ripening of the tomato. The genome is relatively small and many single-gene mutants are available in standard genetic backgrounds, which facilitates research into the control mechanisms. In addition, the parent plant has a relatively short life cycle, it can be transformed genetically and regenerated from tissue culture. Major changes occur in the physiology and biochemistry of a mature green tomato at the onset of ripening. These alterations occur rapidly, are comprehensive, affecting all cell compartments, and fundamentally alter the appearance, the flavour, the texture, the disease-resistance and the survivability or shelf life of the fruit. Table 14.4 gives a brief list of some of the systems that have been studied.

Since the review on tomato fruit composition by Davies and Hobson (1981), more recent studies have emphasized the dynamic aspects of ripening and the nature of the mechanisms controlling the changes (Grierson, 1985, 1986; Brady, 1987; Tucker, 1990; Speirs and Brady, 1992). Although the 'organizational resistance' theory of Blackman and Parija (1928) which explained ripening as a progressive breakdown in cellular compartmentation held sway for so many years – mainly because it is partially correct – evidence gradually accumulated that ripening was a directed process which involved synthesis, as well as degradation. Although some chloroplastic components disintegrate, they are replaced by others to form chromoplasts. The middle lamella partly dissolves, but the plasma membrane remains intact and mitochondrial activity in the cytoplasm persists into senescence. So the wholesale breakdown in cell membranes has not occurred, even by the time tomato fruit are fully ripe (Brady, 1987). The mechanism behind the onset of the very early events in the ripening sequence has still not been elucidated. Secondary events are easier to distinguish – increases in ACC synthase and EFE, in sensitivity of tissues to ethylene, and the synthesis of many other enzymes, such as polygalacturonase (PG) and phytoene synthase.

Table 14.4 Ripening-related changes in tomato fruit.

Change	Mechanism
Colour	Chlorophyll breakdown; disintegration of the light-harvesting complexes and dissolution of the chloroplast lamellae. β-carotene and lycopene accumulation in the plastids as they are converted to chromoplasts.
Texture	Reduction in galactan, araban and polyuronide content of cell walls; solubilization of calcium-pectin complexes, particularly the solubilization and partial depolymerization of polyuronides; loss of electron density in the middle lamella and cell wall erosion observed from light- and electron-microscopy studies.
Flavour, taste and aroma	Decrease in malate and an increase in citrate; Depolymerization and degradation of starch to sugars; Destruction of alkaloids such as α-tomatine; Reduction in polyphenol and polyamine content; Increase in the complexity of the volatile fraction.
Physiological response	Decrease in cytoplasmic volume; Increase in hydraulic conductivity; Redistribution of K^+ between cell compartments; Decrease in phospholipid content.
Enzyme activity	Increase in invertase (EC 3.2.1.26), malic enzyme, β-1:4 glucanase (EC 3.2.1.4), endopolygalacturonase (EC 3.2.1.15), phosphofructokinase (EC 2.7.1.11), ACC synthase (EC 4.4.1.14), EFE and many others.

Some significance in the ripening process has been attached to activity increases in the malic enzyme (Goodenough et al., 1985), fructose-2, 6-bisphosphate concentration (Chalmers and Rowan, 1971; Brady, 1987) and cell wall hydrolases (Fischer and Bennett, 1991). However, these changes are likely to represent the response to ripening initiation rather than be part of the process governing its inception. The first detectable sign of ripening is an increase in the production of ethylene, which occurs one to two days before any visible sign of colour change. The respiratory rise shown by tomato fruit seems to be a response to increased ethylene synthesis. Exactly what brings this about is unclear, but once begun, a chain of events is initiated that leads to ripening.

Closely associated with the respiratory rise is an increase in ATP levels, a surge in protein synthesis and a transient rise in the rate of precursor incorporation into protein (Brady, 1987). The proportion of ribosomes present as polysomes recovered from the cytosol follows the climacteric respiration curve quite well (Speirs *et al.*, 1984), and *in vitro* translation of purified mRNAs indicates that there are at least three groups of mRNAs (Grierson *et al.*, 1985; Biggs *et al.*, 1986). One category, present throughout growth and ripening is thought of as coding for metabolic proteins used in cell maintenance. The second group accumulates during development but diminishes before or at ripening. The final class appears with the onset of ripening, and fits well with the changes in protein synthesis *in vivo* noted by Baker *et al.* (1985).

14.8   NON-RIPENING MUTANTS

A number of tomato ripening mutants have been characterized that show pleiotropic effects on several aspects of ripening, indicating that they affect regulatory processes that control several different events. Although the precise biochemical lesion has not yet been demonstrated for any of these mutants, it is likely that processes involving ethylene synthesis, perception, and signal transduction (refer to Fig. 14.4) are altered in these mutants. A summary of the compositional and physio-logical implications of the introduction of the mutant genes *Never ripe (Nr)*, *ripening inhibitor (rin)* and *non-ripening (nor)* into normally-ripening tomato lines has been published previously (Davies and Hobson, 1981). Two further reviews (Grierson *et al.*, 1987; Tucker and Grierson, 1987) have given more detailed pictures of the biochemistry and molecular biology of lesions that alter the rate of normal development, including some that only affect aspects of pigment biosynthesis. The *rin* and *nor* mutations have proved extremely useful in ripening studies since either of them can virtually stop ripening from taking place (Table 14.5).

Never ripe (Nr) fruit show a weak respiratory rise with a shallow peak in ethylene production; eventually the fruit become orange-red in colour. PG activity is reduced to about 15% of normal, and is confined until late in senescence to the isoform PG-1, usually associated with wild-type fruit during the first few days of ripening (Tucker *et al.*, 1980). A second form, known as PG-2, eventually makes an appearance in very senescent fruit (Crookes and Grierson, 1983).

The *ripening inhibitor (rin)* mutation has a far-reaching effect on the rate of ripening of the mature fruit. Growth and development up to this point is quite normal, and the production of viable seeds is not affected. However, the fruit fail to show peaks in either CO_2 or ethylene, and turn a lemon-yellow colour after several months. They do not accumulate

Table 14.5 Some ripening mutants drastically affecting the rate of ripening.

Mutant	Chromosome on which the gene is located	Effect on the phenotype
Ripening inhibitor (rin)	5	The fruit grow normally and then slowly turn pale yellow in colour. Only very low levels of ethylene are produced. The fruit hardly soften at all and contain a very low activity of PG. They do not ripen when exposed to exogenous ethylene; high oxygen causes a slight pink colour to be produced.
Never ripe (Nr)	9	The fruit slowly assume an orange-red colour. The extent of softening of ethylene, PG and lycopene synthesis is attenuated compared with the wild type.
Alcobaca (alc)	10	According to the genetic background, source of the mutation and time of harvest, the response is variable. In some circumstances, fruit ripen on the plant to a pale red colour, whereas if picked when mature green, they eventually turn yellow and not red. PG activity is very low: high salt conditions cause more extensive ripening and induce a little more PG activity.
Non-ripening (nor)	10	This gene is non-allelic with the Alcobaca mutant, and is more extreme in its action than rin. Ethylene production is very low. Normally, the final colour of the fruit is deep yellow; high salt causes an acceleration in ripening so that the colour then becomes deep orange and they soften somewhat. They contain more PG than Alcobaca, but still much less than 1% of the wild type when ripe.

The variants 'Snowball', 'Spanish Winter' and 'Longkeeper' may well turn out to be mutations all at the same locus as Alcobaca but expressed to differing levels and influenced by the genetic background.

lycopene, and the normal breakdown of plastid proteins is altered (Grierson *et al.*, 1987). Polyuronide solubilization and depolymerization are both inhibited (Seymour *et al.*, 1987a), and the fruit soften a little with extended senescence. The levels of chelator-soluble polyuronide increase from about three months after the first signs of colour change, and PG activity at this time is just detectable. Significantly, the PG gene, which is located on chromosome 10, is intact but hardly expressed as a result of the *rin* mutation on chromosome 5. Normal ripening in this mutant cannot be restored by adding exogenous ethylene.

The *non ripening (nor)* gene is also very effective at blocking ripening. Fruit containing the mutation assume a pale orange colour after some months, and pigmentation is slightly accelerated if the fruit are not picked, but adding ethylene does not improve ripening. However, if plants homozygous for the gene are grown under conditions of high salinity, the fruit are typically reduced in size but they also show partial ripening, so much so that the tomatoes are then edible (Mizrahi *et al.*, 1982; Sharaf and Hobson, 1985). PG activity is not induced, neither is its mRNA, but high salinity promotes invertase activity. Salt-stress on *nor* plants also stimulates the accumulation of mRNA homologous to the cDNA clone pTOM5, which parallels colour development (Davies *et al.*, 1992). The significance of this became clear when it was shown that pTOM5 encodes phytoene synthase, an enzyme important for the synthesis of β-carotene and lycopene (Bird *et al.*, 1991). The incorporation of *rin* or *nor* alleles into standard tomato cultivars has been extensively explored as a way of reducing the loss of fruit firmness during and after ripening, therefore extending the shelf life of the fruit (Richardson and Hobson, 1987). While the time from anthesis to the commercial picking stage is slightly extended and the colour density of the ripe fruit is reduced a little, the rate of deterioration of the harvested fruit is much reduced without loss of consumer appeal (Bedford, 1988).

14.9   CLONING AND CHARACTERIZATION OF RIPENING-RELATED mRNAs

The application of molecular biology techniques to studies of tomatoes has led to major progress in our understanding of gene regulation in tomato. Generation of a cDNA library of mRNA sequences from ripening fruit led to the identification of clones (the pTOM series) for at least 19 mRNAs that increased in abundance during ripening (Slater *et al.*, 1985). A recent compilation of results obtained with over 40 clones from this and other tomato fruit cDNA libraries shows that although the biological function of many of the proteins encoded by these cloned mRNAs remains to be elucidated (Gray *et al.*, 1992), several have been identified. These include those encoding polygalacturonase, ACC synthase, ethylene-forming enzyme, and phytoene synthase. Studies with these and other clones have established that a number of new mRNAs appear at the onset of ripening, as a result of a major increase in transcription of their encoding genes. Sequencing of the clones has allowed the complete amino acid sequences of their corresponding proteins to be predicted, has provided information about the transport of proteins to the cell wall during ripening, and given us an insight into the ways in which the expression of ripening genes is regulated.

14.9.1   Polygalacturonase mRNA

The first ripening-related cDNAs to be sequenced encoded tomato polygalacturonase (EC 3.2.1.15) (Grierson *et al.*, 1986; DellaPenna *et al.*, 1986; Sheehy *et al.*, 1987). This enzyme can be extracted in three distinct isoenzyme forms from tomatoes (PG1, PG2a, PG2b). During ripening there is a major synthesis of these PG isoenzymes, which are all structurally related.

Although different forms of PG may also be found in pollen and in abscission zones, PG1, PG2a and PG2b appear only to be synthesized during ripening, and at no other stage in the life cycle. Following the identification and sequencing of the PG cDNA, the gene itself was isolated and sequenced by Bird *et al.* (1988). The current view is that this is the only PG gene expressed during ripening and it gives rise to three different isoforms of PG, as outlined in Fig. 14.3.

The deduced amino acid sequence of the complete polypeptide encoded by the PG mRNA contains additional residues at the N- and C-terminus that are not present in the mature forms of the enzyme isolated from tomato cell walls. The additional 71 amino acids at the N-terminus (Grierson *et al.*, 1986) are believed to target the newly-synthesized protein to the endoplasmic reticulum (ER) and then to the cell wall. The first 24 amino acids are known to be removed in the ER, where the protein is also glycosylated (Fig. 14.3). The precise role of the remaining 47 N-terminal amino acid residues, and also the 13 amino acids ultimately removed from the C-terminus, is not clear, but they are assumed to be involved in targeting or transport of the mature protein to the cell wall.

The factor(s) responsible for the transcription of PG mRNA during ripening are not clear. Supplying ethylene to mature green fruit stimulates ripening and concomitant synthesis of PG and its mRNA (Maunders *et al.*, 1987). This is inhibited by Ag$^+$, which is believed to interfere with ethylene perception or action (Davies *et al.*, 1988).

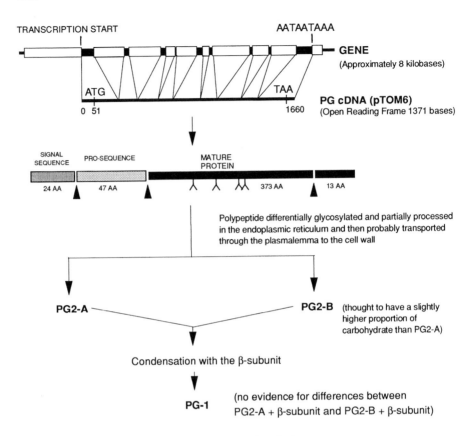

TRANSCRIPTION START

AATAATAAA

ATG

TAA

0 51

1660

GENE
(Approximately 8 kilobases)

PG cDNA (pTOM6)
(Open Reading Frame 1371 bases)

SIGNAL SEQUENCE

PRO-SEQUENCE

MATURE PROTEIN

24 AA

47 AA

373 AA

13 AA

Polypeptide differentially glycosylated and partially processed in the endoplasmic reticulum and then probably transported through the plasmalemma to the cell wall

PG2-A

PG2-B (thought to have a slightly higher proportion of carbohydrate than PG2-A)

Condensation with the β-subunit

PG-1

(no evidence for differences between PG2-A + β-subunit and PG2-B + β-subunit)

Characteristics of the isoforms of tomato polygalacturonase

Component	KDa	PI	Activity retained after 10 min at 50°C compared with original activity	Percentage carbohydrate by weight
PG-1	84 -115	8.6	100%	18.9
PG-2A	43	9.4	50%	11.2
PG2-B	45	9.4?	50%	11.2
β - subunit	38-39	?	?	27.9

Fig. 14.3 The tomato polygalacturonase gene and the pathway of formation of PG1, PG2a and PG2b. (After Bird *et al.*, 1988; Tucker, 1990; Pogson *et al.*, 1991; Fischer and Bennett, 1991.)

However, synthesis of PG mRNA does not occur rapidly in response to ethylene (Lincoln *et al.*, 1987) and, furthermore, inhibiting endogenous ethylene production with antisense genes does not prevent PG synthesis (Hamilton *et al.*, 1990; Oeller *et al.*, 1991).

14.9.2   ACC synthase and EFE mRNAs

Both ACC synthase and EFE are encoded by small multigene families in the tomato. There are at least six genes for ACC synthase and three for EFE (Holdsworth *et al.*, 1988). Members of each gene family are expressed under different physiological conditions or at particular stages in the life cycle, for example to cause enhanced ethylene synthesis during senescence, abscission, ripening and in response to wounding (Kende and Boller, 1981) or infection. cDNA clones for ACC synthase were identified by partially purifying the enzyme, raising monoclonal antibodies, and using these to identify *Escherichia coli* cells transcribing cloned ACC synthase mRNA and translating it into a protein recognized by the antibody (Sato and Theologis, 1989). A less conventional method using an antisense gene was employed to identify a cDNA clone for EFE, which had never previously been purified and for which no molecular probes were available (Hamilton *et al.*, 1990, 1991). This was the first time that an antisense approach had been used to assign a function to a cloned DNA sequence, and gave rise to plants in which EFE, ethylene production, and aspects of ripening were inhibited (Hamilton *et al.*, 1990; Picton *et al.*, unpublished results). Subsequently, antisense genes were also used to inhibit synthesis of ACC synthase and retard ripening (Oeller *et al.*, 1991).

Enhanced expression of genes for ACC synthase and EFE occurs early in ripening and is responsible for elevated ethylene production. The appearance of these mRNAs is independent of ethylene, which indicates that some other factor(s) is required for initiating transcription of these genes and thus causing the rise in ethylene production at the onset of ripening. The further synthesis of ethylene during ripening is autocatalytic ('system 2', refer to section 14.5). It is tempting to speculate that different members of the EFE and ACC synthase gene families may be involved in these two phases of ethylene synthesis.

14.10   POLYGALACTURONASE AND ITS ROLE IN TEXTURE CHANGE

Activation of transcription of the PG gene occurs one to two days after the initiation of the rise in ethylene synthesis that triggers ripening. This leads to the accumulation of PG mRNA in the cytosol and the synthesis of PG protein. There appears to be only one gene for the endo-acting PG isoforms that are synthesized during ripening. The three isoforms that accumulate, PG1, PG2a, PG2b, are believed to be derived from a single mRNA by post-translational processing and glycosylation of the PG polypeptide, or by interaction with other proteins (Fig. 14.3). At first,

PG1 with an M_r of about 100 kDa is the only molecular species detectable, but PG2a and PG2b, rapidly accumulate as ripening proceeds and are the major isoforms in ripe fruit. PG2a (M_r 43 kDa) and PG2b (M_r 45 kDa), are thought to consist of the same polypeptide each being glycosylated to slightly differing extents (Ali and Brady, 1982). PG1 contains components of M_r 43 kDa, 45 kDa and 38 kDa, hence this isoenzyme is thought to be a complex of PG2a or PG2b with a further polypeptide known as the β-chain (Fig. 14.3). Antibodies that have been raised against PG2 were also antigenic against PG1 (Tucker *et al.*, 1980; Ali and Brady, 1982), whereas those generated in a rabbit inoculated with PG2a showed reaction against dissociated PG1 and PG2, but not the β-subunit. Further antibodies were generated by using PG1 as the inoculum, and these recognized both the β-subunit and the PG2-like components of PG1 when dissociated and separated electrophoretically (Pogson *et al.*, 1991). PG1 can thus be broken down to yield PG2 and a β-subunit in a 1:1 ratio (Moshrefi and Luh, 1983). The possibility is that the β-form, present in limited amounts in the cell wall, combines with PG2 after its synthesis in and transport from the cytosol, since these polypeptides are probably transported in an unfolded state. An alternative explanation for the generation of PG1 has been proposed, i.e. that a non-specific 'converter' glycoprotein combines with PG2 during extraction from macerated tomato tissue to produce PG1, whose existence *in vivo* is thus in doubt (Pressey, 1988; Knegt *et al.* 1991). Although the precise relationship between the β-subunit and the 'converter' is not yet entirely clear, it seems that evidence for the actuality of PG1 *in vivo* is substantial. It is probable that during the early stages of ripening the initial amounts of the smaller forms of PG (2a and 2b) combine with the β-units in the cell walls, but as synthesis increases, there are insufficient quantities of β-units and free PG2a and 2b then accumulate. As ripening continues, PG2a and 2b thus become the dominant forms.

Support linking PG1 with the rate of softening in whole fruit during ripening is accumulating (Harman, 1984; Brady *et al.*, 1985). Activity of this larger molecular form of PG is closely associated with the rate at which pectins become solubilized, and DellaPenna *et al.* (1990) suggested that it is also responsible for depolymerization of the pectin chain. The question of the role of the pectic enzymes in tomato fruit softening has been much clarified by the use of techniques in molecular biology. In two complementary approaches to this problem, active PG genes have been inserted into a mutant line in which the natural gene is inhibited, while antisense technology has allowed natural expression of PG to be strongly inhibited in a normal line. The *rin* mutation almost completely prevents PG synthesis in mature fruit (Knapp *et al.*, 1989); the insertion of another PG gene into this line under an inducible promoter resulted in the accumulation of the active enzyme (Giovannoni *et al.*, 1989), but the fruit still did not soften. This supports the view that PG is not the primary determinant of softening. However, it remains a possibility that pleiotropic effects of the *rin* gene preclude PG from bringing about exactly the same changes as in a normal fruit. In direct contrast to this, the technique of antisense RNA technology has been used to inhibit normal PG activity in ripening fruit so that only 1% of the usual activity remains (Smith *et al.*, 1988, 1990a). The reduction largely prevented pectin depolymerization, but had little effect on solubilization of pectin or firmness of the fruit measured by probe-penetration or compression tests. In the cultivar Ailsa Craig, for instance (Schuch *et al.*, 1991), a reduction in PG to less than 1% of its normal level did not lead to a significant alteration in the firmness of the fruit. Despite the lack of effect on fruit firmness, low PG tomatoes transformed with antisense genes were much less susceptible to mechanical damage and cracking (Schuch *et al.*, 1991). In contrast to the results with the Ailsa Craig cultivar, using the cultivar UC82B as the line for manipulation and transformation with PG antisense genes, Kramer *et al.* (1992) made four selections, all of which showed enhanced firmness at maturity and to full ripeness. More recent results with Ailsa Craig have now indicated that low PG does indeed confer improved firmness on fruit throughout ripening (Murray *et al.*, 1993), and this latest finding underlines the difficulties of sampling complex trials and measuring fruit firmness objectively.

All reports so far published relate inhibited PG activity with extended survivability and shelf life of the fruit. However, the major conclusion from experiments where PG levels have been altered is that this enzyme is not the sole, or even the primary, determinant of softening. Consequently, other explanations must be sought for changes in texture that occur during ripening.

The experiments in which PG activity was modified with only a minor effect on firmness has prompted a re-evaluation of the causes of softening. The composition of the fruit cell wall has been extensively reviewed (Tucker and Grierson, 1987; O'Neill *et al.*, 1990) and is described in section 14.1. In essence it consists of protein and three major polysaccharide components, pectin, cellulose and hemicellulose. The pectic substances consist of a α 1-4-linked galacturonan backbone interspersed with 2- and 2-4-linked rhamnosyl residues. Between 50% and 60% of the C-6 carboxyl groups are methoxylated, while calcium ions can form inter- and intra-polymer bridges. A proportion of the rhamnose residues have side-chains of sugars such as galactose, or arabinose. Hemicellulose polymers also contain condensed sugars such as xyloglucans, glucomannans and galactoglucomannans, which may be covalently linked to pectin and hydrogen-bonded to cellulose microfibrils. Cellulose itself consists of linear chains of β-1-4-linked glucosyl residues, which assume a stable microfibrillar structure through hydrogen-bonding (Gross, 1990). Various models of how the components of the cell wall might be arranged into a three-dimensional structure have been critically examined by John and Dey (1986).

Gross (1990) has listed the general reasons behind the association between texture changes in tomatoes and PG activity. In non-ripening mutants, chilling-injured fruit and in tissue infiltrated with silver ions, it can be argued that PG is below the normal level of activity and softening is inhibited. Other studies have demonstrated that PG1 is more closely correlated with loss of firmness than total PG activity. Furthermore, it is also well established that PG isolates can hydrolyse purified tomato cell walls (Koch and Nevins, 1989), and degrade thin sections of unripe tissue (Crookes and Grierson, 1983). On the other hand, there is much evidence that alternative mechanisms affecting softening and tissue degradation exist during fruit development that firstly precede and then complement the action of the pectic enzymes. For instance, the locular tissue surrounding the seeds becomes semi-liquid before PG activity can be detected (Hobson, 1964; Jones *et al.*, 1989), and this probably contributes to a loss of firmness in green fruit. The pre-treatment of cell walls to remove neutral sugar polymers (Wallner, 1978) or the prior action of pectinesterase affected the subsequent susceptibility of the walls to breakdown by PG (Koch and Nevins, 1989), so a sequence of reactions should be more closely correlated with softening than PG activity alone. Some mutants contain normal levels of PG yet still do not soften (Jarret *et al.*, 1984; Giovannoni *et al.*, 1989), and in others where PG activity is low, it is then not well correlated with polyuronide solubilization (Seymour *et al.*, 1987b).

A second pectolytic enzyme, pectinesterase (PE; EC 3.1.1.11) is also found in tomato fruit; changes in activity occur during ripening (Tucker *et al.*, 1982) and it may also be implicated in softening. It is obvious that although PE and PG are present in ripening tomato cells in considerable quantities, demethoxylation, solubilization and molecular weight reduction take place in a strictly controlled manner. Pressey (1986) demonstrated that PG activity was highly dependent upon pH and the ionic strength of the extractant, for the pectic enzymes are strongly adsorbed onto cell wall material. It is possible that PE and PG occupy different sites in the cell wall and middle lamella, thus adding a further control point over their activities (Brady *et al.*, 1987; Rushing and Huber, 1990). While it is impossible with present knowledge to assess the precise role *in vivo* of the various candidate mechanisms that contribute to the loss of firmness towards the end of fruit development, we would like to summarize such evidence as there is available concerning the relationship between physiological and biochemical changes and softening as follows:

1. *Immature green fruit.* Glycosidases remove the side-chains from pectic polymers as a possible preliminary to a subsequent solubilization degradation by PG (Huber, 1983). Autolysis of the locular gel occurs in the absence of PG, with some softening of the fruit (Huber and Lee, 1986) by an unknown mechanism.

2. *Mature green fruit.* At the onset of ripening the autocatalytic phase in ethylene production begins; further galactose is released from cell walls (Kim *et al.*, 1991). Synthesis of PG, solubilization and depolymerization of pectin (Seymour *et al.*, 1987a) and protein release from the cell wall (Hobson *et al.*, 1983) all then occur, but are not necessarily linked. Acidification of the wall material occurs in the wake of PE action, perhaps involving other mechanisms as well; breakdown of cell walls cross-linking by Ca^{2+}, maybe by complexation of Ca^{2+} with citrate (Buescher and Hobson, 1982; Brady *et al.*, 1987) moving from the vacuole. There is evidence that calcium is progressively leached from the cell wall (Jarvis, 1984). Also there is probably a redirection in the incorporation of wall polymers into various types of wall components (Mitcham *et al.*, 1989; Gross, 1990).

3. *Ripe fruit.* Swelling of the middle lamella occurs as the structure disintegrates and allows cell movement. Turgor pressure falls as the cell wall progressively weakens. General (maybe coincidental) relationship between total PG activity and fruit firmness index. Pectin degradation as a result of PG and PE activity appears quite limited at this stage (Seymour *et al.*, 1987a, b).

4. *Over-ripening.* PG continues to rise in activity; with the disintegration of the cells, progressive depolymerization and degradation of the pectic substances results from a breakdown of control mechanisms and a free mixing of enzymes and substrates. Fruit become susceptible to infection by bacteria and fungi. The fruit fall to the ground and the seeds are available for dispersal.

14.11   TOMATO GENETIC ENGINEERING

The control of ripening (refer to section 14.12) must involve specific gene promoters, several of which are being investigated. The availability of fruit-specific, ripening-specific, and ethylene-regulated gene promoters for tomato, and the demonstration that they function in transgenic plants, opens up the possibility of introducing novel genes to alter tomato physiology and biochemistry. This may, in the future, make it possible to introduce genes to improve biochemical properties that affect fruit quality attributes. The advent of antisense gene technology, where the introduction of a second target gene sequence orientated in the inverse direction results in the target gene expression being prevented, has already led to significant developments in this area.

Antisense genes are based on the structure of endogenous genes, and consist of a coding sequence, joined in the inverse orientation, downstream from a suitable gene promoter. When transferred and expressed in the plant, the antisense gene is transcribed by RNA polymerase 2 to produce an antisense RNA, copied from the 'wrong' DNA strand. It is thought that in cells in which the endogenous target gene and the antisense gene are both transcriptionally active, the antisense RNA and the complementary sense mRNA interact to form a RNA–RNA helix, which is rapidly degraded. Antisense genes have been shown to be stably inherited and to act in a gene-dosage-dependent fashion. There is, however, still some doubt as to exactly how they bring about their effects. Whatever the precise mechanism of action of antisense genes, the net result is that the expression of the corresponding endogenous gene is greatly reduced. Using antisense genes, the activity of PG (Sheehy et al., 1988; Smith et al., 1988, 1990b), PE (Hall et al., 1993), ACC synthase (Oeller et al., 1991), ethylene forming enzyme (Hamilton et al., 1990, 1991) and phytoene synthase (Bird et al., 1991) have all been down-regulated in transgenic tomatoes.

The inhibition of PG to less than 1% of the normal level inhibited degradation of pectin in ripening tomatoes without affecting the level of activity of other enzymes such as invertase or PE. Furthermore, the synthesis of coloured carotenoids and ethylene were also normal (Smith et al., 1990b). In some tomato cultivars with reduced PG, a significant increase in firmness was noted (Kramer et al., 1992), although this was not apparent in others (Schuch et al., 1991). The targeted inactivation of the PG gene also rendered the fruit more resistant to splitting, cracking and infection (Gray et al., 1992). Furthermore, the higher pectin chain length and lack of PG may have beneficial effects on the properties of tomato extracts for processing (Schuch et al., 1991). Inhibition of a second wall-modifying enzyme, PE, has also been achieved using antisense genes. This increased the degree of esterification of pectin in ripe fruit (Hall et al., 1993). This may also have some utility for tomato processing.

The genes for the ethylene-forming enzyme and phytoene synthase, which generates phytoene, an essential intermediate in carotenoid production, were both actually identified by introducing antisense genes constructed from ripening-related cDNA clones from tomato into normally ripening lines. This led to the production of tomato fruit with reduced ethylene (Hamilton et al., 1990) and lycopene (Bird et al., 1991) respectively. The reduction in ethylene production to 5% of normal reduced the rate of tomato ripening and this was most pronounced in detached tomatoes and was partially restored by adding ethylene (Picton et al., unpublished). Oeller et al. (1991) generated transgenic tomatoes in which the activity of ACC synthase was inhibited to less than 0.5% of normal using 10 antisense genes. These fruit grew to maturity but never ripened, unless supplied externally with ethylene, when some aspects of ripening were restored. A further method for reducing ethylene production and slowing ripening was developed by Klee *et al.* (1991), who generated transgenic tomatoes in which the substrate for ethylene synthesis, ACC, was alternatively metabolized, greatly reducing ethylene production.

Interestingly, it has been found that a PG gene introduced into tomato in the sense orientation, designed to be transcribed to produce sense mRNA, also down-regulates PG activity (Smith *et al.*, 1990a). This phenomenon of mutual gene inactivation, or co-suppression, is not fully understood. It has been suggested that the effect is brought about by the unexpected generation of antisense RNA from the sense transgene, but other explanations are also possible (Grierson *et al.*, 1991). Experiments with antisense genes have validated some of the features of gene control during ripening outlined in Figure 14.4. Firstly, inhibition of ethylene production slows or prevents ripening, depending on the extent to which ethylene synthesis is inhibited, and this inhibition can be at least partly reversed by adding ethylene. Secondly, inhibiting the expression of genes that act subsequently in the ripening pathway, such as PE, PG or phytoene synthase, all prevent the specific reactions associated with each enzyme, without affecting other aspects of ripening. These results also confirm earlier conclusions (Davies *et al.*, 1988) that expression of some genes can occur even when ethylene perception is inhibited with Ag^+. However, reducing ethylene synthesis in transgenic fruit has shown, unexpectedly, that expression of the PG gene is largely unaffected (Hamilton *et al.*, 1990; Oeller *et al.*, 1991), although PG mRNA does take longer to reach maximum levels in these low-ethylene fruit (Picton *et al.*, manuscript submitted). The accumulation of phytoene synthase mRNA, on the other hand, is significantly reduced (Picton *et al.*, manuscript submitted). Two significant features that require further examination are the indication that an additional factor other than ethylene may also be involved in the regulation of tomato ripening, and that ethylene may stimulate not only the accumulation of mRNAs required for ripening but may also enhance their translation (Picton *et al.*, manuscript submitted).

14.12   CONTROL OF FRUIT-SPECIFIC AND RIPENING-SPECIFIC GENE EXPRESSION

It is obvious that many genes are not constitutively expressed in plants but are restricted to specific organs, such as fruit. Similarly, some genes will only be expressed in fruit during ripening. Signals that control the ripening-specific expression of the PG gene in tomato are located in a 1.4 kilobase (kb) DNA sequence, called the PG promoter, upstream by 51

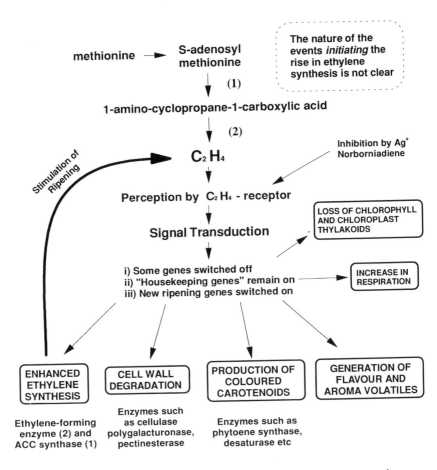

Fig. 14.4 Gene regulation in ripening tomatoes. (Synthesis of some mRNAs and proteins occurs independently of $C_2 H_4$)

base pairs (Fig.14.3) of the coding region. This was demonstrated by Bird *et al.* (1988), who fused this DNA sequence to a reporter gene, encoding the bacterial enzyme chloramphenicol acetyl transferase (CAT), and transferred it to tomato plant cells using *Agrobacterium* Ti plasmids. The transformed cells were regenerated into whole plants, which inherited the chimaeric PG promoter–CAT gene. CAT enzyme activity was not found in roots, stems, leaves, or green fruit but began to appear in the transgenic tomatoes during ripening, in parallel with PG activity. In control experiments, the same CAT gene was fused to the 35S transcription promoter from cauliflower mozaic virus (CaMV) and transgenic tomatoes were regenerated. The CaMV promoter is known to function constitutively in plants and this gene construct gave rise to CAT activity in all organs tested in these plants, which contrasted with the ripening-specific expression conferred by the 1.4 kb region from the PG gene. Quantitative measurements, however, indicated that the chimaeric PG promoter–CAT gene was expressed at a comparatively low level, relative to the normal PG gene. This led to the demonstration that other DNA sequences further upstream and also downstream from the PG gene itself were important in controlling high levels of ripening-specific expression (Nicholass, Smith, Watson, Morris, Bird, Schuch and Grierson, unpublished results).

These results are interpreted to mean that specific *cis*-acting DNA sequences in the vicinity of the PG gene bind to *trans*-acting factors (DNA-binding proteins) that are either synthesized or activated in ripening fruit. These proteins are believed to stimulate the binding of RNA polymerase and the initiation of transcription of the PG gene during ripening. So far, the nature of these protein factors and the DNA sequences they bind to has not been determined in detail for the PG gene. More progress has been made, however, with the E8 gene from tomato, studied by Fischer and his group (Deikman and Fischer, 1988; Lincoln and Fischer, 1988a, b; Cordes *et al.*, 1989).

The E8 gene codes for a protein of unknown function that is expressed specifically during fruit ripening, in response to ethylene. A cloned 4.4 kb tomato genomic fragment containing 2 kb of 5′, upstream sequence, the E8 gene and about 500 bp 3′ to the coding region was tagged with a short segment of DNA from bacteriophage λ and re-inserted into the tomato genome using a Ti plasmid. Expression of this added gene could be compared with the endogenous E8 gene because the transcript from the transgene differed from that of the endogenous gene by the inclusion of the short λ sequence. Accumulation of RNA from the tagged gene occurred in ripening tomatoes, and, like its endogenous counterpart, was stimulated by ethylene (Deikman and Fischer, 1988). Experiments with DNA sequences derived from the 2 kb upstream sequence of the E8 gene have shown the existence of DNA-binding proteins that interact with specific regions of the E8 promoter. One DNA-binding protein in particular, which appears to increase in amount during ripening interacts with a base sequence from 920–936 bp upstream of the transcription start site of the E8 gene. Interestingly, a related DNA sequence is present just upstream of another gene, called

E4, which, although not ripening-specific, is also induced by ethylene (Cordes *et al.*, 1989). A promoter from another gene, known as 2A11, has been shown to direct fruit-specific expression, which begins in the ovules at anthesis. Several positive and negative regulatory elements have been demonstrated to lie within a 1.8 kb 5′, upstream sequence of this gene and a number of DNA-binding proteins from fruit that interact with these sequences have been detected (Van Haaren and Houck, manuscripts submitted).

At the gross level, the technique of antisense DNA technology will allow further insights into the mechanisms in the control of tomato fruit ripening. Impetus is given to these investigations not only because they give information about the genetic control of plant development but also for the commercial possibilities that have been revealed. Nevertheless, public concern over the possible hazards of transgenic tomatoes compared with the advantages to be gained still have to be addressed. Descriptions of the promoter sequences, regions of the genome concerned with organ-specific changes and sites concerned with hormone perception are areas that we see advancing rapidly in the near future. The influences of DNA-binding proteins and regulatory elements surrounding genes will also become obvious soon. In the meantime, the tomato remains a particularly popular research vehicle, and there are many unknowns in the conversion of a mature green tomato into a fully-ripe vegetable-fruit that should even be good to eat!

REFERENCES

Adams, P. (1986) Mineral nutrition. In *The Tomato Crop*, (eds J.G. Atherton and J. Rudich), Chapman & Hall, London, pp. 281–334

Adams, P. (1987) The test of raised salinity. *Grower*, **107**(2), HN23–27

Adams, P. (1990) Effect of salinity on the distribution of calcium in tomato (*Lycopersicon esculentum*) fruit and leaves. In *Plant Nutrition-Physiology and Applications*, (ed M.L. van Beusichem), Kluwer Academic Publishers, Dordrecht, The Netherlands, pp. 473–476

Adams, P. and Ho, L.C. (1989), Effect of constant and fluctuating salinity on the yield, quality and calcium status of tomatoes. *Journal of Horticultural Science*, **64**, 725–732

Ali, Z.M. and Brady, C.J. (1982) Purification and characterisation of the poly-galacturonases of tomato fruits. *Australian Journal of Plant Physiology*, **9**, 155–169

Atherton, J.G. and Rudich, J. (1986) *The Tomato Crop*, Chapman & Hall, London

Atta-Aly, M.A., Saltveit, Jr, M.E. and Hobson, G.E. (1987) Effect of silver ions on ethylene biosynthesis by tomato fruit tissue. *Plant Physiology*, **83**, 44–48

Baker, E.A., Bukovac, M.J. and Hunt, G.M. (1982) Composition of tomato fruit cuticle as related to fruit growth and development. In *The Plant Cuticle*, (eds

D.F. Cutler, K.L. Alvin and C.E. Price), Linnean Society Symposium Series No. 10, 33–44

Baker, J.E., Anderson, J.D. and Hrusochka, W.R. (1985) Protein synthesis in tomato pericarp tissue during ripening characteristics of amino acid incorporation. *Journal of Plant Physiology*, **120**, 167–179

Bathgate, B., Purdom, M.E. , Grierson, D. and Goodenough, P.W. (1985) Plastid changes during the conversion of chloroplasts to chromoplasts in ripening tomatoes. *Planta*, **165**, 197–204

Bedford, L.V. (1988) Sensory quality of fresh tomatoes. Campden Food and Drink Research Association, Technical Memorandum No. 514, Chipping Campden, Glos

Biggs, M.S., Harriman, R.W. and Handa, A.K. (1986) Changes in gene expression during tomato fruit ripening. *Plant Physiology*, **81**, 395–403

Biggs, M.S., Woodson, W.R. and Handa, A.K. (1988) Biochemical basis of high temperature inhibition of ethylene biosynthesis in ripening tomato fruits. *Physiologia Plantarum*, **72**, 572–578

Bird, C.R., Ray, J.A., Fletcher, J.D., *et. al.* (1991) Using antisense RNA to study gene function: inhibition of carotenoid biosynthesis in transgenic tomatoes. *Bio/Techonology*, **9**, 635–639

Bird, C.R., Smith, C.J.S., Ray, J.A., *et al.* (1988) The tomato polygalacturonase gene and ripening specific expression in transgenic plants. *Plant Molecular Biology*, **11**, 651–662

Blackman, F.F. and Parija, P. (1928) Analytic studies in plant respiration. I. *Proceedings of the Royal Society*, **B103**, 412–445

Brady, C.J. (1987) Fruit ripening. *Annual Reviews of Plant Physiology*, **38**, 155–178

Brady, C., McGlasson, W.B. and Speirs, J. (1987) The biochemistry of fruit ripening. In *Tomato Biotechnology* (eds D.J. Nevins and R.A. Jones), Plant Biology, Vol. 4, Alan R. Liss Inc., New York, pp. 279–288

Brady, C.J., McGlasson, W.B., Pearson, J.A., Meldrum, S.J. and Kopeliovitch, E. (1985) Interactions between the amount and molecular forms of poly-galacturonase, calcium and firmness in tomato fruit. *Journal of the American Society for Horticultural Science*, **110**, 254–258

Buescher, R.W. and Hobson, G.E. (1982) Role of calcium and chelating agents in regulating the degradation of tomato fruit tissue by polygalacturonase. *Journal of Food Biochemistry*, **6**, 147–160

Buttery, R.G., Teranishi, R. and Ling, L.C. (1987) Fresh tomato aroma volatiles: a quantitative study. *Journal of Agricultural and Food Chemistry*, **35**, 540–544

Chalmers, D.J. and Rowan, K.S. (1971) The climacteric in ripening tomato fruit. *Plant Physiology*, **48**, 235–240

Cheng, T.S. and Shewfelt, R.L. (1988) Effect of chilling exposure of tomatoes during subsequent ripening. *Journal of Food Science*, **53**, 1160–1162

Crookes, P.R. and Grierson, D. (1983) Ultrastructure of tomato fruit ripening and the role of polygalacturonase isoenzymes in cell wall degradation. *Plant Physiology*, **72**, 1088–1093

Cordes, S., Deikman, J., Margossian, L.J. and Fischer, R.L. (1989) Interaction of a developmentally regulated DNA-binding factor with sites flanking two different fruit-ripening genes from tomato. *Plant Cell* **1**, 1025–1034

Darby, L.A. (1978) Isogenic lines of the tomato 'Ailsa Craig'. *Annual Report of the Glasshouse Crops Research Institute for 1977*, pp. 168–184

Davies, J.N. and Hobson, G.E. (1981) The constituents of tomato fruit – the influence of environment, nutrition and genotype. *Critical Reviews of Food Science and Nutrition*, **15**, 205–280

Davies, K.M., Hobson, G.E. and Grierson, D. (1988) Silver ions inhibit the ethylene-stimulated production of ripening-related mRNAs in tomato. *Plant Cell & Environment*, **11**, 729–738

Davies, K.M., Hobson, G.E. and Grierson, D. (1990) Differential effect of silver ions on the accumulation of ripening-related mRNAs in tomato fruit. *Journal of Plant Physiology*, **135**, 708–713

Davies, K.M., Grierson, D., Edwards, R. and Hobson, G.E. (1992) Saltstress induces partial ripening of the *nor* tomato mutant but expression of only some ripening-related genes. *Journal of Plant Physiology*, **139**, 140–145

Deikman, J. and Fischer, R.L. (1988) Interaction of a DNA binding factor with the 5′ flanking region of an ethylene-responsive fruit ripening gene from tomato. *EMBO Journal*, **7**, 3315–3320

DellaPenna, D., Alexander, D.C. and Bennett, A. B. (1986) Molecular cloning of tomato fruit polygalacturonase: Analysis of polygalacturonase mRNA levels during ripening. *Proceedings of the National Academy of Sciences (USA)*, **83**, 6420–6424

DellaPenna, D., Lashbrook, C.C. Toenjes, K., Giovannoni, J.J., Fischer, R.L. and Bennett, A.B. (1990) Polygalacturonase isoenzymes and pectin depolymerisation in transgenic *rin* tomato fruit. *Plant Physiology*, **94**, 1882–1886

Dinar, M. and Stevens, M.A. (1981) The relationship between starch accumulation and soluble solids content of tomato fruits. *Journal of the American Society for Horticulture Science*, **106**, 515–418

F.A.O. Production Yearbook for 1990 (1991), vol. 43, tables 14, 52, 65, 69–72 and 74, Food and Agriculture Organisation of the United Nations, Rome

Fischer, R.L. and Bennett, A.B. (1991) Role of cell wall hydrolases in fruit ripening. *Annual Reviews of Plant Physiology and Plant Molecular Biology*, **42**, 675–703

Giovannoni, J.J., DellaPenna, D., Bennett, A.B. and Fischer, R.L. (1989) Expression of a chimeric pg gene in transgenic *rin* (Ripening inhibitor) tomato fruit results in polyuronide degradation but not fruit softening. *The Plant Cell*, **1**, 53–63

Goodenough, P.W., Prosser, I.M. and Young, K. (1985) NADP-linked malic enzyme and malate metabolism in ageing tomato fruit. *Phytochemistry*, **24**, 1157–1162

Goodwin, T.W. (1980) *The Biochemistry of the Carotenoids*, vol. 1, 2nd edn, Chapman & Hall, London

Gough, C. and Hobson, G.E. (1990) A comparison of the productivity, quality, shelf-life characteristics and consumer reaction to the crop from cherry tomato plants grown at different levels of salinity. *Journal of Horticultural Science*, **65**, 431–439

Gray, J.E., Picton, S., Shabbeer, J., Schuch, W. and Grierson, D. (1992) Molecular biology of fruit ripening and its manipulation with antisense genes. *Plant Molecular Biology*, (in press)

Grierson, D. (1985) Gene expression in ripening tomato fruit. *CRC Critical Reviews of Plant Science*, **3**, 113–132

Grierson, D. (1986) Molecular biology of fruit ripening. In *Oxford Surveys of Plant Molecular and Cell Biology*, (ed B. J. Miflin), vol. 3, Oxford Press, London, pp. 364–383

Grierson, D., Fray, R., Hamilton, A.J., Smith, C.J.S. and Watson, C.F. (1991) Does co-suppression of sense genes in transgenic plants involve antisense RNA? *Trends in Biotechnology*, **9**, 122–123

Grierson, D. and Kader, A.A. (1986) Fruit ripening and quality. In *The Tomato Crop*, (eds J.G. Atherton and J. Rudich), Chapman & Hall, London, pp. 241–280

Grierson, D. and Tucker, G.A. (1983) Timing of ethylene and polygalacturonase synthesis in relation to the control of tomato fruit ripening. *Planta*, **157**, 174–179

Grierson, D., Purton, M.E., Knapp, J.E. and Bathgate, B. (1987) Tomato ripening mutants. In *Developmental Mutants in Higher Plants*, (eds H. Thomas and D. Grierson), Cambridge University Press, Cambridge, pp. 73–94

Grierson, D., Slater, A., Speirs, J. and Tucker, G.A. (1985) The appearance of polygalacturonase mRNA in tomatoes: one of a series of changes in gene expression during development and ripening. *Planta*, **163**, 263–271

Grierson, D., Tucker, G.A., Keen, J., Ray, J., Bird, C.R. and Schuch, W. (1986) Sequencing and identification of a cDNA clone for tomato poly-galacturonase. *Nucleic Acids Research*, **14**, 8595–8603

Gross, K.C. (1990) Recent developments on tomato fruit softening. *Postharvest News and Information*, **1**, 109–112

Hall, L.N., Tucker, G.A., Smith, C.J.S. *et al.* (1993) Inhibition of pectinesterase. (in press)

Hamilton A.J., Lycett, G.W. and Grierson, D. (1990) Antisense gene that inhibits synthesis of the hormone ethylene in transgenic plants. *Nature, London*, **346**, 284–287

Hamilton, A.J., Bouzayen, M. and Grierson, D. (1991) Identification of a tomato gene for ethylene-forming enzyme by expression in yeast. *Proceedings of the National Academy of Sciences (USA)*, **88**, 7434–7437

Harman, J.E. (1984) Studies on the role of polygalacturonase isoenzymes in tomato fruit softening and ripening. PhD thesis, University of London

Hayase, J., Chung, T.-Y. and Kato, H. (1984) Changes in volatile components of tomato fruits during ripening. *Food Chemistry*, **14**, 113–124

Herrmann, K. (1979) Ubersicht uber die Inhaltsstoffe der Tomaten. *Z. Lebensm. Unters. Forsch.*, **169**, 179–200

Ho, L.C. and Adams, P. (1989) Calcium deficiency – a matter of inadequate transport to rapidly growing organs. *Plants Today*, **2**, 202–207

Ho, L.C. and Hewitt, J.D. (1986) Fruit Development. In *The Tomato Crop*, (eds J.G. Atherton and J. Rudich), Chapman & Hall, London, pp. 201–239

Hobson, G.E. (1964) Polygalacturonase in normal and abnormal tomato fruit. *Biochemical Journal*, **92**, 324–332

Hobson, G.E. (1987) Low-temperature injury and the storage of ripening tomatoes. *Journal of Horticultural Science*, **62**, 55–62

Hobson, G.E. and Kilby, P, (1985) Methods for tomato fruit analysis as indicators of consumer acceptability. *Annual Report of the Glasshouse Crops Research Institute for 1984*, pp. 129–136

Hobson, G.E., Davies, J.N. and Winsor, G.W. (1977) Ripening disorders of tomato fruit. Growers Bulletin No. 4, Glasshouse Crops Research Institute, Littlehampton

Hobson, G.E., Richardson, C. and Gillham, D.J. (1983) Release of protein from normal and mutant tomato cell walls. *Plant Physiology*, **71**, 635–638

Holdsworth, J.J., Schuch, W. and Grierson, D. (1988) Organisation and expression of a wound/ripening-related small multigene family from tomato. *Plant Molecular Biology*, **11**, 81–8

Huber, D.J. (1983) The role of cell wall hydrolases in fruit softening. *Horticultural Reviews*, **5**, 169–219

Huber, D.J. and Lee, J.H. (1986) Comparative analysis of pectins from pericarp and locular gel in developing tomato fruit. In *Chemistry and Function of Pectins*, (eds M.L. Fishman and J.J. Jen), American Chemical Society, Washington, pp. 141–156

Jarvis, M.C. (1984) Structure and properties of pectin gels in plant cell walls. *Plant Cell and Environment*, **7**, 153–164

Jarret, R.L., Tigchelaar, E.C. and Handa, A.K. (1984) Ripening behaviour of the Green Ripe tomato mutant. *Journal of the American Society for Horticultural Science*, **109**, 712–717

Jeffery, D., Smith, C., Goodenough, P., Prosser, I. and Grierson, D. (1984) Ethylene-independent and ethylene-dependent biochemical changes in ripening tomatoes. *Plant Physiology*, **74**, 32–38

John, M.A. and Dey, P.M. (1986) Postharvest changes in fruit cell wall. *Advances in Food Research*, **30**, 139–193

Jones, R.B., Wardley, T.M. and Dalling, M.J. (1989) Molecular changes involved in the ripening of tomato fruit. *Journal of Plant Physiology*, **134**, 284–289

Kende, H. and Boller, T. (1981) Wound ethylene and 1-aminocyclopropane-1-carboxylate synthase in ripening tomato fruit. *Planta*, **151**, 476–481

Khudairi, A.K. (1972) The ripening of tomatoes. *American Scientist*, **60**, 696–707

Kim, J., Gross, K.C. and Solomos, T. (1991) Galactose metabolism and ethylene production during development and ripening of tomato fruit. *Postharvest Biology and Technology*, **1**, 67–80

Kinzer, S.M., Schwager, S.J. and Mutschler, M.A. (1990) Mapping of ripening-related or specific cDNA clones of tomato (*Lycopersicon esculentum*). *Theoretical and Applied Genetics*, **79**, 489–496

Klee, H.J., Hayford, M.B., Kretzmirk, K.A., Barry, G.F. and Kishore, G.M. (1991) Control of ethylene biosynthesis by the expression of a bacterial enzyme in transgenic tomato plants. *The Plant Cell*, **3**, 1187–1193

Klein, J.D. and Lurie, S. (1991) Postharvest heat treatment and fruit quality. *Postharvest News and Information*, **2**, 15–19

Knapp, J., Moureau, P., Schuch, W. and Grierson, D. (1989) Organisation and expression of polygalacturonase and other ripening related genes in Ailsa Craig 'Neverripe' and 'Ripening inhibitor' tomato mutants. *Plant Molecular Biology*, **12**, 105–116

Knegt, E., Vermeer, E., Pak, C. and Bruinsma, J. (1991) Function of the poly-galacturonase converter in ripening tomato fruit. *Physiologia Plantarum*, **82**, 237–242

Koch, J.L. and Nevins, D.J. (1989) Tomato fruit cell wall. *Plant Physiology*, **91**, 816–822

Kramer, M., Sanders, R., Bolkan, H., Waters, C., Sheehy, R.E. and Hiatt, W.R. (1992) Postharvest evaluation of transgenic tomatoes with reduced levels of polygalacturonase: processing, firmness and disease-resistance. *Postharvest Biology and Technology*, **1**, 241–255

Lincoln, J.E. and Fischer, R.L. (1988a) Diverse mechanisms for the regulation of ethylene-inducible gene expression. *Molecular and General Genetics*, **212**, 71–75

Lincoln, J.E. and Fischer, R.L. (1988b), Regulation of gene expression by ethylene in wild-type and *rin* tomato (*Lycopersicon esculentum*) fruit. *Plant Physiology*, **88**, 370–374

Lincoln, J.E., Cordes, S., Read, E. and Fischer, R.L. (1987) Regulation of gene expression by ethylene during *Lycopersicon esculentum* (Tomato) fruit development. *Proceedings of the National Academy of Sciences USA*, **84**, 2793–2797

Lyons, J.M. and Pratt, H.K. (1964) Effect of stage of maturity and ethylene treatment on respiration and ripening of tomato fruits. *Proceedings of the American Society for Horticultural Science*, **84**, 491–500

Martinez, V., Cerda, A. and Fernandez, F.G. (1987) Salt tolerance of four tomato hybrids. *Plant and Soil*, **97**, 233–242

Maunders, M.J., Holdsworth, M.J., Slater, A., Knapp, J.E., Bird, C.R., Schuch, W. and Grierson, D. (1987) Ethylene stimulates the accumulation of ripening-related mRNAs in tomatoes. *Plant Cell and Environment*, **10**, 177–184

McGlasson, W.B., Dostal, H.C. and Tigchelaar, E.C. (1975) Comparison of propylene-induced responses of immature fruit of normal and *rin* mutant tomatoes. *Plant Physiology*, **55**, 218–222

McGlasson, W.B., Last, J.H., Shaw, K.J. and Meldrum, S.K. (1987) Influence of the non-ripening mutants *rin* and *nor* on the aroma of tomato fruit. *HortScience*, 22, 632–634

McMurchie, E.J., McGlasson, W.B. and Eaks, I.L. (1972) Treatment of fruit with propylene gives information about the biogenesis of ethylene. *Nature, London*, **237**, 235–236

Mitcham, E.J., Gross, K.C. and Ng, T.J. (1989) Tomato fruit cell wall synthesis during development and senescence. *Plant Physiology*, **89**, 477–481

Mizrahi, Y., Zohar, R. and Malis-Arad, S. (1982) Effect of sodium chloride on fruit ripening of the nonripening tomato mutants *nor* and *rin*. *Plant Physiology*, **69**, 497–501

Moshrefi, M. and Luh, B.S. (1983) Carbohydrate composition and electrophoretic properties of tomato polygalacturonase isoenzymes. *European Journal of Biochemistry*, **135**, 511–514

Murray, A.J., Hobson, G.E., Schuch, W. and Bird, C.R. (1993) Reduced ethylene synthesis in EFE antisense tomatoes has differential effects on fruit ripening processes. *Postharvest Biology and Technology* (in press)

Mutschler, M.A., Tanksley, S.D. and Rick, C.M. (1987) 1987 linkage map of the tomato, *Lycopersicon esculentum*. *Report of the Tomato Genetics Cooperative*, **37**, 5–34

Oeller, P.W., Min-Wong, L., Taylor, L.P., Pike, D.A. and Theologis, A. (1991) Reversible inhibition of tomato fruit senescence by antisense RNA. *Science*, **254**, 437–439

O'Neill, M., Albersheim, P. and Darvill, A. (1990) The pectic polysaccharides of primary cell walls. In *Methods of Plant Biochemistry*, (ed, P.M. Dey), vol. 2, Academic Press, London, pp. 415–441

Petro-Turza, M. (1987) Flavour of tomato and tomato products. *Food Reviews International*, **2**, 309–351

Picha, D.H. (1986) Effect of harvest maturity on the final fruit composition of cherry and large-fruited cultivars. *Journal of the American Society of Horticultural Science*, **111**, 723–727

Picton, S. and Grierson, D. (1988) Inhibition of expression of tomato-ripening genes at high temperature. *Plant Cell and Environment*, **11**, 265–272

Pogson, B.J., Brady, C.J. and Orr, G.R. (1991) On the occurrence and structure of subunits of endopolygalacturonase isoforms in mature-green and ripening tomato fruits. *Australian Journal of Plant Physiology*, **18**, 65–79

Pressey, R. (1986) Changes in polygalacturonase isoenzymes and converter in tomatoes during ripening. *HortScience*, **21**, 1183–1185

Pressey, R. (1988) Re-evaluation of the changes in polygalacturonases in tomatoes during ripening. *Planta*, **174**, 39–43

Richardson, C. and Hobson, G.E. (1987) Compositional changes in normal and mutant tomato fruit during ripening and storage. *Journal of the Science of Food and Agriculture*, **40**, 245–252

Rick, C.M. (1979) Biosystematic studies in *Lycopersicon* and closely related species of *Solanum*. In *The Biology and Taxonomy of the Solanaceae*, (eds J.G. Hawkes, R.N. Lester and A.D. Skelding), Academic Press, London, pp. 667–677

Rushing, J.W. and Huber, D.J. (1990) Mobility limitations of bound polygalacturonase in isolated cell walls from tomato pericarp tissue. *Journal of the American Society for Horticultural Science*, **115**, 97–101

Salunkhe, D.K., Jadhav, S.J. and Yu, M.H. (1974) Quality and nutritional composition of tomato fruit as influenced by certain biochemical and physiological changes. *Qualitas Plantarum Plant Foods for Human Nutrition*, **24**, 85–113

Sato, T. and Theologis, A. (1989) Cloning the mRNA encoding 1-amino-1-carboxylate synthase, the key enzyme for ethylene biosynthesis in plants. *Proceedings of the National Academy of Sciences (USA)*, **86**, 6621–6625

Sawamura, M., Knegt, E. and Bruinsma, J. (1978) Levels of endogenous ethylene, carbon dioxide and soluble pectin, and activities of pectin methylesterase and polygalacturonase in ripening tomato fruits. *Plant and Cell Physiology*, **19**, 1061–1069

Schuch, W., Kanczler, J., Robertson, D., *et al.* (1991) Fruit quality parameters of transgenic tomato fruit with altered polygalacturonase activity. *HortScience*, **26**, 1517–1520

Seymour, G.B., Harding, S.E. Taylor, A.J., Hobson, G.E. and Tucker, G.A. (1987a) Polyuronide solubilisation during ripening of normal and mutant tomato fruit. *Phytochemistry*, **26**, 1871–1875

Seymour, G.B., Lasslett, Y. and Tucker, G.A. (1987b), Differential effects of pectolytic enzymes on tomato polyuronides *in vivo* and *in vitro*. *Phytochemistry*, **26**, 3137–3139

Sharaf, A.R. and Hobson, G.E. (1985) Effect of salinity on the yield and quality of normal and non-ripening mutant tomatoes. *Acta Horticulturae*, **190**, 175–182

Sheehy, R.E., Kramer, M. and Hyatt, W.R. (1988) Reduction in polygalacturonase activity in tomato fruit by antisense RNA. *Proceedings of the National Academy of Sciences (USA)*, **85**, 8805–8809

Sheehy, R.E., Pearson, J., Brady, C.J. and Hiatt, W.R. (1987) Moiecular characterization of tomato fruit polygalacturonase. *Molecular and General Genetics*, **208**, 30–36

Sisler, E.C. (1982) Ethylene-binding properties of a Triton-X extract of mung bean sprouts. *Journal of Plant Growth Regulation*, **1**, 211–218

Slater, A., Maunders, M.J., Edwards, K., Schuch, W. and Grierson, D. (1985) Isolation and characterisation of tomato cDNA clones for polygalacturonase and other ripening related proteins. *Plant Molecular Biology*, **5**, 137–147

Smith, C.J. S., Watson, C., Ray, J., Bird, C.R., Morris, P.C., Schuch, W. and Grierson, D. (1988) Antisense RNA inhibition of polygalacturonase gene expression in transgenic tomatoes. *Nature*, **334**, 724–726

Smith, C.J. S., Watson, C.F., Bird, C.R., Ray, J., Schuch, W. and Grierson, D. (1990a) Expression of a truncated tomato polygalacturonase gene inhibits expression of the endogenous gene in transgenic tomatoes. *Molecular and General Genetics*, **224**, 477–481

Smith, C.J.S., Watson, C.F., Morris, P.C., Bird, C.R., Seymour, G.B., Gray, J.E., Arnold, C., Tucker, G.A., Schuch, W. and Grierson, D. (1990b) Inheritance and effect on ripening of antisense polygalacturonase genes in transgenic tomatoes. *Plant Molecular Biology*, **14**, 369–379

Sonneveld, C. and Welles, G.W.H. (1984), Growing vegetables in substrates in the Netherlands. Proc. 6th ISOSC Congress on Soilless Culture, Lunteren, 1984, ISOSC, Wageningen, The Netherlands, pp. 613–632

Speirs, J. and Brady, C.J. (1992) Modification of gene expression in ripening fruit. *Australian Journal of Plant Physiology*, **18**, 519–532

Speirs, J., Brady, C.J., Grierson, D. and Lee, E. (1984) Changes in ribosome organisation and messenger RNA abundance in ripening tomato fruits. *Australian Journal of Plant Physiology*, **11**, 225–234

Stevens, M.A. (1986) Inheritance of tomato fruit quality components. In *Plant Breeding Reviews*, (ed. J. Janick), Avi Publishing Co., Inc., Westport, Conn., USA, pp. 273–311

Stevens, M.A. and Rick, C.M. (1986) Genetics and breeding. In *The Tomato Crop*, (eds J.G. Atherton and J. Rudich), Chapman & Hall, London, pp. 35–111

Su, L.Y., McKeon, T., Grierson, D., Cantwell, M. and Yang, S.F. (1984) Development of 1-aminocyclopropane-1-carboxylic acid (ACC) synthase and polygalacturonase activities during maturation and ripening of Tomato fruits in relation to their ethylene production rates. *HortScience*, **19**, 576–578

Taylor, I.B. (1986) Biosystematics of the tomato. In *The Tomato Crop*, (eds J.G. Atherton and J. Rudich), Chapman & Hall, London, pp.1–34

Tucker, G.A. (1990) Genetic manipulation of fruit ripening. *Biotechnology and Genetic Engineering Reviews*, **8**, 133–159

Tucker, G.A. and Brady, C.J (1987) Silver ions interrupt tomato fruit ripening. *Journal of Plant Physiology*, **127**, 163–169

Tucker, G.A. and Grierson, D. (1987) Fruit ripening. In *The Biochemistry of Plants*, (ed D.D. Davies), vol. 12, Academic Press, London, pp. 265–318

Tucker, G.A., Robertson, N.G. and Grierson, D. (1980) Changes in polygalacturonase isoenzymes during the 'ripening' of normal and mutant tomato fruit. *European Journal of Biochemistry*, **112**, 119–124

Tucker, G.A., Robertson, N.G. and Grierson, D. (1982) Purification and changes in activity of tomato pectinesterase isoenzymes. *Journal of the Science of Food and Agriculture*, **33**, 396–400

Varga, A. and Bruinsma, J. (1986) Tomato. In *Handbook of Fruit Set and Development*, (ed. S.P. Monselise), CRC Press, Boca Raton, USA, pp. 461–491

Wallner, S.J. (1978) Postharvest structural integrity. *Journal of Food Biochemistry*, **2**, 229–233

Walls, I.G. (1989) *Growing Tomatoes*, David and Charles, Newton Abbott

Wann, E.V., Jourdain, E.L., Pressey, R. and Lyon, B.G. (1985) Effect of mutant genotypes hp, og^c and dg og^c on tomato fruit quality. *Journal of the American Society of Horticultural Science*, **110**, 212–215

Watkins, C.B., Picton, S. and Grierson, D. (1990) Stimulation and inhibition of expression of ripening-related mRNAs in tomatoes as influenced by chilling temperatures. *Journal of Plant Physiology*, **136**, 318–323

Winsor, G.W. and Adams, P. (1987) *Diagnosis of Mineral Disorders in Plants*, (ed. J.B.D. Robinson), vol. 3, Glasshouse Crops. HMSO London, pp. 168

Yang, S.F. and Hoffman, N.E. (1984) Ethylene biosynthesis and its regulation in higher plants. *Annual Reviews of Plant Physiology*, **35**, 155–189

Yang, R.F., Cheng, T.S. and Shewfelt, R.L. (1990) The effect of high temperature and ethylene treatment on the ripening of tomatoes. *Journal of Plant Physiology*, **136**, 368–372

Glossary of botanical names

COMMON NAME	LATIN NAME	PAGE
Almond	*Prunus amygdalus* Stokes	380
wild	*Prunus fasciculata* A. Gray	380
Apple	*Malus pumila* Mill.	325
Apricot	*Prunus armeniaca* L.	380
Avocado	*Persea americana* Mill.	53
Banana	*Musa* L. sp.	83
Blackthorn (sloe)	*Prunus spinosa* L.	380
Blueberry	*Vaccinium* sp.	356
Carambola (Star fruit)	*Averrhoa carambola* L.	169
Cherimoya	*Annona cherimola* Mill.	167
Cherry	*Prunus avium* L.	380
bird/hagberry	*Prunus padus* L.	380
choke	*Prunus virginiana* L.	380
dwarf	*Prunus pumila* L.	380
evergreen	*Prunus ilicifolia* Walp.	380
Indian	*Prunus chicasa* Michx.	380
Custard apple	*Annona squamosa* L.	167
	Annona reticulata L.	167
Damson	*Prunus institia* L.	380
Feijoa	*Acca sellowiana* Berg.	152
Grape	*Vitis vinifera* L.	189
Grapefruit	*Citrus paradisi* Macfad.	109
Guava	*Psidium guajava* L.	109
Ilama	*Annona diversificola* Saff.	167
Kiwifruit	*Actinidia deliciosa* (A. Chev) C. F. Liang et A. R. Ferguson	235
Lemon	*Citrus limon* (L.) Burm.f.	109
Lime	*Citrus aurantifolia* (Christm.) Swingle	109
Lychee	*Litchi chinensis* Sonn.	109

Mandarin (tangerine)	*Citrus reticulata* Blanco	109
Mango	*Mangifera indica* L.	255
Mangosteen	*Garcinia mangostana* L.	152
Medlar	*Mespilus germanica* L.	325
Muskmelon (cantaloupe, honeydew)	*Cucumis melo* L.	273
Nectarine	*Prunus persica* (L.) Batsch	380
Olive	*Olea europea* L.	379
Orange		
sour	*Citrus aurantium* L.	109
sweet	*Citrus sinensis* L. Osbeck	109
Papaya	*Carica papaya* L.	302
Passion fruit	*Passiflora edulis* Sims	152
Peach	*Prunus persica* Stokes	379
Pear	*Pyrus communis* L.	325
Persimmon	*Diospyros kaki* L.f	152
Plum		
beach	*Prunus maritima* Wangenh.	380
European	*Prunus domestica* L.	380
Pummelo (shaddock)	*Citrus grandis* (L.) Osbeck	109
Quince	*Cydonia oblonga* L.	325
Rambutan	*Nephelium lappaceum* L.	152
Raspberry	*Rubus idaeus* L.	347
Soursop	*Annona muricata* L.	152
Strawberry	*Fragaria ananassa* Duch.	352
Sweetsop (sugar apple)	*Annona squamosa* L.	167
Tamarillo (tree tomato)	*Cyphomandra betacea* (Cav.) Sendt	152
Tomato	*Lycopersicon esculentum* Mill.	405
	Lycopersicon chilense Dun.	407
	Lycopersicon hirsutum Humb. and Bonpl.	406
	Lycopersicon peruvianum (L.) Mill.	407
Watermelon	*Citrullus lanatus* (Thunb.) Mansf.	274

Index